Atomic and Molecular Spectroscopy

Basic Concepts and Applications

Rita Kakkar

CAMBRIDGE
UNIVERSITY PRESS

CAMBRIDGE
UNIVERSITY PRESS

4843/24, 2nd Floor, Ansari Road, Daryaganj, Delhi - 110002, India

Cambridge University Press is part of the University of Cambridge.

It furthers the University's mission by disseminating knowledge in the pursuit of education, learning and research at the highest international levels of excellence.

www.cambridge.org
Information on this title: www.cambridge.org/9781107063884

First published 2015

Printed in India by Thomson Press India Ltd., New Delhi 110001

A catalogue record for this publication is available from the British Library

Library of Congress Cataloging-in-Publication Data
Kakkar, Rita, author.
Atomic and molecular spectroscopy : basic concepts and applications / Rita Kakkar.
pages cm
Includes bibliographical references and index.
Summary: "Elucidates various spectroscopic techniques including atomic spectroscopy, pure rotational spectroscopy, vibrational spectroscopy of diatomic and polyatomic molecules, Raman spectroscopy and electronic spectroscopy"—Provided by publisher.
ISBN 978-1-107-06388-4 (hardback)
1. Atomic spectroscopy. 2. Molecular spectroscopy. I. Title.
QD96.A8K35 2014
539'.60287—dc23
2014020957

ISBN 978-1-107-06388-4 Hardback

To my father
Late Shri Om Prakash Chadha

Contents

List of Figures

List of Tables

Preface

This book is primarily intended for post-graduate students of science, but is simple enough for an undergraduate student to understand. Every chapter begins with simple concepts related to the topic of the chapter, and gradually Quantum Mechanical and Group Theoretical treatments are introduced for a deeper understanding. I have tried to keep the language simple and have introduced new concepts one at a time, so that the reader is not overwhelmed by too many new ideas at the same time.

Spectroscopy is so vast and new concepts are rapidly emerging, so that it is not possible for a book to be complete. This book does not pretend to be complete—but it does try to cover the underlying principles thoroughly so that the reader should not face difficulty in applying these principles to his own problem.

Spectroscopy cannot be understood without a thorough knowledge and understanding of quantum mechanics. Though a number of Quantum Mechanics principles are scattered throughout the book, the reader is advised to first take at least an elementary course on quantum mechanics. The reader must also familiarize himself/herself with Group Theory—at least to the extent of assignment of molecular point groups, calculation of direct products, projection operator techniques, etc. There are a number of excellent texts, namely Cotton, Schonland, Ladd, to name a few. Again, a brief description of these techniques is given as and when required. Though I was advised by a reviewer of the book to include a chapter on Group Theory, it would have added to the volume and cost of the book, because there is nothing I could have removed from the other chapters in order to keep the page number around 400.

One other point—all chapters are arranged in sequence, so that a concept introduced in a certain chapter is used in one of the next chapters—so you cannot expect to understand, say the third chapter, without first reading the first two chapters. I have spent a lot of thought on the arrangement of the chapters—in the short time of one semester, it is extremely important that everything is taught without any repetition. In this respect, though in most books atomic spectroscopy is taught along with electronic spectroscopy, in this book it is the first topic to be covered. The reason is the simplicity of atomic spectroscopy, which has no complications from rotation and vibration, as in the case of molecular electronic spectroscopy. The entire sequence may be different from other books, but in my experience this is the only way I am able to cover the entire syllabus (which also includes NMR and Mössbauer spectroscopy and X-ray diffraction) in 50 hours. After a lot of jugglery during the initial years of teaching the paper, this is the sequence that I found to serve my purpose.

A motivation for writing the book was that, although several excellent books exist on the subject (Banwell, Hollas, Barrow, Chang, to name a few), no single book covers all the topics I wished to include. Many of the texts are also outdated. The level of this book is all the way from Banwell to Hollas and more. The symbols and units used in the book are in conformity with those recommended by IUPAC.

Though every possible effort has been made to avoid errors, in a project of this size, it is inevitable that some errors may have crept in. I shall be thankful to readers if they point them out so that future editions will be error free. Any suggestion or constructive criticism will be highly appreciated.

Rita Kakkar, FRSC

Acknowledgements

This is the best part of the book where I can pen my thoughts without worrying about scientific accuracy. I owe this book to all those who directly or indirectly encouraged me. My first source of motivation was the thousands of students who thronged my spectroscopy class. There can be no better encouragement than the rapt attention of these young souls, their inquisitive questions, coming to me with their doubts after class, their excitement at scoring well in the paper, coming after the NET (National eligibility test) and telling me that they were successful in "cracking" it because of their understanding of spectroscopy and quantum mechanics. For all these students, many of whom have told me that they use my class-notes even for undergraduate teaching, I have brought out this book. As all teachers would agree, there is no satisfaction greater than the high one gets after a well-appreciated class.

To my research group of nearly 40 students, past and present, I owe a lot of thanks for constantly encouraging me to write the book. They are now urging me to write another book on quantum mechanics! To the present lot of students, who had to wait for me to get free from the book before I could check their research papers, thank you for your patience!

I am indebted to all my teachers who made learning fun. I was fortunate to be taught by some of the best teachers, both during my undergraduate and postgraduate studies. For spectroscopy, especially, I was fortunate to be taught by the likes of the late Professors N. K. Ray and V. M. Khanna. Professor N. K. Ray had the knack to make complex quantum mechanics seem so simple, while Professor V. M. Khanna's mastery of spectroscopy was phenomenal. His problem sets were unique and I have included some of his questions in the book. I was also fortunate to have taught three core courses—quantum mechanics, statistical thermodynamics and spectroscopy—in parallel with him and am indebted to him for his guidance. I would also specially mention Professor J. Nagchaudhuri. I do not know where she is at present, but when she left for Kolkata after retirement, she promised to write a book on spectroscopy and remain in touch. That was before the time of the internet and e-mail, and we somehow lost touch. While Professor V. M. Khanna was adept at the theoretical aspects of spectroscopy, she also brought in the experimental aspects. Teaching alongside her, I learnt to include some experimental topics in my teaching. The result is a nice amalgamation of theoretical and experimental spectroscopy, and I sincerely hope that it will be appreciated.

To my publisher's representative, Mr. Gauravjeet Reen of Cambridge University Press, thank you for extending the deadline and your understanding. This soft-spoken gentleman would call me telephonically and ask politely whether this was a good time to talk, and then throw a bombshell by advancing the date of submission of the manuscript. This kept me on my toes. Finally, when the deadline approached, he told me to take my own time, saving me from more sleepless nights.

The acknowledgements would not be complete if I did not appreciate the anonymous reviewers of the book, both national and international. Their comments have definitely helped. I particularly thank one reviewer for suggesting an easy to remember expression for the moment of inertia of linear polyatomic molecules, which I have included in the text. That brings me to my son, Chetan. He was studying in Class IX when I was writing the Rotational Spectroscopy chapter and he just happened to pass by my computer. He asked me why I was deriving the expression for OCS in such a complicated way, when the parallel axis theorem (which he was studying in Physics then) could be used, and I included his derivation in the book.

I dedicate the book to my parents—my late father who always had faith in me and my mother for her excellent advice at all times, her commitment to education—she was a graduate in Economics at a time when women hardly studied. At the age of 90+, she still remembered the poetry we learnt at school and have forgotten. She would advise her children and grand children (from Longfellow's *A Psalm of Life*) "*Be not like dumb, driven cattle! Be a hero in the strife*". Unfortunately, she passed away during production of the book. Without my parents' constant support, I would not have been what I am today. I cannot thank my Dad enough for his support. On looking at my notes, I came across his hand-writing. In my graduation days, our Department library had just one copy of Banwell (the cheap Indian edition had not been published then) and it was issued to students just for a day. Photostat was not so common then, and he actually copied some of the pages for me. I also owe my love for mathematics to him—though I am not as quick as him at doing arithmetic in my head.

Last, but not the least, I would like to express my heartfelt gratitude to my immediate family for putting up with me and my erratic hours during writing of the book. A special word of thanks to my husband, Dr. Subhash Kakkar, for his constant support through thick and thin. To my son Chetan, thank you for making life worth living. Besides the emotional support that the two have provided, both of them have also helped in the preparation of the manuscript—my husband by encouraging me to draw the figures myself, starting by drawing a few of those himself, and my son by providing technical help and advice throughout, the computer wizard that he is.

Rita Kakkar, FRSC

List of Abbreviations

NMR	Nuclear Magnetic Resonance
ESR	Electron Spin Resonance
UV	Ultraviolet
IR	Infrared
MW	Microwave
FWHM	Full Width at Half Maximum
Vis	Visible
FT	Fourier Transform
FFT	Fast Fourier Transform
QM	Quantum Mechanics
PES	Photoelectron Spectroscopy
ESCA	Electron Spectroscopy for Chemical Analysis
XPS	X-ray Photoelectron Spectroscopy
IE	Ionization Energy
COM	Centre of Mass
MO	Molecular Orbital
HOMO	Highest Occupied Molecular Orbital
LUMO	Lowest Unoccupied Molecular Orbital
HMO	Hückel Molecular Orbital
FEM	Free Electron Model
ISC	Intersystem Crossing
VR	Vibrational Relaxation
IC	Internal Conversion
AO	Atomic Orbital
rms	Root Mean Square
LCAO	Linear Combination of Atomic Orbitals

1

Fundamentals of Spectroscopy

La lumière (...) donne la couleur et l'éclat à toutes les productions de la nature et de l'art; elle multiplie l'univers en le peignant dans les yeux de tout ce qui respire.

Light (...) gives colour and brilliance to all works of nature and of art; it multiplies the universe by painting it in the eyes of all that breathe.

Abbé Nollet, 1783

1.1 INTRODUCTION

"Science is spectral analysis. Art is light synthesis", so wrote Karl Kraus, Austrian writer. Light has intrigued both poets and scientists. What exactly is light? What effect does it have on matter? These are questions that have baffled scientists for many years, prompting Einstein in 1917 to say "For the rest of my life I will reflect on what light is". The branch of science that deals with the study of electromagnetic radiation (of which visible light is a part) and its interaction with matter is called *spectroscopy*. The word is derived from the Latin: *spectron* – spectre (ghost or spirit), or the Greek: *σκοπειν* – to see. This literally means that in spectroscopy, you do not look directly at the molecule – the matter – but what you see is its 'ghost' or image. To begin our study, we must, therefore, first discuss the nature of electromagnetic radiation and matter, and then the interaction between the two.

We start this chapter by giving basic formulae and definitions relating to waves, including travelling waves. We then go on to the wave description of electromagnetic radiation and its manifestations, and then discuss the properties emerging from a particulate description of radiation. The entire electromagnetic spectrum, its divisions and sub-divisions, the kind of spectroscopy observed in each region, are the topics of the next section. The populations of energy levels play an important role in the observed intensities. Einstein's coefficients and their interrelation are introduced in this chapter, but the quantum mechanical treatment is reserved for the next chapter. We wind up this chapter with a discussion of line shapes and broadening, followed by a brief introduction to Fourier transform spectroscopy, the almost magical transformation of a time decay to a line width, and the experimental recording of spectra. This chapter sets the foundation for the future chapters. The entire treatment in this chapter is semi-classical, i.e., we recognize that the energy levels of atoms and molecules are quantized, but do not go beyond Bohr's theory. This makes the various processes easier to comprehend, and gives us guidelines for the next chapter, where we bring in quantum mechanics.

1.2 SOME PROPERTIES OF WAVES

A wave is a disturbance that travels and spreads out through some medium. Familiar examples are ripples on the surface of water and vibrations in a string. A wave is characterized by its *frequency*, *wavelength*, *speed*, *direction* and *phase*. These parameters are best understood by imagining an object, such as a ball, moving in a circular path, as shown in Figure 1.1. If, at time $t = 0$, the object is in the horizontal (or 3 o'clock) position, its vertical (y) coordinate is zero. Let the object now start rotating counter-clockwise in a circular path of radius A with an *angular frequency* ω. The angular frequency (ω) is a scalar measure of the rotation rate, measured in radian per second, and is given by $\omega = |\vec{v}| / |\vec{r}|$, where $v = |\vec{v}|$ is the tangential speed at a point about the axis of rotation (measured in m s^{-1}), and $r = |\vec{r}|$ is the radius of rotation. (In this book, all vector quantities are denoted by an arrow above the symbol and $|\vec{r}|$, for example, refers to the magnitude of $\vec{r}$.)

Then, in time t, the object executes an angle $\theta = \omega t$ radian, and its y coordinate is now

$$y(t) = A \sin\theta\,(t) = A \sin\,(\omega t)$$

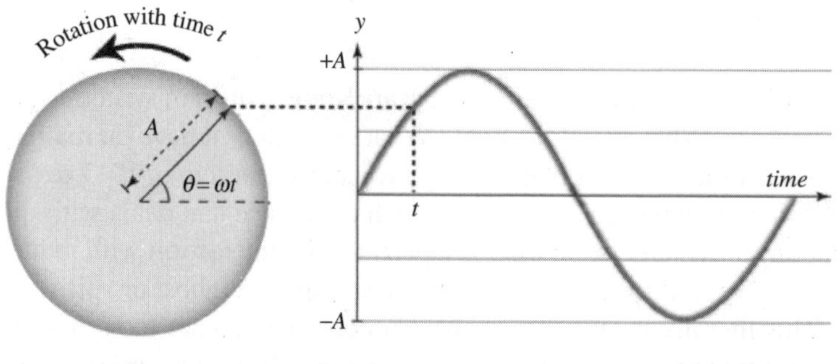

Figure 1.1. Oscillation of a ball in circular motion

When θ reaches $\pi / 2$ (12 o'clock position), the vertical component y achieves its maximum value, i.e., $y = A \sin\,(\pi / 2) = A$. At $\theta = \pi$ (9 o'clock position), y is again zero, and at $\theta = 3\pi/2$ (6 o'clock position), $y = -A$. We can trace $y(t)$ as a function of time as in the panel on the right-hand side of Figure 1.1 until the object reaches its original position when $\theta = 2\pi \equiv 0$. This completes one cycle. Subsequent motion is a repetition of the previous cycle, since $\sin\,(\theta + 2\pi) = \sin\,(\theta)$.

The maximum value of $y(t)$, i.e., A, is called the *amplitude* of the wave. The time taken for completion of a cycle is called the *time period* (T) of the wave and is measured in second (s). A cycle is completed when θ spans 2π; therefore, $\omega T = 2\pi$, or $T = 2\pi / \omega$. Since the pattern repeats itself every T seconds, the number of cycles completed per second is the reciprocal of T. This is called the *frequency* of the wave and is given the symbol ν ($= 1 / T$) with unit s^{-1} (Hz). The maxima in the peaks are called *crests* and the minima are named *troughs* (Figure 1.2). The distance traversed by the object in one cycle, called the *wavelength*, is denoted by the symbol λ.

It is measured as the distance between two successive peaks or crests, troughs or corresponding zero crossings, and is expressed in metre (m).

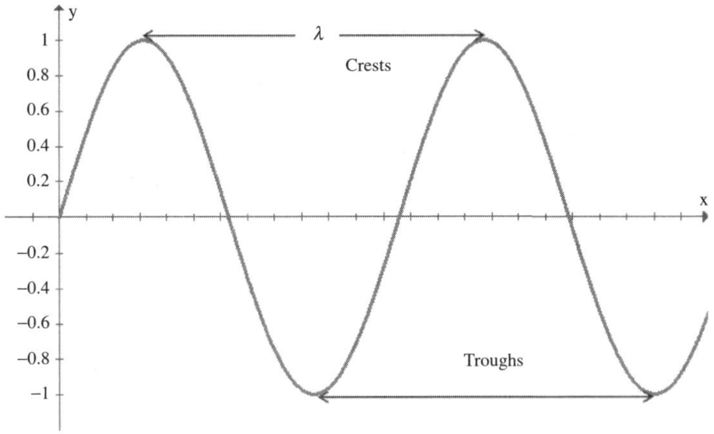

Figure 1.2. The wavelength of a sine wave, λ, measured between crests

The *speed* of light in vacuum is represented by c_0, and is equal to ~3×10^8 m s^{-1} (precisely 299,792,458 m s^{-1}). For media other than vacuum, the speed of light in vacuum, c_0, is replaced by c, the speed in the medium, given by c_0 / n, where n is the refractive index of the medium. Henceforth, we shall use the symbol c for the speed of light in any medium (with $n = 1$ for vacuum). The frequency and wavelength are related by the expression $\lambda\nu = c$; hence, short wavelength radiation has high frequency. The frequency is a more fundamental property than the speed and wavelength of the radiation, and remains constant when radiation propagates through media of different densities, whereas the other two change with the medium.

Another quantity frequently used by spectroscopists is the *wavenumber* ($\tilde{\nu}$), defined as a count of the number of wave crests (or troughs) in a given unit of length: $\tilde{\nu} = \nu / c = 1 / \lambda$. The dimensions of wavenumber are inverse length. The SI unit for the wavenumber is m^{-1}, but the unit cm^{-1} is still in use. For example, light of 400 nm wavelength is given a wavenumber of 25,000 cm^{-1}, rather than 2.5×10^6 m^{-1}. The conversion from cm^{-1} to m^{-1} may be performed by multiplying the wavenumber in cm^{-1} by 100. Alternatively, to express the wavenumber in cm^{-1}, use the value $2.99792458 \times 10^{10}$ cm s^{-1} for c, in place of its value in m s^{-1} to convert wavelength to wavenumber.

Note: Conversion from the wavelength in nm to wavenumber in cm^{-1}:

$$\tilde{\nu}[\text{cm}^{-1}] = \frac{10^7 \text{ cm}^{-1}\text{nm}}{\lambda[\text{nm}]}$$

since 1 nm = 10^{-7} cm.

There is still another term that needs to be defined. The case of a wave starting at zero and immediately increasing (Figure 1.1) is a special one. In general, the oscillation may start from any point, even the highest, as shown in Figure 1.3. In such a case, one may express this wave in terms of the cosine function, since we know that the cosine of zero radian is unity. However, there is another way of expressing this function without changing the general form of our equation. As Figure 1.3 shows, the cosine function is just the sine function, started a little earlier, and the cosine function can be expressed in terms of the sine function just by adding the starting angle. Thus, for the wave shown in Figure 1.3, we may equivalently write

$$y(t) = A\cos\theta(t) \equiv A\sin\left(\theta(t) + \frac{\pi}{2}\right) \tag{1.1}$$

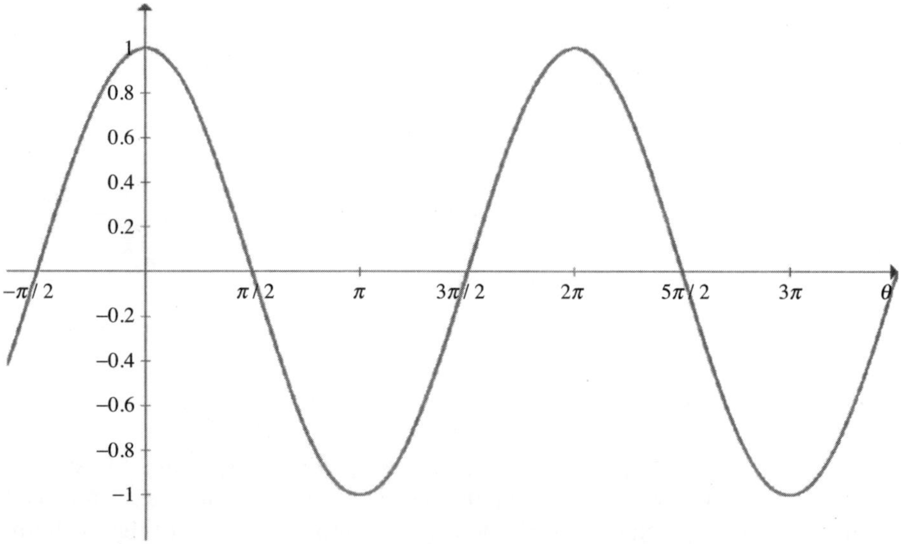

Figure 1.3. A cosine wave expressed as a sine wave with $\phi = \pi / 2$

The angle at which an oscillation starts is referred to as the *phase angle*, or simply the *phase* of an oscillation. It is commonly given a symbol ϕ and is expressed in radian. The general form of the equation is thus $y = A\sin(\theta + \phi') \equiv A\cos(\theta + \phi)$. We have seen that the wave shown in Figure 1.3 can be equivalently expressed in terms of the cosine function, with a phase angle of zero or as a sine function with a phase of $\pi / 2$ [equation (1.1)]. Using the relationship $\cos\theta = \sin\left(\theta + \frac{\pi}{2}\right)$, it is possible to completely describe wave motion in terms of either the sine or cosine functions, just by changing the phase angle.

A trigonometrically more tractable formulation is in terms of the complex exponential $y = Ae^{i(\theta+\phi)}$. Only the real part of this expression has any physical interpretation, and this can be extracted using Euler's formula $y = Ae^{i(\theta+\phi)} = A\left(\cos(\theta+\phi) + i\sin(\theta+\phi)\right)$. Since the cosine function represents the real part of the exponential function, we shall follow the cosine function in our future derivations.

In terms of the various quantities that we have developed, we are now in a position to express $y(t)$ in a variety of forms:

$$\begin{aligned} y(t) &= A\cos(\theta + \phi) = A\cos(\omega t + \phi) = A\cos(2\pi\nu t + \phi) \\ &= A\cos\left(\frac{2\pi t}{T} + \phi\right) = A\cos\left(\frac{2\pi c t}{\lambda} + \phi\right) \end{aligned} \tag{1.2}$$

Of these, the relations in terms of the frequencies (ω and ν) will be used in what follows.

1.2.1 Travelling waves

Light waves are travelling waves and we now discuss the form taken by equation (1.2) for a travelling wave. A travelling wave is any kind of wave that propagates in a single direction (say, z) with negligible change in shape. Hence, it is a function of both spatial (z) and temporal (relating to time, t) variables. We now combine the two dependencies and write

$$y(z, t) = A\cos\phi(z, t) = A\cos(\alpha z + \beta t) \tag{1.3}$$

For this oscillation to move through space, i.e., towards positive z, the point z_0 in space at a time t_0 must move to $z_1 > z_0$ at time $t_1 > t_0$ (Figure 1.4).

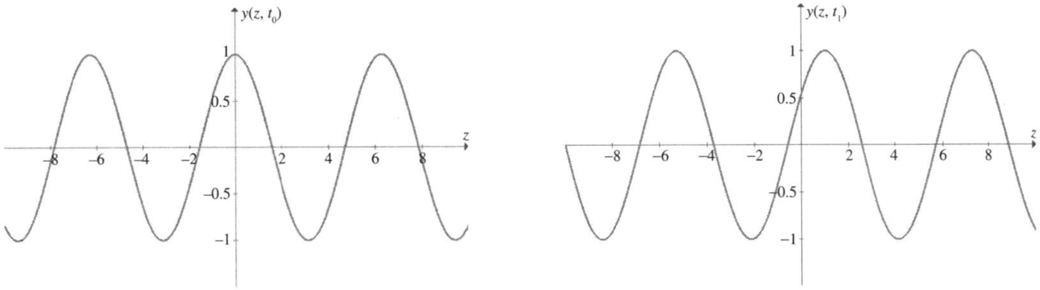

Figure 1.4. 'Snapshots' of a sinusoidal wave at two different times t_0 and $t_1 > t_0$, showing motion of the peak originally at the origin at t_0. The wave is travelling towards $z = +\infty$

The phase of the first wave (t_0) at the origin is zero, but that of the second (at t_1) is negative. Since the wave at location z_1 and time t_1 has the same phase as the wave at location z_0 and time t_0, we can say that:

$$y(z_0, t_0) = y(z_1, t_1)$$

In addition, for the wave to maintain its shape, the phase must be a linear function of z and t; otherwise the wave would compress or stretch out at different locations in space or time. Therefore:

$$\alpha z_0 + \beta t_0 = \alpha z_1 + \beta t_1, \text{ or } \alpha(z_1 - z_0) = -\beta(t_1 - t_0).$$

As discussed, if $t_1 > t_0 \Rightarrow z_1 > z_0$ (i.e., the wave moves towards $z = +\infty$), then α and β must have opposite algebraic signs:

$$\phi(z, t) = |\alpha|z - |\beta|t$$

Since it is the argument of the cosine function, $|\alpha|z - |\beta|t$ has the 'dimensions' of angle (radian). We have already identified $\beta = \omega$ [equation (1.2)], the angular frequency of the oscillation. Similarly, since z has the dimensions of length, α must have the dimensions of rad m^{-1}, i.e., α tells us how many radians of oscillation exist per unit length – the angular spatial frequency of the wave, commonly denoted by k ($= 2\pi / \lambda$): $y_+(z, t) = A\cos(kz - \omega t + \phi)$ – travelling harmonic wave towards $z = +\infty$ with arbitrary phase ϕ. Similarly, for a harmonic wave moving towards $z = -\infty$, $y_-(z,t) = A\cos(kz + \omega t + \phi)$ – travelling harmonic wave towards $z = -\infty$ with arbitrary phase ϕ.

1.3 ELECTROMAGNETIC RADIATION

Up till now, we have not defined $y(t)$; it could be any variable that changes as the wave passes. In the case of light waves, it is their vibrating electric and magnetic fields. Electromagnetic radiation is so named because it consists of mutually perpendicular electric and magnetic fields, normal to the direction of propagation of the wave. The connection of optics with electricity and magnetism was first realized by Maxwell through his equations, which form the basis of all electricity and magnetism. However, towards the end of the nineteenth century, some puzzling phenomena regarding light could not be explained by classical physics, and the quantum theory had to be invoked to explain these. For example, the wave theory of light could explain most phenomena relating to light, such as propagation in a straight line, reflection, refraction, superposition, interference, diffraction, polarization and the Doppler effect, but it could not explain certain other observations regarding blackbody radiation (electromagnetic radiation emitted by a heated object), photoelectric effect (emission of electrons by an illuminated metal) and spectral lines (emission of sharp spectral lines by gas atoms in an electric discharge tube). Young's double slit experiment firmly established the quantum mechanical explanation that both light and matter have dual nature, i.e., they can behave *as either wave or particle, depending on what question you ask: there is a 'wave' aspect and a 'particle' aspect, too*. Thus, the intensity of light depends on the square of the amplitude of the wave, and its energy depends on the frequency of the photon. Since we are more familiar with the wave nature of light, we first characterize electromagnetic radiation as a wave.

1.3.1 Wave nature of light

Light is energy that travels in a straight line in the form of electromagnetic waves. Like all waves, when it encounters an object, light may get absorbed, diffracted, reflected, refracted, scattered or transmitted, depending on the shape and composition of the object, and on the light's wavelength. In fact, many of these processes may occur simultaneously, because objects have uneven compositions or shapes, and beams of light are not monochromatic, i.e., they may include many wavelengths.

Light can be considered as oscillations of an electromagnetic field – characterized by electric ($\vec{E}$) and magnetic ($\vec{B}$) components – perpendicular to the direction of light propagation and to each other (Figure 1.5).

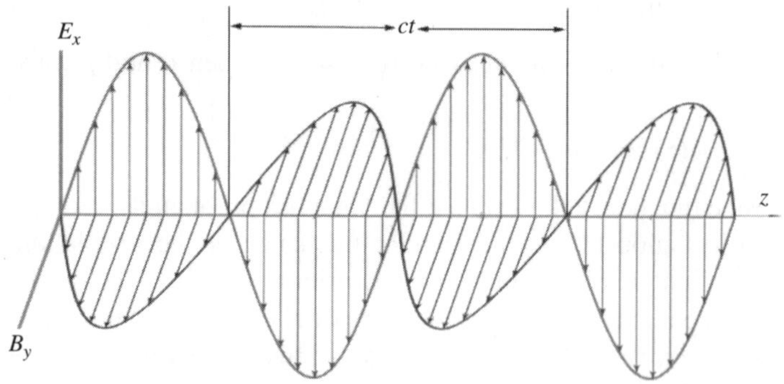

Figure 1.5. A schematic view of an electromagnetic wave propagating along the *z*-axis

It can be shown from Maxwell's relations that the magnitudes of the electric and magnetic fields are related by the wave speed, i.e., $|\vec{E}| = c|\vec{B}|$. The electric and magnetic fields oscillate in the xy plane perpendicular to the direction of propagation (Figure 1.5). As the electric field changes, so does the magnetic field in tandem. The wave represented in this figure has its electric fields aligned in the xz plane and pointing in the x direction. The magnetic fields are aligned in the yz plane and all vectors point in the y direction.

1.3.1.1 Polarization

Electromagnetic radiation has another (and sometimes important to spectroscopists) property: *polarization*. In contrast to sound waves, electromagnetic waves are *transverse waves*, i.e., the oscillations occur perpendicular to the direction of propagation of the wave. For an electromagnetic wave propagating in the z-direction, there are two transverse directions along which the electric and magnetic fields can lie. The $\vec{E}$ vectors could lie either in the x or y directions, or anywhere in the xy plane, i.e., $\vec{E} = \hat{i}E_x + \hat{j}E_y$, where $\hat{i}$ and $\hat{j}$ are unit vectors along the x and y directions, respectively. The polarization of light is defined by the orientation of the wave's electric field. Natural light is generally unpolarized, i.e., its electric field vectors are in all random directions in the xy plane (Figure 1.6). The electromagnetic wave shown in Figure 1.5 is an example of a *plane* polarized light since all the electric field vectors lie in a single plane, the vertical plane (xz), pointing in the x direction. The electric and magnetic fields for this wave are given by

$$\vec{E} = E_0 \cos(kz - \omega t)\hat{i}$$
$$\vec{B} = B_0 \cos(kz - \omega t)\hat{j} = \frac{E_0}{c}\cos(kz - \omega t)\hat{j}$$

This wave has $\vec{E}$ always pointing in the x direction and $\vec{B}$ always pointing in the y direction. A wave like this, where the fields always point along given directions, is also said to be *linearly polarized*. The example wave is linearly polarized in the x direction because the electric field vectors are all in this direction.

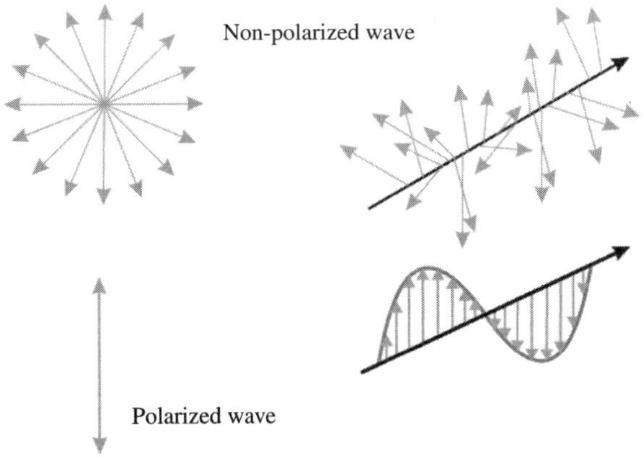

Figure 1.6. Electric field $\vec{E}$ orientation for polarized and non-polarized electromagnetic waves

1.3.2 Particulate nature of radiation

Radiation can be also described in terms of particles of energy, called *photons*. A photon has energy but has neither mass nor charge. Its energy (in Joule) is given as:

$$\varepsilon_{\text{photon}} = h\nu = hc / \lambda = hc\tilde{\nu} \tag{1.4}$$

where h is Planck's constant ($h = 6.626075 \times 10^{-34}$ J s). Equation (1.4) relates the energy of each photon of the radiation to the electromagnetic wave characteristics ($\tilde{\nu}$ and λ). It was Planck who, in 1901, first postulated this relation more as a matter of mathematical necessity to explain blackbody radiation (discussed in Section 1.4), than out of a belief that light consisted of discrete quanta. Einstein, on the other hand, took this mathematical 'trick' and interpreted it literally: if light could only possess energies which were integer multiples of the Planck discrete energy, $E = h\nu$, perhaps light was actually *composed* of discrete packets (photons), each possessing an energy $h\nu$.

By extending Planck's hypothesis of energy quantization to include not just the absorption and emission mechanism but to the light itself, Einstein succeeded in explaining certain puzzling phenomena related to the photoelectric effect. Experiments on the emission of photoelectrons from the surface of metals showed that low-frequency radiation had no effect, but once a threshold frequency was crossed, emission occurred with no time lag. The threshold frequency was dependent on the 'work function' (ionization energy) of the metal. Higher frequencies of the incident radiation led to ejection of electrons with higher kinetic energies. All these observations could be explained if it was assumed that the incident radiation consists of photons of energy $h\nu$, and the maximal kinetic energy of the ejected electrons is given by $E_k = h\nu - \phi$, where ϕ is the work function of the metal, equal to $h\nu_0$, where ν_0 is the threshold frequency.

The photon energy is directly proportional to the wavenumber. However, spectra are usually recorded in terms of the wavelength, so a conversion to energy is required. Putting in the values of h and c, we find that the conversion factor from wavelength to photon energy is

$$hc = 1.986447 \times 10^{-25} \text{ J m} = 1.986447 \times 10^{-16} \text{ J nm}$$

Table 1.1 summarizes some of the characteristics of electromagnetic radiation studied so far, their relationship with wavelength and their common units of measurement.

The internal energies of atoms are small (~10^{-19} J) because of their small size (~10^{-10} m). The individual photon energies are also of the same order and may be more conveniently expressed in *electron volt* (eV), a unit equal to the kinetic energy imparted to an electron when it is accelerated by a potential of 1 V. The conversion factor between electron volt and Joule is numerically equal to the charge on an electron.

$$1 \text{ eV} = 1.60217733 \times 10^{-19} \text{ C} \times 1 \text{ V} = 1.60217733 \times 10^{-19} \text{ J}$$

Combining this with equation (1.4) gives the following conversion factor when the photon energy is expressed in electron volt and wavelength in nanometre.

$$\varepsilon_{\text{photon}}[\text{eV}] = 1240 / \lambda[\text{nm}]$$

Table 1.1. Characteristics of electromagnetic radiation

Characteristic	Common units	Relationship with wavelength
Wavelength	m μm (10^{-6} m) nm (10^{-9} m)	
Wavenumber	m^{-1} cm^{-1}	$\tilde{\nu} = \frac{1}{\lambda}$
Frequency	s^{-1} (Hz) kHz (10^3 s^{-1}) MHz (10^6 s^{-1}) GHz (10^9 s^{-1})	$\nu = \frac{c}{\lambda}$
Speed	m s^{-1}	$c = \nu\lambda$
Energy	Joule (J)	$E = \frac{hc}{\lambda}$

To appreciate the vast range of photon energies, consider that yellow light ($\lambda \approx 600$ nm) has a photon energy of ~2 eV, while X-rays from a copper source, with a wavelength of 0.154 nm, have a photon energy of about 8000 eV. This shows that at least 8000 V is needed to give electrons sufficient energy to produce these X-rays.

The energies discussed above, although small by macroscopic standards, are very significant at the level of individual atoms or molecules. This can be seen by multiplying them by Avogadro's number (N_A), so as to give the energy per mole. The conversion factor from electron volt per molecule to Joule per mole is eN_A, giving

$$1 \text{ eV per molecule} = 96.5 \text{ kJ mol}^{-1}$$

Table 1.2 summarizes the conversion factors amongst various energy units.

Table 1.2. Conversion factors between radiation frequency, wavenumber, photon energy and the corresponding energy per mole

	Corresponding value in			
Unit	**Hz**	**cm^{-1}**	**eV**	**kJ mol^{-1}**
1 Hz	1	3.336×10^{-11}	4.136×10^{-15}	3.990×10^{-13}
1 cm^{-1}	2.998×10^{10}	1	1.236×10^{-4}	1.196×10^{-2}
1 eV	2.418×10^{14}	8066	1	96.49
1 kJ mol^{-1}	2.506×10^{12}	83.60	1.036×10^{-2}	1

Example 1.1 A light bulb of 60 W emits at a wavelength of 0.5 μm. Calculate the number of photons emitted per second.

Solution

The energy of one photon is $\varepsilon_{\text{photon}} = hc / \lambda$; thus, since 60 W = 60 J s^{-1}, the number of photons per second, N, is

$$N = \frac{60\,(\text{J s}^{-1})\,\lambda(\text{m})}{h\,(\text{J s})\,c\,(\text{m s}^{-1})} = \frac{60 \times 0.5 \times 10^{-6}}{6.6256 \times 10^{-34} \times 2.9979 \times 10^{8}} = 1.510 \times 10^{20}\ \text{s}^{-1}$$

Note: A large number of photons is required because Planck's constant h is very small! It is hardly surprising that the quantum nature of this light is not usually apparent.

Example 1.2 Calculate the energy in Joule and the wavenumber in cm^{-1} of:
(a) A photon with wavelength 1 μm
(b) A photon with frequency 6.00 × 10^{14} Hz

Solution

(a) Energy, $E = hc / \lambda$; hence for a wavelength 10^{-6} m:

$$E = (6.626 \times 10^{-34}\ \text{J s}) \times (2.998 \times 10^{8}\ \text{m s}^{-1}) / (10^{-6}\ \text{m}) = 1.986 \times 10^{-19}\ \text{J}$$

Wavenumber, $\tilde{\nu} = 1 / \lambda$; hence $\tilde{\nu} = 1 / (10^{-6}\,\text{m}) = 10^{6}\ \text{m}^{-1} = 10^{4}\ \text{cm}^{-1}$

(b) Energy $E = h\nu$; hence, $E = (6.626 \times 10^{-34}\ \text{J s}) \times (6.00 \times 10^{14}\ \text{s}^{-1}) = 3.98 \times 10^{-19}\ \text{J}$

$\tilde{\nu} = \nu / c = (6.00 \times 10^{14}\ \text{s}^{-1}) / (2.998 \times 10^{8}\ \text{m s}^{-1}) = 2.00 \times 10^{6}\ \text{m}^{-1} = 2.00 \times 10^{4}\ \text{cm}^{-1}$

Example 1.3 Two energy levels are separated in wavenumber by 200 cm^{-1}. Convert this energy to Joule.

Solution

$$E = hc\tilde{\nu} = (6.626 \times 10^{-34}\ \text{J s}) \times (2.998 \times 10^{8}\ \text{m s}^{-1}) \times (20000\ \text{m}^{-1}) = 3.97 \times 10^{-21}\ \text{J}$$

Note: It is essential to convert from cm^{-1} to m^{-1} if we want the final answer in SI units of Joule.

Example 1.4 k_BT, where k_B is Boltzmann's constant, is an important property in chemistry with units of energy. Calculate the value of k_BT at 10 K, 100 K and 300 K, giving your answer in J, kJ mol^{-1} and cm^{-1}.

Solution

At 10 K: $k_BT = (1.381 \times 10^{-23}\ \text{J K}^{-1}) \times (10\ \text{K}) = 1.381 \times 10^{-22}$ J
$= (1.381 \times 10^{-22}\ \text{J}) \times (6.022 \times 10^{23}\ \text{mol}^{-1}) = 0.0832\ \text{kJ mol}^{-1}$
$= (1.381 \times 10^{-22}\ \text{J}) / \{(6.626 \times 10^{-34}\ \text{J s}) \times (2.998 \times 10^{8}\ \text{m s}^{-1})\}$
$= 695\ \text{m}^{-1} = 6.95\ \text{cm}^{-1}$
Likewise, at 100 K, $k_BT = 1.381 \times 10^{-21}\ \text{J} = 0.832\ \text{kJ mol}^{-1} = 69.5\ \text{cm}^{-1}$
and at 300 K, $k_BT = 4.142 \times 10^{-21}\ \text{J} = 2.494\ \text{kJ mol}^{-1} = 208.5\ \text{cm}^{-1}$

Example 1.5 Louis de Broglie proposed (in his doctoral thesis of 1924) that any particle moving with a momentum p can be thought of as having a wavelength given by $\lambda = h/p$. When asked, in his *viva voce* examination, to suggest an experiment that could confirm his hypothesis, he observed that electrons would have a suitable wavelength to be diffracted by crystal lattices, when accelerated to easily attainable velocities. This experiment was promptly carried out by Davisson and Germer, who confirmed de Broglie's hypothesis.

Through what voltage must electrons, initially at rest, be diffracted to first order through an angle of 30°, when transmitted through a crystal with atomic spacing 0.1 nm?

Solution

Bragg's relation is

$$n\lambda = 2d \sin\theta$$

It is given that $n = 1$, $d = 0.1$ nm, $\theta = 30°$. Therefore, $\lambda = 2 \times 0.1 \times 10^{-9} \times 0.5 = 10^{-10}$ m. Since $\lambda = h / p$, $p = h / \lambda = 6.626 \times 10^{-34}\ \text{J s} / 10^{-10}\ \text{m} = 6.626 \times 10^{-24}\ \text{kg m s}^{-1}$.

The kinetic energy E_k can be related to the linear momentum because $E_k = ½\ mv^2$ and $p = mv$. Therefore, $E_k = p^2 / (2m) = (6.626 \times 10^{-24}\ \text{kg m s}^{-1})^2 / (2 \times 9.109 \times 10^{-31}\ \text{kg}) = 2.410 \times 10^{-17}$ J.

A charge q accelerated through a potential V acquires a kinetic energy $E_k = qV$. Therefore, $V = E_k / q = 2.410 \times 10^{-17}\ \text{J} / 1.602 \times 10^{-19}\ \text{C} = 150.4$ V.

Example 1.6 Calculate the momentum of a photon whose wavenumber is 20 cm^{-1}.

Solution

$\tilde{\nu} = 1 / \lambda$; hence, $\lambda = 5 \times 10^{-4}$ m (500 μm).

From the de Broglie expression, $p = h / \lambda = (6.626 \times 10^{-34}\ \text{J s}) / (5 \times 10^{-4}\ \text{m}) = 1.325 \times 10^{-30}\ \text{kg m s}^{-1}$ since 1 J = 1 kg m^2 s^{-2}.

1.4 ELECTROMAGNETIC SPECTRUM

The electromagnetic spectrum spans a large range of frequencies (Figure 1.7). Therefore, it is divided into various regions, starting with the lowest energy radiofrequency region (3 MHz – 3 GHz) to the highest energy γ-rays (>3 × 10^{20} Hz). The units used to describe radiation in various regions of the electromagnetic spectrum are based on convenience. For example, frequency units (Hz) are used to describe the radiofrequency and microwave regions, wavenumber in cm^{-1} for infrared and wavelength in nm for the higher energy radiation. In the radiofrequency region, the photons have sufficient energy only to flip nuclear spins in magnetic fields of a few tesla, and nuclear magnetic resonance (NMR) spectroscopy is studied in this region. In the next higher energy region (the microwave region, 3 GHz – 3000 GHz), the energy is sufficient to cause rotational transitions and electron spin flips (electron spin resonance, ESR). Unlike other forms of spectroscopy, NMR and ESR are resonance techniques, wherein the energy levels are split by interaction with the oscillating magnetic field of the electromagnetic radiation. In the infrared (100 cm^{-1} – 13,000 cm^{-1}) region, vibrational transitions are induced. Valence electron transitions in molecules involve quanta of energy corresponding to the visible and ultraviolet (UV) regions (1000 nm – 10 nm), while core electron transitions are observed with X-rays (10 nm – 0.01 nm). Photoelectron spectroscopy and X-ray diffraction studies are also made using X-rays. Finally, γ-rays (<0.01 nm) are associated with nuclear processes, such as those observed in Mössbauer spectroscopy. The visible part of the electromagnetic spectrum lies between the ultraviolet and infrared regions (between about 400 and 780 nm). It is important to note that higher the frequency (shorter the wavelength), higher the photon energy. Hence, radiowaves are at the long wavelength end of the spectrum and γ-rays are at the short wavelength end of the spectrum.

As shown in Figure 1.7, the region encompassing infrared, UV and visible is further subdivided. The infrared region is divided into far-infrared (33 – 333 cm^{-1}), where rotations of light molecules, particularly the hydrides, phonons of solids, metal–ligand vibrations, as well as ring-puckering and torsional motions of many organic molecules, can be found. The fundamental vibrations of most molecules lie in the mid-infrared region (333 – 3333 cm^{-1}). Finally, the near-infrared region (3333 – 13,000 cm^{-1}) is associated with overtone vibrations

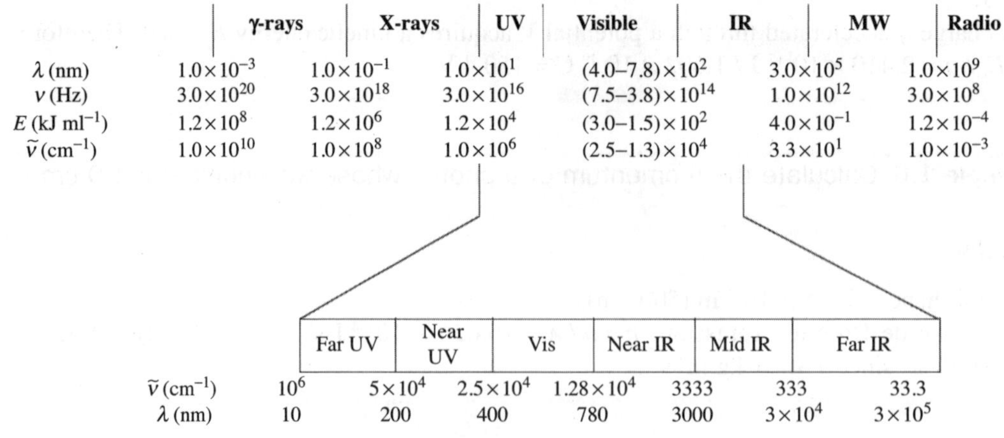

Figure 1.7. The electromagnetic spectrum

and some electronic transitions. Similarly, the visible region is divided into the colours of the rainbow, with red being at the low-energy end (780 nm) and violet at the high-energy end (400 nm). The next region (400 nm – 200 nm) is the near-UV region. Below 200 nm (to about 10 nm), the region is named vacuum-UV because air, particularly the oxygen in it, absorbs in this region, and measurements can only be made if the chamber is evacuated. The regions from 400 nm – 315 nm, 315 – 280 nm and 280 nm – 100 nm are also called UVA, UVB and UVC, respectively. These regions are of interest, as they are the ultraviolet light bands ordinarily encountered from the Sun's emission on Earth.

Example 1.7 In which region of the electromagnetic spectrum do photons with the following wavenumbers lie?
10 cm^{-1}, 2000 cm^{-1}, 20000 cm^{-1}, 40000 cm^{-1}

Solution

From Figure 1.7, these lie, respectively, in the microwave, infrared, visible and ultraviolet regions.

The various types of spectroscopy and associated transitions are summarized in Table 1.3. However, it must be emphasized that there is no strict boundary between regions of the electromagnetic spectrum and the kind of transitions that occur. For example, the microwave region is associated with molecular rotations, but molecular vibrations of light molecules are also observed in this region.

Table 1.3. Types of spectroscopy, showing the typical energies involved and the different types of quantized energy levels probed

Type of radiation	Typical energies		Types of transitions observed
	eV per photon	kJ mol^{-1}	
Radiofrequency	10^{-6}	10^{-4}	NMR
Microwave	$10^{-4} - 10^{-3}$	$0.01 - 0.1$	ESR Molecular rotations
Infrared	$0.01 - 0.1$	$1 - 10$	Molecular vibrations
UV-Visible	$1 - 10$	$10^{2} - 10^{3}$	Bonding electrons
X-rays	$10^{2} - 10^{4}$	$10^{4} - 10^{6}$	Inner-shell electrons
γ-rays	$10^{5} - 10^{6}$	$10^{7} - 10^{8}$	Nuclear internal structure

1.5 BLACKBODY RADIATION

Historically, Planck's study of emission from a blackbody laid the foundation of quantum theory. Blackbody radiation also has practical applications, e.g. in determining stellar temperatures. Consider the cavity shown in Figure 1.8 at a constant temperature, T. Equilibrium exists between the radiation emitted from the cavity and that absorbed by its walls.

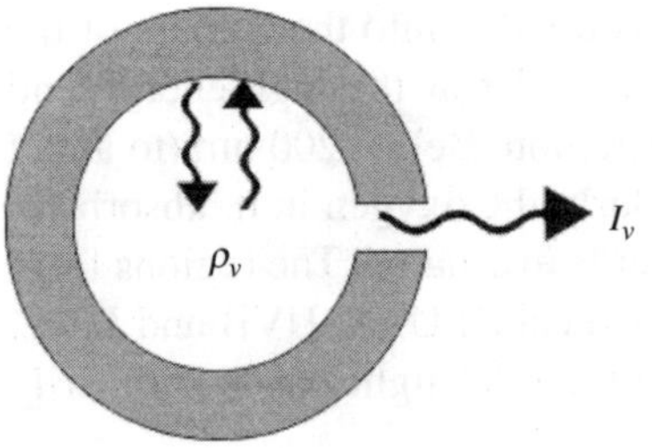

Figure 1.8. Cross-section of a blackbody cavity at a temperature T with a radiation density ρ_v emitting radiation with intensity I_v from a small hole

The energy density of radiation, ρ (in units J m^{-3}) inside the cavity is given by $\rho = \int_0^\infty \rho(\nu)\,d\nu$, where $\rho(\nu) = d\rho / d\nu$ is the frequency distribution of this radiation. Since this represents the radiation density in the frequency interval ν to $\nu + d\nu$, its units are J s m^{-3}. Assuming that the material of the cavity consists of 'oscillators' with quantized energy levels ($nh\nu$, where n is an integer). Planck obtained the following universal function (a function that applies to all shapes and sizes of the cavity, and is also independent of the material of the cavity),

$$\rho(\nu, T) = \left(\frac{8\pi h\nu^3}{c^3}\right)\left(\frac{1}{e^{h\nu/k_BT} - 1}\right) \tag{1.5}$$

named Planck's distribution law. We shall have occasion to use this equation in our derivation of transition probabilities. The intensity is related to the energy density by the formula $I = c\rho$. (Show that this expression is dimensionally correct.)

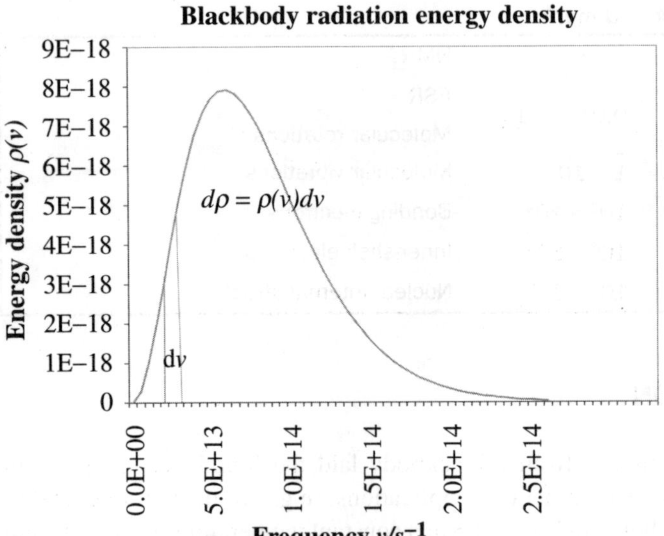

Figure 1.9. The distribution function $\rho(\nu)$ at 1000 K

The first factor in equation (1.5) increases with increasing frequency of the radiation, but the second, temperature dependent factor, decreases with increasing ν because of the exponential dependence on ν in the denominator. At a particular temperature, initially the first factor dominates and there is an increase in the distribution function with frequency, followed by a decrease at higher frequencies due to dominance of the second term. As a result, the function reaches a maximum at some frequency. At 1000 K, the maximum is at a frequency of 7.5×10^{13} s^{-1} (Figure 1.9), which corresponds to the infrared region (Figure 1.7). This maximum shifts to higher frequencies with increasing temperature. Planck attributed this to the increase in the number of oscillators in the higher energy levels with increasing temperature as a result of the Boltzmann distribution. Thus, temperature appears to have a profound effect on the frequency distribution function and the population distribution in the different available energy levels of atomic and molecular systems, and so we shall discuss populations in the next section.

1.6 BOLTZMANN'S POPULATION DISTRIBUTION

It is well known that atomic and molecular systems can exist only in certain well-defined quantum states. In the case of atomic systems, the allowed energy levels depend upon the orbital and spin quantum numbers of the electrons. These are called electronic states, and these are the only states possible for atoms. In addition to electronic energy states, molecules may also have nuclear energy states arising from molecular rotations and vibrations. According to the Boltzmann distribution law, the population N_m of a level m is proportional to $g_m e^{-E_m/k_BT}$:

$$N_m = g_m \frac{N}{q} e^{-E_m/k_BT} = g_m N f_m$$

where E_m is the level energy, g_m is the degeneracy of that level (i.e., the number of states having this energy; larger the number of states associated with an energy level, larger is its population), k_B is the Boltzmann constant, T is the absolute temperature, q is the partition function, N is the total number of molecules in the system and f_m represents the fractional population of the energy level m at thermal equilibrium. At the absolute zero of temperature, the atom does not possess any thermal energy and all the atoms are in the ground state. At other temperatures, the population of the excited states depends upon the ratio E_m / T. As the temperature increases, this ratio becomes smaller and, since the population is proportional to the exponential of the negative of this quantity, the population of the higher energy states increases, but it cannot exceed that of the ground state, unless the upper level is degenerate ($g_m > 1$).

Let us calculate the population of an energy level, having energy 2 kJ mol^{-1} relative to the energy of the ground state at 300 K.

$$\frac{N_m}{N_l} = \frac{g_m e^{-E_m/k_BT}}{g_l e^{-E_l/k_BT}} = \frac{g_m}{g_l} e^{-\Delta E_{lm}/k_BT}$$

$$\Delta E_{lm} = E_m - E_l = \frac{2\times10^3 \text{ J mol}^{-1}}{6.022\times10^{23} \text{ mol}^{-1}} = 3.321\times10^{-21}\text{ J}$$

$$\frac{N_m}{N_l} = e^{-\Delta E_{lm}/k_BT} = \exp\left(-\frac{3.321\times10^{-21}\text{ J}}{1.381\times10^{-23}\text{ J K}^{-1}\times300\text{ K}}\right) = 0.4486,$$

where we have assumed unit degeneracies for the two levels. This means that for every 10,000 molecules in the ground state, there are 4486 at a level 2 kJ mol^{-1} above it.

At 300 K (Example 1.4), k_BT is equivalent to 208 cm^{-1} (~2.5 kJ mol^{-1}). This is small relative to the first excited energy levels of most atoms and ions, which means that very few atoms will be in the excited state. However, in the case of molecules, some vibrational and many rotational energy levels may have energies lower than k_BT, thus allowing a relatively large population in the excited states. Rotational, vibrational and electronic energy intervals are typically of the order 1 cm^{-1}, 1000 cm^{-1} and 25,000 cm^{-1}, respectively. We expect, therefore, a large number of rotational levels to be populated at room temperature, while only the lowest vibrational and electronic energy levels will be occupied at normal temperatures. Although vibrational energy levels are singly degenerate, in the case of rotational energy levels, the degeneracies of the levels also increase with increasing energy, leading to higher populations of the higher energy levels.

Example 1.8 A non-degenerate excited level (g_m = 1) of a molecule lies 10 cm^{-1} above the non-degenerate ground level (g_l = 1).

(a) Calculate the population of the excited level relative to that of the ground state at: T = 1 K, 10 K, 300 K, 10^6 K.

(b) At what temperature will the population of the excited level equal that of the ground level?

(c) Repeat the calculation for the case of a doubly degenerate excited level.

Solution

(a) Use the Boltzmann law:

$$\frac{N_m}{N_l} = \frac{g_m}{g_l}\exp\left(\frac{-\Delta E_{lm}}{k_BT}\right).$$

In this case, $g_l = g_m = 1$ and the energy gap is given by $\Delta E_{lm} = hc\tilde{\nu}$.

$$\frac{N_m}{N_l} = \frac{g_m}{g_l}\exp\left(\frac{-hc\tilde{\nu}}{k_BT}\right) = \frac{g_m}{g_l}\exp\left(\frac{-(6.626\times10^{-34})(2.998\times10^{10})\times\tilde{\nu}}{1.381\times10^{-23}T}\right) = \frac{g_m}{g_l}\exp\left(\frac{-1.438\tilde{\nu}}{T}\right)$$

with $\tilde{\nu}$ in cm^{-1} (it is useful to remember the magnitude of the quantity hc/k_B = 1.438 cm K). Hence,

$$\frac{N_m}{N_l} = \frac{g_m}{g_l}\exp\left(\frac{-hc\tilde{\nu}}{k_BT}\right) = \exp\left(\frac{-1.438\times10}{T}\right)$$

$$= 5.69\times10^{-7} \text{ at } 1\text{ K}$$

$$= 0.237 \text{ at } 10\text{ K}$$

$$= 0.953 \text{ at } 300\text{ K}$$

$$= 0.9999856 \text{ at } 10^6\text{ K}$$

(b) Only at $T = \infty$ (for a two energy level system at equilibrium, if the degeneracies of the levels are equal, the population of the lower level will always be greater than that of the higher energy level).

(c) Now since $g_m = 2$, we have

$$\frac{N_m}{N_l} = \frac{2}{1}\exp\left(-\frac{1.438\tilde{v}}{T}\right)$$

and

$$\frac{N_m}{N_l} = \frac{2}{1}\exp\left(-\frac{1.438\tilde{v}}{T}\right)$$

$$\begin{aligned} N_m / N_l &= 1.137\times 10^{-6} \text{ at } 1\text{ K} \\ &= 0.475 \text{ at } 10\text{ K} \\ &= 1.906 \text{ at } 300\text{ K} \\ &= 1.99997 \text{ at } 10^6\text{ K} \end{aligned}$$

The population of the two levels will be equal when

$$\frac{2}{1}\exp\left(\frac{-hc\tilde{v}}{k_B T}\right) = 1$$

or at

$$T = \frac{-hc\tilde{v}}{k_B \ln(0.5)} = -\frac{14.38}{\ln(0.5)} = 20.7\text{ K}.$$

Example 1.9 For temperatures of 25 °C and 1000 °C, calculate the ratio of molecules in a typical excited rotational, vibrational and electronic energy level to that in the lowest energy level, assuming that the levels are 30 cm^{-1}, 1000 cm^{-1}, and 40,000 cm^{-1}, respectively, above the lowest energy level. Assume that, for the excited rotational level, the rotational quantum number J is 4, and that each level is $(2J + 1)$-fold degenerate, i.e., $g_J = 2J + 1$. Assume that the vibrational and electronic levels are non-degenerate.

Solution

$$\begin{aligned} \frac{N_m}{N_l} = \frac{N_J}{N_0} = \frac{2J+1}{1}\exp\left(-\frac{hc\tilde{v}}{k_B T}\right) &= 9\exp\left(-\frac{1.438\tilde{v}}{T}\right) \\ &= 7.79 \text{ at } 25\text{ °C } (T = 298\text{ K}) \\ &= 8.70 \text{ at } 1000\text{ °C } (1273\text{ K}) \end{aligned}$$

For the vibrational energy level

$$\frac{N_m}{N_l} = \frac{N_v}{N_0} = \exp\left(-\frac{hc\tilde{\nu}}{k_B T}\right) = \exp\left(-\frac{1.438\tilde{\nu}}{T}\right)$$

$$= 8.02 \times 10^{-3} \text{ at } 25\ ^\circ\text{C}$$

$$= 0.323 \text{ at } 1000\ ^\circ\text{C}$$

For the electronic energy level

$$\frac{N_m}{N_l} = \frac{N_e}{N_0} = \exp\left(-\frac{hc\tilde{\nu}}{k_B T}\right) = \exp\left(-\frac{1.438\tilde{\nu}}{T}\right)$$

$$= 1.49 \times 10^{-84} \text{ at } 25\ ^\circ\text{C}$$

$$= 2.38 \times 10^{-20} \text{ at } 1000\ ^\circ\text{C}$$

The above example illustrates the fact that, at room temperature, many excited rotational energy levels are even more populated than the ground state, the excited vibrational levels have a population ~1% of the ground state, but this increases with temperature, while the excited electronic levels have negligible population compared to the ground state. However, at the very high stellar temperatures, even the second level of the hydrogen atom is populated.

1.7 EINSTEIN'S COEFFICIENTS

We have described the nature of electromagnetic radiation and the energies and populations associated with various energy levels of atomic and molecular systems. We shall now discuss what happens when such a system is irradiated with electromagnetic radiation. Consider a system consisting of a lower level l at a constant temperature T, and let there be an excited state m. Since the system is in thermal equilibrium, the number of molecules, N_m, in the upper state is related to those in the lower state, N_l, by the Boltzmann relation

$$N_m = \frac{g_m}{g_l} N_l e^{-h\nu_{lm}/k_B T} \tag{1.6}$$

where k_B is Boltzmann's constant, the degeneracies of levels l and m are g_l and g_m, respectively, and $h\nu_{lm} = E_m - E_l$.

The system is now bathed with radiation density $\rho(\nu, T)$. Transition from the ground state l to the excited state m occurs by absorption of radiation density of the precise frequency ν_{lm}, as required by Bohr's condition. The rate of induced transition is proportional to the population of the lower level and the radiation density at ν_{lm}, and is given by

$$\frac{dN_m}{dt} = B_{lm} N_l \rho(\nu_{lm}) \tag{1.7}$$

where the coefficient B_{lm} ($B_{l\to m}$) is known as the *Einstein coefficient of induced (stimulated) absorption*, and is thus a 'rate constant'. The same radiation can also cause a reverse transition. The rate of induced emission is given by

$$\frac{dN_m}{dt} = -B_{ml} N_m \rho(\nu_{lm}) \tag{1.8}$$

where B_{ml} is the *Einstein coefficient of induced emission* since this emission takes place under the influence of the radiation field. The negative sign indicates that this process reduces the population of the excited state.

In addition, the system in the higher energy state m may spontaneously lose an amount of energy $E_m - E_l = h\nu_{lm}$ to come back to the state l. The *Einstein coefficient of spontaneous emission* is denoted by A_{ml}. The number of molecules passing spontaneously from m to l is given by $N_m A_{ml}$. (Notice that there is no dependence on the radiation density, since this is not a transition caused by the applied radiation.)

$$\frac{dN_m}{dt} = -A_{ml} N_m \tag{1.9}$$

This quantity can be derived from the Dirac radiation theory, but here we will derive it in an indirect manner through relations between the Einstein coefficients of induced absorption and emission and the coefficient of spontaneous emission. Figure 1.10 summarizes the various processes taking place.

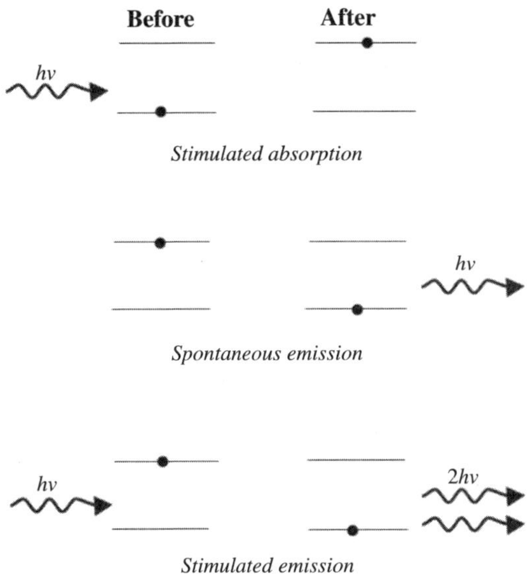

Figure 1.10. Schematic representations of stimulated absorption (top), spontaneous emission (middle) and stimulated emission (bottom) processes in a two-level system

At equilibrium, the number of molecules going from state l to state m must be equal to those going down from state m to state l. Hence, we have the condition

$$N_l B_{lm} \rho(\nu_{lm}) = N_m B_{ml} \rho(\nu_{lm}) + N_m A_{ml} \tag{1.10}$$

Substituting equation (1.6) into equation (1.10), we get

$$N_l B_{lm}\rho(\nu_{lm}) = \frac{g_m}{g_l} N_l e^{-h\nu_{lm}/k_BT}\left(A_{ml} + B_{ml}\rho(\nu_{lm})\right)$$

$$\Rightarrow g_m A_{ml} = \rho(\nu_{lm})\left(g_l B_{lm} e^{h\nu_{lm}/k_BT} - g_m B_{ml}\right) \tag{1.11}$$

The left-hand side of equation (1.11) is independent of the temperature, so it follows that the term proportional to T on the right-hand side must vanish. This can only happen if

$$g_l B_{lm} = g_m B_{ml} \tag{1.12}$$

since this would allow us to factor out the time-dependent term, i.e., equation (1.11) becomes

$$g_m A_{ml} = \rho(\nu_{lm}) g_m B_{ml}\left(e^{h\nu_{lm}/k_BT} - 1\right)$$

The meaning of equation (1.12) is that the rate per atom of induced absorption is equal to the rate per atom of induced emission. Also, the probability of an induced transition is proportional to the degeneracy of the level *to* which the transition is taking place. The equality of the terms independent of T leads to

$$\frac{A_{ml}}{B_{ml}} = \rho(\nu_{lm})\left(e^{h\nu_{lm}/k_BT} - 1\right) \tag{1.13}$$

Applying Planck's radiation law (1.5)

$$\rho(\nu) = \frac{8\pi h\nu^3}{c^3}\left(e^{h\nu/k_BT} - 1\right)^{-1},$$

equation (1.13) becomes

$$\frac{A_{ml}}{B_{ml}} = \frac{8\pi h\nu^3}{c^3} = 8\pi h\tilde{\nu}^3 \tag{1.14}$$

Thus, we can calculate A_{ml} from B_{ml}. In fact, it is enough to determine one Einstein coefficient. The rest follow from equations (1.12) and (1.14).

The implications of equation (1.14) should be clearly appreciated. Since c is the speed of light, the ratio $8\pi h/c^3$ is extremely small. Hence, the importance of A_{ml} clearly depends on the magnitude of ν. At very low values of ν, such as those encountered in the radiofrequency, microwave and infrared regions ($\tilde{\nu}$ < 1000 cm^{-1}), the ratio A_{ml} / B_{ml} is very small, and only processes stimulated by radiation are important. In the ultraviolet and visible regions, where ν is high, spontaneous processes also gain importance.

It is, however, incorrect to compare the coefficients for spontaneous and induced emissions by making use of equation (1.14), because the dimensions of the two quantities in the ratio are not the same. For example, B_{ml} is expressed in m^3 J^{-1} s^{-2}, but A_{ml} is in s^{-1}. Rather, it is more appropriate to use equation (1.13) in the form

$$\frac{A_{ml}}{B_{ml}\rho(\nu_{lm})} = \left(e^{h\nu_{lm}/k_BT} - 1\right) \tag{1.15}$$

for determining the relative importance of spontaneous and induced absorption.

If at $T = 300$ K we examine the microwave region, $\nu_{lm} \approx 10^{10}$ s^{-1}, then

$$\frac{h\nu_{lm}}{k_BT} = \frac{6.626\times10^{-34}\times10^{10}}{1.381\times10^{-23}\times300} = 1.6\times10^{-3}$$

Therefore,

$$\frac{A_{ml}}{B_{ml}\rho(\nu_{lm})} = \left(e^{h\nu_{lm}/k_BT} - 1\right) \approx \frac{h\nu_{lm}}{k_BT} \approx 0.0016$$

where we have used the expansion $e^x = 1 + x + x^2/2 + \ldots$ and retained only the first two terms since $x << 1$. Hence, stimulated emission dominates over spontaneous emission.

In the visible region, $\nu_{lm} \approx 10^{15}$ s^{-1}, $h\nu_{lm}/k_BT \approx 160$ and $A_{ml} >> B_{ml}\,\rho(\nu)$. Thus, in the visible and near infrared regions, spontaneous emission generally dominates over induced emission, unless we can somehow raise the latter by increasing $\rho(\nu)$. However, Planck's distribution (Figure 1.9) shows that $\rho(\nu)$ is generally small at high frequencies because of the presence of the term $\left(e^{h\nu/kT} - 1\right)$ in the denominator of equation (1.5).

According to equation (1.10), at equilibrium, the number of molecules going up is equal to those going down, so we might expect the intensity of absorption to decrease with time. However, since only the induced (stimulated) processes are coherent (the photons are all in phase with one another) with the radiation, the net absorption intensity is given by

$$N_lB_{lm}\rho(\nu_{lm}) - N_mB_{ml}\rho(\nu_{lm}) \tag{1.16}$$

Substituting equation (1.12), this gives

$$N_l\frac{g_m}{g_l}B_{ml}\rho(\nu_{lm}) - N_mB_{ml}\rho(\nu_{lm})$$

$$= B_{ml}\rho(\nu_{lm})\left(N_l\frac{g_m}{g_l} - N_m\right) \tag{1.17}$$

Further substitution of equation (1.6) gives

$$N_l \frac{g_m}{g_l} B_{ml} \rho(\nu_{lm}) - \frac{g_m}{g_l} N_l e^{-h\nu_{lm}/kT} B_{ml} \rho(\nu_{lm})$$
$$= N_l \frac{g_m}{g_l} B_{ml} \rho(\nu_{lm}) \left(1 - e^{-h\nu_{lm}/kT}\right) \tag{1.18}$$

Assuming unit statistical weights ($g_m = g_l = 1$), equation (1.16) shows that if the populations of the two states are equal, there will be no net absorption of radiation. According to equation (1.18), the intensity for a given absorption depends on the factor $h\nu / k_B T$ if $\rho(\nu)$ is kept constant. As we saw in the previous section, this factor is large for vibrational and electronic transitions, leading to high intensities for such transitions.

In the next chapter, we shall explore the quantum aspects of the interaction of radiation and matter. We shall derive an expression for B_{lm} and, consequently, the other Einstein coefficients. At this stage, we only recognize that these processes are taking place and these have a quantum mechanical interpretation. In the next section, we describe an important application of the phenomena discussed so far, i.e., laser action, which is a direct consequence of stimulated emission. Lasers produce *monochromatic coherent* light. These properties have given lasers a large number of applications both in technological and scientific fields.

1.7.1 Lasers

Laser action is essentially a consequence of stimulated emission. In this process, one photon stimulates the emission of another (Figure 1.10); the two photons have the same frequency and are in phase with one another. These two photons can go on to stimulate the emission of more photons, and the resulting cascade can lead to an intense beam (Figure 1.11).

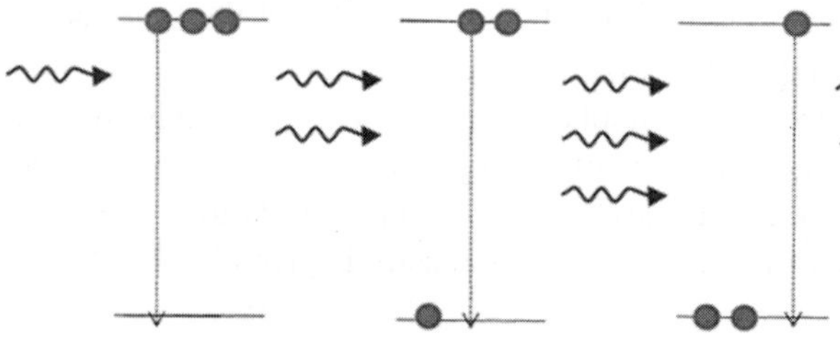

Figure 1.11. Laser action

However, this simple picture is somewhat misleading as we have not accounted for the presence of stimulated absorption and spontaneous emission. Stimulated absorption will absorb the photons, opposing the build-up of photons on which a laser depends. Spontaneous emission will result in the emission of photons of the correct frequency, but since they are not in phase with the other photons, they do not contribute to the laser light.

If we ignore spontaneous emission for the moment, it is clear that we will only have laser action if the rate of stimulated emission *exceeds* the rate of stimulated absorption. For our two levels, the rates of stimulated absorption and stimulated emission are given by equations (1.7) and (1.8), respectively.

Thus, the condition for laser action can be written as:

rate of stimulated emission > rate of stimulated absorption

$$B_{ml}\rho(\nu)N_m > B_{lm}\rho(\nu)N_l$$
$$N_m > N_l$$

where we have assumed that the two levels are non-degenerate so that $B_{ml} = B_{lm}$.

In words, what we have found is that for laser action the population of the upper level must exceed that of the lower level; this is called a *population inversion*. Such a situation will never occur for non-degenerate levels at equilibrium, where the lower level always has the greater population (see Example 1.8). Spontaneous emission also decreases the population of the upper state, further inhibiting laser action.

In practice, a population inversion is created by putting a lot of energy into the system, for example using a flash lamp or an electric discharge; this drives enough molecules or atoms into upper states so that a population inversion is created. Most lasers operate using more than two energy levels: the simplest practical system is one with three levels (Figure 1.12):

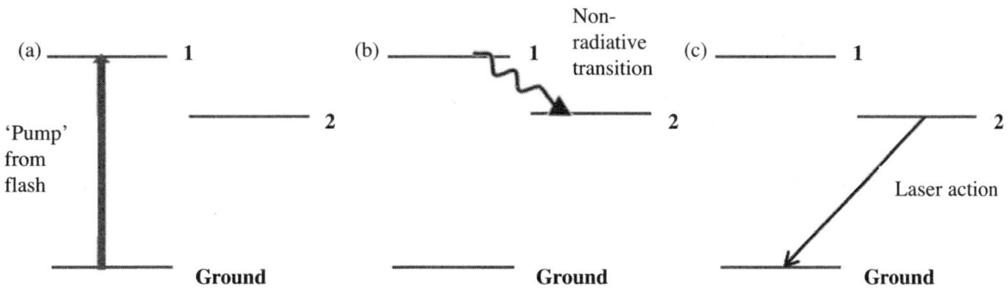

Figure 1.12. A three-level laser system

First, an intense flash of light promotes many molecules or atoms from the ground state to level **1**. Then, by some nonradiative process, the excited molecules in level **1** drop down to level **2**. A nonradiative process is one that does not involve the production of photons: it might take place as a result of collisions, for example. If the rates of the processes are favourable, the result is a population inversion between the ground state and level **2**; laser action can then take place. Designing lasers depends on finding systems in which we can create (and sustain) population inversions. At present, a huge range of lasers is available, emitting light all the way from the IR through to the near UV.

The ruby laser (Figure 1.13) is an example of a three-level system. The energy levels are due to Cr^{3+} ions doped into solid Al_2O_3; the ions have the configuration d^3, and so the ground state is a quartet (we shall learn more about atomic states in Chapter 3). The ions are in a roughly octahedral environment and so the electronic states can be labelled according to the irreducible representations of O_h. The ground state is $^4A_2$; the flash lamp excites this to the (broad) $^4T_2$ level and then there is non-radiative transfer to the ^2E level, generating a population inversion across the $^4A_2-^2E$ levels. Under favourable conditions, laser action can occur (at around 690 nm). We shall see later that this direct transition is spin forbidden so that the ^2E level is long-lived.

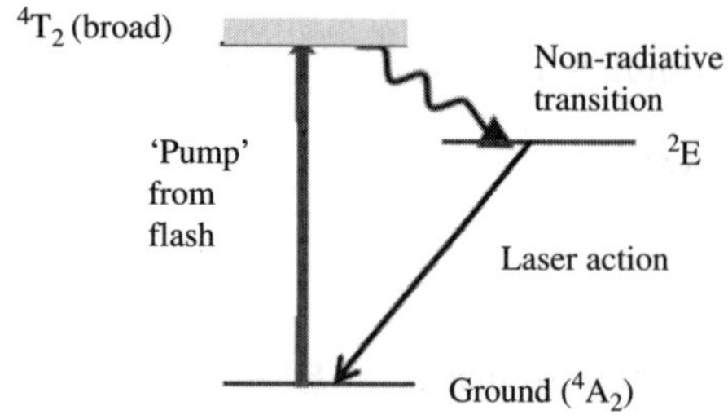

Figure 1.13. The ruby laser

We now consider a consequence of spontaneous emission, because of which the excited state has a finite lifetime and ultimately decays to the ground state. We shall explore the effect of this finite lifetime on the observed spectral lines.

1.8 LINE BROADENING

On the basis of the Bohr frequency condition, i.e., only frequencies corresponding to the difference between two energy levels can be absorbed, $\nu_{lm} = \frac{E_m - E_l}{h}$, one might expect infinitely sharp absorption peaks and this should ideally be the case. This implies that the absorption line should be infinitely narrow like a Dirac delta function $\delta(\nu - \nu_{lm})$. The delta function has the property that the integral over all space (i.e., from $-\infty$ to $+\infty$) gives a single value ν_{lm}. We write this as

$$\int_{-\infty}^{+\infty} \delta(\nu - \nu_{lm})d\nu = \nu_{lm}$$

The function defined is a line at ν_{lm} that is infinitely narrow.

However, in reality, the absorption peaks are not at all sharp. In the case of molecular electronic spectra, the broadening of lines also arises due to the distribution of an electronic transition between different vibrational and rotational transitions. However, even in atomic spectroscopy, where only electronic transitions are possible, there is some line broadening. This is due to a number of reasons, such as natural lifetime broadening, Doppler broadening, pressure or collisional broadening, etc. Line broadening is classified as either *homogeneous*, when all the atoms or molecules experience the same effect, or *inhomogeneous*, in which each atom or molecule is affected differently. In the former class comes natural line broadening, while among the latter is Doppler broadening. We shall see that different line-shapes emerge depending on the line broadening process.

1.8.1 Natural line broadening

Natural line broadening is a consequence of the Heisenberg Uncertainty Principle, which is a fundamental principle. We first discuss this principle.

1.8.1.1 Conjugate variables and uncertainty

Classically, two variables a and b are said to be canonically conjugate if $\frac{\partial H}{\partial a} = \dot{b}$, and $\frac{\partial H}{\partial b} = -\dot{a}$, where H is the Hamiltonian. In these expressions, $\dot{a}$ and $\dot{b}$ represent the time derivatives of the variables a and b, i.e., da / dt and db / dt, respectively. Canonically conjugate variables are governed by the Heisenberg Uncertainty relationship, i.e., $\Delta a \Delta b \geq \hbar/2$. The uncertainties are quantified by the standard deviations of several experimental determinations of the corresponding properties,

$$\Delta a = \sigma_a = \sqrt{\frac{\sum_{i=1}^{N}(a_i - <a>)^2}{N}} = \sqrt{\frac{\sum_{i=1}^{N}\left(a_i^2 - 2<a>a_i + <a>^2\right)}{N}}$$

$$= \sqrt{\left(\frac{\sum_{i=1}^{N} a_i^2}{N} - 2<a>\frac{\sum_{i=1}^{N} a_i}{N} + \frac{N<a>^2}{N}\right)}$$

$$= \sqrt{(<a^2> - <a>^2}$$

and similarly for Δb. Here, the symbol $< a >$ denotes the mean value of a.

Example 1.10 Consider the marks obtained by a class of 10 students: 1, 2, 2, 3, 3, 3, 4, 6, 7 and 9. Compute the average marks and their standard deviations.

Solution

The average mark of the class is given by

$$<x> = \sum_{i=1}^{N} x_i / N = \frac{1+2+2+3+3+3+4+6+7+9}{10} = 4.$$

To calculate the population standard deviation, compute the difference of each data point from the mean, and then square the result of each:

Marks (x_i)	x_i^2	($x_i - <x>$)	($x_i - <x>$)^2
1	1	–3	9
2	4	–2	4
2	4	–2	4
3	9	–1	1
3	9	–1	1
3	9	–1	1
4	16	0	0
6	36	2	4
7	49	3	9
9	81	5	25
Sum = 40	**218**		**58**

Next compute the average of the values in the last column and take the square root $\sqrt{(58/10)}$ = 2.4. Thus, $\sigma_x = \Delta x = 2.4$.
Alternatively, take the average $<x^2>$ by summing x^2 and dividing by 10. From column 2, this is 218 / 10 = 21.8. Then, $\Delta\ x = \sigma_x = \sqrt{(<x^2> - <x>^2} = \sqrt{21.8 - 4^2} = 2.4$.

A large standard deviation indicates that the data points are far from the mean and a small standard deviation implies that they are clustered closely around the mean. For example, each of the three populations {0, 0, 10, 10}, {0, 5, 5, 10} and {4, 5, 5, 6} has the same mean value, 5, but their standard deviations are 2.5, 3.5 and 0.7, respectively. The third population has a much smaller standard deviation than the other two because its values are all close to the mean.

Standard deviation may serve as a measure of uncertainty of a physical measurement. For example, the reported standard deviation of a group of repeated measurements gives the precision of those measurements. For a normal distribution, the probability that the values in a set lie within $<x> \pm \sigma$ is 68%. In statistics, the 68-95-99.7 rule – or three-sigma rule – states that for a normal distribution, nearly all values lie within three standard deviations of the mean, i.e., two standard deviations from the mean account for 95% of the set, while nearly all (99.7%) of the values lie in the range $<x> \pm 3\sigma$.

Example 1.11 Atomic absorbance spectroscopy is used to measure the total iron content in a copper-based alloy. The results of measurement on samples extracted from five separate portions of the alloy are 92.4, 97.3, 102.9, 103.4 and 99.8 mg. What are the mean and standard deviation values for the iron content in the alloy?

Solution

$$\text{Mean} = \frac{92.4 + 97.3 + 102.9 + 103.4 + 99.8}{5} = 99.2 \text{ mg}$$

$$\sigma = 4.0 \text{ mg}$$

Note that the standard deviation will have the same units as the data points themselves.

We are familiar with two conjugate variables, position and momentum. Another pair is time and energy, which also obeys the uncertainty relation $\Delta t \Delta E \geq \hbar / 2$. The ground state has an infinite lifetime, since, unless disturbed, the system will remain forever in this state; hence, the ground state energy is precise. However, unless the lifetime of the excited state is infinite, which can never be the case because of spontaneous emission, there will always be smearing of the energy of all excited states, and hence line broadening (Figure 1.14).

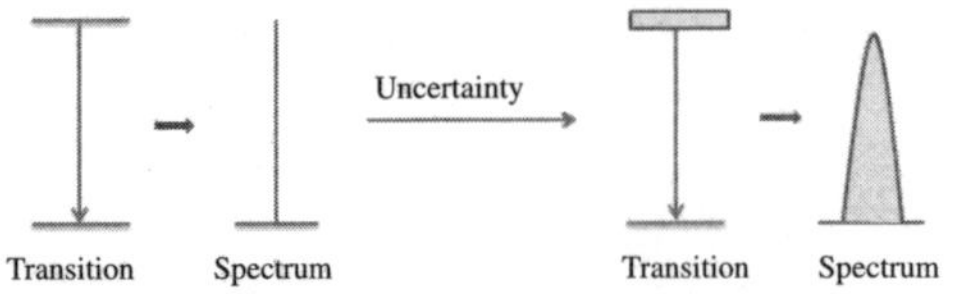

Figure 1.14. Decay of excited state population $N_m(t)$ leads to exponential decay of radiation amplitude, giving a Lorentzian spectrum

We have seen that line broadening results from the finite lifetime of the excited state. Also, even after the radiation is switched off, spontaneous emission persists and leads to the decay of the excited state. We can therefore think of the Einstein coefficient of spontaneous emission, A_{ml}, as a measure of the lifetime of the excited state. Consider molecules in the upper state *m*. As a result of radiative transitions from state *l* to state *m*, the population of the upper level is disturbed and becomes higher than that predicted by the Boltzmann distribution. To bring the system back to thermal equilibrium, spontaneous emission takes place. The photon is emitted in a random direction with arbitrary polarization. It carries away momentum $\frac{h}{\lambda} = \frac{h\nu}{c}$ and the emitting particle (atom, molecule or ion) recoils in the opposite direction. This implies that when a particle changes its energy spontaneously, the emitted radiation is not, as might perhaps be expected, all at the same frequency.

In the absence of radiation, equation (1.9) predicts a first-order decay of the excited state population, i.e., $-\frac{dN_m}{dt} = A_{ml}N_m$ if spontaneous emission is the *only* decay process taking place. Integration gives

$$N_m = N_m^0 e^{-A_{ml}t} \tag{1.19}$$

where $N_m^{\ 0}$ is the population of the excited state at the instant when the radiation is switched off. Thus, the decay of the excited state is exponential, and follows first-order kinetics, with rate constant A_{ml}. The lifetime of the excited state, τ, is the time taken for the population of the excited state to fall to 1 / e of its initial value (where e is the base of natural logarithms). Substitution in equation (1.19) gives $\tau = 1 / A_{ml}$ and equation (1.19) can then be written as $N_m = N_m^0 e^{-t/\tau}$. The spontaneous lifetime of the excited state τ is the average time the atom or molecule can be found in state *m*, since

$$<t> = \frac{\int_0^\infty t N_m^0 e^{-A_{ml}t}\,dt}{\int_0^\infty N_m^0 e^{-A_{ml}t}\,dt} = \frac{1}{A_{ml}} = \tau .$$

This is equal to the uncertainty in the time, $\Delta t = \tau$. This means that A_{ml} determines the lifetime of the excited state and hence the line-width.

1.8.1.2 Damped oscillations

Due to recoil when the photon is ejected from the atom or molecule, the electron is accelerated. According to the classical theory, a charged particle radiates, and hence loses energy, leading to damping of the emitted radiation, whose amplitude also decays exponentially. Prior to Bohr's theory, the electron of an atom was considered held to the nucleus by a Hooke's law kind of force. Thus, if the force constant of the so-called 'spring' is *k*, the acceleration of the electron produces a force $m\ddot{x}$, where $\ddot{x}$ represents the second derivative of *x* with respect to the time. This force is opposed by the restoring force *kx*, i.e.,

$$m\ddot{x} = -kx$$

which is a second-order differential equation with solutions of the kind $A\exp(-\alpha t)$. Substitution in the above equation yields the result

$$m\alpha^2 + k = 0$$

a solution of which is only possible if α is imaginary, say $i\omega_0$ with $\omega_0 = \sqrt{k/m}$, where ω_0 is the natural frequency of the oscillator, so that

$$\ddot{x} + \omega_0^2 x = 0 \tag{1.20}$$

For a *damped oscillator*, the differential equation becomes

$$\ddot{x} + \gamma\dot{x} + \omega_0^2 x = 0$$

where γ is a constant known as the damping constant. Again, substituting $x = A\exp(-\alpha t)$ in the above equation, we get

$$\alpha^2 - \gamma\alpha + \omega_0^2 = 0$$

This is a quadratic equation and its solutions are

$$\alpha = \frac{\gamma \pm \sqrt{(\gamma^2 - 4\omega_0^2)}}{2}$$

We assume that the damping is small, so that $\gamma << \omega_0^2$, and the quantity under the square root is negative. We then expand

$$\alpha = \frac{\gamma}{2} \pm i\omega_0\sqrt{1 - \frac{\gamma^2}{4\omega_0^2}} \approx \frac{\gamma}{2} \pm i\omega_0$$

since $\gamma^2/4\omega_0^2$ is negligible. The solutions are thus

$$\begin{aligned} x &= \frac{1}{2}\left[e^{-\alpha_- t} + e^{-\alpha_+ t}\right] \\ &= \frac{1}{2}\left[e^{-(\gamma/2 - i\omega_0)t} + e^{-(\gamma/2 + i\omega_0)t}\right] \\ &= \frac{1}{2}\left[e^{-\gamma t/2}(e^{i\omega_0 t} + e^{-i\omega_0 t})\right] = e^{-\gamma t/2}\cos\omega_0 t \end{aligned} \tag{1.21}$$

The damping coefficient, γ, is related to the decay of the excited state, and hence to A_{ml}. If spontaneous emission from the upper state m to the lower state l is the only process taking place, i.e., if m is the first excited state and consequently l the ground state, γ is equal to A_{ml}. If this is not the case, γ also includes the effect of emissions from state m to all lower states (all decay channels) and also spontaneous emission from l to all energy states lower than l, which will also contribute to line broadening. If the radiation is still on, its influence should also be considered.

The next step is to see how the exponential decay function (1.21) affects the line-width. Equation (1.21) is a function of the time only, since ω_0 is a constant corresponding to the transition frequency in the absence of line broadening. However, in order to determine the line-width, we need to convert it to a function of frequency. A function in the time domain can be converted to the frequency domain by a Fourier transform, since time and energy (which is directly proportional to the frequency) are conjugate variables. Note that this interconversion is only possible for canonically conjugate variables like energy and time, or position and momentum. We thus have the Fourier transform pair:

$$F(\omega)=\frac{1}{\sqrt{2\pi}}\int_{-\infty}^{\infty} f(t)e^{-i\omega t}\,dt$$

$$f(t)=\frac{1}{\sqrt{2\pi}}\int_{-\infty}^{\infty} F(\omega)e^{i\omega t}\,d\omega \tag{1.22}$$

That is, to convert the exponential decay function (1.21) to a function of the frequency spread, we perform the following Fourier transformation

$$F(\omega)=\frac{1}{\sqrt{2\pi}}\int_{0}^{\infty} e^{-\frac{\gamma t}{2}}(e^{i\omega_0 t}+e^{-i\omega_0 t})e^{-i\omega t}\,dt$$

Since $t > 0$, we have taken the integral limits from 0 to ∞ instead of from $-\infty$ to ∞. This Fourier transform is easily solved, since the integral is just an exponential integral. The solution for this integral is

$$F(\omega)=\frac{1}{\sqrt{2\pi}}\int_{0}^{\infty} e^{-\frac{\gamma t}{2}}(e^{i\omega_0 t}+e^{-i\omega_0 t})e^{-i\omega t}\,dt=-\frac{1}{\sqrt{2\pi}}\left(\left.\frac{e^{\left[-\frac{\gamma}{2}-i(\omega-\omega_0)\right]t}}{(\gamma/2+i(\omega-\omega_0))}\right|_0^{\infty}+\left.\frac{e^{\left[-\frac{\gamma}{2}-i(\omega-\omega_0)\right]t}}{(\gamma/2+i(\omega+\omega_0))}\right|_0^{\infty}\right)$$

The second term is essentially always negligible for spectroscopically important values of ω and ω_0 because its denominator is large. Thus, we need to focus only on the first term:

$$F(\omega)\approx\frac{1}{\sqrt{2\pi}}\left(\frac{1}{(\gamma/2+i(\omega-\omega_0))}\right)$$

for the discussion that follows. When this function is plotted against ω, it peaks at $\omega=\omega_0$.

We multiply and divide by $(\gamma/2-i(\omega-\omega_0))$ in order to rationalize the denominator. This gives

$$F(\omega)=\frac{1}{\sqrt{2\pi}}\left(\frac{1}{(\gamma/2+i(\omega-\omega_0))}\right)\left(\frac{(\gamma/2-i(\omega-\omega_0))}{(\gamma/2-i(\omega-\omega_0))}\right)=\frac{1}{\sqrt{2\pi}}\left(\frac{(\gamma/2-i(\omega-\omega_0))}{(\gamma/2)^2+(\omega-\omega_0)^2}\right)$$

$$=\frac{1}{\sqrt{2\pi}}\left(\frac{\gamma/2}{(\gamma/2)^2+(\omega-\omega_0)^2}-\frac{i(\omega-\omega_0)}{(\gamma/2)^2+(\omega-\omega_0)^2}\right) \tag{1.23}$$

which is a complex function. The real part of this integral is a *Lorentzian* line-shape function. This is the line-shape that is observed for transitions that are dominated by natural line broadening, such as in NMR spectra, etc. The normalized Lorentzian real part of the above function $F(\omega)$ is

$$F(\omega-\omega_0)=\frac{1}{\pi}\left(\frac{\gamma/2}{(\gamma/2)^2+(\omega-\omega_0)^2}\right) \tag{1.24}$$

Note that the integral from $-\infty$ to ∞ of this normalized function is equal to one.

Figure 1.15 plots the exponential decay function (1.21) and the corresponding Lorentzian for three values of γ. The width of the Lorentzian function is controlled by γ; the smaller γ is, the narrower will be $F(\omega-\omega_0)$ and the slower the decay. Thus, in practice, γ is a measure of the width of a spectral line.

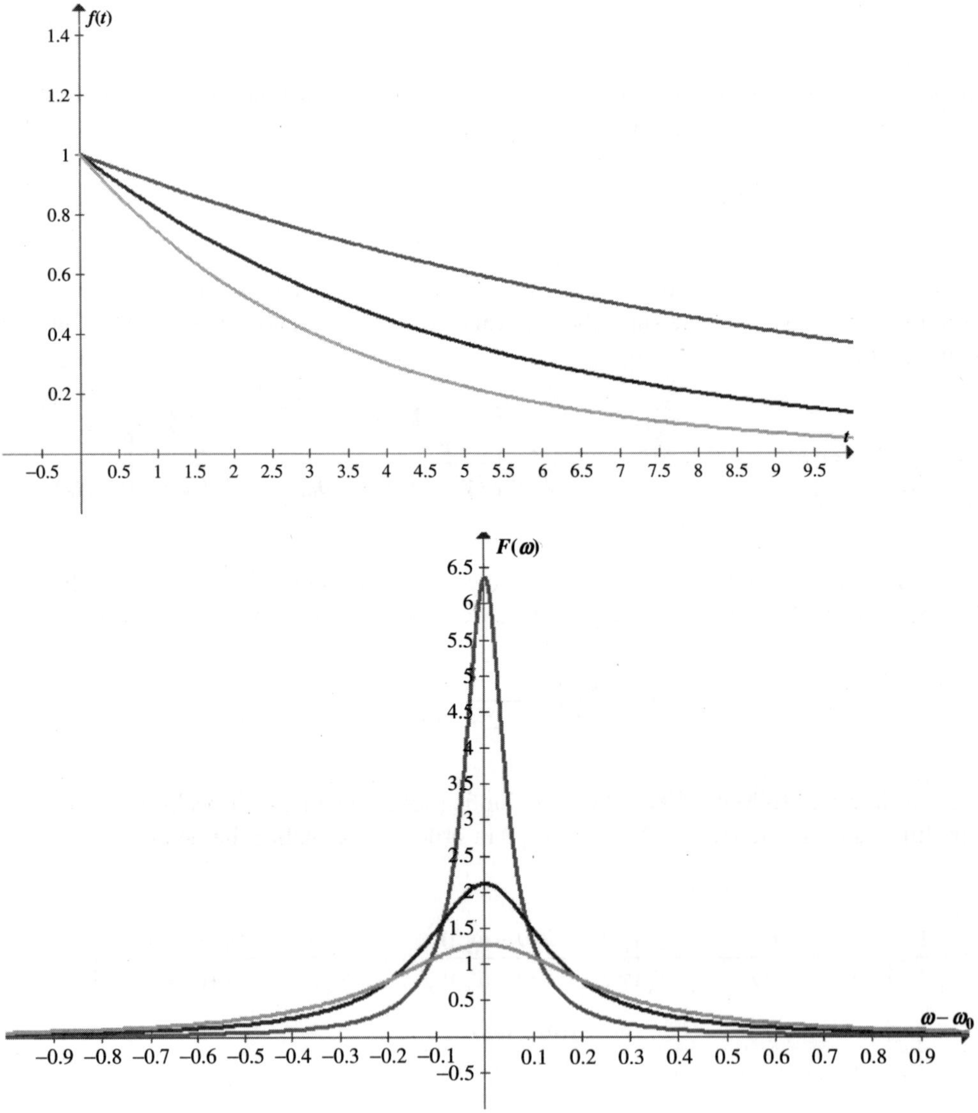

Figure 1.15. The exponential decay function (1.19) and the Lorentzian function, equation (1.23), plotted against $(\omega-\omega_0)$ for various values of γ [0.1 (top), 0.2 (middle) and 0.3 (bottom)]. Notice that the curves become narrower and more peaked as γ decreases

In terms of the frequencies, this function (1.24) is

$$g(\nu-\nu_{lm})=\left(\frac{\gamma}{(\gamma/2)^2+(2\pi)^2(\nu-\nu_{lm})^2}\right) \tag{1.25}$$

since $g(\nu-\nu_{lm})=2\pi F(\omega-\omega_0)$ and $\omega_0=2\pi\nu_{lm}$.

This is the natural line broadening and takes the same form for all atoms and molecules and gives a characteristic Lorentzian shape to the absorption peak (Figure 1.15).

At the peak centre, $\nu=\nu_{lm}$ and consequently $g(\nu-\nu_{lm})=4/\gamma$. The function goes to half this value when $g(\nu-\nu_{lm})=2/\gamma$, i.e., $\nu=\nu_{lm}\pm\frac{\gamma}{4\pi}$. The full width at half-maximum (FWHM) is usually taken as a measure of peak width, which is therefore given by

$$\Delta\nu_{1/2}(\text{Hz})=\frac{\gamma}{2\pi}=\frac{1}{2\pi\tau} \tag{1.26}$$

since $\gamma=1/\tau$. The important result (1.26) is consistent with the Heisenberg time-energy Uncertainty Principle. This formula has been checked experimentally, for example, in the case of the sodium $3^2S_{1/2}\leftarrow 3^2P_{3/2}$ emission transition (one of the sodium *D*-lines) at 589 nm. The experimentally measured lifetime $\tau=16$ ns and the observed homogeneous line-width $\Delta\nu_{1/2}=10$ MHz are consistent with equation (1.26). The Uncertainty Principle therefore requires that if an excited state exists for only τ seconds on average, then the energy level E_m cannot be measured relative to E_l with an accuracy that is greater than $h\Delta\nu_{1/2}$ Joule. Here, we may recall an important property of Fourier transforms: the more concentrated $f(t)$ is, the more spread out its Fourier transform $F(\omega)$ must be (Figure 1.15). In the time-frequency domain, it means that larger the lifetime τ of the excited state, the sharper the line is.

In terms of the FWHM, the Lorentzian line-shape function can be expressed as

$$g(\nu-\nu_{lm})=\left(\frac{\Delta\nu_{1/2}/(2\pi)}{(\Delta\nu_{1/2}/2)^2+(\nu-\nu_{lm})^2}\right) \tag{1.27}$$

Compared to other causes of line broadening, natural line broadening is very small, but is of great theoretical importance, because it is fundamental and cannot be eliminated by any means. This source of broadening is rarely significant in atomic spectroscopy, since a typical lifetime for an atomic energy state is about 10^{-8} s, corresponding to a natural line-width of about 1.6×10^7 Hz (5.3×10^{-4} cm^{-1}), which is very small compared to the transition wavenumber, ~25,000 cm^{-1} ($\Delta\tilde{\nu}/\tilde{\nu}=2\times10^{-8}$). Compare this with a typical rotational state lifetime, which is about 10^3 s, corresponding to a natural line-width of 10^{-4} Hz. Compared to a typical rotational frequency of 10 GHz, this is just 1 in 10^{14}. For the most common Mössbauer isotope, ^{57}Fe, the excited nuclear state lifetime is 2×10^{-7} s and the corresponding line-width is 5×10^{-9} eV. Compared to the Mössbauer γ-ray energy of 14.4 keV, this gives a resolution of 1 in 10^{12}. Though the line-widths are similar for electronic and Mössbauer spectroscopy, the relative line-width is much smaller in the latter owing to the large transition frequency.

Example 1.12 An atom emits a photon with wavelength $\lambda = 700$ nm while decaying from an excited state to the ground state.

(a) Express the energy of the photon in eV.
(b) The lifetime of the excited state $\tau = 7$ ns. Estimate the uncertainty ΔE in the energy of the photon (in eV).
(c) What is the spectral line-width (in wavelength) of the photon?

Solution

(a) $E = hc/\lambda = \dfrac{6.626\times10^{-34}\,\text{J s}\times2.998\times10^{8}\,\text{m s}^{-1}}{700\times10^{-9}\,\text{m}} = 2.838\times10^{-19}\,\text{J} = 2.838\times10^{-19}$ J / 1.602×10^{-19} J eV^{-1} = 1.772 eV

(b) From the Uncertainty relation, $\Delta E\Delta t \approx \hbar$. Therefore, $\Delta E \approx 6.626\times10^{-34}$ J s / ($2\times3.142 \times 7\times10^{-9}$ s) = 1.506×10^{-26} J = 9.403×10^{-8} eV $\approx 10^{-7}$ eV.

(c) Since $E = hc/\lambda$, $\dfrac{dE}{d\lambda} = -\dfrac{hc}{\lambda^2} = -\dfrac{E}{\lambda}$. Therefore, $\dfrac{|\Delta E|}{E} = \dfrac{|\Delta\lambda|}{\lambda}$.

$$|\Delta\lambda| = \frac{|\Delta E|}{E}\lambda = 9.403\times10^{-8}\text{ eV}\times700\times10^{-9}\text{ m}/1.772\text{ eV} = 3.715\times10^{-14}\text{ m} = 3.7\times10^{-5}\text{ nm}.$$

This is a tricky question. Since E and λ are not directly proportional, you cannot simply find $\Delta\nu$ and then use the frequency–wavelength relationship to find $\Delta\lambda$. Since the experimental spectrum is usually a plot of intensity versus wavelength, students generally take the FWHM in terms of $\Delta\lambda$ and then convert these to $\Delta\nu$ by dividing c by $\Delta\lambda$, which is incorrect.

1.8.2 Pressure or collisional broadening

When collisions occur between gas phase molecules, their charge distribution is disturbed, causing an induced dipole that can subsequently absorb or emit radiation. This leads effectively to a broadening of energy levels. If τ_{coll} is the mean time between collisions, and each collision results in a transition between two states, there is a line broadening $\Delta\nu$ of the transition, where

$$\Delta\nu = (2\pi\tau_{coll})^{-1}$$

according to the Uncertainty Principle. Like natural line broadening, this broadening is homogeneous, and usually produces a Lorentzian line-shape because of the similarity in the decay functions. However, for transitions at low frequencies, the line-shape is unsymmetrical. This kind of broadening increases with the pressure.

The time between collisions is related to the attraction between molecules; therefore, line-width investigations are a common technique used to investigate intermolecular forces.

1.8.3 Doppler broadening

For electronic spectra in the visible and UV spectral regions, Doppler broadening sets the limit on resolution. Due to the thermal motion of the atoms, those atoms travelling towards the detector will have transition frequencies different from those of atoms at rest. The reason is the same that an observer hears the whistle of a train moving towards him or her as having a frequency apparently higher than it really is, and lower when it is travelling away from him or her. This effect is known as the Doppler effect. Since different atoms are affected differently because of their differing speeds, the Doppler broadening is inhomogeneous.

The Doppler shift in the frequency is given by the simple form

$$\nu = \nu_{lm}\left(1 \pm \frac{\mathrm{v}}{c}\right) \quad (\nu_{lm} = \text{frequency of an atom at rest}) \tag{1.28}$$

The spread of the line is therefore approximately

$$\Delta\nu_D = 2\nu_{lm}\mathrm{v} / c \tag{1.29}$$

According to the Maxwell–Boltzmann distribution law, the root mean square (rms) velocity is given by

$$\mathrm{v}_{rms} = \sqrt{\frac{3k_B T}{m}} = \sqrt{\frac{3RT}{M}}.$$

For example, for nitrogen molecules at 300 K, the rms speed is 517 m s^{-1}. Substituting this value in equation (1.28), we get

$$\nu = \nu_{lm}\left(1 \pm \frac{517 \text{ m s}^{-1}}{3\times 10^8 \text{ m s}^{-1}}\right) = \nu_{lm}\left(1 \pm 1.72\times 10^{-6}\right)$$

Using a typical frequency of an electronic transition (1×10^{15} Hz) for N_2, we find a total frequency shift of about 3.5×10^9 Hz, which is about 200 times the natural line-width.

Alternatively, measuring the FWHM yields information about the velocity of the molecules from the formula

$$\mathrm{v} = c\Delta\nu_D / 2\nu_{lm} \tag{1.30}$$

Example 1.13

(a) Calculate the natural broadening line-width of the Lyman α -line 2p $\rightarrow$ 1s at $\lambda = 121.6$ nm in the hydrogen atom, given that $A_{ml} = 5 \times 10^8$ s^{-1} at a temperature of $T = 1000$ K in a hydrogen discharge tube.

(b) Convert to velocity via the Doppler formula.

(c) Compare with the Doppler line broadening.

Solution

(a) From equation (1.26),

$$\Delta\nu_{1/2} = \frac{\gamma}{2\pi} = \frac{A_{ml}}{2\pi} = \frac{5\times 10^8}{2\pi}\text{s}^{-1} = 8\times 10^7 \text{s}^{-1}$$

(b) $\lambda = 121.6$ nm corresponds to $\nu_{lm} = 2.5 \times 10^{15}$ s^{-1}.
From equation (1.30), $\mathrm{v} = c\Delta\nu_D / 2\nu_{lm} = 3 \times 10^8 \text{ m s}^{-1} \times 8 \times 10^7 \text{ s}^{-1} / (2 \times 2.5 \times 10^{15} \text{ s}^{-1}) = 4.8 \text{ m s}^{-1}$

(c) The root mean square velocity is given by

$$v_{rms} = \sqrt{\frac{3RT}{M}} = \sqrt{\frac{3\times 8.314\times 1000}{1\times 10^{-3}}} = 4994 \text{ m s}^{-1} \approx 5 \text{ km s}^{-1}$$

The spread of the line is $\Delta\nu_D = 2\nu_{lm} v / c = 2\times 2.5\times 10^{15}\times 5000 / 3.0\times 10^8$ or 8.3×10^{10} s^{-1}.

Example 1.14 If hydrogen atoms on the surface of a blue-white star have an average thermal speed of 2.5×10^4 m s^{-1}, calculate the Doppler broadening of the H_α line, $\lambda = 656.5$ nm.

Solution

From equation (1.30), $\Delta\nu_D = 2\nu_{lm} v / c = 2 \times (3 \times 10^8 \text{ m s}^{-1} / 656.5 \times 10^{-9} \text{ m}) \times 2.5 \times 10^4 \text{ m s}^{-1} / 3 \times 10^8 \text{ m s}^{-1} = 7.6 \times 10^{10} \text{ s}^{-1}$.

The distribution of velocities can be found from the Boltzmann distribution function. The fraction of atoms with velocity v in the direction of the observed light is given by

$$f(v)dv = \sqrt{\frac{m}{2\pi k_B T}}\, e^{-mv^2/2k_BT} dv \tag{1.31}$$

where m is the atomic mass. The distribution of Doppler-shifted frequencies is then

$$g(\nu - \nu_{lm})d\nu = f(v)\frac{dv}{d\nu}d\nu = f(v)\frac{c}{\nu_{lm}}d\nu = \frac{c}{\nu_{lm}}\sqrt{\frac{m}{2\pi k_B T}}\exp\left(-\frac{m}{2k_BT}\left(\frac{c}{\nu_{lm}}\right)^2(\nu-\nu_{lm})^2\right)d\nu \tag{1.32}$$

where we have used equation (1.27). At relatively low temperatures, the distribution is approximately Gaussian (i.e., 'bell-shaped'). (Functions of the type exp $(-\alpha x^2)$ are Gaussian functions.) Thus a spectral line-width influenced by Doppler broadening has a Gaussian line-shape centred on ν_{lm}. The line-width function at resonance (i.e., when $\nu = \nu_{lm}$) is

$$g(\nu_{lm} - \nu_{lm})d\nu = \frac{c}{\nu_{lm}}\sqrt{\frac{m}{2\pi k_B T}}$$

since the exponential function goes to unity. At half this height, the exponential function should be equal to 0.5. Thus,

$$\exp\left(-\frac{m}{2k_BT}\left(\frac{c}{\nu_{lm}}\right)^2(\nu-\nu_{lm})^2\right) = \frac{1}{2}$$

$$\Rightarrow \left(-\frac{m}{2k_BT}\left(\frac{c}{\nu_{lm}}\right)^2(\nu-\nu_{lm})^2\right) = -\ln 2$$

$$\Rightarrow (\nu-\nu_{lm})^2 = \frac{2k_BT\ln 2}{m}\left(\frac{\nu_{lm}}{c}\right)^2 \tag{1.33}$$

or, since the line-width is twice $(\nu - \nu_{lm})$,

$$\Delta\nu_D = \frac{2\nu_{lm}}{c}\sqrt{\frac{2k_B T \ln 2}{m}} = \frac{2\nu_{lm}}{c}\sqrt{\frac{2RT\ln 2}{M}} = 7.16\times 10^{-7}\nu_{lm}\sqrt{T/M} \tag{1.34}$$

with the molecular mass M expressed in g mol^{-1} (u). For example, at 300 K, the Doppler broadening of the helium–neon laser operating at 632.8 nm would be

$$\Delta\nu_D = 7.16\times 10^{-7}\frac{3\times 10^8}{632.8\times 10^{-9}}\sqrt{300/20} \sim 1.3 \text{ GHz}.$$

In terms of the FWHM, the normalized line-shape function (1.32) is

$$g(\nu - \nu_{lm})d\nu = \frac{2}{\Delta\nu_D}\sqrt{\frac{\ln 2}{\pi}}\exp\left(\left(-\frac{2(\nu-\nu_{lm})}{\Delta\nu_D}\right)^2 \ln 2\right)$$

The value of g at resonance (i.e., when $\nu = \nu_{lm}$) is approximately the inverse line-width:

$$g(\nu - \nu_{lm}) = \frac{2}{\Delta\nu_D}\sqrt{\frac{\ln 2}{\pi}} \tag{1.35}$$

Example 1.15 Sodium atoms have contaminated the filament of an incandescent halogen light bulb filament. When the bulb is operating at $T = 3100$ K, assume that the sodium atoms and the optical field are in thermal equilibrium with the filament. Find the ratio of spontaneous emission to stimulated emission in the sodium D_2 (589 nm) line.

Solution

From equation (1.15),

$$\frac{A_{ml}}{B_{ml}\rho(\nu_{lm})} = \left(e^{h\nu_{lm}/k_B T} - 1\right) = \left(e^{hc/\lambda k_B T} - 1\right) = \exp\left(\frac{6.626\times 10^{-34}\times 2.998\times 10^8}{589\times 10^{-9}\times 1.381\times 10^{-23}\times 3100}\right) - 1 = 2637$$

If the atoms were further away from the light bulb, where the light is less intense, the ratio would increase even more. The lesson to be learnt here is that, in thermal light, spontaneous emission is much stronger than stimulated emission.

Example 1.16 Suppose the atoms are instead excited by a sodium laser with 1 W average power and a mode area of 4 mm^2. Find the ratio of spontaneous to stimulated emission, assuming only Doppler broadening.

Solution

First, we must estimate the intensity of the source.

$$I = 1\ \mathrm{W} / 4\ \mathrm{mm}^2 = 2.5 \times 10^5\ \mathrm{J\ m^{-3}}$$

The intensity can be converted to energy density by dividing by c,

$$\rho = I / c = 2.5 \times 10^5 / 3.0 \times 10^8\ \mathrm{m\,s^{-1}} = 8.33 \times 10^{-4}\ \mathrm{J\ m^{-3}}$$

From equation (1.34),

$$\Delta\nu_D = 7.16 \times 10^{-7} \nu_{lm} \sqrt{T / M} = 7.16 \times 10^{-7} \frac{3 \times 10^8\ \mathrm{m\,s^{-1}}}{589 \times 10^{-9}\ \mathrm{m}} \sqrt{3100 / 23} = 4.23 \times 10^9\ \mathrm{s^{-1}}$$

From equations (1.14) and (1.15),

$$\frac{A_{ml}}{B_{ml}\rho(\nu_{lm})} = \frac{8\pi h}{\lambda^3 \rho(\nu_{lm})} = 9.784 \times 10^{-11}$$

This is to be divided by the value of g at resonance [equation (1.35)]

$$\frac{A_{ml}}{B_{ml}\rho(\nu_{lm})} = \frac{8\pi h \nu^3}{c^3 \rho_\nu(\nu_{lm}) g(\nu - \nu_{lm})} = \frac{9.78 \times 10^{-11}}{g(\nu - \nu_{lm})} = 9.78 \times 10^{-11} \left(\frac{4.23 \times 10^9}{2} \sqrt{\frac{\pi}{\ln 2}} \right) = 0.44$$

Lifetime broadening is not typically the dominant broadening mechanism in gases. It is seen from equation (1.34) that the Doppler width $\Delta\nu_D$ is proportional to the transition frequency ν_{lm}, to the square root of the gas temperature T, and is inversely proportional to the square root of the mass. The increase with temperature is due to the larger spread of molecular speeds at higher temperatures. For transitions in the visible or near-UV spectral range at 300 K, the Doppler width is typically within 1 GHz. However, hydrogen atoms and molecules have exceptionally high Doppler widths of around 30 GHz due to their low masses.

Though Doppler broadening appears to be a nuisance as it spreads the frequency profile, it is used to advantage in Mössbauer spectroscopy. It is the main source of broadening in the UV-Vis and infrared regions. Therefore, the experimentally obtained line profiles usually have Gaussian line-shapes. In contrast, for microwave transitions, or when collisional broadening is high, the lifetime broadening becomes larger than the Doppler one, resulting in Lorentzian-type line profiles. At some pressure, the perturbations of rotational energy levels by molecular collisions (pressure broadening) become the limiting factor for resolution.

Besides Doppler broadening, there are other inhomogeneous causes of broadening, such as the different environment or solvent effects experienced by different molecules in crystals or in solution. Besides, Franck–Condon progressions (Chapter 8), which arise from the progression of levels arising from discrete transitions among vibrational levels of the molecule, often have the appearance of Gaussian broadening for molecules in solution. Hence, the Gaussian profile often dominates in solution.

Example 1.17 The copper vapour laser (CVL) operates at $T = 750$ K. Its wavelength is $\lambda = 510.6$ nm and the spontaneous emission lifetime is $\tau = 5 \times 10^{-7}$ s. Determine the form of the line-shape function and its FWHM.

Solution

The homogeneous natural line-width is

$$\Delta\nu_{1/2} = \frac{1}{2\pi\tau} = 3.18 \times 10^5 \text{Hz}$$

The Doppler line-width is, from equation (1.34)

$$\Delta\nu_D = 2.26 \times 10^{-8} \nu_{lm} \sqrt{T/M} = 1.45 \times 10^9 \text{ Hz}$$

Since $\Delta\nu_D >> \Delta\nu_{1/2}$, Doppler broadening dominates this transition. Hence, the line-shape will be Gaussian with a FWHM equal to $\Delta\nu_D$.

Since the Fourier transform of a Gaussian function is also a Gaussian, the Gaussian distribution in frequencies can be thought of as arising from a Gaussian distribution in relaxation times of the excited state, which is usually assumed to arise from an inhomogeneous distribution of the molecules in the sample.

In Figure 1.16, we have compared the normalized Gaussian and Lorentzian functions

$$G(\omega) = \frac{2}{\sqrt{\pi}\gamma} e^{-4(\omega-\omega_0)^2/\gamma^2}$$

$$F(\omega) = \frac{1}{\pi}\left(\frac{\gamma/2}{(\gamma/2)^2 + (\omega-\omega_0)^2}\right) \quad (1.36)$$

plotted against $(\omega - \omega_0)$ for a typical value of the decay constant, $\gamma = 0.10$.

The two lines look similar, but Gaussian lines are broader near the centre of the peak, and Lorentzian lines are broader in the tail portion of the peak. This enables us to recognize the dominating factors contributing to line broadening for a particular transition.

To summarize this section, the rate of spontaneous decay is proportional to the cube of the frequency and hence assumes importance for the high-frequency UV / Vis region and is negligible for spectroscopy in the microwave region. We have seen that faster decays lead to smaller lifetimes and broader line peaks, so the line peaks are also narrower in the microwave region than in the UV / Vis region. However, Doppler broadening tends to become dominant for transitions in the visible region because of the dependence of the Doppler line-width on the transition frequency, and hence the line-shape is Gaussian. In microwave spectroscopy, natural line-widths dominate over Doppler broadening, but the line-widths are determined by pressure broadening, which is the dominant contribution to the line-width in microwave spectroscopy and an important contribution for lines in the IR, and so the line-shape is Lorentzian.

We now turn to the experimental recording of spectra.

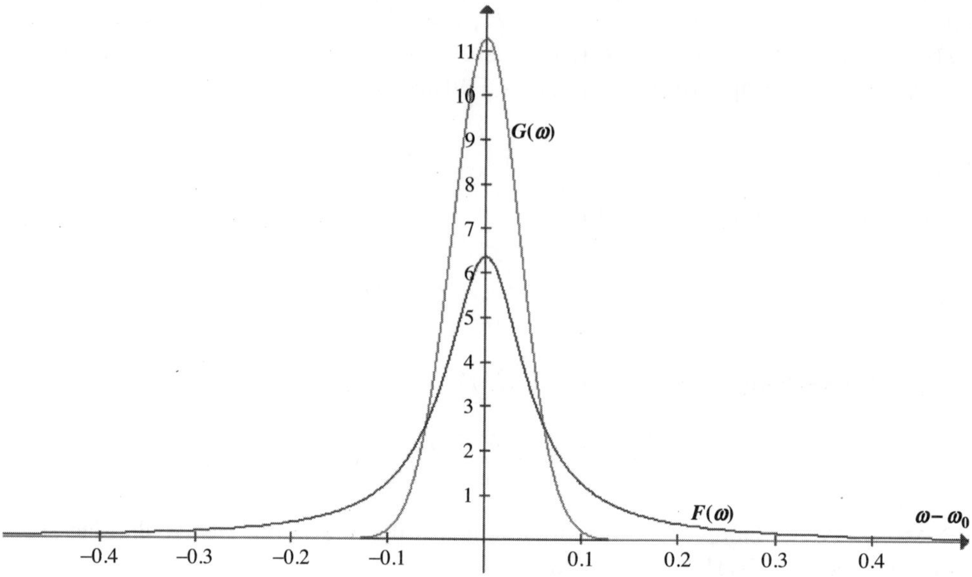

Figure 1.16. Normalized Gaussian and Lorentzian line-shape functions

1.9 LAMBERT–BEER'S LAW

The quantities used in reporting the experimental results can be conveniently expressed in terms of Lambert–Beer's law for an absorbing solute in a non-absorbing solvent. Beer's law states that absorption is proportional to the number, i.e., the concentration, of absorbing molecules (this is only true for dilute solutions), and Lambert's law states that the fraction of radiation absorbed is independent of the intensity of the radiation. Combining these two laws, we can derive Lambert–Beer's Law.

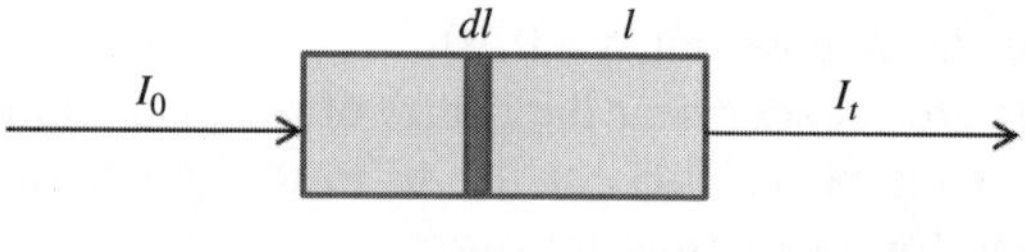

Figure 1.17. The absorption of radiation by a solution

The decrease in intensity, $-dI$, of monochromatic radiation as it penetrates a distance dl, as in Figure 1.17, is, according to Lambert–Beer's law, proportional to the radiation intensity, I, the molar concentration (C), and to the thickness of the absorbing layer dl

$$-dI = \alpha ICdl$$

where α, the proportionality constant, is called the *absorption coefficient*. This law is universal for all kinds of transitions. As we have seen, except in the limit of very high irradiation times, absorption bands usually extend over a range of frequencies due to the finite lifetime of the excited state and other factors contributing to line broadening. An added complication in the case of electronic spectra is that a single absorption band is distributed amongst several vibrational transitions, as we shall see later. Thus, the molar absorption coefficient is a function of the frequencies, i.e., $\alpha(\nu)$. We thus have

$$-dI = \alpha(\nu)ICdl \tag{1.37}$$

Separation of variables and integration with the condition that when $l = 0$, $I = I_0$ gives

$$-\frac{dI}{I} = \alpha(\nu)Cdl$$

$$-\ln\frac{I}{I_0} = \alpha(\nu)Cl \tag{1.38}$$

or

$$I = I_0 e^{-\alpha(\nu)Cl} \tag{1.39}$$

Since it is more convenient to use logarithm to the base 10 than the natural logarithm, the expression (1.39) is usually written as:

$$A = \log\frac{I_0}{I} = \frac{\alpha(\nu)Cl}{\ln 10} = \varepsilon(\nu)Cl \tag{1.40}$$

where $\varepsilon(\nu)$ is the *molar absorption coefficient* or *molar absorptivity*, also called the *molar decadic absorption coefficient.*

The quantity $A = \log(I_0 / I)$ is known as the *absorbance* of the solution, and can be read directly from the spectrum, often as 'absorbance units', though it is dimensionless as it is the logarithm of the ratio of intensities. The ratio I / I_0 is termed the transmittance, and is usually expressed as a percentage.

$$T[\%] = \frac{I}{I_0} \times 100$$

In the experimental study of electronic spectroscopy, one usually deals with wavenumbers ($\varepsilon(\tilde{\nu})$), with unit of cm^{-1} (as stated before, the SI unit m^{-1} is almost never used, and we continue the use of cm^{-1}) instead of ν in s^{-1}. We thus write $\varepsilon(\tilde{\nu})$ in equation (1.40) to emphasize the wavenumber dependence of the molar absorption coefficient. The molar absorption coefficient, $\varepsilon(\tilde{\nu})$, is a more useful quantity than absorbance, because it is independent of concentration and path length, unlike absorbance, which depends upon both. In the SI system, concentration is measured in mol m^{-3} and l in metre, so the units for $\varepsilon(\tilde{\nu})$ (expressed as $\kappa(\tilde{\nu})$) are m^2 mol^{-1} and equation (1.40) can be written as:

$$\log\frac{I_0}{I} = \kappa(\tilde{\nu})C'l \tag{1.41}$$

where C' is expressed in mol m^{-3} and l in metre.

Equation (1.41) is also expressed in molecular dimensions as:

$$I = I_0 e^{-\sigma N x}$$

where N is the number of molecules per m^3, x is the thickness in m; the unit for σ, also known as the *absorption cross section,* is hence m^2. It may be mentioned here that, although σ has the units of area, it should not be interpreted as the area of the absorbing molecule.

However, in this book, we shall continue using equation (1.40) with the concentration expressed in mol dm^{-3} and the path length in cm, so that the unit of $\varepsilon(\tilde{\nu})$ is dm^3 mol^{-1} cm^{-1}, but by convention the units are not quoted.

Example 1.18 Light of wavelength 400 nm is passed through a cell of 1 mm path length containing 10^{-3} mol dm^{-3} of compound X. If the absorbance of this solution is 0.25, calculate the molar absorption coefficient and transmittance.

Solution

Using equation (1.40) and inserting the given data, $A = 0.25$, $C = 10^{-3}$ mol dm^{-3}, $l = 0.1$ cm

$$\varepsilon = \frac{A}{Cl} = \frac{0.25}{10^{-3}\ \text{mol dm}^{-3}\, 0.1\,\text{cm}} = \frac{0.25}{10^{-4}\,\text{mol dm}^{-3}\text{cm}} = 2.5 \times 10^3\ \text{dm}^3\text{mol}^{-1}\text{cm}^{-1}$$

Now

$$A = \log\frac{I_0}{I} = 0.25$$
$$\Rightarrow \frac{I_0}{I} = 1.778$$
$$\Rightarrow \frac{I}{I_0} = 0.56$$

Therefore, transmittance = 56%.

The total intensity of the band is obtained by identifying the band and integrating it graphically to give the *integrated absorption coefficient* $\bar{A} = \int_{\text{band}} \varepsilon(\tilde{\nu})\, d\tilde{\nu}$. A crude, but useful, method is to multiply the FWHM with the molar absorption coefficient at the transition height (denoted by ε_{max}) to approximate the integrated absorption coefficient (Figure 1.18). Since most solution-state spectra are Gaussian shaped, a multiplicative factor 1.06 is used to account for the Gaussian shape.

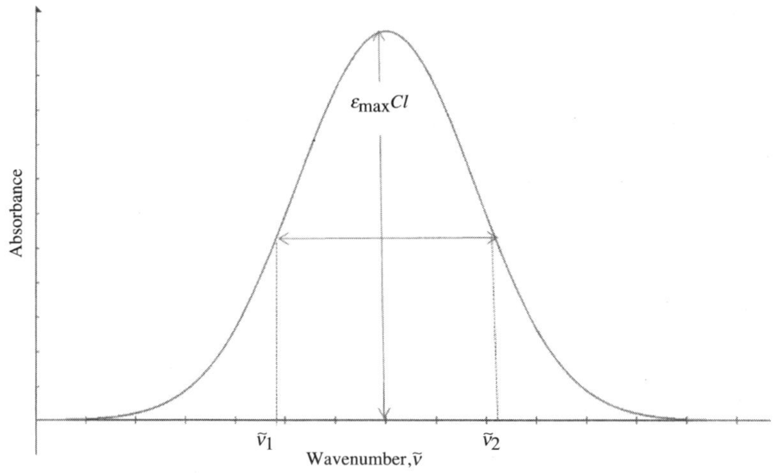

Figure 1.18. An absorption spectrum

Example 1.19 In the long wavelength band of 5-aminoimidazole in water as a solvent, the molar absorption coefficient at the band maximum is 5.4×10^3 dm^3 mol^{-1} cm^{-1} and the bandwidth at half the maximum height is 4047 cm^{-1}. Calculate the integrated absorption coefficient of this band.

Solution

Assuming a Gaussian shape of the absorption curve,

$$\int \varepsilon(\tilde{\nu})d\tilde{\nu} = 1.06 \times (5.4 \times 10^3 \text{dm}^3\text{mol}^{-1}\text{cm}^{-1}) \times (4047 \text{cm}^{-1})$$

$$= 2.3 \times 10^7 \text{dm}^3\text{mol}^{-1}\text{cm}^{-2}$$

Equations (1.37) to (1.41) have been derived strictly for monochromatic radiation. The molar absorption coefficient ε depends not only on the wavelength of excitation but also on the nature of the absorbing material. The value of ε quoted generally in the literature is that of the absorption band maximum, ε_{max} (Figure 1.18). This quantity also varies with the nature of the solvent and temperature. The most intense electronic absorption or fully allowed transition of a compound has an ε_{max} value of the order 10^5 dm^3 mol^{-1} cm^{-1}. Values of ε_{max} below the order 10^3 dm^3 mol^{-1} cm^{-1} correspond to quasi-forbidden or probable transitions and ε_{max} below 100 dm^3 mol^{-1} cm^{-1} indicate forbidden transitions.

1.9.1 Power or saturation broadening

The Boltzmann distribution $\frac{N_m}{N_l} = e^{-\Delta E/k_BT}$ is maintained by exchange of energy through molecular collisions. However, when the difference in the two energy levels is small, the ratio N_m / N_l is close to unity even at room temperature. Now, if this system is irradiated with radiation of high intensity I, the limit $N_m / N_l = 1$ is quickly approached, the Lambert–Beer's Law (1.40) breaks down, and ε no longer remains independent of I. Saturation is said to have occurred.

It is clear that this phenomenon is more common for the low-energy regions of the electromagnetic spectrum, such as the microwave and radiofrequency regions, but high-energy lasers can cause saturation in the higher energy regions, too. Saturation affects the line-shape by flattening out the peak and decreasing its height. This results in what is called *power*, or *saturation, broadening*.

1.10 FOURIER TRANSFORM SPECTROSCOPY

Since we are on the topic of Fourier transforms, we shall discuss an important application of Fourier transforms in spectroscopy. Fourier transform infrared spectra and NMR spectra (FT-IR and FT-NMR) are obtained by observing the time behaviour of a system and then taking its Fourier transform. Fourier transform spectroscopy has now been extended to microwave spectroscopy, too.

Spectra are traditionally recorded by dispersing the radiation and measuring the absorption or emission, one point at a time. In some regions of the spectrum, for which tunable radiation sources are available, the frequency of the source is stepped from ν_n to ν_{n+1} by $\Delta\nu$ and the absorption recorded. The primary attraction of Fourier transform techniques is that all frequencies in the spectrum are detected at once. This property is the so-called multiplex or Fellgett advantage of Fourier transform spectroscopy.

Most Fourier transform measurements at long wavelengths (e.g., FT-NMR) are made by irradiating the system with a short broadband pulse capable of exciting all the frequency components of the system, and then the free induction decay (FID) response is monitored. A simple free induction decay is

$$f(t) = e^{-\alpha t} \cos \omega_0 t \quad t \geq 0, \alpha > 0 \tag{1.42}$$

where ω_0 is some natural frequency of the system (such as the resonance frequency of a proton spin flip). The corresponding transform is

$$F(\omega) = \frac{1}{\sqrt{2\pi}} \left(\frac{\alpha}{\alpha^2 + (\omega - \omega_0)^2} \right)$$

where we have kept only the dominant real part of the equation, as in the derivation of equation (1.24). The real part of $F(\omega)$ is a Lorentzian centred at ω_0 (Figure 1.15). The constant $1/\alpha$ is the lifetime τ of the decay and the FWHM of the Lorentzian is 2α.

As usual, there is a reciprocal relationship between the time domain and the frequency domain. The more rapid the damping (decay) of the signal (i.e., larger α and shorter lifetime τ), the wider is the Lorentzian line-shape function in the spectral domain (Figure 1.15).

Example 1.20 Find the Fourier transform of the finite wave train:

$$f(t) = \begin{cases} \cos \omega_0 t & -\dfrac{N\pi}{\omega_0} < t < \dfrac{N\pi}{\omega_0} \\ 0 & \text{otherwise} \end{cases}$$

A short pulse of the form of $f(t)$ is used in FT-NMR to generate a band of frequencies used to excite the nuclei in a sample. Discuss the relationship between the widths of $f(t)$ and its Fourier transform $F(\omega)$.

Solution

The wave train is an even function of t, so we can take the cosine transform

$$F(\omega) = \left(\frac{2}{\pi}\right)^{1/2} \int_0^{N\pi/\omega_0} \cos\omega_0 t \cos\omega t \, dt$$

$$= \frac{1}{\sqrt{2\pi}} \left[\frac{\sin(\omega-\omega_0)N\pi/\omega_0}{\omega-\omega_0} + \frac{\sin(\omega+\omega_0)N\pi/\omega_0}{\omega+\omega_0} \right]$$

Let us restrict ourselves to positive values of ω. For large values of ω_0, only the first term is important and is shown below

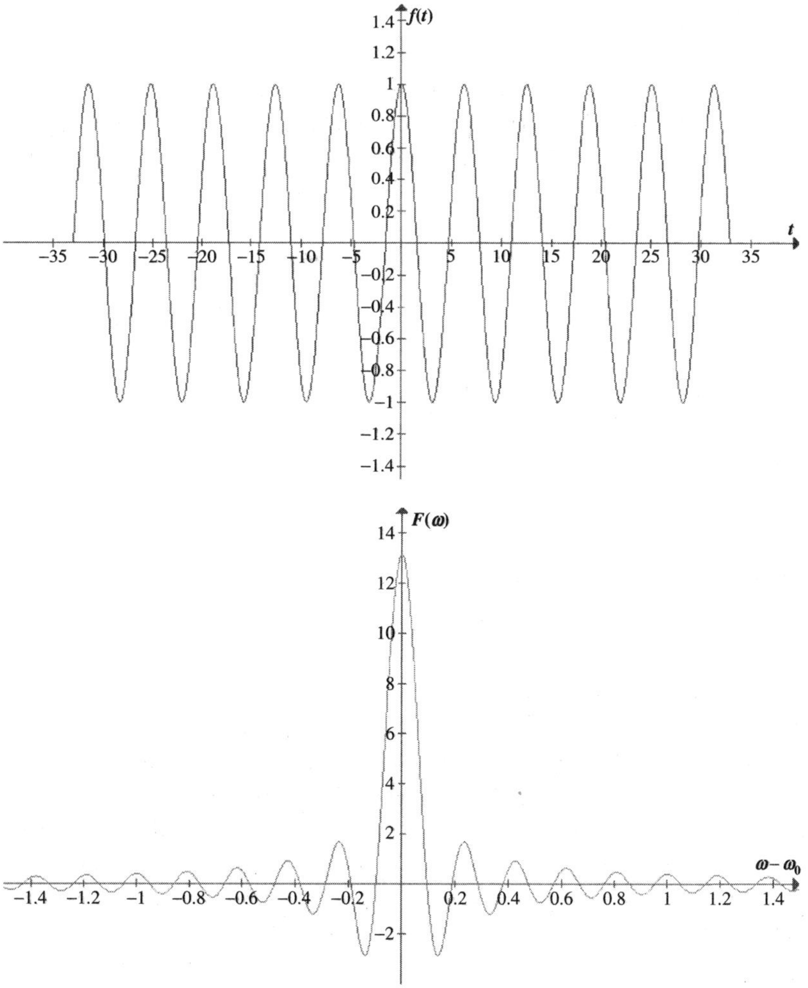

The Fourier transform pair $f(t)$ and $F(\omega)$

The zeroes of $F(\omega)$ occur at the points $(\omega - \omega_0) N\pi / \omega_0 = n\pi$ for $n = \pm 1, \pm 2, \ldots$ or at

$$\frac{\omega}{\omega_0} = 1 \pm \frac{1}{N}, 1 \pm \frac{2}{N}, \ldots$$

Because the contributions outside the central peak of $F(\omega)$ are small (see Figure), we may take $\Delta\omega_0 = \omega_0 / N$ as a measure of the width of $F(\omega)$. The width of the wave train is $\Delta t = 2N\pi / \omega_0$, and so we have the reciprocal relation

$$\Delta\omega \Delta t = 2$$

This result can be written as $\Delta E \Delta t = 2\pi\hbar$ and has a close relationship with the Heisenberg Uncertainty Principle in energy and time, and so we see that our finite wave train does indeed satisfy the Uncertainty Principle.

We shall now show how a Fourier transform can be used to extract frequency information from time signals. This is one of the most important uses of Fourier transforms. Figure 1.19(a) plots $e^{-\alpha t} \cos\omega_0 t$ against time for $\alpha = 0.10$ and $\omega_0 = 1$. Figure 1.19(b) plots the Fourier transform, which clearly shows that only one frequency is involved.

Suppose we now take another function, which is actually a sum of three functions:

$$f(t) = 0.70e^{-0.010t} \cos 1.2t + 1.25e^{-0.025t} \cos 2.0t + 0.75e^{-0.075t} \cos 2.6t \tag{1.43}$$

plotted in Figure 1.20(a). We cannot possibly determine visually how many and what frequencies are involved. If we take the Fourier transform of the function in Figure 1.20(a), however, we obtain

$$F(\omega) = \frac{1}{\sqrt{2\pi}} \left[0.70 \frac{0.010}{0.010^2 + (\omega - 1.2)^2} + 1.25 \frac{0.025}{0.025^2 + (\omega - 2.0)^2} + 0.75 \frac{0.075}{0.075^2 + (\omega - 2.6)^2} \right] \tag{1.44}$$

which is plotted in Figure 1.20(b). This plot clearly shows that three frequencies are involved in the time record. Thus, a Fourier transform extracts the frequencies involved in the time behaviour of the system. Experimentally, the time behaviour is obtained numerically using an extremely efficient algorithm called Fast Fourier Transform (FFT) for the numerical evaluation of Fourier transforms. Prior to the development of FFT, the numerical evaluation of Fourier transforms was rather cumbersome. The ready availability of FFT is one of the reasons why Fourier transform spectroscopy has become so widely used. Another reason that the FT technique has gained importance is its extension to microwave spectroscopy.

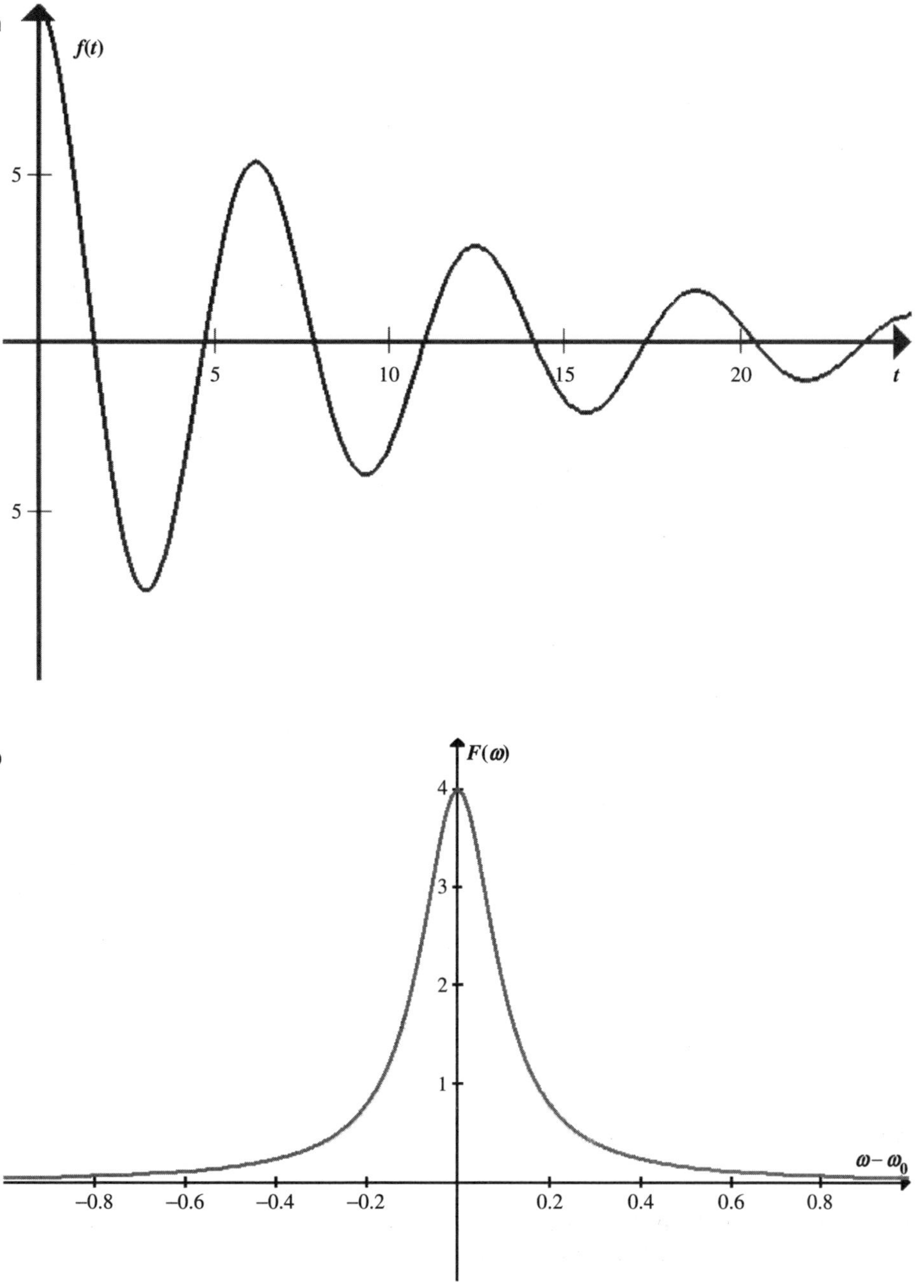

Figure 1.19. (a) The time domain function $e^{-\alpha t}\cos\omega_0 t$ plotted against time for $\alpha = 0.10$ and $\omega_0 = 1$; (b) the corresponding frequency domain function $\frac{1}{\sqrt{2\pi}}\left(\frac{\alpha}{\alpha^2 + (\omega - \omega_0)^2}\right)$ plotted against $(\omega - \omega_0)$

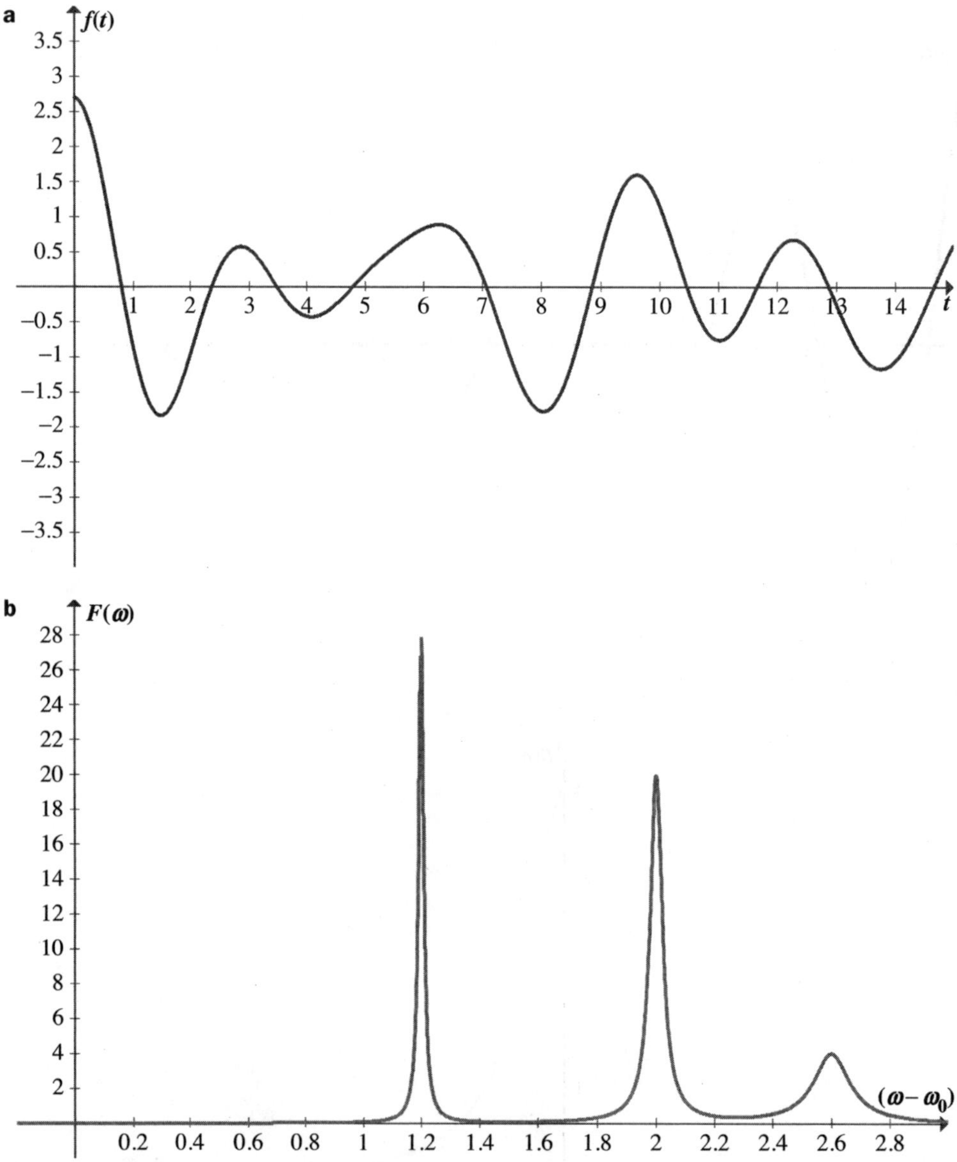

Figure 1.20. (a) The function $f(t)$ given by equation (1.43) plotted versus t; (b) the Fourier transform of $f(t)$, equation (1.44), plotted against $(\omega - \omega_0)$

1.11 SUMMARY

- Waves are characterized by direction, amplitude, wavelength, frequency, angular and linear velocity.
- Like matter, electromagnetic radiation can exhibit dual character–that of waves and that of particles.

- As a wave, it can be considered as a travelling wave having mutually perpendicular electric and magnetic fields, normal to the direction of propagation.
- It can also be considered as consisting of photons carrying energy.
- The electromagnetic radiation comprises a large number of spectral ranges, starting with the low-energy radiofrequency range up to the highest energy γ range where nuclear transitions take place.
- Planck provided the first ideas about quantization of energy levels and a distribution function for frequencies emitted by a blackbody.
- The population difference between the two energy levels undergoing transition is important in determining the intensities of spectral lines.
- Einstein proposed that interaction of radiation with matter leads to stimulated absorption of radiation, but there is also stimulated, as well as spontaneous, emission from the higher energy level.
- Line broadening occurs due to spontaneous emission, resulting in finite lifetime of the excited state.
- If line broadening is due to the natural lifetime of the excited state, the line-shape is Lorentzian.
- Doppler broadening is the main source of line broadening for the high-frequency regions like IR, UV and visible. This leads to Gaussian line-shapes. In solution, inhomogeneous broadening due to interaction of the molecules with solvent also leads to a Gaussian shape.
- Spectra are usually studied in solution and absorbances calculated using the Lambert–Beer law.
- In Fourier transform spectroscopy, the observed time decay of an excited level after applying a pulse of short duration is converted to a frequency function by a Fourier transform.

1.12 EXERCISES

1. The displacement of an object undergoing simple harmonic motion is given by the equation

$$x(t) = 3.00 \sin(8\pi t + \pi/4)$$

where x is in metre, t is in seconds and the argument of the sine function is in radians.
   (a) What is the amplitude of motion?
   (b) What is the frequency of the motion?
   (c) What are the position, velocity, and acceleration of the object at $t = 0$?
2. A particle is oscillating harmonically about the origin along the y-axis with amplitude $A = 0.04$ m and period $T = 2.0$ s.
   (a) What is the angular speed of the corresponding uniform circular motion?
   (b) What is the minimum time required for the particle to travel from $y_1 = -0.01$ m to $y_2 = +0.02$ m?
3. What is the energy (kJ mol^{-1}) associated with ultraviolet radiation of wavelength (a) 184.7 nm, (b) 253.7 nm, (c) 366.0 nm?
4. What is the frequency, the energy (in J and in eV) and the wavenumber (in cm^{-1}) of light at the following wavelengths:

(a) 193 nm (the UV output of an ArF excimer laser)
(b) 600 nm (red light)
(c) 4 μm, in the infrared
(d) 37.5 cm (a typical wavelength used by mobile phones)
(e) 2.922 (radio waves: FM)

5. The energies required to break the C-C bond in ethane, the 'triple bond' in CO, and a hydrogen bond are about 368, 1075 and 17 kJ mol^{-1}. What wavelengths of radiation are required to break these bonds?
6. The maximum kinetic energy of electrons emitted from Na at different wavelengths was measured with the following results:

λ / nm	Maximum kinetic energy / eV
450	0.40
400	0.76
350	1.20
300	1.79

Calculate Planck's constant and the value of the work function from these data.

7. Calculate the de Broglie wavelength for the following cases:
(a) An electron in an electron microscope accelerated with a potential of 100 kV.
(b) A helium atom moving with a speed of 100 m s^{-1}.
(c) A bullet weighing 0.001 kg moving at a speed of 100 m s^{-1}.
Assume that the uncertainty in the speed is 10% and calculate the uncertainty in the position for each of the three cases.
8. Calculate k_BT at 25 °C in J, kJ mol^{-1}, eV and cm^{-1}. A molecule has of the order of k_BT of thermal energy at temperature T. How does this amount of energy compare with the spacing of the lowest two energy levels of the following?
(a) Translational energy levels of N_2 in a laboratory-sized vessel, taken to be a cube with $L = 0.1$ m:
$$E_{n_x,n_y,n_z} = \frac{h^2}{8mL^2}(n_x^2 + n_y^2 + n_z^2) \quad n_x, n_y, n_z = 1, 2, 3, \ldots, [m(\mathrm{N}) = 14.01\ u].$$
(b) Rotational energy levels for N_2 treated as a rigid rotator
$$E_J = hc\tilde{B}J(J+1),\ J = 0, 1, 2, \ldots$$
where the rotational constant $\tilde{B}$ (in cm^{-1}) is defined in terms of the moment of inertia $I = \mu_{red}r^2$, for reduced mass $\mu_{red} = m_1m_2 / (m_1 + m_2)$ and bond length r as $\tilde{B} = h / 8\pi^2 Ic$ [$r(\mathrm{N_2}) = 109.8$ pm].
(c) Vibrational energy levels for N_2 (treated as a simple harmonic oscillator):
$$E_\upsilon = hc\tilde{\nu}\left(\upsilon + \frac{1}{2}\right),\ \upsilon = 0, 1, 2, \ldots$$
where the vibrational wavenumber (related to the vibrational frequency by $\tilde{\nu} = \nu / c$) is given by
$$\tilde{\nu} = \frac{1}{2\pi c}\sqrt{\frac{k}{\mu_{red}}}$$
and k is the bond's force constant [$k(\mathrm{N_2}) = 2294$ N m^{-1}].

9. (a) The population of the $\upsilon = 1$ level relative to the $\upsilon = 0$ ground state of N_2 is given by the Boltzmann distribution:

$$\frac{N_1}{N_0} = \exp\left(-\frac{\Delta E}{k_B T}\right)$$

where $\Delta E = E_1 - E_0$ is the difference in energy between the two levels (which are non-degenerate).
Calculate the ratio N_1 / N_0 at 25 °C and comment on its value.

(b) For the rotational levels, the fact that each *J-level is degenerate* ($g_J = 2J + 1$) must be considered – at higher energies (as J increases), there are more states per level. The Boltzmann distribution gives

$$\frac{N_J}{N_0} = \frac{g_J}{g_0}\exp\left(-\frac{\Delta E_J}{k_B T}\right)$$

where g_J is the degeneracy of the energy level E_J, and $\Delta E_J = E_J - E_0 = E_J$.
Calculate the population of the first and second excited rotational states of N_2 relative to the ground state ($J = 0$).

10. The energy required for the ionization of a certain atom is 5.12 aJ (1 aJ = 10^{-18} J). The absorption of a photon of unknown wavelength ionizes the atom and ejects an electron with velocity 345 km s^{-1}. Calculate the wavelength of the incident radiation.
11. Calculate the energy per photon and the energy per Einstein (mole of photons) for radiation of wavelength (a) 200 nm (ultraviolet), (b) 150 pm (X-ray), (c) 1.00 cm (microwave).
12. A laser used to read CDs emits red light of wavelength 700 nm. How many photons does it emit each second if the power is (a) 0.10 W, (b) 1.0 W?
13. Calculate the de Broglie wavelength of an electron accelerated from rest through a potential difference of (a) 100 V, (b) 1.0 kV, (c) 100 kV.
14. Calculate the energy (in kJ mol^{-1}) and wavenumber of (a) IR photons of wavelength 2.5 μm, (b) green photons of wavelength 532 nm.
15. Calculate the velocity of an electron driven from the surface of potassium by incident light of wavelength 350 nm. The work function of potassium, $\Phi_K = 2.26$ eV.
16. Obtain an expression for the Doppler line-width for a spectral line of wavelength λ emitted by an atom of mass m at a temperature T.
17. The lifetime, τ, of the $3^2P_{3/2} \leftarrow 3^2S_{1/2}$ transition of the sodium atom at 589.0 nm has been measured as 16 ns. Determine the Einstein A and B coefficients, stating the units of your answers.
18. In the Helium–Neon laser (three-level laser), the energy spacing between the upper and lower levels $E_2 - E_1 = 2.26$ eV in the neon atom. After the optical pumping operation stops, at what temperature would the ratio of the population of the upper level E_2 and the lower level E_1 be 0.1?
19. (a) Convert 15,000 cm^{-1} into wavelength (nm) and frequency (Hz).
(b) Convert 450 nm into wavenumber (cm^{-1}) and frequency (Hz).

20. The absorption of radiation energy by a molecule results in the formation of an excited molecule. It would seem that, given enough time, all the molecules in a sample would have been excited and no more absorption would occur. Yet in practice we find that the absorbance of a sample at any wavelength remains unchanged with time. Why?
21. The mean lifetime of an electronically excited molecule is 10^{-8} s. If the emission of the radiation occurs at 610 nm, what are the uncertainties in frequency ($\Delta\nu$) and wavelength ($\Delta\lambda$)? (Hint: Take the ratio $\Delta\nu / \nu \approx \Delta\lambda / \lambda$.)
22. The familiar yellow *D*-lines of sodium are actually a doublet at 587 nm and 589 nm. Calculate the difference in energy (in Joule) between the two lines.
23. The typical energy differences for transitions in the microwave, infrared and electronic spectroscopies are 5×10^{-22} J, 5×10^{-15} J and 1×10^{-18} J, respectively. Calculate the ratio of number of molecules in the two adjacent levels (e.g., the ground level and the first excited level) at 300 K in each case.
24. The resolution of visible and UV spectra can usually be improved by recording the spectra at low temperatures. Why does this work?
25. (a) Convert the following percent transmittance to absorbance: (i) 100%, (ii) 50% and (iii) 0%.
    (b) Convert the following absorbance to percent transmittance: (i) 0.0, (ii) 2.0 and (iii) ∞.
26. (a) A spectrometer is adjusted so that when the cell is empty, the meter indicates 100.0 for a source that is a monochromatic red beam. The cell is 2-cm long. When the cell is filled with 0.01 M MnO_4^-, the meter reading is 78.7. What would be the meter reading for a cell 5-cm long?
    (b) A 1.000 g sample of manganese steel is dissolved in acid, and the manganese is oxidized to MnO_4^-. The resulting solution is made up to 100.0 mL in a volumetric flask, and some of the solution is used to fill the 2-cm cell described in (a). The meter reading is 91.6. What is the manganese content of the steel?

2

Theoretical Principles

Dad	I understand that light is emitted when an atom decays from an excited state to a lower energy state.
Feynman	That's right.
Dad	And light consists of particles called photons.
Feynman	Yes.
Dad	So the photon 'particle' must be inside the atom when it is in the excited state.
Feynman	Well no.
Dad	Well how do you explain that the photon comes out of the atom when it was not there in the first place.
Feynman (Physics Teacher 1969)	I'm sorry. I don't know I can't explain it to you.

2.1 INTRODUCTION

As was stated in Chapter 1, spectroscopy is a study of the interaction of matter with radiation. In order to understand their mutual interaction, we must therefore characterize both radiation and matter. The starting point is quantum mechanics, because the origin of quantum mechanics was based on unexplained phenomena related to both the nature of electromagnetic radiation and that of matter. In fact, quantum mechanics owes its birth to some puzzling results of spectral and other experiments. We may say that "spectroscopy is quantum mechanics in action." We therefore begin our study of quantum mechanics with a recapitulation of its postulates. An understanding of the postulates is also essential for our development of the time-dependent perturbation theory and also forms the basis for the connection between theory and experiment. Indeed, spectroscopy is the experimental verification of the laws of quantum mechanics.

Quantum mechanics is based on a few postulates (from the Latin *postulatum*), meaning a truth that does not need any further proof, because it is obvious by itself, and its success lies in the fact that no experiment till date has proved it wrong. There is no agreement of how many axioms one needs to describe the machinery of quantum mechanics; however, the six given below is an acceptable number. The first three postulates, as we shall see, make up the mathematical background of quantum mechanics, and the fourth and fifth supply the association between the mathematics introduced by the first three and the results of a measurement process. The sixth is the antisymmetry principle, which is the basis of Pauli's Exclusion Principle.

As a general rule, the statement of each postulate will be followed by some comments. The significance of words within the postulates will be explained at the right time, namely, as and when they are introduced.

2.2 THE POSTULATES OF QUANTUM MECHANICS EXPLAINED

We now state the postulates of quantum mechanics, particularly in relation to spectroscopy.

Postulate 1

a The state of a dynamical system of N particles is described as fully as possible by an appropriate state function $\Psi(q_1, q_2, \ldots, q_{3N}, t)$. This is a function of all the $3N$ coordinates of the system and time (and the N spin coordinates).

b The state function Ψ has the property that $\Psi^* \Psi d\tau$ is the probability of finding particle **1** between q_1 and $q_1 + dq_1$, particle **2** between q_2 and $q_2 + dq_2$, …, particle N between q_N and $q_N + dq_N$, at a specified time t. This is the Born interpretation of the wave function. Since $\Psi^*\Psi d\tau$ represents a probability, the probability that the particle is *somewhere* is unity, we have the normalization condition $\int \Psi^* \Psi d\tau = 1$, where $d\tau$ is the volume element.

Note: Ψ must be well behaved, i.e., it must be single valued, finite and continuous for all values of the position. It should be well bounded and go to zero at the boundaries.

Postulate 2

The wave function that represents the state of the system changes in time according to the time-dependent Schrödinger equation

$$\hat{H}(\vec{q}, \vec{p}, t)\Psi(\vec{q}, t) = i\hbar \frac{d\Psi(\vec{q}, t)}{dt} \tag{2.1}$$

where $\hat{H}$ is called the Hamiltonian operator and is a function of position, $\vec{q}$, the momentum, $\vec{p}$, and the time t.

Consider the case in which the Hamiltonian $\hat{H}$ does not contain the time explicitly. Such a system is called a *conservative* system. In such a system, the sum of the kinetic and potential energies of the system remains constant with time. A conservative system is, therefore, an isolated system and is not acted upon by external forces. In such systems, the potential energy, V, depends only on the position and is independent of time. $\Psi(q_1, q_2, \ldots, q_{3N}, t)$ may then be expressed as a product of a time-independent function $\psi(q_1, q_2, \ldots, q_{3N})$ and a time-dependent function $F(t)$, or simply,

$$\Psi(\vec{q}, t) = \psi(\vec{q})F(t) \tag{2.2}$$

where $\vec{q} = (q_1, q_2, \ldots, q_{3N})$ represents the position vector.

Substituting in equation (2.1) gives

$$\hat{H}(\vec{q}, \vec{p})\big(\psi(\vec{q})F(t)\big) = i\hbar \frac{d}{dt}\big(\psi(\vec{q})F(t)\big)$$

or

$$F(t)\hat{H}(\vec{q},\vec{p})\psi(\vec{q}) = i\hbar\psi(\vec{q})\frac{dF(t)}{dt}$$

because $\hat{H}(\vec{q},\vec{p})$, which is independent of time, does not have any effect on $F(t)$, and, similarly, $\psi(\vec{q})$ is independent of time, and thus the $\frac{\hat{d}}{dt}$ operator operates only on $F(t)$.

If we divide throughout by $\Psi(\vec{q},t) = \psi(\vec{q})F(t)$, we obtain

$$\frac{1}{\psi(\vec{q})}H(\vec{q},\vec{p})\psi(\vec{q}) = \frac{i\hbar}{F(t)}\frac{dF(t)}{dt}. \tag{2.3}$$

In this equation, the right-hand side contains only t, and the left-hand side depends only on the coordinates $\vec{q}$. The two sides of the equation must, however, remain equal, no matter what values of t or the coordinates are used, and hence each side must be separately equal to a constant, say E. Hence, the left-hand side gives

$$\hat{H}(\vec{q},\vec{p})\psi(\vec{q}) = E\psi(\vec{q})$$

which is the familiar time-independent Schrödinger equation.

The right-hand side of equation (2.3) can be rearranged to

$$\begin{aligned}
&\frac{dF(t)}{dt} = \frac{F(t)}{i\hbar}E \\
&\Rightarrow \frac{dF(t)}{F(t)} = \frac{E}{i\hbar}dt \\
&\Rightarrow \int_0^t \frac{dF(t')}{F(t')} = \frac{E}{i\hbar}\int_0^t dt' \\
&\Rightarrow \ln F(t')\big|_0^t = \frac{E}{i\hbar}t'\big|_0^t \\
&\Rightarrow \ln\frac{F(t)}{F(0)} = \frac{E}{i\hbar}t \\
&\Rightarrow \frac{F(t)}{F(0)} = e^{\frac{E}{i\hbar}t} = e^{-iEt/\hbar}
\end{aligned}$$

or

$$F(t) = F(0)e^{-iEt/\hbar}$$

The constant $F(0)$ in this expression is simply $\psi(\vec{q})$, the wave function at time $t = 0$. Hence, the state function $\Psi(\vec{q},t)$ for a particle is separable into the product of a space-dependent function $\psi(\vec{q})$ and a time-dependent function $F(t)$ $\left[\text{i.e., } \Psi(\vec{q},t) = \psi(\vec{q})F(t) = \psi(\vec{q})e^{-iEt/\hbar}\right]$ for conservative systems.

The probability density at a point $\vec{q}$ is given by

$$\begin{aligned}
\Psi^*(\vec{q},t)\Psi(\vec{q},t) &= \psi^*(\vec{q})e^{iEt/\hbar}\psi(\vec{q})e^{-iEt/\hbar} \\
&= \psi^*(\vec{q})\psi(\vec{q})
\end{aligned}$$

where the complex conjugate of $\Psi(\vec{q},t)$ is represented by $\Psi^*(\vec{q},t)$. Such states are called *stationary* states because the corresponding probability density has no time dependence. Since $F(t)$ has no significant effect on energy or particle distribution, we can ignore it while dealing with stationary states. The average position $<q>$ is also unaffected by the time factor, and hence all quantities are independent of time for stationary states.

Postulate 3

For every observable property of a system, there exists a corresponding linear Hermitian operator and the physical properties of the observable can be inferred from the mathematical properties of the associated operator.

In this book, we shall represent all quantum mechanical operators with a caret (hat or ^) symbol above them. To obtain the operator for a given observable, we write down the classical expression for the quantity in terms of the position and momentum and replace the position by the corresponding operator (e.g., x by $\hat{x}$, which represents multiplication by x) and the momentum by its operator, i.e., $\hat{p}_x = -i\hbar\frac{d}{dx}$, where $\hbar = h/2\pi$.

For example, the kinetic energy

$$E_k = \frac{1}{2}m\text{v}^2 = \frac{(m\text{v})^2}{2m} = \frac{p^2}{2m}$$

Therefore, the corresponding operator

$$\hat{T} = \frac{1}{2m}\left(-i\hbar\frac{d}{dx}\right)\left(-i\hbar\frac{d}{dx}\right) = -\frac{\hbar^2}{2m}\frac{d^2}{dx^2}$$

in one dimension, and

$$\hat{T} = -\frac{\hbar^2}{2m}\left(\frac{\partial^2}{\partial x^2} + \frac{\partial^2}{\partial y^2} + \frac{\partial^2}{\partial z^2}\right) = -\frac{\hbar^2}{2m}\nabla^2$$

in three dimensions. Here ∇^2 is the Laplacian operator.

Example 2.1 Find the operator for $\hat{L}_z$, the z-component of the angular momentum.

Solution

Classically, $\vec{L} = \vec{r} \times \vec{p}$, where $\vec{r}$ is the position vector and $\vec{p}$ the momentum. Accordingly,

$$\vec{L} = \begin{bmatrix} \hat{i} & \hat{j} & \hat{k} \\ x & y & z \\ p_x & p_y & p_z \end{bmatrix} = \hat{i}(yp_z - zp_y) - \hat{j}(xp_z - zp_x) + \hat{k}(xp_y - yp_x) = \hat{i}L_x + \hat{j}L_y + \hat{k}\hat{L}_z$$

Therefore,

$$\hat{L}_Z = xp_y - yp_x = x\left(-i\hbar\frac{\partial}{\partial y}\right) - y\left(-i\hbar\frac{\partial}{\partial x}\right) = i\hbar\left(y\frac{\partial}{\partial x} - x\frac{\partial}{\partial y}\right)$$

Note: $\hat{i}$, $\hat{j}$ and $\hat{k}$ are unit vectors in the Cartesian *x*, *y* and *z* directions.

In quantum mechanics, the operators $\hat{A}$ and $\hat{B}$ associated with two canonically conjugate dynamical variables *a* and *b* must satisfy the equation $\left|[\hat{A},\hat{B}]\right| = i\hbar$, where $[\hat{A},\hat{B}]$ represents the commutator of the operators $\hat{A}$ and $\hat{B}$, and is equal to $\hat{A}\hat{B} - \hat{B}\hat{A}$. In terms of the commutator, it can be shown that $\Delta a \Delta b \equiv \sigma_a \sigma_b \geq \left\{\left|\frac{1}{2i}\left[\hat{A},\hat{B}\right]\right|\right\}$.

For example, in classical mechanics, the position *x* and the *x* component of the momentum, p_x, are conjugate variables. Inserting the corresponding operators, the commutator

$$[\hat{x},\hat{p}_x]\psi = x\left(-i\hbar\frac{d}{dx}\right)\psi - \left(-i\hbar\frac{d}{dx}\right)x\psi$$

$$= -i\hbar x\left(\frac{d\psi}{dx}\right) + i\hbar\left(xd\frac{d\psi}{dx} + \psi\right) = i\hbar\psi$$

Therefore

$$[\hat{x},\hat{p}_x] = i\hbar$$

and, according to the Uncertainty Principle, $\Delta x \, \Delta p_x \geq \frac{\hbar}{2}$.

Postulate 4

Suppose that $\hat{A}$ is a linear Hermitian operator corresponding to an observable and that there is a set of identical systems in state ψ, which is an eigenfunction of $\hat{A}$, i.e., $\hat{A}\psi = a\psi$, where *a* is a real number. Then, if an experimentalist makes a series of measurements of the quantity corresponding to $\hat{A}$ on different members of the set, he/she will always get the same result *a*. It is only when ψ and $\hat{A}$ satisfy the eigenvalue equation that an experiment will give the same result on each measurement.

Consider the uncertainty Δa associated with a single experimental measurement of *a*. The mean value, $<a>$, of a large number of experimental determinations of *a* is

$$<a> = \frac{\int \psi^* \hat{A}\psi d\tau}{\int \psi^* \psi d\tau} = \int \psi^* \hat{A}\psi d\tau,$$ assuming ψ is normalized, i.e., $\int \psi^* \psi d\tau = 1$.

Since $\hat{A}\psi = a\psi$ and the constant *a* commutes with ψ, we can write

$$\int \psi^* \hat{A}\psi d\tau = a\int \hat{\psi}^* \psi d\tau = a$$

Therefore, $\langle a\rangle = a$ since $\int \psi^* \psi d\tau = 1$. Likewise, $\langle a^2\rangle = \int \psi^* \hat{A}^2 \psi d\tau = a\int \psi^* \hat{A} \psi d\tau = a^2$. The uncertainty Δa can now be obtained from

$$\Delta a = \sqrt{(\langle a^2\rangle - \langle a\rangle^2)} = 0$$

In other words, the result of a precise experimental measurement of $\hat{A}$ in the state ψ will certainly be the value a.

Postulate 5

Given an operator $\hat{A}$ and a set of identical systems characterized by a function ψ that is not an eigenfunction of $\hat{A}$, a series of measurements of the property corresponding to $\hat{A}$ on different members of the set will not give the same result. Rather, a distribution of results will be obtained, the average of which will be given by

$$\langle a\rangle = \frac{\int \psi^* \hat{A} \psi d\tau}{\int \psi^* \psi d\tau}$$

This is the so-called *mean-value theorem.* If the ψ's are normalized, the denominator becomes unity. However, it must be emphasized that any result of a measurement of a dynamical variable is one of the eigenvalues of the corresponding operators. The question is "What is the probability of obtaining a particular eigenvalue in a measurement?"

Suppose ψ is a state which is not an eigenfunction of the operator $\hat{A}$. This function can be expanded in terms of the eigenfunctions ψ_n of the operator

$$\psi = \sum_{n=1}^{\infty} c_n \psi_n .$$

This expression can also be written as $\psi = \sum_{m=1}^{\infty} c_m \psi_m$ since m, n or any other variable in such an expression is just a running variable to keep count of the number of terms in an expansion.

To obtain a particular coefficient, c_m, multiply both sides of this equation by ψ_m^* and integrate.

$$\int \psi_m^* \psi d\tau = \int \psi_m^* \sum_{n=1}^{\infty} c_n \psi_n d\tau = c_m$$

since the ψ_n form an orthonormal set (recall that the eigenfunctions of an operator having different eigenvalues are orthogonal to each other and can be normalized to form an orthonormal set), and so $\int \psi_m^* \psi_n d\tau = \delta_{mn}$.

Note: δ_{mn} is the Krönecker delta function, which equals unity for $m = n$ and zero for all $m \neq n$.

The mean value $\langle a\rangle$ is given by

$$\langle a\rangle = \int \psi^* \hat{A} \psi d\tau = \int \left[\sum_{m=1}^{\infty} c_m^* \psi_m^*\right] \hat{A} \left[\sum_{n=1}^{\infty} c_n \psi_n\right] d\tau = \sum_{m=1}^{\infty}\sum_{n=1}^{\infty} c_m^* c_n \int \psi_m^* \hat{A} \psi_n d\tau$$

Since ψ_n and ψ_m are all eigenfunctions of the operator $\hat{A}$, $\hat{A}\psi_n = a\psi_n$. Hence,

$$<a> = \sum_{m=1}^{\infty}\sum_{n=1}^{\infty} c_m^* c_n \int \psi_m^* \hat{A}\psi_n d\tau = \sum_{m=1}^{\infty}\sum_{n=1}^{\infty} c_m^* c_n a_n \int \psi_m^* \psi_n d\tau = \sum_{n=1}^{\infty} c_n^* c_n a_n \int \psi_n^* \psi_n d\tau = \sum_{n=1}^{\infty} |c_n^2| a_n$$

Comparing with the expression for the average value $<a> = \sum_{n=1}^{\infty} p_n a_n$, we have $p_n = c_n^* c_n = |c_n|^2$, i.e., the probability of obtaining a value a_n in a measurement is the square of the Fourier coefficient of the expansion in terms of the eigenfunctions.

If ψ is normalized,

$$\int \psi^* \psi d\tau = 1 \Rightarrow \int \left[\sum_{m=1}^{\infty} c_m^* \psi_m^*\right]\left[\sum_{n=1}^{\infty} c_n \psi_n\right] d\tau = \sum_{m=1}^{\infty}\sum_{n=1}^{\infty} c_m^* c_n \int \psi_m^* \psi_n d\tau = 1$$

Since $\int \psi_m^* \psi_n d\tau = \delta_{mn}$, $\sum_{n=1}^{\infty} |c_n^2| = 1$, as should be the case if the $|c_n^2|$ are to represent probabilities.

Postulate 6

The total wave function must be antisymmetric with respect to the interchange of all coordinates of one fermion with those of another. These coordinates must include the electron spin.

The Pauli Exclusion Principle is a direct result of this *antisymmetry principle*. We shall develop this postulate in our study of atomic spectroscopy in the next chapter.

2.3 TIME-DEPENDENT PERTURBATION

Thus far, we have seen that the Schrödinger equation (2.1), i.e., $\hat{H}\Psi = i\hbar \frac{d\Psi}{dt}$, yields solutions of the kind, $\Psi_n(\vec{q}, t) = \psi_n(\vec{q})e^{-iE_n t/\hbar}$, corresponding to allowed energy levels E_n's of stationary states [equation (2.2)]. We now apply the postulates to develop time-dependent perturbation theory, since we are now looking at systems absorbing radiation with energies varying with time.

Let us consider two solutions Ψ_l and Ψ_m of the general Schrödinger equation (2.1). Since Ψ_l and Ψ_m are individually eigenfunctions of the same linear operator $\hat{H}$, any linear combination of these, i.e., $c_l\Psi_l + c_m\Psi_m$, also represents a valid solution of equation (2.1) by the principle of superposition. Here c_l and c_m are constants independent of time.

We now want to treat *transitions* between these two quantum states, which are driven by a *time-dependent* Hamiltonian. This moves us from the realm of *statics* to *dynamics*. Since the state of this system will now depend on time, we shall use time-dependent perturbation theory to describe the system. The perturbation causes a change in the energy of the system, and its new Hamiltonian is $\hat{H} + \hat{H}'$, where $\hat{H}'$, the *perturbation* Hamiltonian, describes the interaction between the system and the radiation field. In this case, $\hat{H}$ is time-independent and $\hat{H}'$ is time-dependent.

We shall consider the case in which $\hat{H}'$ is relatively small, i.e., $\hat{H}' << \hat{H}$, so that its only effect is to cause transitions between the stationary states described by $\hat{H}$, without otherwise disturbing the stationary state wave functions and their energies. This amounts to applying first-order perturbation theory, since we are using the zeroth-order wave functions.

We consider two states of the system, *l* and *m*. As a result of the perturbation, transitions can occur between these two states. The system can now be described by the wave function

$$\Psi = c_l(t)\Psi_l(\vec{q}, t) + c_m(t)\Psi_m(\vec{q}, t) \tag{2.4}$$

where c_l and c_m are functions of time. This is equivalent to expressing the perturbed wave function Ψ in terms of a linear combination of unperturbed wave functions, i.e., in general,

$$\Psi = \sum_{n=1}^{\infty} c_n(t)\Psi_n(\vec{q}, t) \tag{2.5}$$

Therefore, equation (2.1), i.e., the Schrödinger equation for the *perturbed state*, should now become

$$(\hat{H} + \hat{H}')\left(\sum_n c_n(t)\Psi_n(\vec{q}, t)\right) = i\hbar \frac{d}{dt}\left(\sum_n c_n(t)\Psi_n(\vec{q}, t)\right) \tag{2.6}$$

This can be resolved as

$$\hat{H}\left(\sum_n c_n(t)\Psi_n(\vec{q}, t)\right) + \hat{H}'\left(\sum_n c_n(t)\Psi_n(\vec{q}, t)\right)$$
$$= i\hbar\left(\sum_n \frac{dc_n(t)}{dt}\Psi_n(\vec{q}, t)\right) + i\hbar\left(\sum_n c_n(t)\frac{d\Psi_n}{dt}(\vec{q}, t)\right) \tag{2.7}$$

Since each Ψ_n is individually a solution of equation (2.1), i.e., $\hat{H}\Psi_n = i\hbar\frac{d\Psi_n}{dt}$, and the c_n's of the unperturbed system are independent of time, the first term on the left-hand side and the last term on the right-hand side of equation (2.7) are equal. This leaves

$$\hat{H}'\left(\sum_n c_n(t)\Psi_n(\vec{q}, t)\right) = i\hbar\left(\sum_n \frac{dc_n(t)}{dt}\Psi_n(\vec{q}, t)\right) \tag{2.8}$$

To obtain $dc_m(t)/dt$, we multiply both sides of equation (2.8) by $\Psi_m^*(\vec{q}, t)$ and integrate over all space to get

$$\int \Psi_m^*(\vec{q}, t)\hat{H}'\left(\sum_n c_n(t)\Psi_n(\vec{q}, t)\right)d\tau = i\hbar\int \Psi_m^*(\vec{q}, t)\left(\sum_n \frac{dc_n(t)}{dt}\Psi_n(\vec{q}, t)\right)d\tau$$
$$= i\hbar\sum_n \frac{dc_n(t)}{dt}\int \Psi_m^*(\vec{q}, t)\Psi_n(\vec{q}, t)d\tau$$

since the coefficients depend only on time and are independent of space. Using the orthonormal property of Ψ_n's ($\int \Psi_m^*(\vec{q}, t)\Psi_n(\vec{q}, t) = \delta_{mn}$) to eliminate all the terms in the summation in the last expression, except when $n = m$, we have

$$\int \Psi_m^*(\vec{q}, t)\hat{H}'\sum_n c_n(t)\Psi_n(\vec{q}, t)d\tau = i\hbar\frac{dc_m(t)}{dt}$$
$$\Rightarrow \frac{dc_m(t)}{dt} = \frac{1}{i\hbar}\int \Psi_m^*(\vec{q}, t)\hat{H}'\sum_n c_n(t)\Psi_n(\vec{q}, t)d\tau \tag{2.9}$$

Expanding $\Psi_n(\vec{q},t) = \psi_n(\vec{q})e^{-iE_nt/\hbar}$ and inserting $\Psi_m^*(\vec{q},t) = \psi_m^*(\vec{q})e^{iE_mt/\hbar}$, we obtain

$$\frac{dc_m(t)}{dt} = \frac{1}{i\hbar}\sum_n c_n(t)\int \psi_m^* e^{iE_mt/\hbar}\hat{H}'\psi_n e^{-iE_nt/\hbar}\,d\tau = \frac{1}{i\hbar}\sum_n c_n(t)H'_{mn}e^{-i(E_n-E_m)t/\hbar} \tag{2.10}$$

where $H'_{mn} = \int \psi_m^*\hat{H}'\psi_n d\tau$ is the integral over the time-independent functions as usual and it is understood that the wave functions are space dependent. This is a set of first-order differential equations involving all the coefficients $c_n(t)$.

To obtain $c_m(t)$, we integrate equation (2.10):

$$c_m(t) - c_m(0) = \frac{1}{i\hbar}\int_0^t \sum_n c_n(t')H'_{mn}e^{-i(E_n-E_m)t'/\hbar}dt' \tag{2.11}$$

This is a rather disappointing result, because it says that, in order to evaluate a coefficient c_m, we need the whole set of coefficients. Fortunately, simplification is possible if we take into account that the perturbation is applied for a short time, so that its total effect is small. Let us now assume that the system starts at $t = 0$ in a state Ψ_l and that the perturbation H' is turned on at time $t = 0$ and off at time t. We pick t short enough so that the total probability of a transition out of the state l is very small, and then the coefficients $c_n(t)$ can be taken equal to $c_n(0)$ plus first-order corrections. Then at time $t = 0$,

$$c_l(0) = 1;\ c_n(0) = 0,\ n \neq l$$

We assume *that this is still true a very short time after t = 0*, when radiation is applied on the system and transition is occurring. Therefore, the term on the right-hand side of equation (2.9) is zero for all $n \neq l$, and the *initial rate of transition*, which is given by $\frac{dc_m(t)}{dt}$, is equal to

$$\frac{dc_m(t)}{dt} = \frac{1}{i\hbar}c_l(t)H'_{ml}e^{-i(E_l-E_m)t/\hbar}$$

or

$$\frac{dc_m(t)}{dt} = \frac{1}{i\hbar}H'_{ml}e^{-i(E_l-E_m)t/\hbar}.$$

since $c_l(t) \approx 1$. Therefore,

$$c_m(t) = c_m(0) + \frac{1}{i\hbar}\int_0^t H'_{ml}e^{-i(E_l-E_m)t'/\hbar}dt' = \frac{1}{i\hbar}\int_0^t H'_{ml}e^{-i\omega_{lm}t'}dt' \tag{2.12}$$

where

$$\omega_{lm} = \frac{E_l - E_m}{\hbar} = \frac{2\pi}{h}(E_l - E_m) = 2\pi\nu_{lm} \tag{2.13}$$

and ν_{lm} is the frequency of the transition. The special coefficient (for $n = l$), $c_l(t)$, is given by

$$c_l(t) = 1 + \frac{1}{i\hbar}\int_0^t H'_{ll}dt' \tag{2.14}$$

since $c_l(t) = 1$ at $t = 0$. If the perturbation is turned off at time t, then

$$c_m(t' > t) = \frac{1}{i\hbar}\int_0^t H'_{ml} e^{-i\omega_{lm}t'} dt' \tag{2.15a}$$

$$c_l(t' > t) = 1 + \frac{1}{i\hbar}\int_0^t H'_{ll} dt' \tag{2.15b}$$

We can recognize equation (2.15a) as the Fourier transform integral (see also Chapter 1). Since the probability of finding the system in a state m is proportional to $|c_m(t)|^2$, we may now state: *The probability of finding the system in a state m, $|c_m^2|$ is proportional to the square of the Fourier transform of the perturbation.*

2.3.1 The interaction of electromagnetic radiation with a molecular system

As yet, we have not qualified the nature of the perturbation, i.e., $\hat{H}'$, and the preceding results are general, applicable to all kinds of time-dependent perturbation. We now consider the specific case when the perturbation is in the form of electromagnetic radiation. The electromagnetic field of the radiation is represented by an oscillating electric field $\vec{E}$ and an oscillating magnetic field $\vec{B}$, both of which are mutually perpendicular and normal to the direction of propagation of the electromagnetic radiation (Figure 1.5). However, the magnitudes of the electric and magnetic fields are very different and are related by $|\vec{E}| = c|\vec{B}|$. Hence, the interaction with a charged particle is almost entirely due to $\vec{E}$, that with $\vec{B}$ being negligible, because the magnitude of interaction with a particle of charge 'q' is $q|\vec{E}|$ in the former case and $q|\vec{B}|$ in the latter, and $|\vec{E}| >> |\vec{B}|$ because of the large magnitude of the speed of light. We may therefore ignore the interaction with the magnetic field. However, magnetic dipole transitions do occur and are important, especially if the transition is forbidden by electric dipole selection rules, but they are beyond the scope of this book. Throughout the book, we shall focus only on electric dipole transitions because they are ~10^5 times more intense than magnetic dipole transitions and ~10^7 times more intense than nuclear quadrupole transitions.

In the case of a molecule, the charge is represented by a dipole, and the interaction with the electric field is given by $-\vec{\mu}.\vec{E}$, where $\vec{\mu}$ is the dipole moment of the molecule. For simplicity, we shall consider only the x-component of the dipole moment of the molecule and the x-component of the electric vector, E_x, and then generalize the result to all three dimensions. The x-component of the electric dipole moment can be written as

$$\mu_x = \sum q_i x_i$$

where the summation extends to all the charged particles (electrons and nuclei) of charge q_i in the system. This quantity will vary under the perturbation. The oscillating electric field of the electromagnetic radiation can disturb the potential energy of the molecule and allow it to escape from the initial stationary state, here assumed to be characterized by the state l. The electric field of the radiation oscillates at the point occupied by the molecule with a frequency ν. For example, the x-component of the electric field of the radiation can be described at the position occupied by the molecule by an equation which is usually written (see Chapter 1) as

$$E_x = E_x^0 \cos(kz - \omega t)$$

where E_x^0 represents the amplitude of the electric field and we have ignored the initial phase. The direction of propagation of the wave is z. If the origin is taken at the molecule ($z = 0$) and the wavelength is greater than the size of the molecule (< 1 nm) so that the electric field strength is uniform at different parts of the molecule (in fact, with the exception of γ-rays and X-rays, which do not concern us here, this condition is satisfied by all electromagnetic radiations that we shall encounter; see Figure 1.7), this equation becomes

$$E_x = E_x^0 \cos(\omega t) = E_x^0 \cos(2\pi \nu t)$$

The perturbation Hamiltonian $\hat{H}'$ is now given by

$$\hat{H}' = -\mu_x E_x = -\mu_x E_x^0 \cos 2\pi \nu t$$

It is here more convenient to use the exponential form for the time dependency of the electric field and write

$$E_x = E_x^0 \frac{(e^{2\pi i \nu t} + e^{-2\pi i \nu t})}{2}$$

This gives $H' = -\mu_x E_x = -\frac{\mu_x E_x^0}{2}\left(e^{2\pi i \nu t} + e^{-2\pi i \nu t}\right)$ and the corresponding perturbation operator is

$$\hat{H}' = -\frac{\hat{\mu}_x E_x^0}{2}\left(e^{2\pi i \nu t} + e^{-2\pi i \nu t}\right) \tag{2.16}$$

Combining with equations (2.13) and (2.15a) gives

$$\begin{aligned} c_m(t' > t) &= \frac{1}{i\hbar}\int_0^t H'_{ml} e^{-i\omega_{lm} t'}\, dt' \\ &= -\frac{1}{2i\hbar}\int_0^t \left(\int_{-\infty}^{+\infty} \psi_m^* \hat{\mu}_x E_x^0 \left(e^{2\pi i \nu t} + e^{-2\pi i \nu t}\right)\psi_l\, dx\right) e^{-i(E_l - E_m)t'/\hbar}\, dt' \end{aligned} \tag{2.17}$$

Separation of the space-dependent and time-dependent parts yields

$$c_m(t) = -\frac{E_x^0}{2i\hbar}\left(\int_{-\infty}^{+\infty} \psi_m^* \hat{\mu}_x \psi_l\, dx\right)\int_0^t \left(e^{i(E_m - E_l + h\nu)t'/\hbar} + e^{i(E_m - E_l - h\nu)t'/\hbar}\right) dt' \tag{2.18}$$

The space integral over x is usually represented by $\left|\vec{M}_{xlm}\right|$ and referred to as the *transition moment integral*, having the same unit as dipole moment (C m), i.e., $\left|\vec{M}_{xlm}\right| = \int_{-\infty}^{+\infty} \psi_m^* \hat{\mu}_x \psi_l\, dx$. The symbol $\left|\vec{M}_{xlm}\right|$ signifies the x-component of the transition dipole moment for a transition from the l state to the m state. Though an important quantity and the basis of all selection rules, we shall defer a discussion on it for the moment and dwell on the time dependence.

Thus equation (2.18) can be rewritten as

$$c_m(t) = -\frac{E_x^0}{2i\hbar}\left|\vec{M}_{xlm}\right|\int_0^t \left(e^{i(E_m - E_l + h\nu)t'/\hbar} + e^{i(E_m - E_l - h\nu)t'/\hbar}\right) dt'$$

Integration of the time-dependent function yields

$$c_m(t) = -\frac{E_x^0}{2i\hbar}\left|\vec{M}_{xlm}\right|\frac{\hbar}{i}\left(\left.\frac{e^{i(E_m-E_l+h\nu)t'/\hbar}}{E_m-E_l+h\nu}\right|_0^t + \left.\frac{e^{i(E_m-E_l-h\nu)t'/\hbar}}{E_m-E_l-h\nu}\right|_0^t\right)$$

$$= \frac{E_x^0}{2}\left|\vec{M}_{xlm}\right|\left(\frac{e^{i(E_m-E_l+h\nu)t/\hbar}-1}{E_m-E_l+h\nu} + \frac{e^{i(E_m-E_l-h\nu)t/\hbar}-1}{E_m-E_l-h\nu}\right) \tag{2.19}$$

Since $c_m^*(t)c_m(t)$ represents the probability of an $l \rightarrow m$ transition, it is important to establish under what conditions it is significant. Both terms in parenthesis have numerators of the type $e^{i\theta} - 1$. On taking the complex conjugate and multiplying, they take the form

$$(e^{-i\theta}-1)(e^{i\theta}-1) = 2 - 2\cos\theta = 4\sin^2\frac{\theta}{2} \tag{2.20}$$

Since $\sin\theta$ cannot exceed unity, the numerator of each term in the parenthesis of equation (2.19) cannot exceed 2. Further, since $E_x^0\left|\vec{M}_{xml}\right|$ is usually small for a single frequency, the magnitude of each term depends on the denominator and will be large only if the denominator is close to zero.

Let us consider under what conditions this happens. If we are considering absorption spectra, the final state is higher in energy than the initial state, and $E_m - E_l > 0$. Since $h\nu$ is also positive, the denominator of the first term is a sum of two positive quantities and hence large in magnitude. In the second term, the denominator is $E_m - E_l - h\nu$ and is hence small. In contrast, for emission, the denominator is small for the first term and large for the second. We may thus ignore the first term for absorption spectra and the second when considering emission spectra. We, therefore, find that transition to state m from state l is significant only when $|E_m - E_l| \approx h\nu$. This is nothing but the Bohr condition for the frequency of the radiation absorbed or emitted.

Let us consider the specific case of absorption spectra. In this case

$$c_m(t) \approx \frac{E_x^0}{2}\left|\vec{M}_{xlm}\right|\left(\frac{e^{i(E_m-E_l-h\nu)t/\hbar}-1}{E_m-E_l-h\nu}\right) \tag{2.21}$$

or

$$\left|c_m(t)\right|^2 = c_m{}^*(t)c_m(t) \approx (E_x^0)^2\left|\vec{M}_{xlm}\right|^2\left(\frac{\sin^2\dfrac{(E_m-E_l-h\nu)t/\hbar}{2}}{(E_m-E_l-h\nu)^2}\right) \tag{2.22}$$

Putting $\omega = \dfrac{E_m - E_l - h\nu}{\hbar}$, we can write for the probability that a particle which started out in state ψ_l will be found in state ψ_m at time t, called the *transition probability*,

$$P_{m\leftarrow l}(t) \equiv \left|c_m(t)\right|^2 = \frac{(E_x^0)^2\left|\vec{M}_{xlm}\right|^2}{\omega^2\hbar^2}\sin^2\frac{\omega t}{2} \tag{2.23}$$

So far we have considered only monochromatic radiation, i.e., radiation of a single frequency. For a range of frequencies, this expression has to be integrated over the entire frequency range. However, only a small frequency range near the Bohr frequency $h\nu_{lm} = E_m - E_l$ will make an effective contribution. To see this, let us plot the function $\frac{\sin^2(\omega t/2)}{(\omega/2)^2}$ as a function of ω (Figure 2.1). Since $\lim_{\theta\to 0}\frac{\sin\theta}{\theta} = \lim_{\theta\to 0}\cos\theta = 1$ from l'Hôpital's rule, $\lim_{\theta\to 0}\frac{\sin^2(\omega t/2)}{(\omega t/2)^2} = 1$ or $\frac{\sin^2(\omega t/2)}{(\omega/2)^2} \to t^2$ as $\omega \to 0$. Because of the property of the sine function, this function goes to zero whenever $\omega t/2 = n\pi$, where n is an integer, or

$$\omega = \frac{2n\pi}{t}, n = 0, \pm 1, \pm 2, ...$$

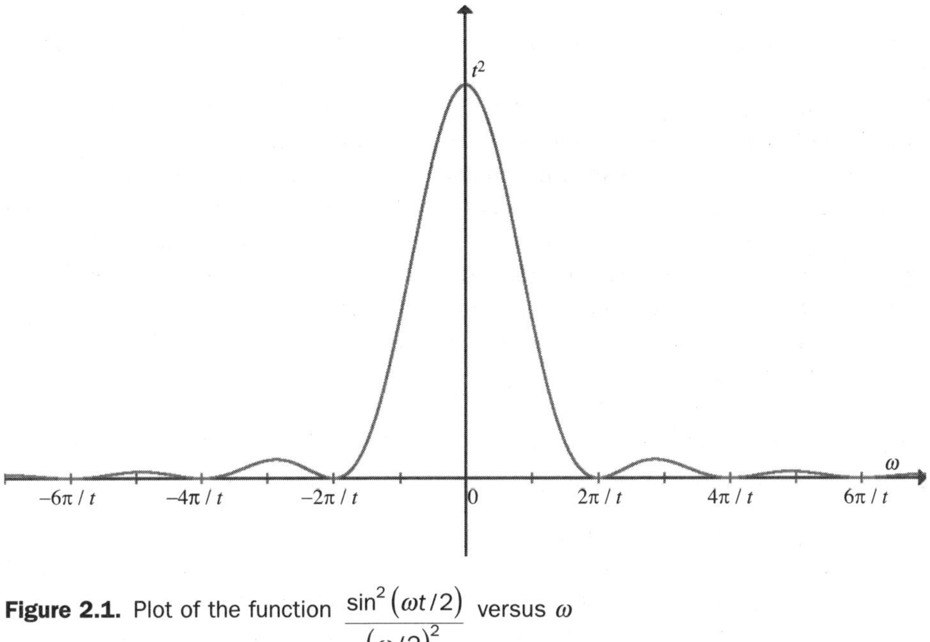

Figure 2.1. Plot of the function $\frac{\sin^2(\omega t/2)}{(\omega/2)^2}$ versus ω

From the figure, it is obvious that the height of the central peak at $\omega = 0$, i.e., when $E_m - E_l = h\nu$, is t^2, and its full base is $4\pi/t$. Approximating the shape of the peak as triangular, the area under the curve is given by the product of the height of the peak and half the base width, which may be taken as $\Delta\omega = 2\pi/t$. Since this area ($\sim 2\pi t$) represents the probability of transition to state m when $\omega \approx 0$ or $E_m - E_l \approx h\nu$, we conclude that the transition probability increases linearly with time.

With increasing time of irradiation, the function of Figure 2.1 gets more sharply peaked. In the limit that $t \rightarrow \infty$, the function $\dfrac{\sin^2(\omega t/2)}{(\omega/2)^2}$ tends to a delta function, $2\pi t\delta(\omega)$. Substitution in equation (2.23) gives a transition probability that varies linearly with time

$$P_{l\rightarrow m}(t\rightarrow\infty)=\frac{(E_x^0)^2\left|\vec{M}_{xlm}\right|^2}{4\hbar^2}2\pi t\delta(\omega) \tag{2.24}$$

The significance of the result is that the transition rate, w_{lm}, which is the transition probability per unit time, is independent of time.

$$w_{lm}=\frac{P_{lm}(t)}{t}=\frac{(E_x^0)^2\left|\vec{M}_{xlm}\right|^2}{4\hbar^2}\left[\delta(\nu')+\delta(\nu'')\right] \tag{2.25}$$

where $\nu' = \nu_{lm} - \nu$ and $\nu'' = \nu_{lm} + \nu$ and we have used the substitution $\delta(\omega)=\dfrac{1}{2\pi}\delta(\nu')$ since $\delta(kx)=\dfrac{1}{|k|}\delta(x)$. The two delta functions in the above expression account for the contributions of stimulated absorption and emission. Equation (2.25) is also called the *Fermi Golden Rule* of spectroscopy. It says that the long time limit results in the familiar Bohr frequency condition: $h\nu_{lm} = E_m - E_l$, i.e., the energy of the photon must match the energy difference of the two states. Since w_{lm} is independent of time, we conclude that the transition rate in this limit is constant. This is the expected behaviour of light sources of ordinary intensity. For instance, the amount of visible light absorbed by a coloured substance, which is proportional to the transition rate, remains constant in time upon illumination with a source of given intensity.

For shorter irradiation times, the peak is broad. The uncertainty in energy of the system is $\Delta E = \hbar\Delta\omega = \hbar\dfrac{2\pi}{t} = \dfrac{h}{t}$, or $\Delta E\Delta t \approx h$, and thus $\Delta\nu\Delta t \approx 1$. There is a paradox here: If monochromatic radiation is applied to the system for a time t, the system sees radiation of width $\Delta\nu = \dfrac{1}{t}$, which is certainly not monochromatic! For example, a pulse of radiation 10 ns long has an intrinsic width of at least 100 MHz, and we can cause a transition even if the photon frequency does not exactly match the transition frequency. Since the other peaks of Figure 2.1 allow for transitions to take place at even higher frequency shifts than the $\pm\pi/t$ considered here for the main peak, though with small probability, the uncertainty is even higher. Thus, the uncertainty can be written as $\Delta E\Delta t \geq h$ and $\Delta\nu\Delta t \geq 1$, which is approximately the Uncertainty Principle for a state having lifetime Δt. The latter is to be expected: we cannot specify the frequency of light if the pulse duration is less than one period. The lifetime of the excited state also limits the duration of the interaction with the radiation, which cannot exceed the lifetime of the excited state. This results in a spread of frequencies inversely proportional to the lifetime of the excited state (see Chapter 1). Even if all other factors contributing to line broadening are eliminated, the lifetime broadening remains, and it represents the minimum line-width for a spectroscopic transition.

In this derivation, we have used first-order perturbation theory and neglected higher order corrections. For a first-order perturbation to be valid, the perturbation should be smaller than the uncertainty in energy, i.e., $|H'_{ml}| << \frac{\hbar}{t}$ or $t << \frac{\hbar}{|H'_{ml}|} = \frac{\hbar}{|E_x^0 \vec{M}_{xlm}|}$. This implies that, since the total probability of a transition to state m is proportional to t, the irradiation time should be small enough to make the total probability small, but, at the same time, it should not be too small, as that would broaden the peak, since $\Delta\omega$, the width of the peak is given by $\Delta\omega \approx 2\pi / t$. Moreover, to justify neglect of the first term of equation (2.19), we need to have $2\frac{E_m - E_l}{\hbar} >> \Delta\omega$, since the absorption and emission peaks appear at $\omega \pm \frac{E_m - E_l}{\hbar}$ and their separation is thus $2\frac{E_m - E_l}{\hbar}$. To prevent their overlap, the separation should be larger than the peak widths. We therefore require that $t >> \frac{h}{2(E_m - E_l)} = \frac{1}{2\nu}$. Combining these two conditions, we obtain $(E_m - E_l) >> \pi|H'_{ml}| = \pi|E_x^0 \vec{M}_{xlm}|$.

To take into account the spread of frequencies from the source, we integrate equation (2.23) over all frequencies, which gives

$$|c_m(t)|^2 = \frac{(E_x^0)^2 |\vec{M}_{xlm}|^2}{\hbar^2} \int_{-\infty}^{+\infty} \frac{\sin^2 \frac{\omega' t}{2}}{\omega'^2} d\nu' \tag{2.26}$$

where $\nu' \equiv \nu_{lm} - \nu = \omega' / 2\pi$. Substitution for ω' in equation (2.26) gives

$$\begin{aligned} |c_m(t)|^2 &= \frac{(E_x^0)^2 |\vec{M}_{xlm}|^2}{\hbar^2} \frac{1}{2\pi} \int_{-\infty}^{+\infty} \frac{\sin^2 \frac{\omega' t}{2}}{\omega'^2} d\omega' \\ &= \frac{(E_x^0)^2 |\vec{M}_{xlm}|^2}{4\hbar^2} t \end{aligned} \tag{2.27}$$

where we have used the standard integral $\int_{-\infty}^{+\infty} \frac{\sin^2 a\theta}{\theta^2} d\theta = \pi|a|$.

Therefore,

$$|c_m(t)|^2 = \frac{(E_x^0)^2 |\vec{M}_{xlm}|^2}{4\hbar^2} t \tag{2.28}$$

2.4 EINSTEIN'S COEFFICIENT OF INDUCED ABSORPTION

We first extend equation (2.28) to all three dimensions. We assume that the radiation is isotropic, i.e., the average field strengths in all three directions are equal and independent of each other, so that

$$(E_x^0)^2 + (E_y^0)^2 + (E_z^0)^2 = (E_0)^2$$

$$(E_x^0)^2 = (E_y^0)^2 = (E_z^0)^2 = \frac{(E_0)^2}{3}$$

The dipole moment is also a vector quantity, and hence $\mu^2 = \mu_x^2 + \mu_y^2 + \mu_z^2$. Therefore, the transition dipole moment is also

$$\left|\vec{M}_{lm}\right|^2 = \left|\vec{M}_{xlm}\right|^2 + \left|\vec{M}_{ylm}\right|^2 + \left|\vec{M}_{zlm}\right|^2$$

Adding the three directions, we then have

$$\begin{aligned} \left|c_m(t)\right|^2 &= \frac{(E_x^0)^2\left|\vec{M}_{xlm}\right|^2}{4\hbar^2}t + \frac{(E_y^0)^2\left|\vec{M}_{ylm}\right|^2}{4\hbar^2}t + \frac{(E_z^0)^2\left|\vec{M}_{zlm}\right|^2}{4\hbar^2}t \\ &= \frac{(E_0)^2\left(\left|\vec{M}_{xlm}\right|^2 + \left|\vec{M}_{ylm}\right|^2 + \left|\vec{M}_{zlm}\right|^2\right)t}{12\hbar^2} = \frac{(E_0)^2\left|\vec{M}_{lm}\right|^2 t}{12\hbar^2} \end{aligned} \quad (2.29)$$

The probability of the transition is clearly dependent on the strength of the electric field. However, the energy density is a more convenient measure of the amount of radiation than the electric field, which is difficult to measure. In order to couch equation (2.29) into a more useful form for comparison with experimental quantities, we define certain new quantities related to electromagnetic waves.

We have seen that electromagnetic radiations carry energy through their electrical and magnetic fields. The *intensity* is defined as the average power transmitted per unit area, perpendicular to the direction of flow. It is usually denoted by I and its SI units are W m^{-2}. It is related to the energy density ρ (expressed in J m^{-3}) of the radiation field and the speed of flow c (in m s^{-1}) by the equation $I = \rho c$. For electromagnetic fields, the energy density in vacuum is the sum of the contributions due to the electric and magnetic fields

$$\rho = \rho_E + \rho_B = \frac{1}{2}\left(\varepsilon_0 E_0^2 + \frac{B_0^2}{\mu_0}\right),$$

Here, $\mu_0 = 4\pi \times 10^{-7}$ H m^{-1} ($\equiv$ N s^2 C^{-2}) is the permeability of free space, and $\varepsilon_0 = 8.85 \times 10^{-12}$ F m^{-1} ($\equiv$ C^{2}N^{-1} m^{-2}) is the permittivity of free space, and H and F are the Henry and Farad units of inductance and capacitance, respectively. The energy density is almost entirely due to the electric field, the contribution of the magnetic field being negligible in comparison. The energy density due to the electric field is therefore $\rho_E = \frac{1}{2}\varepsilon_0 E_0^2 \equiv \int \rho(\nu, T)d\nu$, where $\rho(\nu, T)$ is the energy density of radiation per frequency interval at temperature T.

In terms of the incident spectral radiation energy density, the transition probability becomes

$$\left|c_m(t)\right|^2 = \frac{\rho_\nu\left|\vec{M}_{lm}\right|^2 t}{6\varepsilon_0\hbar^2}$$

where we have assumed that $\rho(\nu, T)$ has a constant value, ρ_ν, at temperature T in the frequency range under consideration, since it is a slowly varying function of the frequency. Hence, the rate of transition,

$$w_{lm} = \frac{d|c_m|^2}{dt} = \frac{\rho_\nu |\vec{M}_{lm}|^2}{6\varepsilon_0 \hbar^2} \tag{2.30}$$

which is independent of time. Thus, the time-dependence of the coefficients of ψ_m for a state is seen to depend on the energy density of the radiation and the square of the transition dipole moment.

The rate of change of the system as a result of the perturbing effect of the radiation is usually written in terms of Einstein's coefficient of induced absorption (Chapter 1), defined as

$$\frac{d|c_m|^2}{dt} = B_{lm}\rho_\nu \tag{2.31}$$

Comparison with equation (2.30) yields

$$B_{lm} = \frac{|\vec{M}_{lm}|^2}{6\varepsilon_0 \hbar^2} = \frac{8\pi^3}{3h^2(4\pi\varepsilon_0)}|\vec{M}_{lm}|^2 \tag{2.32}$$

This is the form Einstein's coefficient of induced absorption takes when the radiation density in equation (2.31) is expressed in frequency (Hz) units. If it is expressed in wavenumber, the expression in (2.32) has to be divided by c, the speed of light. Similarly, in terms of the angular frequency, it has to be multiplied by 2π. However, we shall follow the form given in equation (2.32), with the implicit understanding that the radiation density is expressed in frequency units.

We have now seen that the rate of the transition depends on the Einstein coefficient of induced absorption. We have also derived in Chapter 1 the equations connecting the three Einstein coefficients. This implies that all three Einstein coefficients can be determined from equation (2.32). What remains is to connect this theoretical quantity to the quantities observed experimentally.

It should be mentioned here that, though we treated it classically in Chapter 1, spontaneous emission is a quantum effect and can only be explained if we recognize that if the matter that is interacting with the electromagnetic field has quantized energy levels ('first quantization'), the same should be true for the radiation ('second quantization'). Thus, there is a zero-point energy associated with electromagnetic radiation. Zero-point fluctuations of the electromagnetic field, called vacuum noise, are responsible for initiating spontaneous emission. The combined system of atom and electromagnetic field undergoes a transition from an electronic excitation to a photonic excitation, resulting in spontaneous emission.

This is best described in Dirac's own words in 1927: "The light quanta has the peculiarity that it apparently ceases to exist when it is in one of its stationary states, namely the zero state.... When a light quanta is absorbed it is said to jump into this zero state and when one is emitted it can be considered to jump from the zero state to one in which it is physically in evidence, so that it appears to have been created. Since there is no limit to the number of light quanta that may be created in this way we must suppose that there are an infinite number of light quanta in the zero state."

Detailed calculations show that only half of the decay of the excited state is due to stimulated emission from the zero-point field, the other half being the loss of the energy of the oscillating charge due to the influence of its own 'self-field', i.e., the radiation reaction field, which damps the radiation field. Thus, spontaneous emission is not really 'spontaneous' in the sense that it is induced by the zero field, but it is not induced by the applied field.

2.4.1 Comparison with experimental quantities

We now compare the theoretical quantities derived here with experimental quantities. As was seen in Chapter 1, the Lambert-Beer law is used to obtain experimental spectra. In order to compare the Lambert-Beer law equation (1.39) with the theoretical quantities obtained in the previous section, we must somehow express the former in terms of the radiation density and the latter in terms of the concentration C. We had seen that the rate of transfer of molecules from state l to state m is given by equations (2.31) and (2.32) as

$$w_{lm} = \frac{d|c_m|^2}{dt} = B_{lm}\rho_\nu = \frac{8\pi^3}{3h^2(4\pi\varepsilon_0)}|\vec{M}_{lm}|^2 \rho_\nu$$

Each transfer decreases the energy of the radiation by $h\nu_{lm}$. If there are N molecules per dm^3 in the initial state l (essentially equal to the total number of molecules for electronic spectroscopy) of the sample, the decrease in intensity (or energy) of the radiation as it passes a cross section of thickness dl is $B_{lm}\rho_\nu h\nu_{lm}Ndl$, i.e.,

$$-dI = \frac{8\pi^3}{3h^2(4\pi\varepsilon_0)}|\vec{M}_{lm}|^2 \rho_\nu h\nu_{lm}Ndl \tag{2.33}$$

Moreover, N, the number of molecules per unit volume, is related to the molar concentration as $N = N_A C$, where N_A is Avogadro's number. Substituting both these quantities in the above equation gives,

$$-dI = \frac{8\pi^3}{3h^2(4\pi\varepsilon_0)}|\vec{M}_{lm}|^2 \rho_\nu h\nu_{lm}N_A Cdl \tag{2.34}$$

Furthermore, since the intensity is defined as the energy flowing through a cross section of unit area in one second, it is related to the energy density by $I = c\rho = c\int\rho(\nu)\,d\nu$, where c is the speed of light. As we noted previously, $\rho(\nu)$ is a very slowly varying function of the frequency and can be taken as a constant, ρ_ν, its value at the transition frequency. Moreover, since the band extends over a range of frequencies, identification of the theoretical quantities must be made with the total absorbance rather than that for a given frequency, since the theoretical quantity is for the entire band. Substitution in the differential form of Beer's law, equation (1.37), gives

$$-dI = \rho_\nu Cc\left(\int_{\text{band}} \alpha(\nu)\,d\nu\right)dl$$

Comparison with equation (2.34) gives

$$\int_{\text{band}} \alpha(\nu)d\nu = \frac{B_{lm}h\nu_{lm}N_A}{c} = \frac{8\pi^3 N_A}{3hc(4\pi\varepsilon_0)}\nu_{lm}|\vec{M}_{lm}|^2$$

where ν_{lm} is the centre of the absorption band or band maximum of the absorption band.

$$\int_{\text{band}} \alpha(\nu)d\nu = c\ln 10 \int_{\text{band}} \varepsilon(\tilde{\nu})d\tilde{\nu} = \frac{8\pi^3 N_A}{3hc(4\pi\varepsilon_0)}\nu_{lm}|\vec{M}_{lm}|^2 = \frac{B_{lm}N_A h\nu_{lm}}{c}$$

$$\bar{A} = \int \varepsilon(\tilde{\nu})\, d\tilde{\nu} = \frac{B_{lm} N_A h \nu_{lm}}{c^2 \ln 10} = \frac{8\pi^3 N_A}{3hc(4\pi\varepsilon_0)\ln 10}\tilde{\nu}_{lm}\left|\vec{M}_{lm}\right|^2 \qquad (2.35)$$

The important result to note is that equation (2.35) connects the macroscopic experimental quantity, the integrated absorption coefficient, $\bar{A}$, with the microscopic theoretical quantity $\left|\vec{M}_{lm}\right|^2$, which can be obtained from the wave functions of the initial and final states. Equation (2.35) thus offers a bridge between the empirical and the fundamental. If $\bar{A}$ is to be expressed in $dm^3\ mol^{-1}\ cm^{-2}$, a unit in which it is frequently expressed, $\tilde{\nu}_{lm}$ is in cm^{-1} and the transition dipole moment has the SI unit C m, this expression needs to be multiplied by another factor 10^3. Insertion of the values of the fundamental constants (N_A in mol^{-1}, $1/(4\pi\varepsilon_0)$ in $N\ m^2\ C^{-2}$) gives

$$\bar{A} = 9.784 \times 10^{60}\, \tilde{\nu}_{lm}\left|\vec{M}_{lm}\right|^2 \qquad (2.36)$$

where the constant term has the unit $dm^3\ mol^{-1}\ cm^{-1}\ C^{-2}\ m^{-2}$.

Equation (2.35) shows that the integrated absorption intensity is proportional to the wavenumber of the transition and the square of the transition dipole moment. Thus, knowing the wave function and performing the integration

$$\left|\vec{M}_{lm}\right| = \int_{-\infty}^{+\infty} \psi_m^* \hat{\mu} \psi_l \, d\tau$$

we can evaluate the integrated absorption coefficient and, conversely, we can calculate $\left|\vec{M}_{lm}\right|$ if the integrated absorption coefficient is known, by using the relation

$$\left|\vec{M}_{lm}\right| = \sqrt{\bar{A}/(9.784 \times 10^{60}\,\tilde{\nu}_{lm})} = 3.197 \times 10^{-31}\sqrt{\bar{A}/\tilde{\nu}_{lm}} \qquad (2.37)$$

where the integrated absorption intensity and the frequency are expressed in $dm^3\ mol^{-1}\ cm^{-2}$ and cm^{-1}, respectively, and the transition dipole moment is in C m units. When comparing with theoretical quantities, it sometimes becomes necessary to express the integrated absorption intensity in $km\ mol^{-1}$, the unit most frequently encountered in computational chemistry software. Since the conversion factor is $1\ km\ mol^{-1} = 100\ dm^3\ mol^{-1}\ cm^{-2}$, the factor in equation (2.36) becomes 9.784×10^{58}.

Example 2.2 Do a unit analysis and verify the constant in equation (2.35).

Solution

Verifying units,

$$\frac{8 \times 3.142^3 \times 6.022 \times 10^{23}\,\mathrm{mol^{-1}}}{3 \times 6.626 \times 10^{-34}\,\mathrm{J\ s} \times 2.998 \times 10^{10}\ \mathrm{cm\ s^{-1}} \times 4 \times 3.142 \times 8.854 \times 10^{-12}\,\mathrm{C^2 s^2 kg^{-1} m^{-3}} \times 2.303}$$

$$= 9.784 \times 10^{57}\frac{\mathrm{mol^{-1}}}{\mathrm{C^2 m^{-1} cm}} = 9.784 \times 10^{57}\frac{\mathrm{dm^3 mol^{-1} cm^{-2}}}{\mathrm{C^2 m^{-1} cm^{-1} (0.1m)^3}} = 9.784 \times 10^{60}\frac{\mathrm{dm^3 mol^{-1} cm^{-2}}}{\mathrm{C^2 m^2 cm^{-1}}}$$

Multiplication by the wavenumber (cm^{-1}) and square of the transition dipole moment (in $C^2\ m^2$) gives the integrated absorption coefficient in $dm^3\ mol^{-1}\ cm^{-2}$.

Example 2.3 Given that a protein with 18 residues has ε_{max} of 8000 $dm^3\ mol^{-1}\ cm^{-1}$ and a spectral width of 40 cm^{-1} for the amide I band, estimate the transition dipole moment $|\vec{M}_{01}|$ for a single residue in Debye. Given that the amide I band has a frequency of about 1600 cm^{-1}.

Solution

From equation (2.37),

$$\left|\vec{M}_{01}\right|^2 = 1.022\times10^{-61}\bar{A}/\tilde{\nu}_{lm} = 1.022\times10^{-61}\frac{8000\ \mathrm{M^{-1}cm^{-1}}40\ \mathrm{cm^{-1}}}{1600\ \mathrm{cm^{-1}}}\frac{1}{18\ \text{oscillators}}$$

$$= 1.136\times10^{-60}\,\mathrm{C^2m^2}$$

$$\therefore \left|\vec{M}_{01}\right| = 1.066\times10^{-30}\,\mathrm{C\,m}\times\frac{D}{3.336\times10^{-30}\,\mathrm{C\,m}} = 0.319\ \mathrm{D}$$

2.5 EINSTEIN'S COEFFICIENTS OF INDUCED AND SPONTANEOUS EMISSION

For reverse transitions, i.e., induced emission,

$$\left|\vec{M}_{ml}\right| = \int_{-\infty}^{+\infty}\psi_l^*\hat{\mu}\psi_m\,d\tau$$

Since $\hat{\mu}$ is a quantum mechanical operator, and all quantum mechanical operators are Hermitian, $\left|\vec{M}_{ml}\right| = \left|\vec{M}_{lm}\right|$.

Combining equations (1.12) and (1.14) with (2.32) gives

$$A_{ml} = \frac{8\pi h\nu^3}{c^3}B_{ml} = \frac{64\pi^4\nu^3}{3hc^3(4\pi\varepsilon_0)}\frac{g_l}{g_m}\left|\vec{M}_{lm}\right|^2 \quad (2.38)$$

from which $\Delta\nu$ can be calculated from the Heisenberg Uncertainty Principle as

$$\Delta\nu = \frac{16\pi^3\nu^3}{3hc^3(4\pi\varepsilon_0)}\frac{g_l}{g_m}\left|\vec{M}_{lm}\right|^2 \quad (2.39)$$

This is the natural line width. The dependence of $\Delta\nu$, the frequency spread, on ν^3 results in a much larger spread for an excited electronic state, typically 30 MHz, than for an excited rotational state, typically 10^{-4} – 10^{-5} Hz, because of the much higher frequency for an electronic excitation.

We have also seen that the Einstein coefficient of induced absorption, B_{lm}, can be related to experimental quantities through equation (2.35). Since A_{ml} is also related to B_{lm} through equation (2.38), we may combine these two to get A_{ml} in terms of the integrated absorption coefficient:

$$A_{ml} = \frac{(\ln 10)}{N_A}\frac{g_l}{g_m}\frac{8\pi v^2}{c}\int_{\text{band}}\varepsilon(\tilde{v})\,d\tilde{v}$$
$$= \frac{2.303}{N_A}8\pi\tilde{v}^2 c\frac{g_l}{g_m}\int_{\text{band}}\varepsilon(\tilde{v})\,d\tilde{v} \tag{2.40}$$

from which the lifetime of the excited state can be calculated [equation (1.19)] as

$$\tau = \frac{N_A g_m}{8\pi\tilde{v}^2 c(2.303)g_l\int_{\text{band}}\varepsilon(\tilde{v})\,d\tilde{v}} \tag{2.41}$$

Inserting the values of all the constants (and putting $g_m = g_l$), we get a simple result

$$\tau = \frac{3.5\times10^8}{\tilde{v}^2\int_{\text{band}}\varepsilon(\tilde{v})\,d\tilde{v}} \tag{2.42}$$

with $\tilde{v}$ expressed in cm^{-1}. This is an important result. The shape of the absorption curve allows us to calculate the lifetime of the excited state. Equation (2.42) implies that an excited state resulting from an allowed transition has a shorter lifetime.

Example 2.4 A typical strong absorption band due to an allowed transition has $\int_{\text{band}}\varepsilon(\tilde{v})d\tilde{v} = 10^8$ and a band maximum at ~300 nm, or ~30,000 cm^{-1}. Calculate the lifetime of the excited state, assuming both states are singly degenerate.

Solution

The emission from the excited state reached by this absorption should, according to the above analysis, have a lifetime, due to the emission process, of

$$\tau = \frac{3.5\times10^8}{(30{,}000)^2\times10^8} \approx 3.9\times10^{-9}\,\text{s} = 3.9\text{ ns}$$

This is a typical lifetime of an electronic excited state.

Example 2.5 The lifetime of the 3 $^2P_{1/2} \rightarrow$ 3 $^2S_{1/2}$ transition of the sodium atom at 589.6 nm is measured to be 16.4 ns.

(a) What are the Einstein *A* and *B* coefficients for this absorption?

(b) What is the transition dipole moment?

(c) What is the emission line width, assuming that it is determined by lifetime broadening?

Solution

(a) The A_{ml} coefficient is given by

$$A_{ml} = 1/\tau = 6.10 \times 10^{7}\ s^{-1}$$

$$\frac{A_{ml}}{B_{ml}} = \frac{8\pi h \nu^{3}}{c^{3}}$$

Rearrangement gives

$$B_{ml} = \frac{A_{ml}}{8\pi h \tilde{\nu}^{3}}$$

From the transition wavelength, $\tilde{\nu} = 1/\lambda = 1/(589.6 \times 10^{-7}\ cm) = 1.696 \times 10^{4}\ cm^{-1}$. Insertion of these values gives

$$B_{ml} = 2.159 \times 10^{29}\ J^{-1}\ m^{3}\ s^{-2}$$

From relation (1.12), $B_{lm} = B_{ml}$ since the degeneracies of the two levels are the same, as their J values are identical (½).

(b) Rearrangement of equation (2.32) gives for $|\vec{M}_{lm}|$

$$|\vec{M}_{lm}| = \hbar\sqrt{6\varepsilon_0 B_{lm}} = 2.10 \times 10^{-29}\ C\ m$$

(c) The emission line-width

$$\Delta\nu = 1/(2\pi\tau) = 9.7 \times 10^{6}\ s^{-1}$$

2.6 THE BASIS OF SELECTION RULES

The results of the previous section provide a quantitative relationship between the absorption of radiation by a given sample and the transition moment of the absorbing species. The rate of transition is found to be directly proportional to the square of the transition moment integral. Let us first understand the significance of this quantity. The transition dipole moment is given by $|\vec{M}_{lm}| = \int_{-\infty}^{+\infty} \psi_m^{*} \hat{\mu} \psi_l \, d\tau$ and in order to evaluate it, we require the wave functions of both the ground and excited states. This is difficult, because quantum mechanics only allows us to calculate the exact wave functions for one-electron systems. For all the rest, one has to resort to some approximation. Even then, the task is formidable for most molecules.

Fortunately, even though we may not be able to calculate the wave functions, we may still be able to make qualitative predictions of which transitions can be induced by radiation and thus lead to the absorption of radiation, and which transitions cannot be induced, and therefore fail to absorb radiation. Such general statements are called *selection rules*, and for a given system, they can be deduced by deciding under what circumstances the transition moment integral $|\vec{M}_{lm}|$ becomes necessarily zero by using the symmetry properties of the ground and excited states and the dipole moment operator. The transition is then said to be *forbidden*, and should not, according to the approximate theory, occur at all. Actually, forbidden transitions do occur, and, as might have been expected, calculation by more refined theories gives small, but non-vanishing, values for the intensities. Observed intensities of forbidden transitions are generally quite small, much smaller than those of allowed transitions (for which $|\vec{M}_{lm}|$ gives non-vanishing intensities). For example, $n \rightarrow \pi^*$ transitions, forbidden by the orbital selection rules, do occur, albeit with low intensity. The nonbonding electrons lie in the plane of the molecule, while the π^* orbitals are out of plane, preventing a mutual interaction. However, out-of-plane vibrations allow these interactions to occur. Such transitions are termed vibronic (vibration + electronic).

It is a well-known fact of mathematics that $\int f(x)\,dx$ represents the area under the curve when $f(x)$ is plotted against x, areas below the abscissa being counted negative. We are familiar with fundamental definitions of *odd* and *even* functions. A function is odd if $f(x) = -f(-x)$, and even if $f(x) = f(-x)$. Not surprisingly, polynomials consisting only of odd powers of x are odd, and those consisting of only even powers of x are even. Polynomials containing both even and odd powers, such as $x^2 + x$, are neither even nor odd (on inversion, the function becomes $x^2 - x$), and the following discussion does not apply to them.

Let us first examine an even function $f(x)$; for example, $f(x) = x^2$, shown in Figure 2.2(a); such a function is symmetric in x. The *shaded* area under the parabola can be expressed by the integral

$$\int_{-a}^{+a} f(x)\,dx = \int_{-a}^{0} f(x)\,dx + \int_{0}^{a} f(x)\,dx = \int_{a}^{0} f(-x)d(-x) + \int_{0}^{a} f(x)\,dx = \int_{0}^{a} f(-x)\,dx + \int_{0}^{a} f(x)\,dx$$

$$= \int_{0}^{a} f(x)\,dx + \int_{0}^{a} f(x)\,dx$$

$$= 2\int_{0}^{a} f(x)\,dx$$

where we have used $f(-x) = f(x)$. This result is obvious since the areas under the two halves of the curve are both above the abscissa and are equal in magnitude.

Next, let us examine a function antisymmetric in x. A plot of such a function, $f(x) = x^3$, is shown in Figure 2.2(b), where it can be seen that, for every positive value of x, there is a positive y value, and for every negative value of x, a corresponding negative value of y. Hence $f(x) = -f(-x)$; such a function is called an *odd* function. Integration of this function gives

$$\int_{-a}^{+a} f(x)\,dx = \int_{-a}^{0} f(x)\,dx + \int_{0}^{a} f(x)\,dx = \int_{a}^{0} f(-x)d(-x) + \int_{0}^{a} f(x)\,dx = \int_{0}^{a} f(-x)d(x) + \int_{0}^{a} f(x)\,dx$$

$$= -\int_{0}^{a} f(x)d(x) + \int_{0}^{a} f(x)\,dx$$

$$= 0$$

since $f(-x) = -f(x)$. This is again obvious since the (shaded) areas under the curve for positive and negative x are equal, but above and below the abscissa, respectively, and thus cancel. This illustrates one of the key concepts of odd functions: *the integral of an odd function is zero if it is evaluated by limits that are symmetric across the origin.*

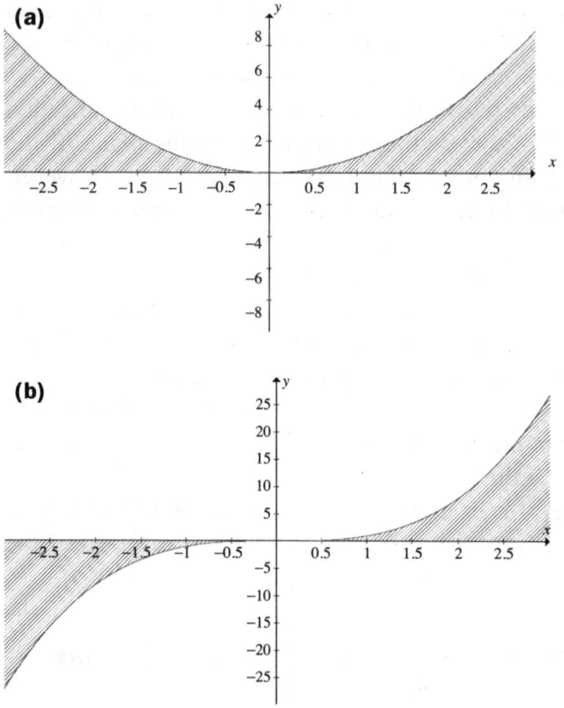

Figure 2.2. Graph of (a) x^2; (b) x^3

The product of an even and an odd function is always odd, whereas the products of two even or two odd functions are always even. The integral over all space of an odd function always vanishes identically, whereas the integral over all space of an even function may or may not vanish, but generally does not.

We now apply these ideas to the deduction of selection rules of transitions.

2.7 OVERVIEW OF SELECTION RULES

Within a given approximation, for example, the dipolar approximation, radiative transitions occur only if the transition probability is non-zero, i.e., $\left|\vec{M}_{lm}\right| \neq 0$. Even if $\left|\vec{M}_{lm}\right| = 0$, higher order moments may lead to non-zero transition probabilities, and these transitions may be quadrupole allowed, etc. However, their probabilities are much lower, and they give rise to much weaker lines than dipole allowed transitions. We shall therefore ignore these. Jumps that are likely to occur are called *allowed* transitions; those which are unlikely are said to be *forbidden*.

The integral $\left|\vec{M}_{lm}\right|$ can be expanded into a sum of integrals, each involving only one coordinate (i.e., one of the components of $\vec{\mu}$), and it is necessary only to demonstrate that the integral vanishes for all components to show that a transition is forbidden. Provided the molecule has some elements of symmetry, it is possible to demonstrate that $\left|\vec{M}_{lm}\right|$ vanishes for some transitions. It is often sufficient to consider the symmetry of the three functions that appear in the expression for the transition dipole moment, ψ^*_m, $\hat{\mu}$ and ψ_l. For example, let us look at the x component of the transition dipole moment, $\left|\vec{M}_{xlm}\right| = \int_{-\infty}^{+\infty} \psi^*_m \hat{\mu}_x \psi_l d\tau$. This is a triple product, which should be even for the integral to survive. The x-component of the dipole moment, $\mu_x = \sum q_i x_i$, behaves like x, i.e., it is an odd function since it changes sign on inversion. Since the overall product should be an even function, this is possible only if the product of the two wave functions is also odd, i.e., if one of the wave functions is odd and the other even. This is equivalent to the statement that only levels of different symmetry can combine. Of course, this classification only applies to centrosymmetric molecules. The selection rule can be conveniently expressed as $+ \leftrightarrow -$, but $+ \not\leftrightarrow +$ and $- \not\leftrightarrow -$, which is the basis of the well-known Laporte selection rule for electronic transitions. The Laporte rule is a spectroscopic selection rule that states that electronic transitions that conserve *parity*, i.e., g (gerade) $\rightarrow$ g, or u (ungerade) $\rightarrow$ u are forbidden. *Parity* indicates whether the wave function changes its sign upon the inversion transformation $\vec{r} \rightarrow -\vec{r}$.

Let us apply these symmetry considerations to atomic and molecular spectra. The s and d orbitals are symmetric (g) because applying an inversion does not change the sign of the wave function, while the p and f orbitals are antisymmetric (u) (Figure 2.3). The dipole moment operator is also u. In other words, if a molecule is centrosymmetric, transitions within a given set of p or d orbitals (i.e., those that only involve a redistribution of electrons within a given subshell) are forbidden. However, this rule is violated in the case of transition metals, whose complexes are coloured due to d $\leftrightarrow$ d transitions, though they are Laporte forbidden for centrosymmetric octahedral complexes. Hence these transitions are weak. On the other hand, for tetrahedral complexes, where there is no centre of symmetry, these transitions are more intense.

Also, s $\leftrightarrow$ s and s $\leftrightarrow$ d transitions are forbidden, but s $\leftrightarrow$ p transitions are allowed. However, though allowed by the Laporte selection rule, s $\leftrightarrow$ f transitions are also forbidden. In Chapter 3, we shall show that detailed quantum mechanical treatment allows us to deduce the orbital selection rule, $\Delta l = \pm 1$. This illustrates an important pitfall – symmetry tells us which transitions are forbidden, but this does not imply that the others are allowed. They may be forbidden for reasons other than symmetry. In the present case, since a photon has an intrinsic angular momentum of one unit, it can alter the orbital angular momentum by only one unit.

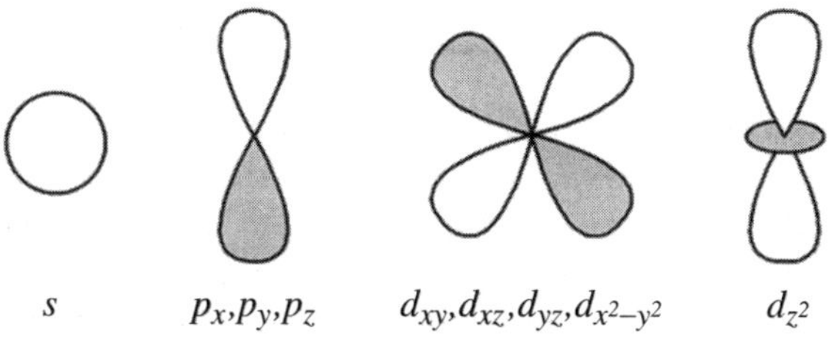

Figure 2.3. Symmetries of atomic orbitals

The spin selection rule $\Delta s = 0$ may be similarly deduced. The photon has an angular momentum of one unit and is hence a boson. The electron is however a fermion since its spin angular moment is half-integral. Therefore, it is not possible for the photon to change the spin of the electron, and hence we have the spin selection rule, which is among the strictest of all selection rules. The use of symmetry for deducing selection rules shall become clearer in subsequent chapters, where more examples will be discussed.

Let us look at the allowed decays from the first four Bohr levels of hydrogen (Figure 2.4). This is commonly known as a Grotrian diagram and depicts the allowed transitions in an atom.

The levels with the same l (all the levels in the same column) have spectral series assigned to them. These series are labelled s, p, d and f, abbreviations for **s**harp, **p**rincipal, **d**iffuse and **f**undamental. The principal series (main branch) contains all transitions between states with $l = 1$ and the ground state. Since the ground state is almost exclusively occupied at room temperature, the *principal* series (1s $\rightarrow$ np) dominates the *absorption* spectra. Two subordinate series (side branches) starting at states $l = 0$ (ns $\rightarrow$ n'p) and $l = 2$ (nd $\rightarrow$ n'p) have the level $l = 1$ (n'p) in common as the lower level. They appear in the emission spectrum as *sharp* and *diffuse* subordinate series. The notation s, p, d and f, for an electron with $l = 0$, 1, 2 or 3, respectively, is based on this notation for the spectral series. For higher values of l, i.e., $l = 4$, 5, 6 etc., the notation is continued alphabetically as g, h, i, etc.

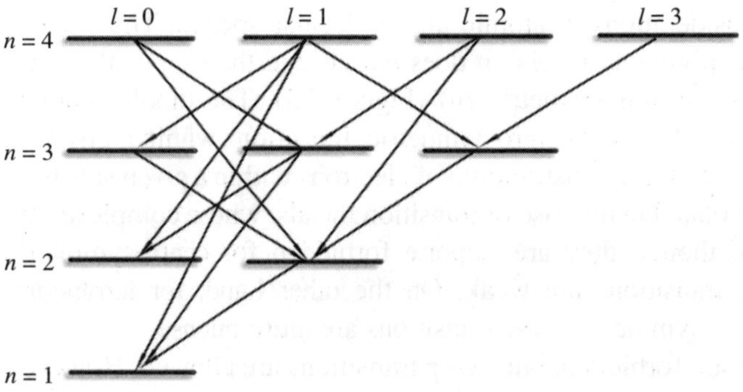

Figure 2.4. Allowed decays for the first four Bohr levels of hydrogen atom

The magnitude of the transition dipole moment, $\left|\vec{M}_{lm}\right| = \int_{-\infty}^{+\infty} \psi^*_m \hat{\mu} \psi_l \, d\tau$, depends on the spatial overlap $\psi^*_m \psi_l$ between the excited and ground state wave functions. Therefore, there needs to be a significant spatial overlap between the two wave functions throughout the molecule for the transition moment that couples the two states to be large. The magnitude also depends on the dipole moment, which is a product of the electronic charge and position. Because of the position vector, $\vec{r}$, in the dipole moment ($\vec{\mu} = -e\vec{r}$), good overlap at sites that are at a large distance from the nucleus will have enhanced contributions to the transition dipole moment. Molecules with good spatial overlap between two states that are not of the same parity will have large transition dipole moments and strong (allowed) transitions in the electronic spectrum. Conversely, if the above criteria are not met, then the transitions will be weak.

Allowed transitions in the visible region typically have A_{ml} coefficients in the range 10^6–10^8 s^{-1}. Forbidden transitions in this region have A_{ml} coefficients below 10^4 s^{-1}. These probabilities decrease as the wavelength of the jump increases. Consequently, levels that can decay by allowed transitions in the visible region have lifetimes generally shorter than 1 μs, but similar forbidden transitions have lifetimes in excess of 10–100 μs. Although no jump turns out to be absolutely forbidden, some jumps are so unlikely that levels whose electrons can only fall to lower levels by such jumps are very long lived. Levels with lifetimes in excess of 1 hour have been observed under laboratory conditions. Levels which can only decay slowly, and usually only by forbidden transitions, are said to be *metastable*.

An example of a transition that is forbidden by all selection rules – electric dipole, magnetic dipole and electric quadrupole – is the radiative transition between the 2s state and the ground state 1s of the hydrogen atom. For this transition, electric dipole transition is forbidden by parity, and electric quadrupole and all other higher multipole transitions are forbidden by angular momentum conservation. It is clear that the 2s orbital can be populated by decays from 3p and 4p, but once there, the electron is 'stuck' because there is no lower energy p state. This is an example of a metastable state. It is characterized by a lifetime which is much larger than that of the other states, e.g., 2p. Eventually, a metastable state will decay by collisions or so-called forbidden transitions (but which are actually allowed by other selection rules). In this case, the atom finds a way to return to the ground state by simultaneous emission of two photons. Its lifetime of $1/7$ s is, however, extremely long compared to the lifetime of the 2p state, 1.6×10^{-9} s.

In Chapter 3, we introduce atomic spectroscopy and derive these selection rules rigorously. Atomic spectroscopy is the most convenient starting point, since the study of spectroscopy originated from the investigation of the emission spectra of excited atoms and we are more familiar with atomic structure. Moreover, atomic spectroscopy is free from the added complications of simultaneous rotational and vibrational fine structure.

2.8 SUMMARY

- Application of first-order time-dependent perturbation theory to the interaction of electromagnetic radiation with matter reveals that
  - the width of the central peak is proportional to $1/t$ and the height to t^2 and the Uncertainty Principle is satisfied.
  - only transitions that satisfy the Bohr condition have high probability.
- The transition probability is proportional to the square of the transition dipole moment.
- Transitions are governed by selection rules.
- The symmetry properties of the wave functions of the initial and final states often give a clue to whether a transition between the two states is allowed or forbidden.

2.9 EXERCISES

1. (a) The mean time for a spontaneous 2p → 1s transition is 1.6×10^{-9} s, while the mean time for a spontaneous 2s → 1s transition is as long as 0.14 s. Explain.
   (b) Explain in words why the rate of spontaneous emission has to increase with frequency.
   (c) The transition dipole moment is not the same as the dipole moment. Explain clearly the distinction between the two.

2. For a rotational transition, an estimate of the Einstein B coefficient can be obtained from equation (2.32) by substituting the permanent dipole moment μ of the molecule for the transition dipole moment, i.e.,

$$B_{lm} = \frac{8\pi^3\mu^2}{3h^2(4\pi\varepsilon_0)}$$

Use this expression to estimate the B coefficient for the rotational $J = 0 \rightarrow 1$ transition in HCN, which has a dipole moment of 2.98 Debye (convert first to the SI unit C m) and rotational constant $\tilde{B} = 1.477\ \text{cm}^{-1}$.

Hence, determine a value for the Einstein A coefficient, the lifetime of the upper state and the natural line-width of the transition. Comment on these results. For simplicity, ignore the effects of degeneracy.

3. Consider the calcium resonance transition at $\lambda = 422.7$ nm. The upper level (4^1P) has a radiative lifetime of 4.5 ns. The lower level is 4^1S. For our purpose, we may ignore the other possible decay routes, e.g., from 4^1P to 3^1D. Evaluate the Einstein coefficients and the transition dipole moment values for the transition. For atomic transitions, 4^1S refers to a ^{1}S state with principal quantum number 4, and 4^1P similarly to a ^{1}P state with $n = 4$. The degeneracies of the S and P states are 1 and 3, respectively.
4. The emission spectrum of hydrogen consists of several series of sharp emission lines in the ultraviolet (Lyman series), in the visible (Balmer series) and in the infrared (Paschen series, Brackett series, etc.) regions of the spectrum.
   (a) What feature of the electronic energies of the hydrogen atom explains why the emission spectrum consists of discrete wavelengths rather than a continuum of wavelengths?
   (b) Account for the existence of several series of lines in the spectrum. What quantity distinguishes one series of lines from another?
   (c) Draw an electronic energy level diagram for the hydrogen atom and indicate on it the transition corresponding to the line of lowest frequency in the Balmer series.
   (d) What is the difference between an emission spectrum and an absorption spectrum? Explain why the absorption spectrum of atomic hydrogen at room temperature has only the lines of the Lyman series.

3

Atomic Spectroscopy

Twinkle, twinkle little star,
I don't wonder what you are,
For by spectroscopic ken,
I know that you are hydrogen.
—Anonymous

3.1 INTRODUCTION

We first take up atomic spectroscopy because it was an attempt to explain the emission spectrum of the hydrogen atom that led Bohr to postulate his theory. In all the three techniques that are applied in atomic spectroscopy, viz. atomic absorption, emission and fluorescence, the processes of excitation and subsequent decay to the ground state are involved, and the corresponding energies are measured and used for analytical purposes. We begin with a discussion of Bohr's theory of hydrogen and hydrogen-like atoms due to historical reasons.

3.2 THE BOHR THEORY OF THE HYDROGEN ATOM

Bohr's interpretation of the behaviour of an electron was based on the rather arbitrary assumption that it moves in a circular orbit about the nucleus in such a way that its *angular momentum is an integral multiple of* $\hbar$. This statement and the ordinary rules of dynamics and electrostatics led to allowed orbits for the electron, each allowed orbit having a certain energy.

The angular momentum postulate requires that

$$m_e vr = n\hbar, n = 1, 2, 3, ... \tag{3.1}$$

where m_e is the mass of the electron, v its velocity and r the radius of its orbit.

If it is postulated that the wave associated with an electron in a Bohr orbit should form a standing wave around the nucleus, the condition that $2\pi r = n\lambda$ is imposed, where n is an integer signifying the number of waves on the circumference, and $2\pi r$ is the circumference of the electron orbit. This stipulation, together with de Broglie's wavelength relation, $\lambda = h / m_e v$, leads, interestingly, to the same requirement, i.e., $2\pi r = nh / m_e v$ or $m_e vr = nh / 2\pi = n\hbar$, as arbitrarily proposed by Bohr.

The coulomb force of attraction between an electron of charge $-e$ revolving around a proton of charge $+e$ in an orbit of radius r is balanced by the centripetal force

$$\frac{e^2}{4\pi\varepsilon_0 r^2} = \frac{m_e v^2}{r} \tag{3.2}$$

Here ε_0 is the permittivity of the vacuum, which is numerically equal to $8.854187816... \times 10^{-12}$ F m^{-1}.

Elimination of v between equations (3.1) and (3.2) gives

$$r = \frac{\varepsilon_0 h^2}{\pi m_e e^2} n^2 = a_0 n^2 \tag{3.3}$$

for the radii of the quantized orbits. The radius of the smallest orbit ($n = 1$) is called the *first Bohr radius* and is denoted by a_0. Its numerical value is

$$a_0 = \frac{\varepsilon_0 h^2}{\pi m_e e^2} = 52.9 \text{ pm} \tag{3.4}$$

The various orbits are then given by the sequence a_0, $4a_0$, $9a_0$, ... according to equation (3.3).

The total energy of a given orbit is given by the sum of the kinetic and the potential energy (which is negative), namely,

$$E = \frac{1}{2} m_e v^2 - \frac{e^2}{4\pi\varepsilon_0 r}$$

Substituting equation (3.2), one finds

$$E = -\frac{e^2}{8\pi\varepsilon_0 r} \tag{3.5}$$

This is the classical value for the energy of a bound electron. If all values of r were allowed, the electron could have any negative value of the energy. However, according to equation (3.3) derived from Bohr's postulates, the r values are quantized. The resulting values of the energy are

$$E_n = -Ry / n^2 \tag{3.6}$$

where the quantity Ry, the Rydberg unit of energy, is $m_e e^4 / 8\varepsilon_0^2 h^2 = 2.17987 \times 10^{-18}$ J ($\equiv$ 13.60589253 eV), and n is the principal quantum number. This is also the binding energy of the electron in the ground state, $n = 1$, because this is the energy required to separate the electron from the nucleus.

In general, an electron in a lower energy state n_l, can undergo a transition to a higher state n_m ($n_m > n_l$) with absorption of energy $\Delta E = E_m - E_l$. In terms of wavenumber,

$$\tilde{\nu} = \Delta\varepsilon = \varepsilon_m - \varepsilon_l = -\frac{R_\infty}{n_m^2} - \left(-\frac{R_\infty}{n_l^2}\right) = R_\infty \left(\frac{1}{n_l^2} - \frac{1}{n_m^2}\right) \tag{3.7}$$

In equation (3.7), all quantities are expressed in wavenumber units ($\varepsilon = E/hc$), $\tilde{\nu}$ is the wavenumber of the transition and R_∞ is the Rydberg constant in wavenumber ($R_\infty = Ry/hc = m_e e^4/8c\varepsilon_0^2 h^3 = 1.0973731534 \times 10^7\ \text{m}^{-1} = 1.0973731534 \times 10^9\ \text{cm}^{-1}$).

If we allow n_l to take the values 1, 2, …, absorption of energy will raise the electron into the states with quantum number n_m = 2, 3, …; transitions from n_l = 2 carry the electron by absorption into the states with n_m = 3, 4, …, and so on. An initial value of the principal quantum number therefore gives rise to a *series* of absorption lines in the spectrum. Six such series have been recognized for the hydrogen atom, corresponding to the initial values of n_l = 1, 2, …; they are known, respectively, as the Lyman, Balmer, Paschen, Brackett, Pfund and Humphreys series.

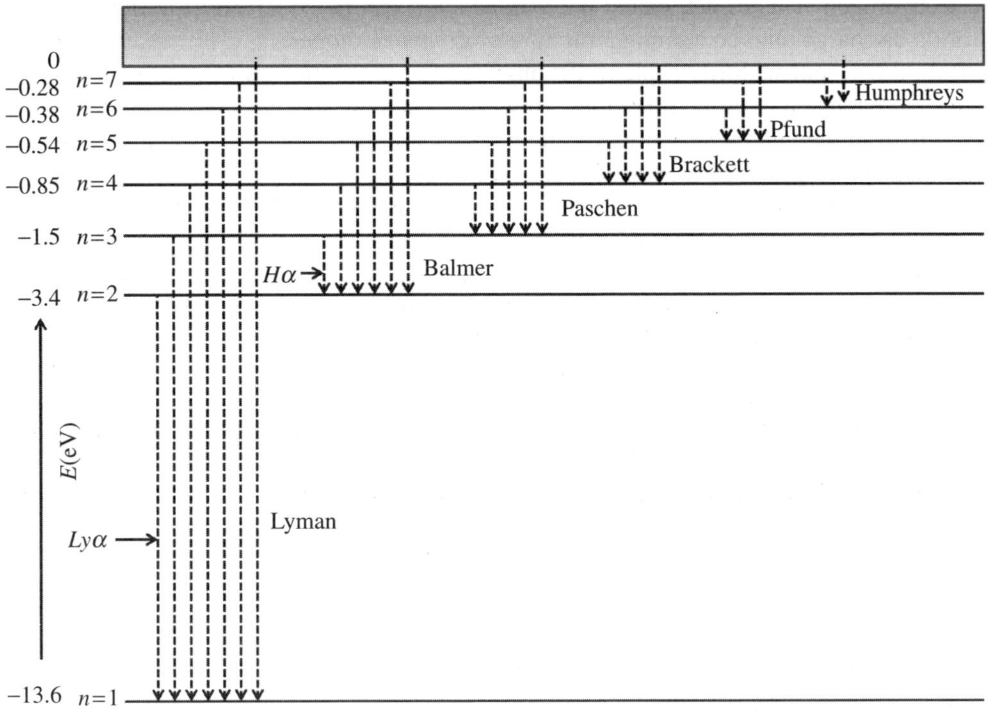

Figure 3.1. The hydrogen spectral series

An identical spectral line will be produced in emission if the electron falls from state n_m to state n_l (Figure 3.1). It should be mentioned that each line series discussed above shows a *continuous absorption* or *emission* to high wave numbers of the convergence limits. The convergence limit represents the situation where the atomic electron has absorbed just sufficient energy from radiation to escape from the nucleus with zero velocity. It can, however, absorb more energy than this and hence escape with higher velocities and, since the kinetic energy of an electron moving in free space is *not* quantized, *any* energy above the ionization energy can be absorbed. Hence the spectrum in this region is continuous.

The value of the Rydberg constant given above (R_∞) is for a nucleus of infinite mass. Since the nucleus of the hydrogen atom has a finite mass, the electron does not revolve around a stationary nucleus. Rather, both revolve around their common centre of mass, making this system a two-particle case. This can be reduced to a one-particle problem by replacing the mass of the electron in the expression for the Rydberg constant by the reduced mass, μ_e, given by

$$\mu_e = \frac{m_e M}{m_e + M},$$

where M is the mass of the nucleus. The Rydberg constant for the hydrogen atom is thus more accurately given by $R_H = \mu_e e^4 / 8\varepsilon_0^2 ch^3$. For the hydrogen atom, the nucleus consists of only a proton, and the ratio M/m_e is equal to 1836. Its isotope, deuterium, has a neutron as well in its nucleus, and hence the value of this ratio is about twice that of hydrogen. Its Rydberg constant is also slightly different from that of hydrogen, which causes a shift in its transition frequencies relative to hydrogen. This effect, called the *isotope shift*, can be observed as a 'doubling' of lines from a discharge tube containing a mixture of the two isotopes.

The Bohr theory satisfactorily explains the observed low-resolution absorption and emission spectra of hydrogen and hydrogen-like ions. However, when the spectral lines of the hydrogen spectrum are examined under very high resolution, they are found to consist of closely spaced doublets. This splitting, called fine structure, was one of the first experimental evidences for electron spin. There are certain other shortcomings of the Bohr model: It fails to describe why certain spectral lines are brighter than others, i.e., there is no mechanism for calculating *transition probabilities*. It also fails to explain why the electron does not radiate while travelling in the circular orbit, as required by classical electromagnetic theory. Bohr's theory is difficult to apply to multi-electron atoms and cannot be applied at all to molecules.

Example 3.1 Imagine a sample of atomic hydrogen exposed to blackbody radiation characteristic of temperature T. Use equation (1.15) for the ratio of the rate of spontaneous emission to that of stimulated emission for two levels separated by energy $h\nu_{lm}$:

$$\frac{A_{ml}}{B_{ml}\rho_\nu} = \left(e^{h\nu_{lm}/k_B T} - 1\right)$$

to calculate the above ratio for atomic hydrogen at 1000 K for the case of:

(a) a transition between $2^2S_{1/2}$ and $3^2P_{1/2}$;

(b) a transition between $5^2S_{1/2}$ and $6^2P_{1/2}$.

Comment on the values you obtain.

Solution

As we shall soon see, $3^2P_{1/2}$ means a $P_{1/2}$ term in which the electron has a principal quantum number, n, of 3. The degeneracy of the levels is $(2J + 1)$, where J is the subscript, but, for simplicity, we shall ignore degeneracy.

The transition wavenumbers are given by equation (3.7).

(a) $\tilde{\nu} = 109737\left(\frac{1}{2^2} - \frac{1}{3^2}\right) = 15241\ \text{cm}^{-1}$; $\nu = 4.569 \times 10^{14}\ \text{s}^{-1}$; Hence, $A_{ml}/B_{ml}\rho_\nu = 3.321 \times 10^9$

(b) $\tilde{\nu} = 109737\left(\frac{1}{5^2} - \frac{1}{6^2}\right) = 1341\ \text{cm}^{-1}$; $\nu = 4.021\times10^{13}\ \text{s}^{-1}$; Hence, $A_{ml}/B_{ml}\rho_{\nu} = 5.884$

A_{ml} is therefore more important only for transitions involving the lower energy levels of a hydrogen atom.

The spectra of hydrogen-like ions such as He^+, Li^{2+}, …, are also well accounted for by equation (3.6), which is now given by

$$\varepsilon_n = -\frac{Z^2 R_X}{n^2} = -\frac{Z^2 \mu_e e^4}{8\varepsilon_0^2 c h^3 n^2} \tag{3.8}$$

where Z is the atomic number of the ion, and the energy is expressed in wavenumber units.

Before taking up multi-electron atoms, we take a closer look at the hydrogen atom and hydrogen-like ions and their spectra. In order to better understand the remaining chapter, a brief recapitulation of quantum mechanics is in order.

3.3 HYDROGEN AND HYDROGEN-LIKE ION SPECTRA

The hydrogen atom and hydrogen-like ions are spherical in shape and, hence, it is most convenient to express the Hamiltonian and wave functions in spherical polar coordinates (Figure 3.2). The Hamiltonian may be written as

$$\hat{H} = -\frac{\hbar^2}{2\mu_e}\nabla^2 - \frac{Ze^2}{4\pi\varepsilon_0 r} \tag{3.9}$$

where the two terms are, respectively, the kinetic and potential energy of the electron. The Laplacian operator ($\nabla^2_{r\theta\phi}$) in spherical polar coordinates is given by

$$\nabla^2_{r\theta\phi} = \frac{1}{r^2}\frac{\partial}{\partial r}\left(r^2\frac{\partial}{\partial r}\right) + \frac{1}{r^2\sin\theta}\frac{\partial}{\partial\theta}\left(\sin\theta\frac{\partial}{\partial\theta}\right) + \frac{1}{r^2\sin^2\theta}\frac{\partial^2}{\partial\phi^2} \tag{3.10}$$

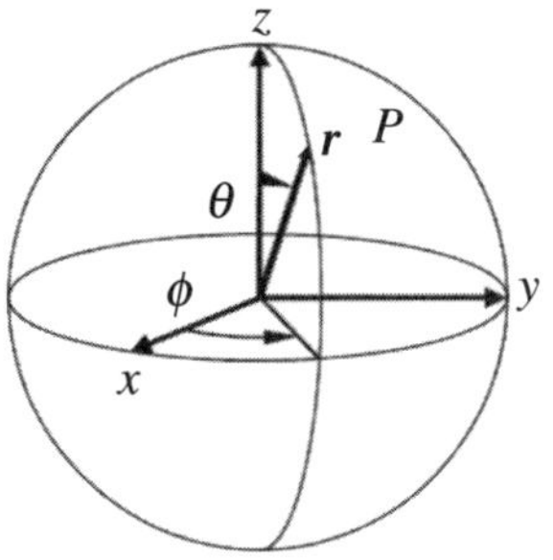

Figure 3.2. Spherical polar coordinates in terms of Cartesian coordinates, $x = r\sin\theta\cos\varphi$, $y = r\sin\theta\sin\varphi$ and $z = r\cos\theta$

This constitutes a central field problem, since the potential energy is a function of the distance of the electron from the nucleus only, i.e., $V(r)$, and there is no angular dependence. The $\hat{L}^2$ operator, which is the operator for the square of the orbital angular momentum, is given by

$$\hat{L}^2 = -\hbar^2\left(\frac{1}{\sin\theta}\frac{\partial}{\partial\theta}\left(\sin\theta\frac{\partial}{\partial\theta}\right)+\frac{1}{\sin^2\theta}\frac{\partial^2}{\partial\phi^2}\right) \tag{3.11}$$

Comparison of the Hamiltonian operator in equation (3.9) with the Laplacian operator in equation (3.10) shows that the angular momentum square operator is identical to the angular part of the Hamiltonian, and hence the two operators commute. Further, the angular momentum square operator obeys the following commutation rules:

$$\begin{aligned}
&\left[\hat{L}^2,\hat{L}_x\right]=\left[\hat{L}^2,\hat{L}_y\right]=\left[\hat{L}^2,\hat{L}_z\right]=0\\
&\left[\hat{L}_x,\hat{L}_y\right]\neq 0\\
&\left[\hat{L}_x,\hat{L}_z\right]\neq 0\\
&\left[\hat{L}_y,\hat{L}_z\right]\neq 0
\end{aligned} \tag{3.12}$$

We may therefore say that the x-, y- and z-components of the orbital angular momentum are all *complementary variables*. Knowing one of the three precludes knowledge of the other two. They are not all simultaneously well defined. By convention, the known component is taken as the z-component.

Detailed quantum mechanical solution of the hydrogen atom problem leads to three quantum numbers, n (principal), l (azimuthal) and m (magnetic), signifying the total energy, total angular momentum and z-component of the angular momentum, respectively.

The total wave function $\psi_{nlm}(r, \theta, \phi)$ may be expressed as the product of a radial (r-dependent) part $R_{nl}(r)$ and the spherical harmonics $Y_{lm}(\theta, \phi)$, where the spherical harmonics themselves are products of the θ- and φ-dependent parts, i.e.,

$$Y_{lm}(\theta,\phi) = \Theta_{lm}(\theta)\Phi_m(\phi) = N_{l,|m|}P_l^{|m|}(\cos\theta)e^{im\phi} \tag{3.13}$$

Here, the $P_l^{|m|}(\cos\theta)$ are a class of special functions, the *associated Legendre polynomials*, and $N_{l,|m|}$ denote their normalization constants, given by

$$N_{l,|m|} = \left(\frac{2l+1}{4\pi}\frac{(l-|m|)!}{(l+|m|)!}\right)^{1/2}$$

Some of the first associated Legendre polynomials are listed below

$$P_0^0(x)=1;\quad P_1^0(x)=x;\quad P_2^0(x)=\frac{1}{2}(3x^2-1);\quad P_3^0(x)=\frac{1}{2}(5x^3-3x)$$

where $x = \cos\theta$. It is obvious that the Legendre functions are odd for odd l and even for even l.

The spherical harmonics are eigenfunctions of the $\hat{L}^2$ and $\hat{L}_z$ operators with eigenvalues $l(l+1)\hbar^2$ and $m\hbar$, respectively. The energy is given in terms of the principal quantum number by the same expression (3.8) as that obtained by Bohr. Substitution of $n = 1$, $l = 0$ and $m = 0$ gives for the 1s orbital of hydrogen and hydrogen-like ions,

$$\psi_{1s} = \sqrt{\frac{Z^3}{a_0^3 \pi}} e^{-Zr/a_0} \tag{3.14}$$

In the early 1920s, an experiment performed by Stern and Gerlach discovered two important new aspects of electrons. First, in addition to being charged, electrons behave like tiny bar magnets. They also have a tiny intrinsic amount of angular momentum, equal to $\hbar/2$. This quantity is called *spin*, and all known elementary particles have non-zero spin. Electrons are called *spin-½* particles and are thus fermions (elementary particles with half-integral spins). The second surprising discovery was *how much* the path of the electrons is deflected in the presence of a magnetic field. If electrons were like ordinary magnets with random orientations, they would show a continuous distribution of paths. However, what was observed was quite different. The electrons were deflected by the same amount, either up or down, in approximately equal numbers. This indicated that the z-component of the electron's spin is *quantized*: It can take only one of two discrete values. We say that the spin is either *up* or *down* in the z direction.

We may express these findings in quantum mechanical terms as follows: For a single electron, the spin quantum number $s = ½$. Like any other angular momentum, this means that the magnitude of its square is quantized and is given by

$$\left|\vec{s}^{\,2}\right| = s(s+1)\hbar^2 = \frac{1}{2}\left(\frac{1}{2}+1\right)\hbar^2 = \frac{3}{4}\hbar^2 \tag{3.15}$$

Also, the spin angular momentum obeys commutation rules similar to the orbital angular momentum (3.12). Since $s = ½$, its component in the z direction can take on two values +½ (spin-up or α) and –½ (spin-down or β) and we have the additional relations

$$\begin{aligned}
\hat{s}^2|\alpha\rangle &= \frac{3}{4}\hbar^2|\alpha\rangle \\
\hat{s}^2|\beta\rangle &= \frac{3}{4}\hbar^2|\beta\rangle \\
\hat{s}_z|\alpha\rangle &= \frac{1}{2}\hbar|\alpha\rangle \\
\hat{s}_z|\beta\rangle &= -\frac{1}{2}\hbar|\beta\rangle
\end{aligned} \tag{3.16}$$

The splitting of lines in the hydrogen spectrum can now be explained as follows: The orbital motion of the electron in the electric field of the nucleus produces a magnetic field at the site of the electron that couples to the magnetic moment of the electron's spin. This spin–orbit coupling is a relativistic effect. The total angular momentum $\vec{j}$ is given by

$$\vec{j} = \vec{l} + \vec{s} \tag{3.17}$$

The j quantum number can take on values (positive ones only) permitted by the Clebsch–Gordan series:

$$j = |l + s|, |l + s - 1|, \ldots, |l - s| \tag{3.18}$$

Since, generally, $s \leq l$, the number of values of j is $2s + 1$, which is called the *multiplicity* of a state. If the multiplicity equals 1, 2, 3, …, the term is called singlet, doublet, triplet, …, respectively. Note that these names are traditionally kept for the number $2s + 1$ even if $l < s$, although the number of terms is then $2l + 1$.

For the ground state of the hydrogen atom, with the electron configuration, $1s^1$,

$l = 0$ (s orbital)

$s = ½$ (one electron)

so

$j = |l + s| = ½$

$|l - s|$ is also $|-½| = ½$, so there is only one allowed value of j. Note that lower case letters (l, s, j) are used for the angular momenta, which is the convention for a single electron.

As another example, for a d^1 configuration:

$s = ½$, hence $(2s + 1) = 2$, $l = 2$

The permitted values of j are $j = 2 \pm ½ = 5/2$ and $3/2$, as the vector addition below shows.

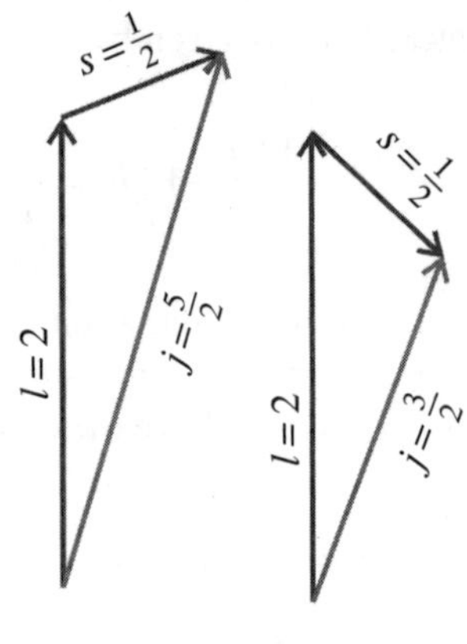

Hence for a single electron, $j = l \pm ½$, since s can only have the value ½. For a given electron configuration, an atom can, therefore, assume different states depending on how the orbital and spin angular momenta of the electrons are coupled. Before we proceed further, we quantitatively derive the selection rules for atomic transitions.

3.3.1 Transition probabilities

As stated in Chapter 2, all transitions are governed by selection rules. In the specific case of a hydrogen atom, we had deduced, on symmetry and angular momentum conservation considerations alone, that the l selection rule is $\Delta l = \pm 1$. We now deduce the angular momentum selection rules quantum mechanically.

The probability of transition from state l to state m is proportional to $|\vec{M}_{lm}|^2$, where $|\vec{M}_{lm}|$ is given by $\int \psi_m^* \hat{\mu} \psi_l \, d\tau$, and $\hat{\mu}$ is the operator corresponding to the dipole moment vector, which is defined as $\vec{\mu} = \sum_{i=1}^{n} e\vec{r}_i$ for an arbitrary atom. Here e is the electronic charge and $\vec{r}_i$ is the distance of the ith electron from the nucleus. The summation extends over all n electrons. For a single electron, as in the hydrogen atom, $\vec{\mu} = e\vec{r}$. The dipole moment can be expressed in terms of its Cartesian components, $\vec{\mu} = \hat{i}\mu_x + \hat{j}\mu_y + \hat{k}\mu_z$, where $\hat{i}$, $\hat{j}$ and $\hat{k}$ are unit vectors along the x-, y- and z-axes. Since $\vec{r} = \hat{i}x + \hat{j}y + \hat{k}z$, substitution of $x = r\sin\theta\cos\phi$, $y = r\sin\theta\sin\phi$, $z = r\cos\theta$ gives

$$\vec{\mu} = \mu_0(\hat{i}\sin\theta\cos\varphi + \hat{j}\sin\theta\sin\varphi + \hat{k}\cos\theta)$$
$$\mu_0 = \int \psi_{\text{el}}^{\ *}\,\hat{\mu}\,\psi_{\text{el}}\, d\tau_{\text{el}} \tag{3.19}$$

For the sake of simplicity, let us assume that the radiation is z polarized. Then, for the transition $l \to l'$, $m \to m'$, the transition moment integral becomes

$$\left|\vec{M}_{ll'}\right| = \int_0^{2\pi}\int_0^{\pi} Y_{l'm'}^{*}(\theta,\phi) e\hat{r}\cos\theta Y_{lm}(\theta,\phi)\sin\theta\, d\theta d\phi \tag{3.20}$$

where the integral over θ extends from 0 to π, the volume element being given by $\sin\theta\, d\theta$, while the integral over ϕ is from 0 to 2π. Inserting equation (3.13), we get

$$\begin{aligned}\left|\vec{M}_{ll'}\right| &= \int_0^{2\pi}\int_0^{\pi}\left(N_{l',|m'|}P_{l'}^{|m'|}(\cos\theta)e^{im'\phi}\right)^{*}\mu_0\cos\theta N_{l,|m|}P_l^{|m|}(\cos\theta)e^{im\phi}\sin\theta\, d\theta d\phi\\ &= \int_0^{2\pi}\int_0^{\pi} N_{l',|m'|}P_{l'}^{|m'|}(\cos\theta)e^{-im'\phi}\mu_0\cos\theta N_{l,|m|}P_l^{|m|}(\cos\theta)e^{im\phi}\sin\theta\, d\theta d\phi\\ &= N_{l',|m'|}N_{l,|m|}\int_0^{\pi} P_{l'}^{|m'|}(\cos\theta)\mu_0\cos\theta P_l^{|m|}(\cos\theta)\sin\theta\, d\theta\int_0^{2\pi} e^{-im'\phi}e^{im\phi}d\phi\end{aligned} \tag{3.21}$$

Solving first the ϕ equation, we get

$$\begin{aligned}\int_0^{2\pi} e^{-im'\phi}e^{im\phi}d\phi &= \int_0^{2\pi} e^{i(m-m')\phi}d\phi = \left.\frac{e^{i(m-m')\phi}}{i(m-m')}\right|_0^{2\pi} = \frac{1}{i(m-m')}\left[e^{i(m-m')2\pi} - 1\right]\\ &= \frac{1}{i(m'-m)}\left[\cos 2\pi(m-m') + i\sin 2\pi(m-m') - 1\right]\\ &= 0\end{aligned} \tag{3.22}$$

for the case $m' \neq m$, since $\cos(n\pi) = (-1)^n$ when n is an integer. In equation (3.22), the m's are integers, and hence $2\,(m - m')$ is even and the cos term always equals +1. The sine term is equal to zero since $\sin 2n\pi$ is zero for all integer values of n. For the special case, $m' = m$, the integral is simply equal to 2π. Thus, we may write

$$\int_0^{2\pi} e^{-im'\phi}e^{im\phi}d\phi = \int_0^{2\pi} e^{i(m-m')\phi}d\phi = 2\pi\delta_{mm'} \tag{3.23}$$

where the Krönecker delta

$$\delta_{mm'} = \begin{cases} 1 \text{ if } m = m' \\ 0 \text{ if } m \neq m' \end{cases} \tag{3.24}$$

and so we have the first selection rule $\Delta m = 0$ for z-polarized radiation. For molecules with no unpaired electrons, the selection rule for m is not apparent in the spectrum unless there is an applied external field, since all m values for a given l are degenerate in the absence of an applied field. In a similar fashion, we obtain the selection rule $\Delta m = \pm 1$ for x- and y-polarized radiation (more complicated though), and so the consolidated selection rule is $\Delta m = 0, \pm 1$ for unpolarized radiation.

Returning to z-polarized radiation, substitution in (3.20) of the recursion relation for associated Legendre polynomials,

$$\cos\theta P_l^{|m|}(\cos\theta) = \frac{(l+|m|)}{2l+1} P_{l-1}^{|m|}(\cos\theta) + \frac{(l-|m|+1)}{2l+1} P_{l+1}^{|m|}(\cos\theta) \tag{3.25}$$

with $m' = m$ gives

$$\left|\vec{M}_{ll'}\right| \propto \int_0^\pi P_{l'}^{|m|}(\cos\theta)\left[\frac{l+|m|}{2l+1} P_{l-1}^{|m|}(\cos\theta) + \frac{l-|m|+1}{2l+1} P_{l+1}^{|m|}(\cos\theta)\right]\sin\theta\, d\theta$$

$$\propto \int_0^\pi \left[P_{l'}^{|m|}(\cos\theta)\frac{l+|m|}{2l+1} P_{l-1}^{|m|}(\cos\theta) + P_{l'}^{|m|}(\cos\theta)\frac{l-|m|+1}{2l+1} P_{l+1}^{|m|}(\cos\theta)\right]\sin\theta\, d\theta \tag{3.26}$$

the proportionality constant being $N_{l',|m'|} N_{l,|m'|} 2\pi\mu_0$. The associated Legendre polynomials are orthonormal, i.e.,

$$\int_0^{2\pi}\int_0^{\pi} Y_{l'm'}^{*}(\theta,\varphi) Y_{lm}(\theta,\phi)\sin\theta\, d\theta\, d\phi = \delta_{ll'}\delta_{mm'}$$

$$\int_0^\pi P_{l'}^{|m|}(\cos\theta) P_l^{|m|}(\cos\theta)\sin\theta\, d\theta = \frac{2(l+m)!}{(2l+1)(l-m)!}\delta_{l'l} \tag{3.27}$$

The first term in the integral (3.26) is non-zero only when $l' = l - 1$ and the second term is non-zero only when $l' = l + 1$. In all other events, the integrals involving products of two different values of l vanish. Thus, the combined selection rule for the change in orbital angular momentum during atomic transitions is $l' = l \pm 1$, or

$$\Delta l = \pm 1 \tag{3.28}$$

Before we proceed further, we note that there is no restriction on Δn. We thus have the combined selection rules $\Delta n = \pm 1, \pm 2, \pm 3, \ldots$ $\Delta l = \pm 1$, $\Delta m = 0, \pm 1$ (applicable only in the presence of an external field) for transitions of a single electron. Transitions within the same configuration are forbidden, i.e., *either n or l* must change for one electron, or, in other words, the electron must move to another orbital.

Note that these selection rules are derived for electric dipole transitions only. The selection rules for magnetic dipole and electric quadrupole transitions are different and will not be considered here.

Thus, according to our electric dipole selection rules, the 1s electron can be promoted to the 2p orbital ($\Delta n = +1$, $\Delta l = +1$) by absorption of radiation, giving rise to the $(2p)^1$ configuration. For this configuration,

$$n = 2, l = 1, s = \tfrac{1}{2}, j = |l + s|, |l + s - 1|, \ldots, |l - s| = \tfrac{3}{2}, \tfrac{1}{2}$$

i.e., we obtain two states from this configuration with $j = \tfrac{3}{2}$ and $\tfrac{1}{2}$, having slightly different energies.

Transitions to both j states are allowed by the selection rule, $\Delta j = 0, \pm 1$ (which we state without derivation). Hence the 2p ← 1s transition (it is conventional to write spectroscopic transitions with the *upper* level written first) is split into a doublet, as observed experimentally. This coupling is very weak for hydrogen and the first few hydrogen-like ions; the doublets and compound doublets can be resolved only with spectrometers of the highest resolving power.

3.4 MULTI-ELECTRON ATOMS

The situation gets more complex as the number of electrons increases. However, when determining the possible states, it is sufficient to consider only shells and subshells that are not completely filled because the angular momenta of the electrons in a filled subshell always add up such that the total orbital, the total spin, and the total angular momentum for this subshell are all zero.

For multi-electron systems, it is not possible to represent the state in terms of the quantum numbers n, l, m and s used for single electrons. Instead, the *state* is characterized by a term, given by the term symbol $^{2S+1}L_J$, where the value of the multiplicity $(2S + 1)$ is written as the upper left superscript and L is denoted by the corresponding capital letter as follows:

Total Orbital Momentum

L	0	1	2	3	4	5
	S	P	D	F	G	H

For the $(1s)^1$ configuration of the hydrogen atom, the term symbol is thus $^2S_{1/2}$, since the total orbital angular momentum is zero, corresponding to an S state, the multiplicity of the state is $2S + 1 = 2$ (it is a doublet) and the J value is $\tfrac{1}{2}$. Note the use of capital letters to denote the state symbol, as opposed to small letters for the electron configuration $(1s)^1$. It is important to distinguish between an *electron configuration* and a *state.* The former is a listing of the number of electrons in each subshell and shows the occupation scheme of atomic orbitals. In the example above, the ground state electron configuration of hydrogen is $(1s)^1$, and its state is represented by $^2S_{1/2}$. Similarly, for the configuration $(d)^1$ discussed above, the state is ^2D and the J values are $\tfrac{3}{2}$ and $\tfrac{5}{2}$. In multi-electron atoms, the term symbol is a more convenient representation than the electron configuration, as it contains all the relevant information in a concise way.

We first discuss the spectra of alkali metals, which, like hydrogen and hydrogen-like ions, have an $(ns)^1$ electron configuration.

3.4.1 Alkali metal spectra

All alkali metals have a single electron in the outer s orbital, and hence their spectra bear a superficial resemblance to the hydrogen spectrum. There are two major differences, however. The first is that the energy of an electron no longer depends on the principal quantum number alone. The radial distribution curves in Figure 3.3 show that the s electrons have the greatest *penetrating power* and experience the greatest attraction to the nucleus. The p and d electrons are more *shielded*, and hence the energy of the electron depends on both the n and l quantum numbers, with $E_{ns} < E_{np} < E_{nd}$. The other difference is the amount of splitting of the spectral lines, which increases with the atomic number.

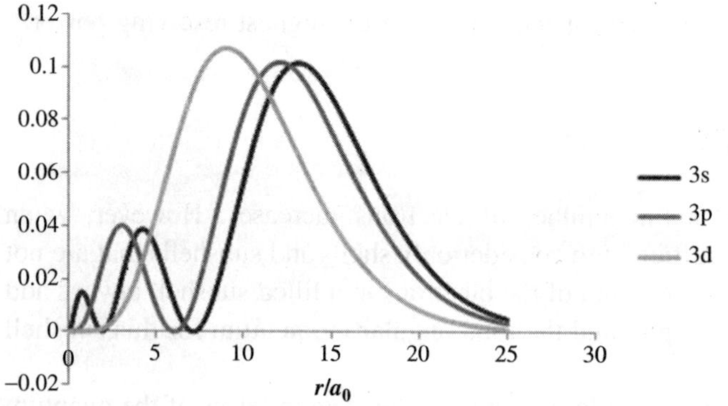

Figure 3.3. Radial distribution curves for the 3s, 3p and 3d orbitals of hydrogen

These effects are taken into account by using an effective principal quantum number $n_{eff}(n, l)$ or a quantum defect $\Delta(n, l)$. For example, the energy terms of the sodium atom are given by

$$E_{nl} = -\frac{R_{Na}hc}{n_{eff}^2} = -\frac{R_{Na}hc}{(n-\Delta)^2} \tag{3.29}$$

The quantum defect Δ increases with increasing atomic number from lithium to caesium in the alkali metals group, but decreases with increasing orbital quantum number l. It does not depend much on the principal quantum number n, as can be seen in the following values for the quantum defect for sodium $\Delta(n, l)$: {1.37 (3s), 1.36 (4s), 1.35 (5s)}, {0.88 (3p), 0.87 (4p), 0.86 (5p)}, {0.010 (3d), 0.011 (4d), 0.013 (5d)}, and {0.000 (4f) and –0.001 (5f)}.

Based on these values, the Grotrian diagram for sodium is shown in Figure 3.4. The s, p, d and f labels remain the same as for hydrogen.

The emission spectrum is dominated by the 3p → 3s transition corresponding to the well-known sodium D lines. As for hydrogen, the upper level $(3p)^1$ is split into two states,

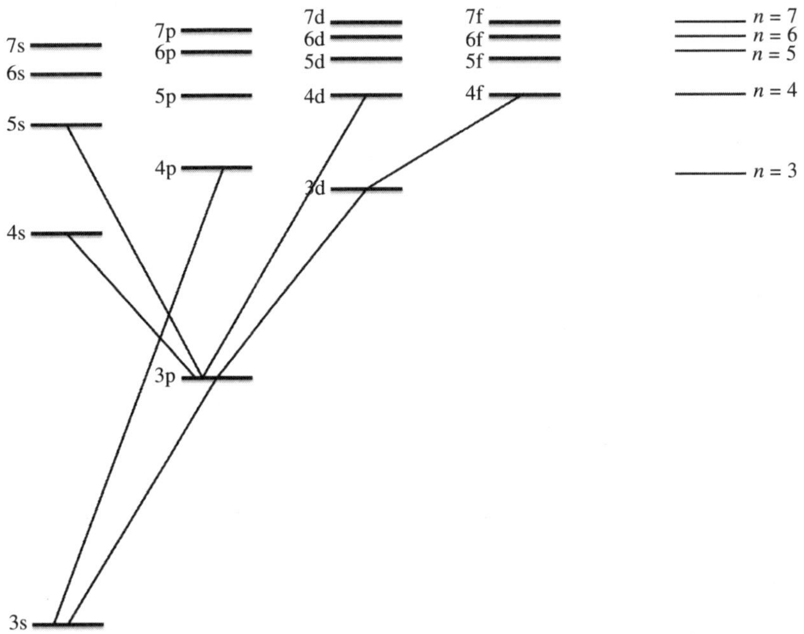

Figure 3.4. Grotrian diagram for sodium

$3^2P_{3/2}$ and $3^2P_{1/2}$ ($L = 1$, $S = ½$, $J = ½$, $3/2$). The 3 written before the state symbol represents the principal quantum number in order to distinguish it from hydrogen. To decide which of these states is lower in energy, we turn to Hund's rules of maximum multiplicity.

1. States having higher S values lie lower in energy.
2. When two states have the same S values, the one having the higher L value lies lower in energy.
3. For orbitals that are less than half-filled, states having lower J values lie lower in energy (normal ordering); for orbitals that are more than half-filled, states having higher J values lie lower in energy (inverted ordering). For half-filled orbitals, $L = 0$, and hence $J = S$ and there is only one value of J.

In other words, of all possible terms, those with the largest spin multiplicity ($2S + 1$) have lowest energy; of all terms with the same S, the one with the lowest energy is that with the largest value of L. For configurations that are less than half-filled, the term with the lower J is lower in energy. Thus, for the excited $(n\text{p})^1$ configurations of hydrogen and alkali metals, the $^2P_{1/2}$ level is lower in energy, since there is only one electron in the p orbital out of a maximum possible occupancy of six (less than half-filled).

Unlike the hydrogen atom case, the doublet in the alkali metal spectra can be easily resolved. For example, the doublet $3^2P_{3/2,1/2} \leftarrow 3^2S_{1/2}$ of sodium (Figure 3.5) appears at 585.99 and 588.99 nm, corresponding to the well-known yellow sodium D lines. Similarly, the colour of the lithium doublet is red, and that for potassium is lilac, in conformity with the increasing spacing with the increasing atomic number. As for hydrogen (Chapter 2), this transition belongs to the principal series.

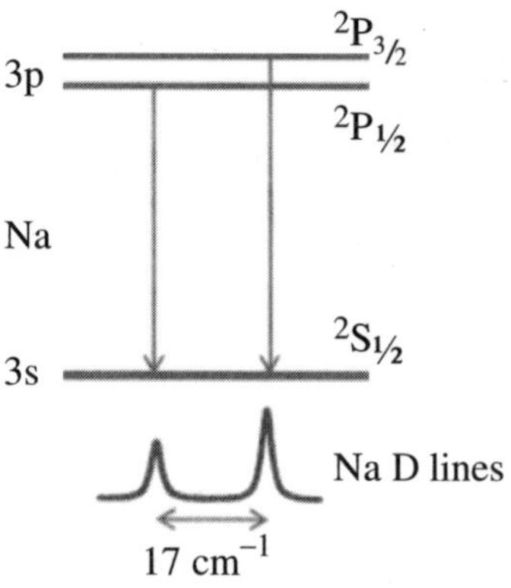

Figure 3.5. The 3p → 3s transition of sodium (*D* lines)

Example 3.2 How many lines will be observed for 4f → 3d (^{2}F to ^{2}D) transitions?

Solution

For the ^{2}D state, the allowed J values are $\frac{3}{2}$ and $\frac{5}{2}$, and for the ^{2}F state, the allowed values are $\frac{5}{2}$ and $\frac{7}{2}$. Application of the selection rules ($\Delta L = \pm 1$, $\Delta J = 0, \pm 1$) leads to the following allowed transitions and thus a triplet will be observed.

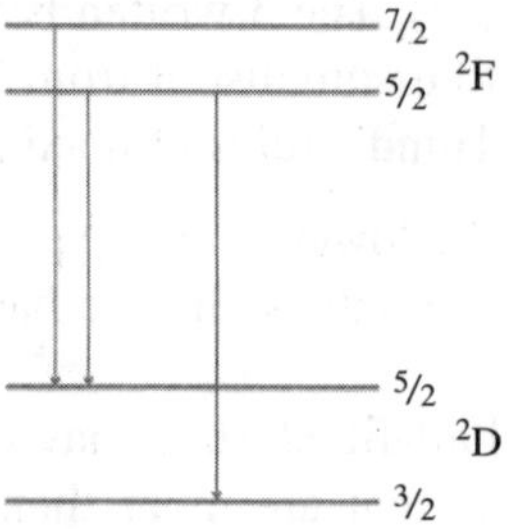

A similar application of the selection rules to the 3d → 2p (3^2D → 2^2P) emission convinces us that this too should be a triplet. Also note that for single outermost electron atoms, such as alkali metals and hydrogen, the orbital term symbol is the capital letter corresponding to the orbital containing the unpaired electron.

3.4.2 Two-electron systems

When more than one valence electron is present, the situation increases in complexity. Consider the simplest two-electron system, the helium atom. The Hamiltonian operator becomes

$$\begin{aligned}\hat{H} &= -\frac{\hbar^2}{2\mu_e}\nabla_1^2 - \frac{\hbar^2}{2\mu_e}\nabla_2^2 - \frac{2e^2}{4\pi\varepsilon_0 r_1} - \frac{2e^2}{4\pi\varepsilon_0 r_2} + \frac{e^2}{4\pi\varepsilon_0 r_{12}} \\ &= \hat{H}^0 + \hat{H}_{ee} \end{aligned} \tag{3.30}$$

where the subscripts '1' and '2' refer to electrons 1 and 2, and the last term signifies the coulombic repulsion between the two electrons ($\hat{H}_{ee}$). Due to the presence of this term, it is not possible to get an exact solution for helium atom (or any other multi-electron atom) since this term depends on the coordinates of both electrons. It is usual to take the other terms (which amount to two helium ions) as the zero-order Hamiltonian ($\hat{H}^0$) and treat the inter-electronic repulsion term as a perturbation. The first-order perturbation correction to the zero-order energy is given by

$$E^{(1)} = \left\langle \psi^{(0)} \middle| \hat{H}_{ee} \middle| \psi^{(0)} \right\rangle$$

where the $\psi^{(0)}$ are the zero-order wave functions, which are eigenfunctions of the zero-order Hamiltonian. We consider first what might be an excited configuration $(n\mathrm{s})^1(n'\mathrm{s})^1$ of two non-equivalent electrons, for reasons which will become apparent later on. For both electrons, $l = 0$ and $s = {}^1/_2$, but the n are different. Since now there are two electrons, their spin angular momenta couple. As usual, the permitted values of the total spin quantum number for the combined system are given by the Clebsch-Gordan series as follows:

$$S = |s_1 + s_2|, |s_1 + s_2 - 1| \ldots |s_1 - s_2| = |\tfrac{1}{2} + \tfrac{1}{2}|, \ldots |\tfrac{1}{2} - \tfrac{1}{2}| = 1, 0$$

In a similar fashion, the permitted values of the total orbital angular momentum are

$$L = |l_1 + l_2|, |l_1 + l_2 - 1| \ldots |l_1 - l_2| = |0 + 0|, \ldots |0 - 0| = 0$$

The two states are therefore $^3\mathrm{S}$ and $^1\mathrm{S}$. The J values are 1 and 0 ($L + S$), respectively, leading to $^3\mathrm{S}_1$ and $^1\mathrm{S}_0$ states.

Repeating the procedure for the ground state electronic configuration of helium $(1\mathrm{s})^2$ leads to the same two states since the l and s values for the two electrons are the same as before: They differ only in their n quantum numbers. However, the $^3\mathrm{S}$ term cannot exist for the $(1\mathrm{s})^2$ configuration, since this would require a state with the s_i of both electrons aligned in the same direction in the same orbital (both electron spins α or β), which is contrary to the Pauli Exclusion Principle. The Pauli Exclusion Principle arises from the sixth postulate of quantum mechanics (Chapter 2), the antisymmetry principle, which requires that the total wave function of a fermion (such as an electron) must be antisymmetric with respect to electron exchange. Let us see how this principle manifests itself for the ground and first excited states of helium.

If we describe the zero-order wave function for the $(1\mathrm{s})^1(2\mathrm{s})^1$ configuration of helium as $\varphi_1 = 1\mathrm{s}(1)2\mathrm{s}(2)$, meaning that the first electron is in the 1s orbital and the second in the 2s orbital, it defies the principle of indistinguishability of electrons, since we have labelled the two electrons. If we define a permutation operator, $\hat{P}(1,2)$, which exchanges the two electrons, we find that φ_1 is not an eigenfunction of this operator, since

$$\hat{P}(1,2)\varphi_1 = \hat{P}(1,2)(1\mathrm{s}(1)2\mathrm{s}(2)) = 1\mathrm{s}(2)2\mathrm{s}(1) = \varphi_2 \tag{3.31}$$

Similarly,

$$\hat{P}(1,2)\varphi_2 = \varphi_1 \tag{3.32}$$

Since $[\hat{P}(1,2), \hat{H}] = 0$, i.e., the Hamiltonian operator is invariant under electron exchange (if you replace **1** by **2** and **2** by **1** in the Hamiltonian operator for the helium atom, it remains unchanged), we look for a function ψ that is simultaneously an eigenfunction of both $\hat{P}(1, 2)$ and $\hat{H}$. We recognize that two permutations of a wave function should regenerate it. Thus,

$$\hat{P}^2(1, 2)\psi = \psi$$

Therefore, the only possible eigenfunctions of the $\hat{P}(1, 2)$ operator are ± 1. The eigenfunction corresponding to the +1 eigenvalue is termed symmetric because the wave function does not change sign on electron exchange, while the one with the −1 eigenvalue is termed antisymmetric.

We look for linear combinations of our two functions φ_1 and φ_2 that satisfy these requirements. Since both φ_1 and φ_2 are eigenfunctions of the zero-order Hamiltonian with the same eigenvalue $E^0 = E_{1s} + E_{2s}$, any linear combination of these two functions is also an eigenfunction. An obvious choice is the sum and difference of the two. Thus, we take the two normalized linear combinations,

$$\psi_1 = \frac{1}{\sqrt{2}}(\varphi_1 + \varphi_2)$$
$$\psi_2 = \frac{1}{\sqrt{2}}(\varphi_1 - \varphi_2) \tag{3.33}$$

where the first is symmetric with respect to electron exchange, and the second is antisymmetric, since

$$\hat{P}(1,2)\,\psi_1 = \frac{1}{\sqrt{2}}\hat{P}(1,2)(\varphi_1 + \varphi_2) = \frac{1}{\sqrt{2}}(\varphi_2 + \varphi_1) = \psi_1$$
$$\hat{P}(1,2)\,\psi_2 = \frac{1}{\sqrt{2}}\hat{P}(1,2)(\varphi_1 - \varphi_2) = \frac{1}{\sqrt{2}}(\varphi_2 - \varphi_1) = -\psi_2 \tag{3.34}$$

using equations (3.31) and (3.32). We may be tempted to conclude that ψ_2 is the required antisymmetric wave function, but there is more to the picture. The antisymmetry principle states that *the total wave function must also include the spin component.*

We can write four possible wave functions for the two spin states α and β. These are $\alpha(1)\alpha(2)$, $\alpha(1)\beta(2)$, $\beta(1)\alpha(2)$ and $\beta(1)\beta(2)$. The first and last functions are symmetric with respect to electron exchange, but the other two are neither symmetric nor antisymmetric with respect to electron exchange, just like our orbital functions φ_1 and φ_2. Taking a cue from our handling of the spatial wave functions, we take the two linear combinations $\frac{1}{\sqrt{2}}[\alpha(1)\beta(2) \pm \beta(1)\alpha(2)]$, where the positive combination is a symmetric function and the other is antisymmetric. We now have three symmetric and one antisymmetric spin functions. These have to be combined with our two spatial functions, one of which is symmetric and the other antisymmetric. According to Pauli's exclusion principle, the total wave function of an electron must be antisymmetric with respect to electron exchange (Postulate 6). We know that the product of a symmetric function with an antisymmetric one is antisymmetric, and the product of two symmetric or two antisymmetric products is symmetric. Hence, we must combine our symmetric orbital function with the antisymmetric spin function, and the antisymmetric orbital function with the symmetric spin functions. There are three such functions; so we recognize that the antisymmetric orbital function belongs to a triplet state and has a spin degeneracy of three. Similarly, the symmetric orbital function is a singlet state.

We now have four total wave functions

$$\frac{1}{2}(1s(1)2s(2)+1s(2)2s(1))(\alpha(1)\beta(2)-\alpha(2)\beta(1))$$

$$\frac{1}{\sqrt{2}}(1s(1)2s(2)-1s(2)2s(1))\begin{cases}\alpha(1)\alpha(2)\\ \beta(1)\beta(2)\\ \frac{1}{\sqrt{2}}(\alpha(1)\beta(2)+\alpha(2)\beta(1))\end{cases} \tag{3.35}$$

We can see that the symmetric spatial wave function is the singlet $^1S_0$ state and the antisymmetric spatial wave functions correspond to the $^3S_1$ state. Since the spatial parts are also different for the singlet and triplet wave functions, it follows that their energies are different. Operation of the Hamiltonian (3.30) on the two wave functions yields the result that the energies are given by $E^0 + J \pm K$, where the plus sign refers to the singlet wave function, J and K are, respectively, the coulomb and exchange integrals, both of which are positive in sign because they signify repulsions, which means that the triplet state is lower in energy. We can see that the triplet wave function in equation (3.35) vanishes whenever the two electrons are at the same position, and hence it is lower in energy because it avoids interelectronic repulsions. Figure 3.6 shows the possible transitions for singlet and triplet helium.

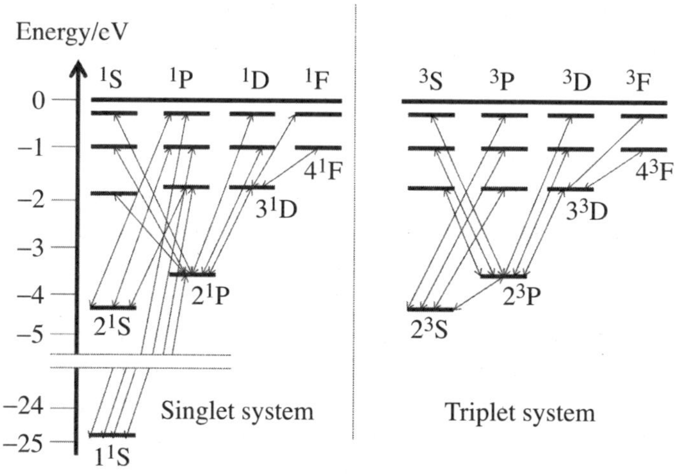

Figure 3.6. Singlet and triplet helium. The 1^1S-state has the energy −24.77 eV, while the 2^3S-state has energy −5.00 eV. The transitions shown here apply to one electron and the other remains in the 1s state

What happens to the triplet wave function if the two electrons are equivalent, as in a $(1s)^2$ or $(2s)^2$ configuration? The orbital part of the triplet wave function now becomes $\frac{1}{\sqrt{2}}(1s(1)1s(2)-1s(2)1s(1))$, which vanishes. The $^3S_1$ state simply does not exist for equivalent electrons. Thus, Pauli's exclusion principle only allows the $^1S_0$ state for the $(1s)^2$ and $(2s)^2$ configurations.

We now have two spin states, singlet and triplet, for the $(1s)^1(2s)^1$ configuration and require a spin selection rule. As stated in the previous chapter, the electrons are fermions, while the photons are bosons, and hence cannot induce a change in the electron spin because this would violate the law of spin conservation. Thus, we have $\Delta S = 0$. Let us confirm this selection rule by looking at a transition from a singlet state to another singlet state. We note that the α and β functions are normalized and orthogonal to each other. We thus have for a transition from a singlet state to a singlet state:

$$
\begin{aligned}
\left|\vec{M}_{ss'}\right| &\propto \frac{1}{2}\left(\int\int(\alpha(1)\beta(2)-\alpha(2)\beta(1))(\alpha(1)\beta(2)-\alpha(2)\beta(1))d\sigma_1\,d\sigma_2\right)\\
&= \frac{1}{2}\left(\int\int(\alpha(1)\beta(2)\alpha(1)\beta(2)-\alpha(1)\beta(2)\alpha(2)\beta(1)-\alpha(2)\beta(1)\alpha(1)\beta(2)+\alpha(2)\beta(1)\alpha(2)\beta(1))d\sigma_1 d\sigma_2\right)\\
&= \frac{1}{2}\begin{pmatrix}\int\alpha(1)\alpha(1)\,d\sigma_1\int\beta(2)\beta(2)\,d\sigma_2-\int\alpha(1)\beta(1)\,d\sigma_1\int\beta(2)\alpha(2)\,d\sigma_2\\ -\int\beta(1)\alpha(1)\,d\sigma_1\int\alpha(2)\beta(2)\,d\sigma_2+\int\beta(1)\beta(1)\,d\sigma_1\int\alpha(2)\alpha(2)\,d\sigma_2\end{pmatrix}\\
&= \frac{1}{2}(1\times1-0\times0-0\times0+1\times1)=1
\end{aligned}
\tag{3.36}
$$

where we have used the orthonormal properties of the spin functions. Here σ_1 and σ_2 represent the spin volume elements. Thus, this is a spin-allowed transition. It may be similarly shown that a transition from a singlet $\frac{1}{\sqrt{2}}(\alpha(1)\beta(2)-\alpha(2)\beta(1))$ to any of the triplet wave functions in equation (3.35) is forbidden. We thus have the spin selection rule $\Delta S = 0$.

Transitions to both the excited states arising from the $(1s)^1(2s)^1$ configuration are forbidden by the orbital selection rule, since there is no change in the l quantum number of either of the two electrons. Transition to $^3S_1$ is also forbidden by the spin selection rule.

As a second example, consider the configuration $(np)^1(n'p)^1$, e.g., $(2p)^1(3p)^1$ of two non-equivalent p electrons. Here, $l_1 = l_2 = 1$, and hence $L = 2$, 1 and 0 corresponding to D, P and S states. As in the previous example, S can have the values 1 and 0. Therefore, the possible terms of the $(np)^1(n'p)^1$ configuration are ^3D, ^1D, ^3P, ^1P, ^3S and ^1S. The total degeneracy of each term is given by $(2S + 1)(2L + 1)$, which is 3×5, 1×5, 3×3, 1×3, 3×1 and 1×1, respectively, for the six states, making a total of 36, i.e., 6×6, since there are six ways each electron can be placed in a p orbital. Applying Hund's rules, the ground state is ^3D. The total number of terms in this state is $(2S + 1)(2L + 1) = 3 \times 5 = 15$. The possible J values are $|L+S|...|L-S|$, i.e., 3, 2, 1. Since the degeneracy of each J value is equal to $2J + 1$, the total number of states adds up to $7 + 5 + 3 = 15$, as required. As the orbitals are less than half-filled, the ground state is $^3D_1$.

^{3}D terms
$L = 2, S = 1$

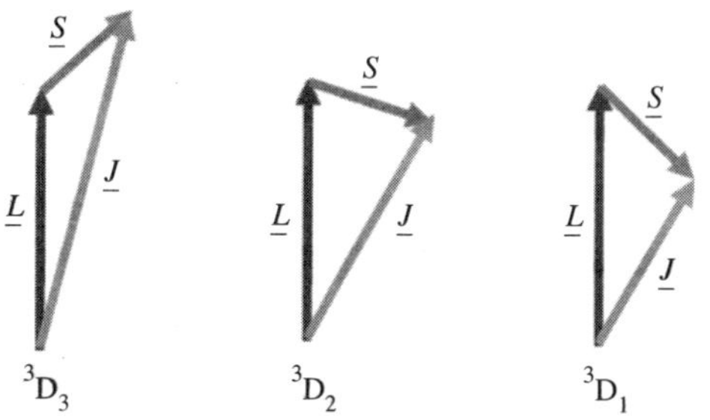

Example 3.3 Find all the states arising from a (2p)(3d) electron configuration.

Solution

The possible values of the S quantum number arise from a Clebsch–Gordan series.
$S = |s_1 + s_2|, |s_1 + s_2 - 1|, \ldots |s_1 - s_2|$ for two electrons, i.e., $S = 1, 0$

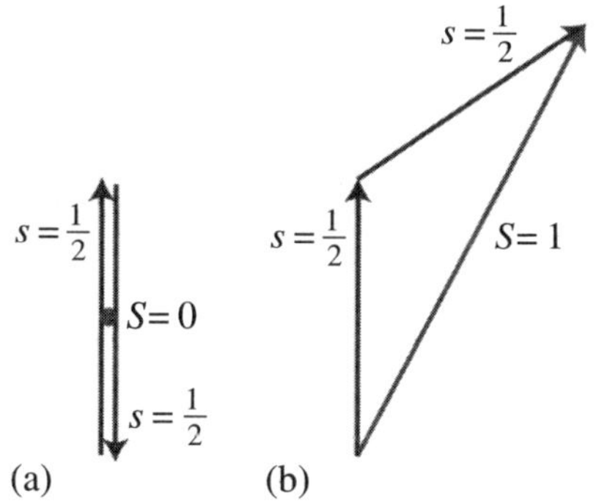

Coupling of two electron spins to give $S = 0, 1$

Similarly, the possible values of the L quantum number are $L = |l_1 + l_2|, |l_1 + l_2 - 1|, \ldots |l_1 - l_2|$ for two electrons, i.e., 3, 2, 1

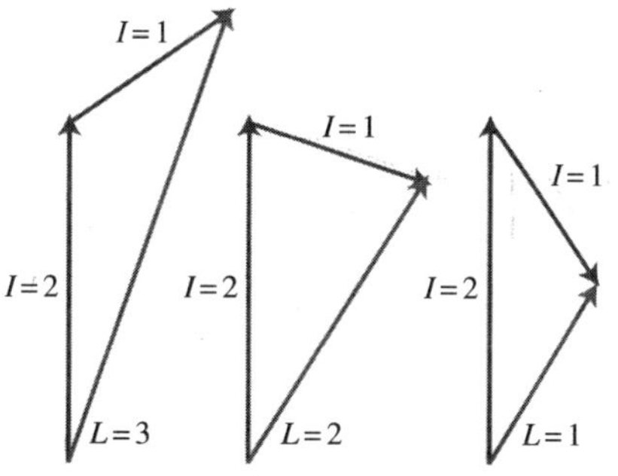

Coupling of orbital angular momenta for $l_1 = 1$, $l_2 = 2$. The possible states are therefore ^{3}D, ^{1}D, ^{3}P and ^{1}P.

As seen above, it is fairly simple to derive the possible states for non-equivalent electrons, but the situation becomes complicated for equivalent electrons. For the helium atom, we saw that when the two electrons are equivalent, not all the terms are possible, due to limitations imposed by the exclusion principle. Thus, the configuration $(ns)^2$ admits only the term ^{1}S, and for the $(np)^2$ configuration, the only possible terms are ^{1}S, ^{1}D and ^{3}P. We must therefore devise a procedure to weed out the terms not conforming to the Pauli exclusion principle in the case of equivalent electrons. Two methods are described below.

3.4.2.1 Term symbols by writing microstates

In the $(np)^2$ configuration, corresponding to $l = 1$, the quantum numbers of the electrons are (n, l, m, m_s) and (n, l, m', m_s'). To comply with Pauli's exclusion principle, the two sets of quantum numbers must differ either in m or m_s or both, since n and l are common. Since $l = 1$, the possible values for m and m' are +1, 0 and −1 and those for m_s are $+^1/_2$ and $-^1/_2$. We shall find the terms which conform to the exclusion principle by explicitly writing down the permitted microstates. To simplify the notation, the spin values will be designated as plus (+) or minus (−) and the common quantum numbers n and l will be omitted.

The number of possible combinations of quantum numbers is given by the combination $^6C_2$ or 15. The allowed values for (m, m_s, m', m_s') are arranged in their common values of $M_L = m + m'$ and $M_S = m_s + m_s'$ in Table 3.1.

Table 3.1. The allowed values for (m, m_s, m', m_s')

$\sum m$ \ $\sum m_s$	+1	0	−1
+2		$(\overset{+}{1},\overset{-}{1})$	
+1	$(\overset{+}{1},\overset{+}{0})$	$(\overset{+}{1},\overset{-}{0})$ $(\overset{-}{1},\overset{-}{0})$	$(\overset{-}{1},\overset{-}{0})$
0	$(\overset{+}{1},\overset{+}{-1})$	$(\overset{+}{1},\overset{-}{-1})$ $(\overset{-}{1},\overset{+}{-1})$ $(\overset{+}{0},\overset{-}{0})$	$(\overset{-}{1},\overset{-}{-1})$
−1	$(\overset{+}{0},\overset{+}{-1})$	$(\overset{+}{0},\overset{-}{-1})$ $(\overset{-}{0},\overset{+}{-1})$	$(\overset{-}{0},\overset{-}{-1})$
−2		$(\overset{+}{-1},\overset{-}{-1})$	

The configurations are analyzed according to a method developed by J. C. Slater in 1929. Figure 3.7(a) shows the number of different wave functions corresponding to each value of M_L and M_S.

M_L
↑

$$\begin{pmatrix} & \boxed{1} & \\ \boxed{1}\ \boxed{2}\ \boxed{1} \\ \boxed{1}\ \boxed{3}\ \boxed{1} \\ \boxed{1}\ \boxed{2}\ \boxed{1} \\ & \boxed{1} & \end{pmatrix} - \begin{pmatrix} \boxed{1} \\ 1 \\ 1 \\ 1 \\ \boxed{1} \end{pmatrix} = \begin{pmatrix} \boxed{1}\ \boxed{1}\ \boxed{1} \\ \boxed{1}\ \boxed{2}\ \boxed{1} \\ \boxed{1}\ \boxed{1}\ \boxed{1} \end{pmatrix} \begin{pmatrix} \boxed{1}\ \boxed{1}\ \boxed{1} \\ \boxed{1}\ \boxed{1}\ \boxed{1} \\ \boxed{1}\ \boxed{1}\ \boxed{1} \end{pmatrix} = (1) \to M_S$$

(a)　　(b)　　(c)　　　(d)　　　(e)

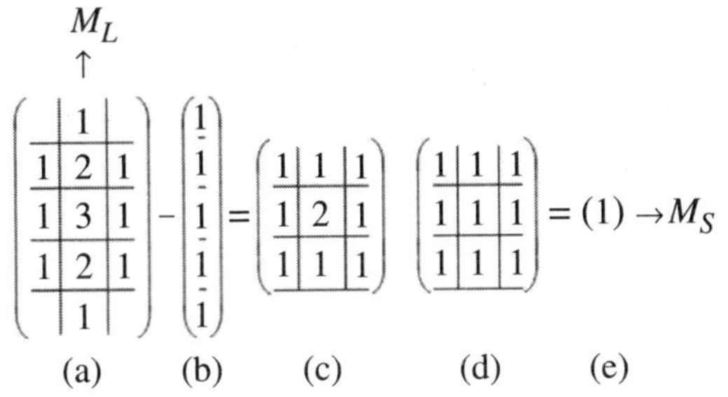

Figure 3.7. (a) Two equivalent p electrons; (b) ^1D; (c) remaining terms; (d) ^3P; (e) ^1S

From Figure 3.7(a), it is clear that the largest value of $M_L = \sum m = 2$ and this must correspond to $L = 2$ or a D term. This term is associated only with $M_S = \sum m_s = 0$, and therefore it must correspond to $S = 0$, making up a ^1D term. This state accounts for five wave functions with $M_L = 2, 1, 0, -1, -2$ and $M_S = 0$. These five wave functions are depicted in Figure 3.7(b). Subtracting these from (a) leaves (c). In (c), the largest value of $M_L = 1$, corresponding to a P term. Since $M_L = 1$ is associated with $M_S = +1, 0, -1$, it corresponds to ^3P, which has a degeneracy of $(2S + 1)(2L + 1) = 9$ terms having $M_L = +1, 0, -1$ and $M_S = +1, 0, -1$. Subtraction of (d) from (c) leaves (e), which is a wave function with $M_L = 0$, $M_S = 0$, which must correspond to a term with $L = 0$ and $S = 0$ or ^1S. Thus, the only possible terms of the $(np)^2$ configuration are ^1S, ^1D and ^3P. The substates are $^1S_0$, $^1D_2$, $^3P_2$, $^3P_1$ and $^3P_0$. The total degeneracy is $1 \times 1 + 1 \times 5 + 3 \times 3 = 15$ ($^6C_2$). Hence, from an initial 36 states, we are left with only 15 which satisfy Pauli's exclusion principle. Note that the $L + S$ values for the three allowed states are 0, 2 and 2, respectively. In fact, it is possible to show that only the states for which $L + S$ is even satisfy the Pauli exclusion principle.

Applying Hund's rules to the carbon atom $(2p)^2$, we obtain the following states in order of increasing energy: $^3P_0 < {}^3P_1 < {}^3P_2 < {}^1D_2 < {}^1S_0$. A useful approximation is that an 'electron hole' behaves like an electron for constructing spectroscopic terms, so $(p)^5$ has the same term as $(p)^1$, $(p)^4$ the same as $(p)^2$ and so on. Also, for half-filled orbitals, $L = 0$ and $J = S$, and so there is only one value of J. Thus, for the $(p)^3$ configuration, there is only one state, $^4S_{3/2}$. Oxygen with $(2p)^4$ has the same states as carbon, but the relative order of the states with different J is opposite, according to Hund's third rule because the p orbital is more than half-filled in oxygen.

The writing of all microstates and tabulating the M_L and M_S values is tedious but instructive. It helps to convince us that a term does not correspond to one microstate, but to an array of microstates. Even ^1S, with $M_L = 0$, $M_S = 0$, does not correspond to a unique microstate, but to an array of microstates belonging only to this term. For $(p)^2$, there are three microstates giving this combination. However, the writing down of microstates becomes much more tedious for $(d)^2$. Fortunately, there is another method, spin factoring, which can be applied.

3.4.2.2 Spectral terms by spin factoring

In the spin-factoring method, we obtain the 'partial terms' for each of the spin sets (the electrons with all $m_s = +\frac{1}{2}$ or $-\frac{1}{2}$ form a spin set). The partial terms of the spin sets (designated

α and β) are multiplied to generate the complete terms. Since an empty, a half-filled, or a filled complete set of orbitals contributes nothing to the orbital angular momentum, corresponding to an S term, an empty spin set or a complete spin set (half-filled orbitals) also gives an S partial term. Thus, $(s)^0$ (empty) or $(s)^1$ (completely filled because there can be only one α electron in an s orbital) both give an S partial term.

One electron in an orbital gives a term with the corresponding label, $(s)^1 \rightarrow$ S, $(p)^1 \rightarrow$ P, $(d)^1 \rightarrow$ D, $(f)^1 \rightarrow$ F, etc., so a spin set (α or β) consisting of one electron gives these same partial terms. For example, for $(p)^1$, there are three possible ways the electron can be placed in the three orbitals:

m	1	0	−1	M_L
	↑			1
		↑		0
			↑	−1

The three M_L values constitute a P term. One vacancy in a spin set (one hole) gives the same partial term as for one electron, P for $(p_\alpha)^1$ or $(p_\alpha)^2$, D for $(d_\alpha)^1$ or $(d_\alpha)^4$, F for $(f_\alpha)^1$ or $(f_\alpha)^6$, and so on. The total number of α electrons possible in a p orbital is three, so $(p_\alpha)^3$ also constitutes an S term.

If we write the microstates for the spin set $(d_\alpha)^2$, we get 10 microstates.

Microstate		$M_L = \Sigma m$	**No. of terms**
(2, 1)		3	1
(2, 0)		2	1
(2, −1)	(1, 0)	1	2
(2, −2)	(1, −1)	0	2
(1, −2)	(0, −1)	−1	2
(0, −2)		−2	1
(−1, −2)		−3	1

Consider the first column. We place the first electron in $m = 2$. The second electron cannot have $m = 2$ because we cannot place two α electrons in the same orbital, but it can be in any one of the other m states 1, 0, −1, or −2, giving total M_L values of 3, 2, 1 or 0. Similarly, if the first electron is placed in $m = 1$, the second electron can be in any of the states $m = 0, -1$ or -2 (we have already considered the possibility $m = 2$ and the (2, 1) and (1, 2) states are identical because electrons are indistinguishable). The highest value of M_L is 3, constituting an F term ($L = 3$). An F term has seven components corresponding to $M_L = 3, 2, 1, 0, -1, -2$ and -3. This leaves us with three microstates having M_L values 1, 0 and −1, corresponding to a P state. Thus, the partial terms for two d electrons of one spin set are F and P. The $(d_\alpha)^3$ configuration is the hole counterpart of $(d_\alpha)^2$ and generates the same terms. The partial terms for $(f)^2$, $(f)^3$, etc., can be generated in the same way (Table 3.2). We need only the terms to the left of the solid line since those to the right are obtained by the hole formalism. The partial terms for various numbers of electrons in a spin set are given for orbitals up to f in Table 3.2.

Once we have the partial terms in Table 3.2, it is fairly easy to obtain the states for any configuration. As a first example, let us use the partial terms to derive the spectral terms for the

Table 3.2. Partial terms arising from the occupancy of a single spin set (α or β)

Orbital set	Orbital occupancy (electrons or holes)				
	0	1	2	3	4
s	S	S			
p	S	P	P	S	
d	S	D	PF	PF	D
f	S	F	PFH	SDFGI	SDFGI

$(p)^2$ configuration. The possible spin configurations are $(p_\alpha)^1(p_\beta)^1$, $(p_\alpha)^2(p_\beta)^0$ and $(p_\alpha)^0(p_\beta)^2$. We obtain the partial terms for each spin configuration from the table and take the *product* of the partial terms. The product of partial terms with L values L_1 and L_2 includes all integral L values from $|L_1 + L_2|$ through $|L_1 - L_2|$. For example, the electron configuration $(p_\alpha)^1(p_\beta)^1$ gives all integral L values from $(1+1)$ to $(1-1)$, i.e., 2, 1 and 0, or D, P and S states. For $(p)^2$, we have

Configuration	Degeneracy	Product	Terms	Spin	
$(p_\alpha)^2(p_\beta)^0$	${}^3C_2 \times 1 = 3$	$P \times S \rightarrow (L = 1 + 0 = 1)$	P	$M_S = +1$	^3P
$(p_\alpha)^0(p_\beta)^2$	$1 \times {}^3C_2 = 3$	$S \times P \rightarrow (L = 0 + 1 = 1)$	P	$M_S = -1$	^3P
			D		^1D
$(p_\alpha)^1(p_\beta)^1$	${}^3C_1 \times {}^3C_1 = 9$	$P \times P \rightarrow (L = \lvert 1+1 \rvert, \ldots, \lvert 1-1 \rvert = 2, 1, 0)$	P	$M_S = 0$	^3P
			S		^1S

In the first line, we observe a P state with $M_S = 1$. It must be a component of a ^3P state. The other M_S values are -1 and 0, and we eliminate these. The remaining terms are a D and an S state, both with $M_S = 0$. Thus, these are ^1D and ^1S states. We can check the total degeneracy of these terms $(3 \times 3 + 1 \times 1 + 1 \times 5)$ against the 15 microstates.

Similarly, for $(d)^2$, for the first configuration $(d_\alpha)^2(d_\beta)^0$, the term $(d_\alpha)^2$ corresponds to P and F (Table 3.2) and $(d_\beta)^0$ to S. The product $(P + F) \times S = P + F$ and the spin multiplicity is 3 since $M_S = 1$. Therefore, there are two states ^3P and ^3F. We eliminate the other components corresponding to $M_S = 0$ and -1 for these terms from the $(d_\alpha)^1(d_\beta)^1$ and $(d_\alpha)^0(d_\beta)^2$ configurations, leaving ^1G, ^1D and ^1S terms.

Configuration	Degeneracy	Product	Terms	Spin	
$(d_\alpha)^2(d_\beta)^0$	${}^5C_2 \times 1 = 10$	$(P + F) \times S \rightarrow L = 1, 3$	P	$M_S = +1$	^3P
			F		^3F
$(d_\alpha)^0(d_\beta)^2$	$1 \times {}^5C_2 = 10$	$S \times (P + F) \rightarrow L = 1, 3$	P	$M_S = -1$	^3P
			F		^3F
$(d_\alpha)^1(d_\beta)^1$	${}^5C_1 \times {}^5C_1 = 25$	$D \times D \rightarrow$	G		^1G
		$(L = \lvert 2+2 \rvert, \ldots \lvert 2-2 \rvert = 4, 3,$	F		^3F
		$2, 1, 0$	D	$M_S = 0$	^1D
			P		^3P
			S		^1S

Thus, we have ^3P, ^3F, ^1S, ^1D and ^1G terms arising from a $(d)^2$ configuration. According to Hund's rules, the ground state is ^3F. You can check that the total degeneracy of these terms is 45. Similarly, the terms corresponding to the other $(d)^n$ configurations can be obtained, as given in Table 3.3.

Table 3.3. Terms for $(3d)^n$ free ion configurations

Configuration	No. of quantum states	No. of energy levels	Ground term	Excited terms
$(d)^1$, $(d)^9$	10	1	^{2}**D**	-
$(d)^2$, $(d)^8$	45	5	^{3}**F**	^3P, ^1G, ^1D, ^1S
$(d)^3$, $(d)^7$	120	8	^{4}**F**	^4P, ^2H, ^2G, ^2F, $2 \times {}^2D$, ^2P
$(d)^4$, $(d)^6$	210	16	^{5}**D**	^3H, ^3G, $2 \times {}^3F$, ^3D, $2 \times {}^3P$, ^1I, $2 \times {}^1G$, ^1F, $2 \times {}^1D$, $2 \times {}^1S$
$(d)^5$	252	16	^{6}**S**	^4G, ^4F, ^4D, ^4P, ^2I, ^2H, $2 \times {}^2G$, $2 \times {}^2F$, $3 \times {}^2D$, ^2P, ^2S

3.4.3 Ground state terms

If only the ground state is required, the procedure is fairly simple. We begin with an example.

Example 3.4 Without deriving all spectral terms, obtain the ground state levels for Cr^{2+} $(d)^4$, Mn^{2+} $(d)^5$ and Ni^{2+} $(d)^8$.

Solution

A simple graphical method for determining just the ground term alone for free ions uses a 'fill in the boxes' arrangement. The microstate with maximum M_S and M_L must be one of the microstates of the array for the ground state term having maximum S, and for this S, maximum L. This can be achieved by filling the orbitals individually, starting with the highest m, and adding one electron to each orbital before any is doubly occupied, in conformity with Hund's rules. To calculate S, simply sum the *unpaired* electrons using a value of ½ for each. To calculate L, use the labels for each column to determine the value of m for that box; then add all the individual box values together to get M_L as the absolute value. Only unpaired electrons need to be considered.

Configuration	Configuration with maximum M_S and M_L (for maximum M_S)					M_L	M_S	Ground state term	Ground state level
	m = 2	1	0	−1	−2				
Cr^{2+} $(d)^4$	↑	↑	↑	↑		2	2	^5D	$^5D_0$
Mn^{2+} $(d)^5$	↑	↑	↑	↑	↑	0	$5/2$	^6S	$^6S_{5/2}$
Ni^{2+} $(d)^8$	↑↓	↑↓	↑↓	↑	↑	3	1	^3F	$^3F_4$

The ground state terms arising from different number of unpaired d electrons are summarized in Table 3.3. Note that $(d)^n$ gives the same terms as $(d)^{10-n}$. Note that for 5 electrons with 1 electron in each box, the total value of M_L is zero, and hence L is zero. This is why L for a $(d)^1$ configuration is the same as that for $(d)^6$. The other thing to note is the idea of the 'hole' approach. The overall result shown in Table 3.3 is: four configurations $\{(d)^1, (d)^4, (d)^6, (d)^9\}$ give rise to D ground terms, four configurations $\{(d)^2, (d)^3, (d)^7, (d)^8\}$ give rise to F ground terms and the $(d)^5$ configuration (half-filled) gives an S ground term.

3.4.4 Quantitative treatment of the coupling of orbital and spin angular momenta for multi-electron atoms

Though the above treatment of spin–orbit coupling gives satisfactory agreement of the calculated magnetic moments with the experimentally determined ones for most atoms, the agreement is not so good for the heavier atoms. In our previous discussion for the helium atom, we had not considered the spin–orbit interaction in the Hamiltonian (3.30). However, for any meaningful treatment of multi-electron atoms, this term has to be taken into account. Adding this term to the Hamiltonian gives

$$\hat{H} = \hat{H}^0 + \hat{H}_{ee} + \hat{H}_{s-o} \tag{3.37}$$

where the respective terms are the zero-order term, the electron–electron repulsion term and the spin–orbit interaction term. The last two terms are treated using perturbation theory. We had seen that the squares of the orbital angular momentum ($\hat{L}^2$), spin angular momentum ($\hat{S}^2$) and total angular momentum ($\hat{J}^2$) operators commute with $\hat{H}^0 + \hat{H}_{ee}$, but this is no longer the case when the spin–orbit interaction term is added. Further treatment therefore depends on the relative magnitudes of the last two terms. The cases we have considered so far are those for which $\hat{H}_{s-o} << \hat{H}_{ee}$, and hence the former was ignored. In such cases, the spin–orbit coupling scheme is referred to as *LS* or Russell–Saunders coupling and is generally applicable to atoms with atomic numbers less than 30. In case the reverse is true, another kind of coupling scheme, known as *jj* coupling, is applicable. We first consider the case where the coulombic term $\hat{H}_{ee}$ is dominant and the spin–orbit interaction is small, i.e., *LS* or Russell–Saunders coupling prevails, as in the cases discussed above.

3.4.4.1 Russell–Saunders or *LS* coupling

In this case, since $\hat{H}_{s-o} << \hat{H}_{ee}$, we first solve the Schrödinger equation for $\hat{H}^0 + \hat{H}_{ee}$ as before and then consider $\hat{H}_{s-o}$ as a perturbation. A relativistic treatment of the electron yields the spin–orbit interaction term

$$H_{s-o} = A\vec{L}.\vec{S} \tag{3.38}$$

where A is a constant referred to as the spin–orbit coupling constant. Note that the spin–orbit interaction is proportional to the dot product $\vec{L}.\vec{S}$. Since $\vec{J} = \vec{L} + \vec{S}$, $\vec{J}^2$ is given by

$$\left|\vec{J}\right|^2 = \left|\vec{L} + \vec{S}\right|.\left|\vec{L} + \vec{S}\right| = \vec{L}^2 + \vec{S}^2 + 2\vec{L}.\vec{S} \tag{3.39}$$

and therefore

$$\vec{L}.\vec{S} = \frac{1}{2}\left[\vec{J}^2 - \vec{L}^2 - \vec{S}^2\right] \tag{3.40}$$

The L, S and J remain good quantum numbers since the spin–orbit coupling introduces only a small perturbation and the atoms remain spherically symmetrical. The spin–orbit coupling Hamiltonian can now be expressed in terms of the 'good' quantum numbers of LS spin–orbit coupling

$$\hat{H}_{s-o} = A\hat{L}.\hat{S} = \frac{A}{2}\left[\hat{J}^2 - \hat{L}^2 - \hat{S}^2\right] \tag{3.41}$$

and the energy correction is given by

$$\Delta E_{s-o} = \frac{A}{2}\hbar^2\left[J(J+1) - L(L+1) - S(S+1)\right] \tag{3.42}$$

The spin–orbit interaction, therefore, causes splitting of spectroscopic lines in atomic spectra, and the total energy of a state depends on the value of the total angular momentum quantum number J.

The energy separation between energy levels of a multiplet also depends on J:

$$\begin{aligned}\Delta E_{J\to J+1} &= \frac{A}{2}\hbar^2\left[(J+1)(J+2) - L(L+1) - S(S+1) - (J(J+1) - L(L+1) - S(S+1))\right]\\ &= A\hbar^2(J+1)\end{aligned} \tag{3.43}$$

This relationship is called the Landé interval rule: The separation between two adjacent levels within a term is proportional to the larger of the two J-values involved. This rule can be used to determine the value of J.

Example 3.5 Use the result from first-order perturbation theory to determine the energy of the fine structure levels in the ^{3}P-state from a (p)^2-configuration.

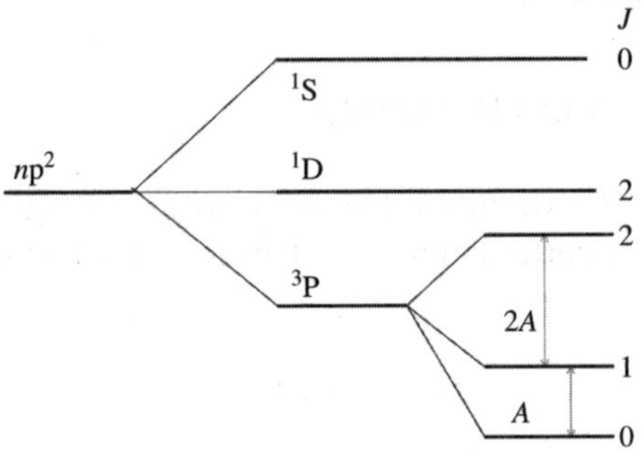

Solution

We have the following values of the quantum numbers: $L = 1$, $S = 1$, $J = 0, 1, 2$. Inserting L and S in equation (3.42) gives the formula

$$\Delta E_{s-o} = A(^3\text{P})\frac{\hbar^2}{2}[J(J+1)-2-2]$$

with the following energies:

J	$\Delta E_{s-o}/\hbar^2$
2	A
1	$-A$
0	$-2A$

The energy gap between adjacent J levels is the order of 50–800 cm^{-1}, much less than the spacing between the main spectroscopic terms, which is the order of 10,000 cm^{-1}. For carbon, the separation between the $^3P_0$ and $^3P_1$ terms is 16.4 cm^{-1}, and that between $^3P_0$ and $^3P_2$ is 43.5 cm^{-1} (Figure 3.8). According to the Landé interval rule, the separation between $^3P_2$ and $^3P_1$, and $^3P_1$ and $^3P_0$ should be in the ratio 2:1, which is only approximately obeyed (1.65:1).

The spin–orbit coupling constant is different for different terms and can be of either sign. It follows that $A > 0$ for less than half-filled subshells, while $A < 0$ for more than half-filled subshells, according to Hund's third rule. For example, for oxygen, which has a (p)^4 electron configuration, and hence the same ^3P ground state as carbon, the spacing between the $^3P_2$ and $^3P_1$ terms is 158 cm^{-1}, and that between $^3P_2$ and $^3P_0$ is 227 cm^{-1}. The ratio of the spacings between $^3P_2$ and $^3P_1$, and $^3P_1$ and $^3P_0$ is then 158/69 (= 2.29), which is close to the expected

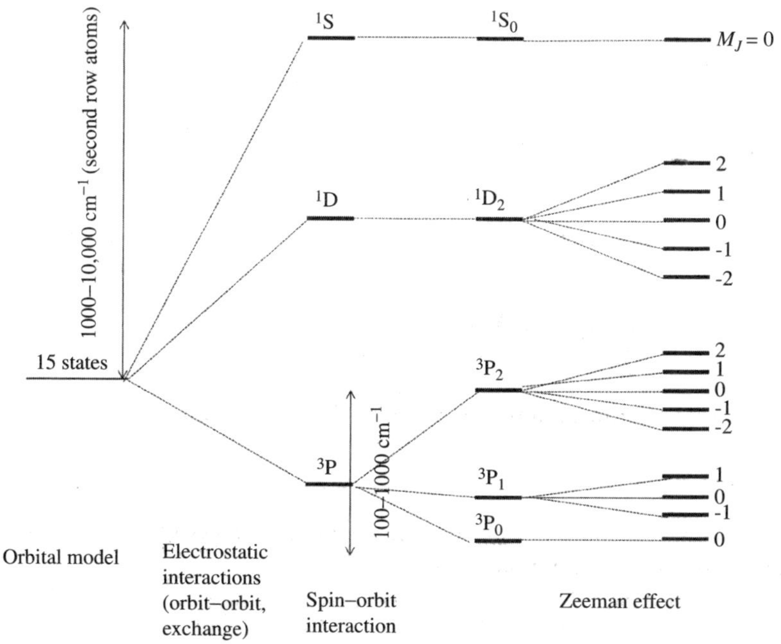

Figure 3.8. Schematic energy level structure of the (2p)^2 configuration in LS coupling

value of 2.0. Note that the amount of splitting increases as the number of electrons (and hence the electron repulsion) increases, and that the energy order is reversed, i.e., $^3P_2$ is lowest in energy since the spin–orbit coupling constant is negative for more than half-filled states.

Figure 3.8 shows schematically the interactions by which the 15 states of the ground state configuration of carbon are split. The strong electrostatic interactions (including exchange) lead to a splitting into three terms ^3P, ^1D and ^1S. The weaker spin–orbit interaction splits the ground ^3P term into three components, $^3P_0$, $^3P_1$ and $^3P_2$. Each term component can be further split into the $(2J + 1)$ M_J levels by an external magnetic field, an effect known as the Zeeman effect (Figure 3.8), which we shall discuss later.

In summary, if only the $\hat{H}^0$ term is retained in the atomic Hamiltonian operator, the terms in a given configuration are degenerate (Figure 3.8). If the electron–electron Coulomb interaction term $\hat{H}_{ee}$ is added, the different terms get different energies due to the stronger interelectronic repulsion in the singlet state. Finally, when the spin–orbit interaction $\hat{H}_{s-o}$ is also included, the degeneracy of the levels in the terms is lifted.

3.4.4.2 Term symbols for Russell–Saunders coupling

As stated above, the Russell–Saunders coupling scheme applies when there is strong coupling among the orbital angular momenta (and the spin angular momenta) themselves and weak spin–orbit coupling. It is a good approximation for light atoms. Given the electron configuration, the number of terms is determined by all possible combinations of L and S obtained when adding the individual l_i and s_i, under consideration of the Pauli principle.

Spin–spin coupling

The overall spin S arises from adding the individual spins together and is a result of coupling of spin quantum numbers for the separate electrons.

$$s_1 + s_2 + \ldots = \sum_i s_i = S$$

Orbit–orbit coupling

The orbital angular momenta of all electrons, l_i, couple similarly to yield a total orbital angular momentum L.

$$l_1 + l_2 + \ldots = \sum_i l_i = L$$

The total orbital angular momentum quantum number L defines the energy state for a system of electrons.

Spin–orbit coupling

Coupling occurs between the resultant spin and orbital momenta of an electron, which gives rise to J, the total angular momentum quantum number. We write

$$\vec{J} = \vec{L} + \vec{S}$$

The possible values of the quantum numbers are

$$L = (l_1 + l_2 + ...), (l_1 + l_2 + ... - 1), ..., 0$$
$$S = (s_1 + s_2 + ...), (s_1 + s_2 + ... - 1), ..., 0 \text{ or } \tfrac{1}{2}$$

and J is given by the Clebsch–Gordan series

$$J = |L + S|, |L + S - 1|, ..., |L - S|$$

where L and S are their maximum values $\left(\sum_i l_i \text{ and } \sum_i s_i\right)$. The Russell–Saunders term symbol that results from these considerations is given by $^{2S+1}L_J$. It is implicit in the use of this notation that, since spin–orbit coupling is small, the energy of a state depends mostly on the values of S and L, the different values of J merely provide a 'fine structure'.

The total angular momentum J of an atom is very important because it determines (among other things) the magnetic properties of the atom and the transition probabilities in radiative processes. Like any other angular momentum, the total angular momentum obeys the commutation rules given for the orbital angular momentum (3.12).

The z-component of J is determined by the quantum number M_J such that

$$J_z = M_J \hbar$$

with

$$M_J = \pm|J|, \pm|J - 1|, ...$$

For each electronic configuration, there are several possible corresponding values of J, each associated with a different energy of the atom.

3.4.4.3 Selection rules for Russell–Saunders coupling

As for any other kind of spectroscopy, not all transitions are allowed. The selection rules are given as follows:

1. Transitions involve change of n and l quantum numbers of one electron. Transitions between states that require the change in quantum numbers of more than one electron are forbidden. This is the *Laporte* selection rule.
2. The change in l of the electron involved in the transition satisfies the following relation: $\Delta l = \pm 1$; $\Delta m = 0, \pm 1$
3. The *parity* must change. The parity of a configuration is *even* or *odd* according to whether $\sum_i l_i$ is even or odd, the sum being taken over all electrons (in practice only those in open subshells need be considered).
   The parity of a term can also be specified in the term symbol by adding an 'o' as upper right superscript if the term parity is odd, e.g., $^3P^o$. In practice, parity can be simply determined from the electron configuration by $(-1)^{\sum_i l_i}$ where the sum is taken over the quantum numbers l_i (note that it is not a vector sum over the $\vec{l}_i$). A minus sign refers to

odd parity, a plus sign to even parity. Parity is not always indicated in the term symbol if the situation is clear, although it actually plays an important role for the selection rules.

4. $\Delta J = 0, \pm 1$ ($0 \leftrightarrow 0$ not allowed) and $\Delta M_J = 0, \pm 1$ ($\Delta M_J = 0$ not allowed if ΔJ is also 0) The possibility of $J = 0 \rightarrow J' = 0$ is forbidden by the law of conservation of angular momentum, since it can be shown that a photon has an intrinsic angular momentum of one unit. Hence, it is impossible for there to be a transition $J = 0 \rightarrow J' = 0$ with emission (or absorption) of a photon carrying one unit of angular momentum. Also $\Delta M_J = 0$ is not possible with $\Delta J = 0$, because in this case again the conservation of angular momentum requires a change in J_z with a corresponding change in M_J.

Atomic states are always eigenstates of parity and $\hat{J}^2$, so the selection rules can be regarded as absolutely valid in electric dipole transitions. These are the rigorous selection rules applicable to all coupling schemes.

In specific coupling schemes, further selection rules apply. In the case of ideal *LS* coupling, we also require:

1. $\Delta S = 0$ and $\Delta M_S = 0$ (frequently violated if spin–orbit coupling is large). This rule follows from conservation of the total spin in a transition.
2. $\Delta L = 0, \pm 1$ (but $0 \leftrightarrow 0$ is not allowed) and $\Delta M_L = 0, \pm 1$. This is a consequence of the conservation of angular momenta.
3. $\Delta l_i = \pm 1$ if only the *i*th electron is involved in transition. This rule follows from the parity change rule since the parity of an atom is the product of parities of separate electron wave functions, $(-1)^{l_i}$.

However, since *LS* coupling is only an approximation, these rules are themselves approximate.

As the atomic number increases, the electron becomes more relativistic, and spin–orbit coupling of each electron takes place in preference to spin–spin and orbit–orbit coupling with other electrons. This scheme is referred to as *jj* coupling.

3.4.4.4 *jj* Coupling

The other coupling scheme is *jj* coupling, which is opposite to Russell–Saunders coupling, i.e., there is strong coupling between each electron's orbital and spin angular momenta. In this case ($\hat{H}_{s-o} > \hat{H}_{ee}$), we first solve the Schrödinger equation for $\hat{H}^0 + \hat{H}_{s-o}$ and then consider $\hat{H}_{ee}$ as a perturbation. The configurations of $\hat{H}^0$ are split by $\hat{H}^0 + \hat{H}_{s-o}$ into terms with fixed combinations of $nl_i j_i$. The perturbation caused by $\hat{H}_{ee}$ lifts the degeneracy in J of these terms. We write

$$\vec{j}_1 = \vec{l}_1 + \vec{s}_1$$
$$\vec{j}_2 = \vec{l}_2 + \vec{s}_2$$
$$\vec{j}_1 + \vec{j}_2 + \ldots = \sum_i \vec{j}_i = \vec{J}$$

For a given electron configuration, the term is then defined by the set of j_i for all electrons in unfilled subshells. Since the only possible value of s is ½, each electron can have values $j_i = l_i \pm ½$. The term symbol is written as $(j_1\, j_2)$, and L and S are not defined, as they are no longer 'good' quantum numbers. For example, Pb has an electron configuration $(6p)^2$ (other subshells filled) and is well described by *jj* coupling. Each electron can have j values of $1 \pm ½$ since $l = 1$ for a p electron. Thus, for $(6p)^2$ we get three terms $\left(\frac{1}{2}\,\frac{1}{2}\right)$, $\left(\frac{1}{2}\,\frac{3}{2}\right)$ and $\left(\frac{3}{2}\,\frac{3}{2}\right)$. Each term consists of different levels specified by the different possible values of J resulting from the vector summation of the $\vec{j}_i$. For example, the possible values of J for $\left(\frac{1}{2}\,\frac{1}{2}\right)$ are $\left|\frac{1}{2}+\frac{1}{2}\right|\ldots\left|\frac{1}{2}-\frac{1}{2}\right| = 1, 0$ and the possible terms are $\left(\frac{1}{2}\,\frac{1}{2}\right)_{1,0}$, $\left(\frac{1}{2}\,\frac{3}{2}\right)_{2,1}$ and $\left(\frac{3}{2}\,\frac{3}{2}\right)_{3,2,1,0}$, where the subscript stands for J. As for Russell–Saunders terms, the levels have to be consistent with the Pauli principle so that we end up with the levels $\left(\frac{1}{2}\,\frac{1}{2}\right)_0$, $\left(\frac{1}{2}\,\frac{3}{2}\right)_{2,1}$, $\left(\frac{3}{2}\,\frac{3}{2}\right)_{2,0}$, the others being forbidden. The ground state of Pb is the level $\left(\frac{1}{2}\,\frac{1}{2}\right)_0$.

This coupling applies mainly to heavy atoms beyond the alkaline earth metals. In the case of many transition metals, the coupling is neither purely *LS* nor purely *jj*, but is intermediate between the two. Usually, the coupling is closer to one of these two model schemes, so we take that as the model for the transition metal and the other effect as a perturbation.

3.4.5 Intermediate coupling

If $\hat{H}_{ee} \cong \hat{H}_{s-o}$, neither *LS* coupling nor *jj* coupling are good approximations and we are in the intermediate coupling regime. In this case, we have to treat $\hat{H}_{ee} + \hat{H}_{s-o}$ as a perturbation. A good test for the applicability of *LS* coupling is Landé's interval rule (3.43). For example, in the first excited term of Mg-I, a $^3P_0$ term, the separation between the $J = 2$ and $J = 1$ levels is 4.07 mm^{-1}, while the separation between $J = 1$ and $J = 0$ is 2.01 mm^{-1}. The interval rule is approximately satisfied, showing that the term conforms closely to *LS*.

We have already stated that every J level has a degeneracy of $2J + 1$, as determined by the possible values of M_J, which is lifted in the presence of an external field. For the ground state of the carbon atom, for example, the $^3P_2$ term is 5-fold degenerate, $^3P_1$ term 3-fold degenerate and the $^3P_0$ term 1-fold degenerate. It can be seen that $5 + 3 + 1 = 9$, the total degeneracy of the ^3P term $[(2S + 1)(2L + 1)]$. This final degeneracy is removed in the presence of a magnetic field, the *Zeeman effect*, or electric field (*Stark effect*), which gives 15 different energy levels for the $(p)^2$ configuration (Figure 3.8).

3.4.6 Interaction with external fields

3.4.6.1 The Zeeman effect

The Zeeman effect is the name given to the splitting of spectral lines in an external magnetic field. Pieter Zeeman observed a line triplet instead of a single spectral line at right angles to a magnetic field and a line doublet parallel to the field. Later, when more complex splittings were observed, they were called *anomalous Zeeman splittings*, and Zeeman's observations became known as the *normal Zeeman effect*. Later findings established that the so-called anomalous Zeeman effect is actually the rule and the normal Zeeman effect the exception.

On application of an external magnetic field, the magnetic moment of the atom interacts with the magnetic field, and a new perturbation term, $\hat{H}' = -\vec{\mu}_J . \vec{B}$, is introduced. Here $\vec{\mu}_J$ is the magnetic moment of the atom and $\vec{B}$ the strength of the applied magnetic field (or, rather, the magnetic flux density). The average magnetic moment of an atom is proportional to $\vec{J}(\vec{\mu}_J \propto |\vec{J}|)$, which means that a substance composed of single atoms will be paramagnetic if $J \neq 0$ and diamagnetic if $J = 0$.

The *normal* Zeeman effect is observed only for transitions between atomic states with total spin $S = 0$. The total angular momentum $\vec{J} = \vec{L} + \vec{S}$ of the state is then a pure orbital angular momentum ($J = L$). For the corresponding magnetic moment, $\vec{\mu}_L$ we can simply say that

$$\vec{\mu}_L = -\frac{e}{2m_e}\left|\vec{L}\right| = -\frac{e\hbar}{2m_e}\sqrt{L(L+1)} = -\mu_B\sqrt{L(L+1)} \tag{3.44}$$

where μ_B is a constant called the Bohr magneton, which is equal to $9.2740154 \times 10^{-24}$ J T^{-1}.

In the presence of an external magnetic field, the energy levels of a state are split into $2L + 1$ sublevels, corresponding to the different possible orientations of $\vec{L}$ with respect to the magnetic field. This adds a perturbation term $-\hat{\mu}_L.\hat{B}$ to the Hamiltonian. The effective energy of this magnetic interaction E_{int} is given by

$$E_{int} = -\vec{\mu}_L.\vec{B} \tag{3.45}$$

The component of the orbital magnetic moment in the direction of the magnetic field μ_z is given by the magnetic quantum number

$$\mu_z = -\frac{e\hbar}{2m_e}M_L = -\mu_B M_L \tag{3.46}$$

where M_L is the component of L in the direction of the field (z).

Thus

$$E_{int} = \mu_B B M_L \tag{3.47}$$

which implies that more negative M_L values are lower in energy.

Let us consider a specific transition of Cd in the absence and presence of a magnetic field. The electronic configuration of Cd in the ground state is [Kr]$(4d)^{10}(5s)^2$, which is similar to He, and hence the ground state is a singlet. In the emission spectrum (as in a Cd lamp), four lines are observed in the visible: violet (467.9 nm), blue (480.1 nm), green (508.7 nm) and red (643.8 nm). We shall specifically consider the last (red) 5d $\rightarrow$ 5p ($^1D_2 \rightarrow {}^1P_1$) transition corresponding to a frequency of 465.7 THz (Figure 3.9). Since the splittings of the two levels are identical, only three lines will be observed, corresponding to $\Delta M_L = 0, \pm 1$, according to the selection rules.

Applying equation (3.47), the separation between adjacent levels is $\mu_B B$. The lines corresponding to $\Delta M_L = \pm 1$ are labelled σ lines and are linearly polarized perpendicular to the field when viewed at right angles to it, while those corresponding to $\Delta M_L = 0$ are labelled π lines and are plane polarized in a direction parallel to the field. Zeeman's observations are therefore explained. This is called the normal Zeeman effect.

However, for the majority of transitions, many more lines than the three predicted by the normal Zeeman effect are observed. This is called the *anomalous Zeeman effect* for historical reasons, although this is more frequently observed, because the existence of spin was not known at the time of Zeeman's observations. This effect is observed whenever any of the states

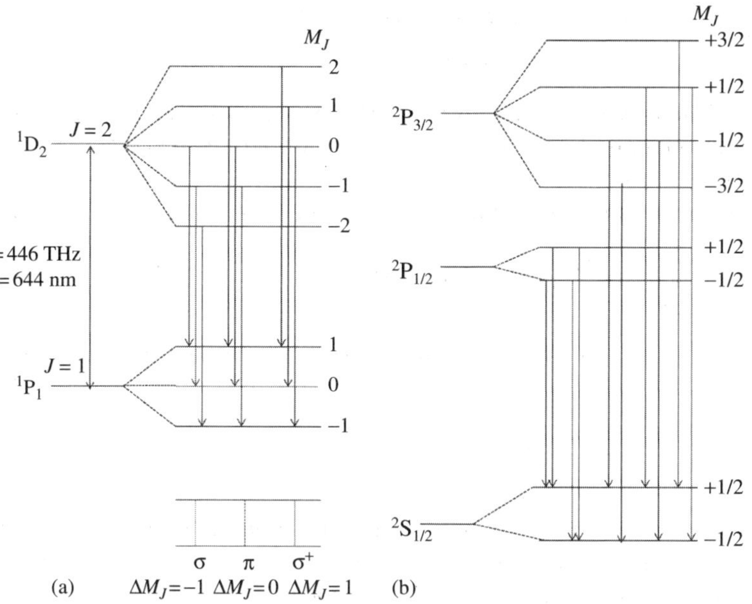

Figure 3.9. (a) Normal Zeeman effect for a d → p transition of Cd. The field splits the degenerate M_J levels by $\Delta E = \mu_B B M_J$. Note that in this case $J = L$; (b) Anomalous Zeeman effect for the D lines of Na. Here, the levels are split by $\Delta E = \mu_B B M_J g_J$. In both cases, transitions can occur if $\Delta M_J = 0, \pm 1$

undergoing transitions has non-zero spin, which results in different splitting for the upper and lower states. This effect is due to the interaction between the external magnetic field and the magnetic moment associated with the electron spin and orbital motion. It is found experimentally and can be proved from quantum electrodynamics that the magnetic moment due to the spin is approximately twice that due to orbital motion, and the analogous equation to (3.44) for the electron spin is

$$\mu_S = -g_e \mu_B \sqrt{S(S+1)} \tag{3.48}$$

where g_e is called the Landé g factor for a free electron and has the value 2.002319304386, which is usually approximated to 2.

For an electron that has both orbital and spin angular momenta, the energy splitting then depends on the quantum numbers, L, S and J, and more complicated structures are observed in the corresponding spectra (Figure 3.9(b)). The magnetic moment is proportional to $\vec{J}$, i.e., $\vec{\mu}_J \propto |\vec{J}|$, and the combined effect of the orbital and spin angular momenta is given by

$$\mu_J = -g_J \frac{e\hbar}{2m_e} \sqrt{J(J+1)} = -g_J \mu_B \sqrt{J(J+1)} \tag{3.49}$$

where g_J is the Landé g factor given by

$$g_J = 1 + \frac{J(J+1) + S(S+1) - L(L+1)}{2J(J+1)} \quad (3.50)$$

If $S = 0$, then $J = L$, and substitution in equation (3.50) gives $g_J = 1$. Therefore, equation (3.49) becomes identical to equation (3.44). We then have the normal Zeeman effect. If $L = 0$, then $J = S$ and g_J becomes equal to 2.

The component of the magnetic moment in the direction of the magnetic field μ_J is given by

$$\mu_J = -\frac{e\hbar}{2m_e} g_J M_J = -\mu_B g_J M_J \quad (3.51)$$

where M_J is the component of J in the direction of the field. The effective energy of this magnetic interaction E_{int} is given by

$$E_{\text{int}} = -\vec{\mu}_J . \vec{B} = g_J \mu_B B M_J \quad (3.52)$$

The g_J values for the $^2P_{3/2}$, $^2P_{1/2}$ and $^2S_{1/2}$ states involved in the sodium D lines are $4/3$, $2/3$ and 2, respectively. From Figure 3.9(b), it is apparent that the sodium D_1 line is split into a quartet because of the unequal splitting of the two states involved in the transition. Likewise, it can be shown that the D_2 line is split into a sextet.

Example 3.6 The magnetic field of the sun and stars can be determined by measuring the Zeeman splitting of spectral lines. Suppose that the sodium D_1 line emitted in a particular region of the solar disk is observed to be split into the four-component Zeeman effect (see Figure 3.9(b)). What is the strength of the solar magnetic field B in that region if the wavelength difference $\Delta\lambda$ between the shortest and longest wavelength is 0.022 nm? The wavelength of the sodium D_1 line is 589.6 nm.

Solution

The figure below shows that the longest wavelength (shortest energy) transition corresponds to the $M_J = -1/2$ component of the $3^2P_{1/2}$ level to the $M_J = 1/2$ component of the $3^2S_{1/2}$ level. The calculated g_J values for the initial and final levels are $2/3$ and 2, respectively. Applying equation (3.52) shows that the net shift in the energy of the transition with respect to the zero field line is

$$\Delta E_{\text{long}} = -\left(\frac{2}{3}.\frac{1}{2}\mu_B B + 2.\frac{1}{2}\mu_B B\right) = -\frac{4}{3}\mu_B B$$

A similar calculation for the shortest wavelength transition ($M_J = 1/2 \rightarrow M_J = -1/2$) shows a shift in the opposite direction, i.e., $\Delta E_{\text{short}} = \frac{4}{3}\mu_B B$. The net difference in energies between the two lines is then twice this, i.e., $\Delta E = \frac{8}{3}\mu_B B$.

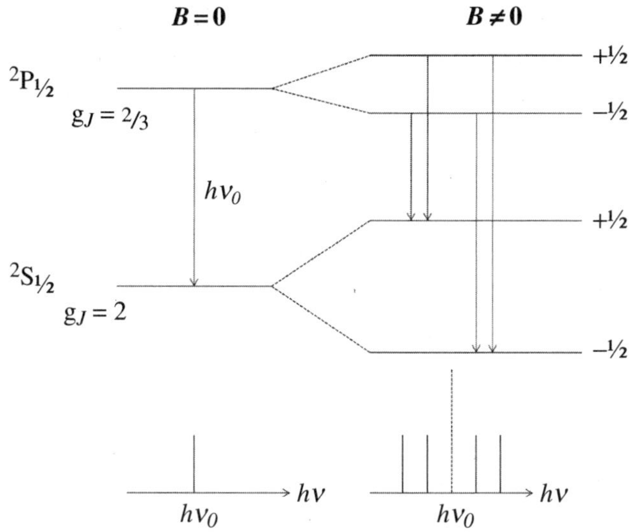

We are given that the wavelength difference is 0.022 nm. Since $E = hc/\lambda$, differentiation of this expression with respect to λ yields

$$\frac{dE}{d\lambda} = -\frac{hc}{\lambda^2}$$

or

$$\Delta\lambda = -\frac{\lambda^2}{hc}\Delta E = \frac{\lambda^2}{hc}\frac{8}{3}\mu_B B$$

since $\Delta E_{\text{long}} - \Delta E_{\text{short}} = -\frac{8}{3}\mu_B B$. Hence B is given by

$$B = \frac{3}{8}\frac{hc}{\mu_B}\frac{\Delta\lambda}{\lambda^2}$$

We are also given that the wavelength of the D_1 line is 589.6 nm. Substitution of the various quantities gives

$$B = \frac{3}{8}\frac{(6.626\times10^{-34}\,\text{J s})\times(2.998\times10^{8}\,\text{ms}^{-1})}{(9.274\times10^{-24}\,\text{J T}^{-1})}\frac{(0.022\times10^{-9}\,\text{m})}{(589.6\times10^{-9}\,\text{m})^2} = 0.51\,\text{T}$$

In comparison, the Earth's magnetic field is much smaller and averages around 50 μT.

Example 3.7 Calculate the Zeeman splitting of the $(7s)^1(6s)^1\ ^3S_1$ to $(6p)^1(6s)^1\ ^3P_1$ transition in Hg, corresponding to the blue line at 435.8 nm. Also, determine the number of lines in the $(7s)^1(6s)^1\ ^3S_1$ to $(6p)^1(6s)^1\ ^3P_2$ and $(7s)^1(6s)^1\ ^3S_1$ to $(6p)^1(6s)^1\ ^3P_0$ transitions.

Solution

From equation (3.50), the g_J values corresponding to the $^3S_1$ and $^3P_1$ states are 2 (as expected for a pure spin state) and 3/2, respectively. The spectral line at $\lambda = 435.8$ nm will be split into several lines by

$$\Delta E = \left(\frac{3M_J}{2} - 2M_{J'}\right)\mu_B B$$

where M_J are the allowed values of J for the $^3P_1$ state and $M_{J'}$ are those for $^3S_1$. Since $J = 1$ in both cases, both M_J and $M_{J'}$ can take on the values 0, ±1. All possibilities are listed in the table below:

Allowed transitions $(7s)^1(6s)^1\ ^3S_1$ to $(6p)^1(6s)^1\ ^3P_1$ in Hg

$M_{J'}$	M_J	ΔM_J	Polarization	Energy shift (in units of $\mu_B B$)
1	1	0	π	$-1/2$
1	0	1	σ	-2
1	−1	2	Forbidden	-
0	1	−1	σ	$3/2$
0	0	0	π	0
0	−1	1	σ	$-3/2$
−1	1	−2	Forbidden	–
−1	0	−1	σ	2
−1	−1	0	π	$1/2$

Hence, six distinct lines should be observed for this transition. Similar arguments for the $(7s)^1(6s)^1\ ^3S_1 \rightarrow (6p)^1(6s)^1\ ^3P_2$ transition show a nine-line pattern, while the $(7s)^1(6s)^1\ ^3S_1 \rightarrow (6p)^1(6s)^1\ ^3P_0$ transition shows a triplet with a spacing of $\pm 2\mu_B B$.

3.4.6.2 The Stark effect

In the presence of an electric field $\vec{E}$, the atomic levels will also be split and complete spectra observed.

3.4.6.3 Paschen-Back effect

When the strength of the applied magnetic field becomes greater than the mutiplet splitting, we have the so-called Paschen-Back effect. In this region, the orbital and spin angular momenta are no longer coupled to each other but are coupled separately to the magnetic field. In this case, only the normal Zeeman effect is observed even when doublets or higher spin states are present. It is important to note that although in principle ESR and NMR can be observed at any magnetic field, the usual experiments are always performed in the Paschen-Back region.

Before we end this chapter, we describe an experimental technique to verify what we have learnt recently. Photoelectron spectroscopy allows us a direct look at the orbital energies. Here we will describe photoelectron spectroscopy as applied to atoms. Most of the principles of photoelectron spectroscopy will become clear in this chapter. In future chapters, we shall discuss its application to molecular systems.

3.5 PHOTOELECTRON SPECTROSCOPY

When a high-energy photon strikes an electron in an atom or molecule, ionization may result (Figure 3.10). The kinetic energy of the ejected electron is equal to the photon energy minus the ionization energy (IE):

$$E_k = h\nu - \text{IE} \tag{3.53}$$

In *photoelectron spectroscopy* (PES), high-energy monochromatic radiation, typically the 58.4 nm (He-I) emission of a He discharge lamp, is shined on the gaseous sample to be studied. Electrons ejected from the sample are counted as a function of their kinetic energy. That is, the photoelectron spectrometer produces a graph of electrons ejected versus electron kinetic

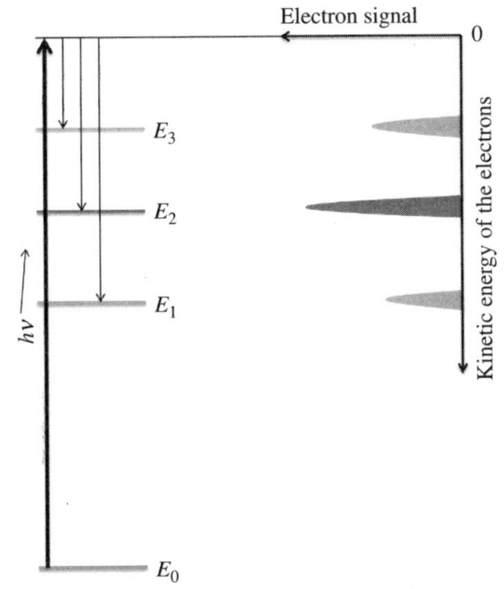

Figure 3.10. Principle of conventional PES. Ionization of molecules with monochromatic radiation of fixed frequency leads to the photoejection of electrons (photoelectric effect). From the measurement of the electron kinetic energy distribution, the positions of ionic energy levels E_1, E_2, E_3, ..., can be determined with respect to an energy level E_0 of the neutral molecule

energy. Knowing the values of $h\nu$ and E_k in equation (3.53), one can plot the data as electrons ejected versus ionization energy. Such a plot is called a photoelectron spectrum. The 58.4 nm photon carries an energy of 21.2 eV and so can ionize electrons with an ionization energy less than 21.2 eV. Typically, these include just the more loosely bound valence electrons. By using a higher energy source, more tightly held electrons can be ionized. For example, the Cr Kα X-ray emission line (5414.7 eV) is commonly employed in a method known as ESCA (Electron Spectroscopy for Chemical Analysis), which is analogous to PES, and is also known as XPS (X-ray photoelectron spectroscopy). Development of this technique led to award of the Nobel Prize to the Swedish physicist Kai Siegbahn in 1981.

As an instructive example, consider the complete photoelectron spectrum of Ar, shown in Figure 3.11, a composite of PES and ESCA results. The electronic configuration of Ar is $(1s)^2(2s)^2(2p)^6(3s)^2(3p)^6$, which defines a $^1S_0$ ground state. That is, there is zero orbital and zero spin angular momentum. When an electron is ejected from the outermost orbital (3p), the atom is left with the electronic configuration $(1s)^2(2s)^2(2p)^6(3s)^2(3p)^5$, which defines $^2P_{3/2}$ and $^2P_{1/2}$ levels, of which the $^2P_{3/2}$ level is lower in energy according to Hund's third rule. In the photoelectron spectrum of Ar, we assign the lowest IE doublet near 16 eV to the transitions $3^1S_0 \rightarrow 3^2P_{3/2}$ (15.759 eV) and $3^1S_0 \rightarrow 3^2P_{1/2}$ (15.937 eV). The next lowest energy ionization occurs if a 3s electron is removed from the neutral atom, yielding a $3^2S_{1/2}$ level. This gives rise to a peak at 29.24 eV. In a similar manner, the complete spectrum of Ar can be readily and satisfyingly explained.

Koopmans' theorem, which states that the ionization energy IE_i is equal in magnitude to the orbital energy (with electron–electron repulsions included), is implicit in the preceding discussion. This theorem allows us to connect the theoretical orbital energies with the observed ionization energies. Koopmans' 'theorem' is really an approximation. For example, although there is only one '3p orbital energy' for Ar, there are two levels resulting from the two physically distinguishable ways that the orbital and spin angular momenta may be arranged in the configuration $(3p)^5$. The orbital and spin angular momenta may be either parallel or

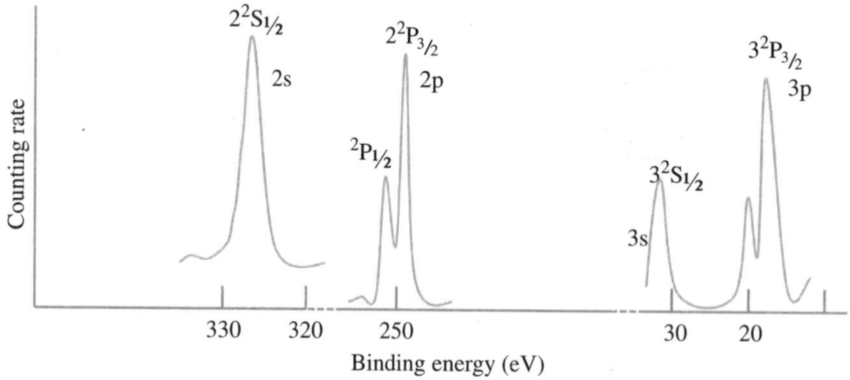

Figure 3.11. Photoelectron spectrum of Ar excited by Kα X-radiation [adapted from Siegbahn et al. (1969)]

antiparallel, resulting in a total momentum, J, of $\frac{3}{2}$ and $\frac{1}{2}$ in this case. To extract the energy of an electron in the 3p orbital from the two peaks near 16 eV would require some additional theory. (It is also an approximation, albeit a good one, to associate a particular ^2P state with the 3p orbital. Any other configuration which gives rise to a ^2P state will contribute to some extent to the ^2P state in question.)

One other feature of photoelectron spectra worth noting is that under ideal circumstances, the area under a peak is proportional to the degeneracy of the orbital from which the ionization takes place. For example, in comparing the areas under the 2s and 2p peaks of Ar at 326.3 and 248.52 eV, respectively, the two peaks resulting from the 2p ionization together have about three times the area of the 2s peak. Further, within the 2p region, the $^2P_{3/2}$ peak has twice the area of the $^2P_{1/2}$ peak because the degeneracy of a state characterized by a certain value of J is $2J + 1$ (For $^2P_{3/2}$, $J = \frac{3}{2}$, and $2J + 1 = 4$. For $^2P_{1/2}$, $J = \frac{1}{2}$, and $2J + 1 = 2$). This is the most direct way to assign the order of the two levels $^2P_{3/2}$ and $^2P_{1/2}$ and is experimental confirmation of Hund's third rule.

3.6 SUMMARY

- There are two principal coupling schemes: the Russell–Saunders (or *LS*) coupling and *jj* coupling.
- The *LS* coupling scheme is applicable to light atoms. In this coupling scheme, L and S are good quantum numbers, i.e., the angular momenta of the individual electrons combine to the total orbital angular momentum L and to the total spin S of the atom. The state of an atom is characterized by a set of quantum numbers, L, S and J.
- The state of a configuration with the same L and S constitutes a *term*. Each term of a configuration has a different energy, which depends on the value of L. Every term contains $(2L + 1)(2S + 1)$ states.
- States of a term having the same L and S values, but different J values have practically the same energy and constitute a multiplet, the multiplicity being $2J + 1$. Different values of J provide a 'fine structure'. The splitting of an *LS* term according to the value of J is a spin–orbit effect.
- Multiplicity occurs when several levels are close together. Normally $S \leq L$; therefore, the total number of different states associated with an *LS* term (called the multiplicity of the term) is $2S + 1$.
- The $2J + 1$ states in a level are degenerate except when a specific direction in space is preferred, for example, on superimposing an external field.
- In the case of magnetic fields, these states are referred to as Zeeman sublevels.
- For atoms with large atomic number Z, the *jj*-coupling scheme is more appropriate where the l_i and s_i of every electron are first combined to the total angular momentum of the electron j_i so that L and S lose their meaning.
- Photoelectron spectroscopy can provide information about orbital energies.
- In this chapter, we have ignored effects of nuclear spin coupling and other effects, which can be studied from more advanced books.

3.7 EXERCISES

1. (a) What is the orbital angular momentum of an electron in the orbitals (i) 4*d*, (ii) 2*p*, (iii) 3*p*?
   (b) State the orbital degeneracy of the levels in a hydrogen atom that have energy (i) $-hcR_H$, (ii) $-hcR_H/9$, (iii) $-hcR_H/25$.
2. The following wavenumbers represent the energies of some of the observed transitions in the Balmer series of the emission spectrum of atomic hydrogen: 15233.2, 20564.7, 23032.5, 24373.1 cm^{-1}. What are the principal quantum numbers (*n*) for the upper and lower states of each of the transitions?
3. The Humphreys series is a group of lines in the spectrum of atomic hydrogen. It begins at 12368 nm and has been traced to 3281.4 nm. What are the transitions involved? What are the wavelengths of the intermediate transitions?
4. The energies of some transitions observed in the *absorption* spectrum of an unknown one-electron ion (e.g., He^+, Li^{2+}, Be^{3+}) are 1646254.1, 1560886.3, 1316965.2 cm^{-1}. Identify the ion.
5. Give two reasons why the 1s → 2s transition for a hydrogen atom is forbidden.
6. What is an atomic term symbol? What information does the term symbol $^3F_4$ provide about the angular momentum of the atom?
7. Write the possible term symbols for the following atoms using the Russell–Saunders coupling scheme: (a) Li, (b) Na, (c) Sc and (d) Br
8. Calculate the total degeneracy of a term defined by the term symbol ^4G.
9. What are the permitted *J* values for the following terms?
   (a) ^6S (b) ^1F (c) ^2H (d) ^4P (e) ^3D
10. Determine the allowed wave functions consistent with Pauli's principle (including the spin) for the excited state $(1s)^1(2s)^1$ configuration of helium, and give their term symbols. Show why one of the terms does not survive for the $(1s)^2$ and $(2s)^2$ configurations.
11. (a) Show that the singlet → singlet transitions are allowed by the electric dipole selection rules.
    (b) Show that the transition from a singlet to the three components of a triplet state in equation (3.35) is forbidden.
12. Write the term symbols for all states arising from the excited $(2p)^1(3p)^1$ configuration of carbon. For the ground state $(2p)^2$, the Pauli principle forbids some of the states. Determine the ones that survive and then write down the ground state for carbon and oxygen, justifying your selection.
13. Obtain the ground state levels for Fe^{2+}, Cr^{3+}, N and O.
14. Derive the spectral terms for the $(p)^2$ and $(d)^2$ configurations by (a) writing out the microstates, (b) spin factoring.
15. What is the ground state configuration of Na? What is the first excited state configuration? To what atomic levels do these configurations rise? The yellow Na '*D*' line represents emission from the first excited configuration to the ground configuration. The *D* line is actually a doublet with components at 589.15788 and 586.75537 nm. Label each transition with the appropriate symbol indicating initial and final atomic levels. What is the energy separation (in cm^{-1}) between the two levels of the excited configuration?

16. (a) Using the selection rules for atomic spectra, state which of the following transitions are allowed:
    (i) $^2S_{1/2} \rightarrow {}^2P_{3/2}$
    (ii) $^2S_{1/2} \rightarrow {}^2P_{1/2}$
    (iii) $^2S_{1/2} \rightarrow {}^2D_{3/2}$
    (iv) $^1P_1 \rightarrow {}^1D_2$
    (v) $^3P_1 \rightarrow {}^3P_2$
    (vi) $^3S_1 \rightarrow {}^3S_1$
    (b) With the additional selection rule for ΔM_J, show the Zeeman splitting of the spectral lines in (a).
17. The term symbols for an $(nd)^8$ configuration are ^1S, ^1D, ^1G, ^3P and ^3F. Calculate the values of J associated with each of these term symbols. Which term represents the ground state?
18. Derive the possible term symbols (ignoring spin–orbit coupling) for the following configurations:
    (a) B: $(1s)^2(2s)^2(2p)^1$
    (b) C*: $(1s)^2(2s)^2(2p)^1(3s)^1$
    (c) C*: $(1s)^2(2s)^2(2p)^1(3p)^1$
    (d) Ti^+: $[Ar](3d)^1$
    Explain briefly why there are fewer possible term symbols for the ground state configuration of carbon, $(1s)^2(2s)^2(2p)^2$ than for the excited state in (c) above.
19. The following table lists the terms arising from the ground state configurations of some second row atoms. For each term, list the possible spin–orbit levels, and (using Hund's rules) predict the ground state level for each atom.

Element	**Configuration**	**Terms arising**
C	$(1s)^2(2s)^2(2p)^2$	^1D; ^3P; ^1S
N	$(1s)^2(2s)^2(2p)^3$	^4S; ^2D; ^2P
O	$(1s)^2(2s)^2(2p)^4$	^1D; ^3P; ^1S
F	$(1s)^2(2s)^2(2p)^5$	^2P

20. A number of possible transitions in the beryllium atom are listed below. Which are fully allowed? For those not allowed, which selection rule would have to be broken? Comment on any mechanism that exists to cause such a breakdown. One of the term symbols is impossible – which one and why?
    $(2s)(5s)\ (^1S_0) \rightarrow (2s)(5d)\ (^1D_2)$
    $(2s)(5s)\ (^3S_1) \rightarrow (2s)(2p)\ (^1P_1)$
    $(2s)(5s)\ (^1S_0) \rightarrow (2s)^2\ (^1S_0)$
    $(2s)(5p)\ (^3P_1) \rightarrow (2s)(3s)\ (^3S_1)$
    $(2s)(5p)\ (^1P_1) \rightarrow (3s)(4s)\ (^1S_0)$
    $(2s)(3p)\ (^3P_1) \rightarrow (3p)^2\ (^3D_2)$
    $(2s)(3p)\ (^3P_0) \rightarrow (3p)(4p)\ (^3D_2)$
    $(2s)(3p)\ (^3P_0) \rightarrow (3p)(4p)\ (^3P_0)$
21. (a) Write down the allowed values of the total angular momentum quantum number J for an atom with spin S and orbital quantum number L.

   (b) Write down the *L*, *S* and *J* quantum numbers for the states described as $^2S_{1/2}$, $^3D_2$ and $^5P_3$
   (c) Determine if any of these states is (are) impossible, and, if so, explain why.
22. (a) Sketch the radial distribution function of the 3s, 3p and 3d orbitals of Na and show in your graph where the inner shell electrons lie. Explain why these orbitals have different energies.
   (b) A transition from the 3s to 3p level in Na is at 16961 cm^{-1}. Transitions from this 3p level to the d level form a series of lines, of which the first three are at 8752, 16214 and 19386 cm^{-1}. Sketch the energy level diagram for the transitions involved and deduce the ionization energy (in cm^{-1}) of Na in its ground state.
   (c) Explain the fact that the 3s → 3p transition referred to in part (b) is split into a doublet separated by 17 cm^{-1} and under high resolution the p to d transitions are each seen to consist of three lines.
23. Calculate the energy of the spectroscopic lines associated with the 3p ← 3s transitions for (a) Na in the absence of an external magnetic field, (b) Na atoms perturbed by an external magnetic field *B*.
24. The $(3s)^2(3p)^1(4p)^1$ configuration of an excited state of the Si atom leads to a state with orbital angular momentum $L = 1$ and which is split by spin–orbit coupling into the levels shown below. Deduce possible term symbols for the three levels and indicate the dipole allowed transitions which would arise in absorption from these levels when the 4p electron is excited into a 5s orbital.

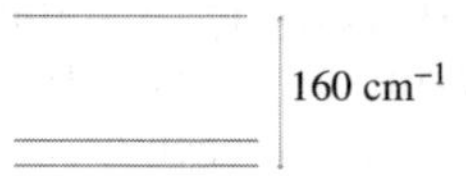

25. (a) Draw up a table showing the microstates allowed by the Pauli exclusion principle for an $(nd)^2$ electronic configuration and hence derive the permitted term symbols assuming Russell–Saunders coupling.
   (b) Use the angular momentum coupling arguments to show that the levels derived in *jj* coupling for an *npnf* configuration are $\left(\frac{1}{2}\,\frac{5}{2}\right)_{3,2}$, $\left(\frac{1}{2}\,\frac{7}{2}\right)_{4,3}$, $\left(\frac{3}{2}\,\frac{7}{2}\right)_{5,4,3,2}$ and $\left(\frac{3}{2}\,\frac{5}{2}\right)_{4,3,2,1}$.
26. Explain the following observations:
   (a) The four lowest energy electronically excited states of He (at low resolution) lie at 159850 cm^{-1}, 166271 cm^{-1}, 169083 cm^{-1} and 171129 cm^{-1} above the ground state. The absorption spectrum of He, however, displays a strong line at only one of these energies.
   (b) He gas shows negligible absorption of radiation in the infrared, but strong absorption bands are observed at 4858 cm^{-1} and 9233 cm^{-1} when the gas is excited in an electric discharge.
   (c) A $^1D_2 \rightarrow {}^1P_1$ transition observed in the emission spectrum of Cd consists of a single line. In the presence of a magnetic field, three lines are observed. For the $^2P_{1/2} \rightarrow {}^2S_{1/2}$ emission line in Na the single line is observed to split into four lines when a magnetic field is applied.
27. Write down the possible term symbols for atoms with the electronic configurations:
   (a) $(1s)^2$
   (b) $(2s)^1$
   (c) $(2s)(3s)$

(d) (2p)(3d)
(e) (2p)(3p)
(f) $(2p)^2$

28. Make a table of energies of hydrogen's $n = 3$ levels when placed in (a) weak and (b) strong magnetic fields. What transitions are possible to the $n = 1$ level?
29. How many fine structure levels is the 4f level of hydrogen split into? Rank these levels with respect to their energies. What happens to these levels when the atom is placed in a weak magnetic field?
30. State the ground state electron configuration and magnetic dipole moment of hydrogen ($Z = 1$) and sodium ($Z = 11$).
31. (a) The magnetic moment of silver atom is only 1 Bohr magneton although it has 47 electrons. Explain.
    (b) What is the magnetic moment of an atom in the $^3P_0$ state?
    (c) What is the ratio of intensities of spectral lines in the hydrogen spectrum for the transitions

$$2^2P_{1/2} \rightarrow 1^2S_{1/2} \text{ and } 2^2P_{3/2} \rightarrow 1^2S_{1/2}?$$

32. State, with reasons, which of the following transitions are forbidden for electric dipole transitions:

$$^3D_1 \rightarrow {}^2F_3$$
$$^2P_{3/2} \rightarrow {}^2S_{1/2}$$
$$^2P_{1/2} \rightarrow {}^2S_{1/2}$$
$$^3D_2 \rightarrow {}^3S_1$$

33. For the $^2P_{3/2} \rightarrow {}^2S_{1/2}$ transition of an alkali atom, sketch the splitting of the energy levels and the resulting Zeeman spectrum for atoms in a weak external magnetic field (express your results in terms of the frequency v_0 of the transition in the absence of an applied magnetic field).
34. The spacings of adjacent energy levels of increasing energy in a calcium triplet are 30×10^{-4} and 60×10^{-4} eV. What are the quantum numbers of the three levels? Write down the levels using the appropriate spectroscopic notation.
35. An atomic transition line with a wavelength of 350 nm is observed to be split into three components in a spectrum of light from a sun spot. Adjacent components are separated by 1.7 pm. Determine the strength of the magnetic field in the sun spot.
36. Calculate the energy spacing between the components of the ground state energy level of hydrogen when split by a magnetic field of 1.0 T. What frequency of electromagnetic radiation could cause a transition between these levels? What is the specific name given to this effect?
37. Consider the transition $^2P_{1/2} \rightarrow {}^2S_{1/2}$ for sodium in the magnetic field of 1.0 T. Sketch the splitting.
38. To excite the mercury line at 546.1 nm, an excitation potential of 7.69 V is required. If the deepest term in the mercury spectrum lies at 84,181 cm^{-1}, calculate the numerical values of the two energy levels involved in the emission of 546.1 nm.
39. (a) Explain the origin of spin–orbit coupling in atoms. How is the term symbol modified to indicate the resulting energy levels?
    (b) What are Hund's rules? Give a physical justification for each of them, and explain how they are used to predict the lowest energy level of an atom.

(c) The following table lists the terms arising from the ground state configurations of the second row atoms. For each term, list the possible spin–orbit levels, and (using Hund's rules) predict the ground state level of each atom.

Element	Configuration	Terms arising
Li	$(1s)^2(2s)^1$	^2S
Be	$(1s)^2(2s)^2$	^1S
B	$(1s)^2(2s)^2(2p)^1$	^2P
C	$(1s)^2(2s)^2(2p)^2$	^1D, ^3P, ^1S
N	$(1s)^2(2s)^2(2p)^3$	^4S, ^2D, ^2P
O	$(1s)^2(2s)^2(2p)^4$	^1D, ^3P, ^1S
F	$(1s)^2(2s)^2(2p)^5$	^2P
Ne	$(1s)^2(2s)^2(2p)^6$	^1S

How is Hund's third rule applied to determine the lowest level of the lowest energy term from a configuration with an *exactly* half-full subshell?

40. (a) What are the selection rules for a radiative transition in a many-electron atom?
(b) How many fine structure components would be observed in the emission line $n = 4 \rightarrow n = 3$ of the hydrogen atom, if the effects of spin–orbit coupling were fully resolved in the spectrum? Illustrate the transitions on a Grotrian diagram.
(c) A line in the spectrum of potassium has been identified as arising from the transition $4^2P \leftarrow 3^2D$. Upon closer inspection, it is found to consist of three lines at wavenumbers 8494.13, 8496.45 and 8554.17 cm^{-1}. Draw a diagram representing these transitions, given that the splitting is regular for the ^2P term and inverted for the ^2D term. Calculate the separation of the individual J components of the states involved.

41. The wavenumbers of the first two lines in each of the principal ($3^2S \leftarrow n^2P$, $n \geq 3$), sharp ($3^2P \leftarrow n^2S$, $n \geq 4$) and diffuse ($3^2P \leftarrow n^2D$, $n \geq 3$) series of atomic sodium are given below:

Series	$\tilde{\nu}$ / cm^{-1}	
Principal	16952.2	30263.0
Sharp	8786.2	16246.9
Diffuse	12218.3	17595.2

(a) Draw an energy level diagram illustrating these transitions.
(b) The ionization energy of sodium from its ground state is 495.80 kJ mol^{-1}. Calculate the value of the effective principal quantum number, n_{eff}, for the ground state, Na 3^2S, and the quantum defect, Δ. Take $R_{Na} = R_{\infty}$.
(c) In a similar manner, complete the other entries in the following table for Na atoms:

State	3^2S	4^2S	5^2S	3^2P	4^2P	3^2D	4^2D
n_{eff}		2.643	3.648	2.116			3.989
Δ							

(d) Compare and interpret the values of n_{eff} for the S, P and D states. In particular, how does Δ depend on l and why?

42. Calculate the Zeeman splitting in reciprocal centimetre between the three spectral lines of a transition between a $^1S_0$ term and a $^1P_1$ term in the presence of a magnetic field of 2 T. How does the splitting compare with typical optical transition wavenumbers, such as those for the Balmer series of hydrogen?
43. When 58.4 nm light from a helium discharge lamp is directed into a sample of krypton, electrons are ejected with a velocity of 1.59×10^6 m s^{-1}. The same radiation releases electrons from rubidium vapour with a speed of 2.45×10^6 m s^{-1}. What are the ionization energies of the two species?
44. Make a list showing the order of increasing energy of all of the transitions expected in the photoelectron spectrum of nitrogen in which N^+ is the product.

4

Pure Rotational Spectroscopy

Light brings us the news of the Universe.
The Universe of Light (1933), 1.
—Sir William Bragg

4.1 OVERVIEW OF MOLECULAR SPECTRA

In the previous chapter, we had seen that, except for dissociation continua, the spectra of atoms consist of sharp lines. This is because the only transitions that are possible in atoms are between the different allowed electronic energy states. Molecular spectra, on the other hand, consist of bands, in which a densely packed line structure can be seen under high resolution. This is because, besides transitions of electrons between different molecular orbitals, molecules can also change their energies in two other ways, namely by changes in their rotational and vibrational energies. Just like the electronic states, the rotational and vibrational energies are quantized and hence exist as discrete energy levels.

In the study of molecular spectroscopy, it is customary to separate the total energy into the electronic, vibrational and rotational components, i.e., $E = E_{el} + E_{vib} + E_{rot}$. The separation of the electronic energy from the other terms is essentially the Born–Oppenheimer approximation. The nuclei and electrons of a molecule are subject to the same forces, but the nuclei are much more massive than the electrons, which move more rapidly. Hence, we may consider the nuclei to be stationary when considering the motion of the electrons. Considering a diatomic molecule, we may write the exact Hamiltonian in terms of the kinetic energies of the electrons and nuclei and the potential energies of their mutual interactions as

$$\hat{H} = \hat{T}_{el} + \hat{V}_{el-el} + \hat{T}_n + \hat{V}_{n-n} + \hat{V}_{el-n} \tag{4.1}$$

In the absence of the last term, it would be possible to separate the Hamiltonian into a sum of two independent terms depending on the electronic and nuclear coordinates. If the Hamiltonian is separable into independent terms, it is possible to write the total wave function as a product of the wave functions of the independent coordinates, and a separation of variables in the Schrödinger equation can lead to independent equations, which, when solved, give the energy of each term. The total energy is then the sum of energies obtained from each equation.

The Born–Oppenheimer approximation assumes that we can study the behaviour of electrons in a field of frozen nuclei. The approximate electronic Hamiltonian is then

$$\hat{H}_{\rm el} = \hat{T}_{\rm el} + \hat{V}_{\rm el-el} + \hat{V}_{{\rm el}-n} = -\frac{\hbar^2}{2m_e}\sum_i \nabla_i^2 + \sum_{i<}\sum_j \frac{e^2}{4\pi\varepsilon_0 r_{ij}} - \sum_\alpha\sum_i \frac{Z_{\alpha i}e^2}{4\pi\varepsilon_0 r_{\alpha i}} \tag{4.2}$$

with fixed nuclear positions for calculating the last term. Here, the symbols i and j refer to electrons and the summations run from 1 to n, the total number of electrons in the system. The symbol α refers to the nuclei and the summation runs from 1 to N, the number of nuclei. In equation (4.2), r_{ij} refers to the inter-electronic distance and $r_{\alpha i}$ is the distance of the ith electron from nucleus α. The $E_{\rm el}$ obtained by solving the electronic Schrödinger equation $\hat{H}_{\rm el}\psi_{\rm el} = E_{\rm el}\psi_{\rm el}$ is used as an effective potential in the nuclear Hamiltonian, which now also includes the internuclear repulsion term:

$$\hat{H}_n = -\frac{\hbar^2}{2}\sum_\alpha \frac{1}{M_\alpha}\nabla_\alpha^2 + E_{\rm el}(R_{\alpha\beta}) + \sum_{\alpha<}\sum_\beta \frac{Z_\alpha Z_\beta}{R_{\alpha\beta}} \tag{4.3}$$

The nuclear equation can be further separated into the vibrational, rotational and translational components, leading to $\psi = \psi_{\rm el}\psi_n = \psi_{\rm el}\psi_{\rm vib}\psi_{\rm rot}\psi_{\rm trans}$ and $E = E_{\rm el} + E_{\rm vib} + E_{\rm rot} + E_{\rm trans}$. Translation involves movement of the molecule as a whole through space without any change in the relative positions of the nuclei; in rotational motion, the molecule rotates through its centre of mass without any change in the relative positions of the nuclei. It is only in vibrational motion that the nuclei change their relative positions, which influences the effective potential in the nuclear Hamiltonian (4.3) because of variation in $R_{\alpha\beta}$.

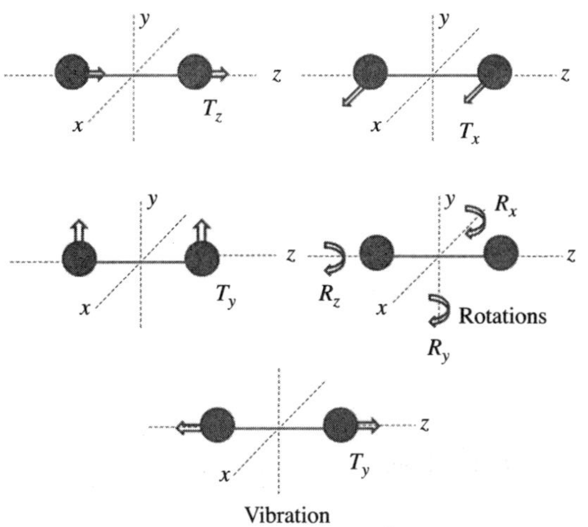

Figure 4.1. Translational, rotational and vibrational motion of a diatomic molecule. Note that rotation about the internuclear axis (R_z) changes neither the angular momentum nor the energy

For a diatomic molecule, equation (4.3) is a function of only one internuclear distance, $R_{\alpha\beta}$, and the equation is a function of six coordinates, three for each nucleus. If we transform our coordinate system to three coordinates of the centre of mass, x_{COM}, y_{COM} and z_{COM}, and three internal coordinates of the molecule, r, θ and φ (this is also a central field problem), the equation separates into two, one involving the centre of mass coordinates,

$$-\frac{\hbar^2}{2M}\nabla^2_{\mathrm{COM}}\Psi_{\mathrm{COM}} = E_{\mathrm{trans}}\Psi_{\mathrm{COM}} \tag{4.4}$$

which gives the translational energy. Here, M is the total mass of the molecule. This is simply the particle-in-a-box Schrödinger equation and yields the translational energy, $E_{\mathrm{trans}} = \frac{h^2}{8M}\left(\frac{n_x^2}{a^2}+\frac{n_y^2}{b^2}+\frac{n_z^2}{c^2}\right)$. Here, a, b and c are the dimensions of confinement and n_x, n_y and n_z are the translational quantum numbers. Because molecules are generally confined in large spaces compared to electrons in atoms, the translational quantum numbers at room temperature are very large (~10^9) and the spacing between consecutive energy levels is too small to be experimentally detectable. Hence, we can ignore translational spectroscopy because of the near continuous translational energy levels.

This leaves the nuclear energy as the sum of vibrational and rotational energies. In general, a molecule containing N atoms has $3N$ degrees of freedom, of which three, involving the centre of mass coordinates, are translational. In addition, there are three rotational degrees of freedom (linear molecules have only two rotational degrees of freedom), and the rest ($3N - 6$ / $3N - 5$) are vibrational degrees of freedom. The rotational energy spacings are much smaller than vibrational energy spacings (Figure 4.2). This is because vibration involves distortion of the molecule, which requires more energy than rotation. These large differences in energies provide a justification for the Born–Oppenheimer approximation.

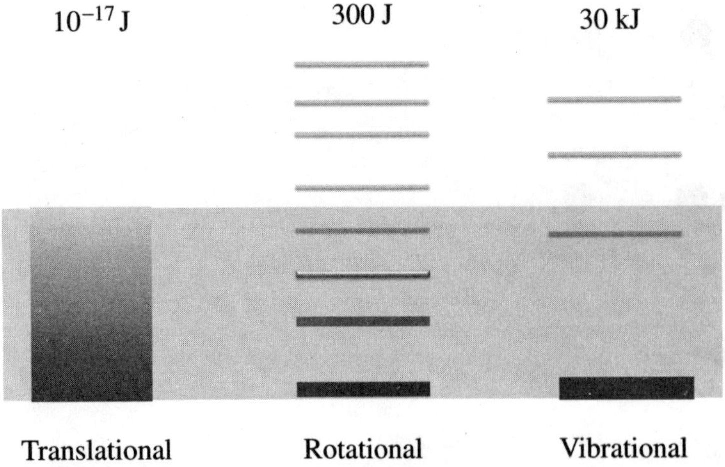

Figure 4.2. Relative spacings of translational, rotational and vibrational energy levels

Example 4.1 A typical diatomic molecule rotates at a rate of 10^{11} rev s^{-1}. How many vibrations does it undergo during one revolution? (Take $k = 1000$ N m^{-1}, $\mu_{red} = 10\ u$).

The frequency of vibration is given by

$$\nu_{osc} = \frac{1}{2\pi}\sqrt{\frac{k}{\mu_{red}}} = \frac{1}{2\times 3.1416}\sqrt{\frac{1000\ \text{N m}^{-1}}{10\ \times 1.6605\times 10^{-27}\ \text{kg}}} \approx 4\times 10^{13}\ \text{s}^{-1}.$$

Hence, more than a 100 vibrations take place during a single rotation.

The difference in timescales for these types of motions is the basis of the Born–Oppenheimer approximation – rotation and vibration can be treated separately as independent motions; specifically, the bond length for rotation can be taken as an average of the internuclear separation over each vibrational period. However, strictly speaking, the separation of the vibrational and rotational energies as two independent entities is not entirely justified. A rapidly rotating molecule undergoes centrifugal distortion, increasing the bond length and influencing vibrations. Similarly, the bond length of a vibrating molecule is constantly varying, and this influences its rotational energy levels. Nevertheless, in what follows, we shall first consider the two motions as independent, and then consider the influence of one upon the other. Thus, in considering rotational motion, we shall at first consider the molecule as a rigid rotator (or rotor), with a fixed bond length. Later, we will consider non-rigidity, i.e., the effect of centrifugal distortion. Similarly, we shall consider the idealized harmonic oscillator model for molecular vibrations in the next chapter and then study the influence of anharmonicity.

We start with the lowest energy region of the electromagnetic spectrum where molecular spectra are observed, i.e., the microwave region. The energy emitted or absorbed in this region is just sufficient to alter the rotational energy of the molecules. Since this is the lower energy end of the electromagnetic spectrum, no simultaneous vibration or any other transition can take place in this region, and so this is a convenient starting point. We shall first take up the simple system of a diatomic molecule, but the treatment that follows is equally applicable to other linear molecules as well. We shall describe the extrapolation to other linear systems later on in this chapter. Further, initially we describe a system of rotating masses classically, and then introduce the quantum restrictions imposed due to finite molecular dimensions, and how these enable one to determine the molecular properties. As seen in Chapter 2, quantum mechanical operators are most conveniently derived from the corresponding classical expressions. This is going to be our general strategy in the rest of the book as well. We shall see that the rotational energies are quantized and depend on the size and shape of the molecule. This enables one to determine these properties of the molecule from a study of its rotational spectrum.

4.2 THE RIGID DIATOMIC MOLECULE (THE RIGID ROTOR)

We start with the rigid rotor, the simple model for a rotating diatomic molecule, shown in Figure 4.3. Masses m_1 and m_2 are joined by a rigid bar (the bond) whose length is $r_e = r_1 + r_2$.

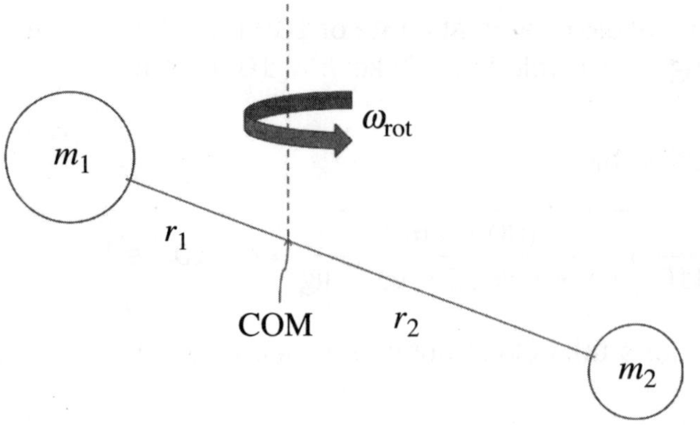

Figure 4.3. Rotational motion

The molecule rotates end over end about a point COM, the centre of mass, which is defined by the moment of balancing equation

$$m_1 r_1 = m_2 r_2 \tag{4.5}$$

The moment of inertia of a system of masses about an axis is defined by $\sum_i m_i r_i^2$, where r_i is the perpendicular distance of the *i*th mass m_i from the axis. Hence, the moment of inertia about a rotational axis passing through COM (Figure 4.3) is

$$\begin{aligned} I &= m_1 r_1^2 + m_2 r_2^2 \\ &= (m_1 r_1) r_1 + (m_2 r_2) r_2 \\ &= (m_2 r_2) r_1 + (m_1 r_1) r_2 \\ &= (m_1 + m_2) r_1 r_2 \end{aligned} \tag{4.6}$$

But

$$\begin{aligned} m_1 r_1 &= m_2 r_2 = m_2 (r_e - r_1) \\ m_1 r_1 + m_2 r_1 &= m_2 r_e \\ \therefore (m_1 + m_2) r_1 &= m_2 r_e \\ r_1 &= \frac{m_2 r_e}{m_1 + m_2} \end{aligned} \tag{4.7}$$

Similarly,

$$r_2 = \frac{m_1 r_e}{m_1 + m_2} \tag{4.8}$$

Substituting in (4.6), we get

$$I = (m_1 + m_2) \frac{m_2 r_e}{m_1 + m_2} . \frac{m_1 r_e}{m_1 + m_2} = \frac{m_1 m_2}{m_1 + m_2} r_e^2 = \mu_{red} r_e^2 \tag{4.9}$$

where μ_{red} is the reduced mass of the system. In order to avoid confusion with dipole moment, which also has the same symbol *mu,* though it is a vector ($\vec{\mu}$), the reduced mass has been written with the subscript '*red*'.

Since the particles are at a fixed distance from each other, the potential energy does not vary as the rotor moves through space, and V is a constant, which we may take as zero. Thus, the energy is entirely kinetic, given by

$$E_k = \frac{1}{2}m_1 v_1^2 + \frac{1}{2}m_2 v_2^2$$

Here, v is the *tangential* velocity along the circular path taken by each atom around the molecular centre of mass. The velocity is the length of the path divided by the time taken to complete one revolution, which is the inverse of the frequency of rotation, ν_{rot}, so that we have $v_1 = 2\pi r_1 \nu_{rot} = r_1\omega_{rot}$, where $\omega_{rot} = 2\pi\nu_{rot}$ is the angular velocity in rad s^{-1}. Likewise, $v_2 = r_2\omega_{rot}$. Note that the linear velocities for the two masses are different, but their angular velocities are the same.

$$E_k = \frac{1}{2}m_1 r_1^2\omega_{rot}^2 + \frac{1}{2}m_2 r_2^2\omega_{rot}^2 = \frac{1}{2}\Big(\underbrace{m_1 r_1^2 + m_2 r_2^2}_{=I}\Big)\omega_{rot}^2 = \frac{1}{2}I\omega_{rot}^2 \tag{4.10}$$

The angular momentum $\vec{L}$ is given by $\vec{L} = \vec{r} \times \vec{p}$, where $\vec{p}$ is the linear momentum defined by $\vec{p} = m\vec{v}$. Therefore, L, the magnitude of $\vec{L}$, is given by

$$L = \left|\vec{L}\right| = r_1 p_1 + r_2 p_2 = r_1 m_1 v_1 + r_2 m_2 v_2 = m_1 r_1^2\omega_{rot} + m_2 r_2^2\omega_{rot} = I\omega_{rot} \tag{4.11}$$

Therefore,

$$E_k = \frac{1}{2}I\omega_{rot}^2 = \frac{L^2}{2I} \tag{4.12}$$

Equation (4.12) gives the connection between *angular momentum* and kinetic energy of rotation. The relation is analogous to that between *linear momentum* and kinetic energy for translational motion ($E_k = p^2/2m$).

We now need to describe the system quantum mechanically. Equations (4.9), (4.11) and (4.12) show the relationship between I, L and E_{rot}, respectively (since the rotational energy is entirely kinetic) in classical mechanics. We expect the relationship to persist in quantum mechanics, so that the quantization of the angular momentum and energy will go hand-in-hand. The system is a two-particle system. Separation out of the translational motion from the total Hamiltonian leads to the mass m being replaced by μ_{red}, the reduced mass of the system. This is equivalent to shifting the origin to one atom (Figure 4.4).

To obtain the Hamiltonian for the system, we substitute the operator for $\hat{L}^2$ in equation (4.12). The same result may be obtained alternatively. The $V = 0$ is a special case of a central force field ($V(r)$), so we use spherical polar coordinates, as for the hydrogen atom, to deal with the motion. Expressed in spherical polar coordinates, the Laplacian operator is

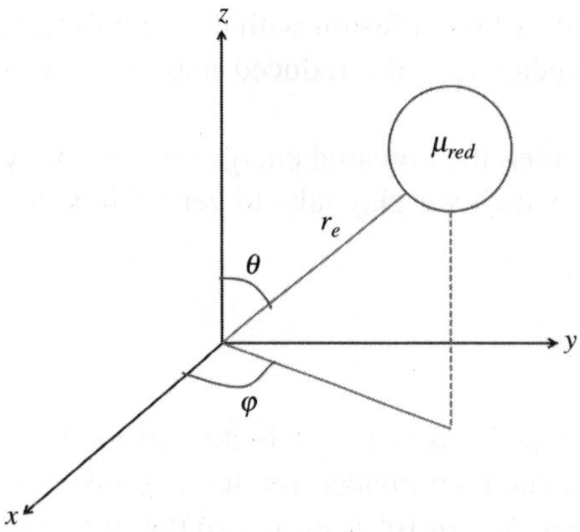

Figure 4.4. Diatomic molecule reduced to a one-body problem with mass μ_{red}

$$\nabla^2_{r\theta\varphi} = \frac{1}{r^2}\frac{\partial}{\partial r}\left(r^2\frac{\partial}{\partial r}\right) + \frac{1}{r^2 \sin\theta}\frac{\partial}{\partial\theta}\left(\sin\theta\frac{\partial}{\partial\theta}\right) + \frac{1}{r^2\sin^2\theta}\frac{\partial^2}{\partial\varphi^2} \tag{4.13}$$

and so, in spherical polar coordinates, the Schrödinger equation, $\hat{H}(r,\theta,\varphi)\psi(r,\theta,\varphi) = E\psi(r,\theta,\varphi)$ is

$$\left\{-\frac{\hbar^2}{2\mu_{red}}\left[\frac{1}{r^2}\frac{\partial}{\partial r}\left(r^2\frac{\partial}{\partial r}\right) + \frac{1}{r^2\sin\theta}\frac{\partial}{\partial\theta}\left(\sin\theta\frac{\partial}{\partial\theta}\right) + \frac{1}{r^2\sin^2\theta}\frac{\partial^2}{\partial\varphi^2}\right] + V(r,\theta,\varphi)\right\}\psi(r,\theta,\varphi)$$
$$= E\psi(r,\theta,\varphi) \tag{4.14}$$

In our rigid rotor example,

$$V(r,\theta,\varphi) = 0 \quad (r = r_e)$$
$$V(r,\theta,\varphi) = \infty \quad (r \neq r_e),$$

i.e., r is held constant at r_e, and $r \neq r_e$ is not permitted.

Therefore, $\psi(r,\theta,\varphi)$ becomes $\psi(r_e,\theta,\varphi)$ and $\frac{\partial}{\partial r}\psi(r_e,\theta,\varphi) = 0$, since r is constant. The Schrödinger equation can therefore be written as

$$-\underbrace{\frac{\hbar^2}{2\mu_{red}r_e^2}}_{I}\left[\frac{1}{\sin\theta}\frac{\partial}{\partial\theta}\left(\sin\theta\frac{\partial}{\partial\theta}\right) + \frac{1}{\sin^2\theta}\frac{\partial^2}{\partial\varphi^2}\right]\psi(r_e,\theta,\varphi) = E_{\text{rot}}\psi(r_e,\theta,\varphi) \tag{4.15}$$

Comparison with the angular momentum square operator (3.11) yields the simplified equation

$$\frac{\hat{L}^2}{2I}\psi(r_e,\theta,\varphi) = E_{\text{rot}}\psi(r_e,\theta,\varphi) \tag{4.16}$$

Since r_e in $\psi(r_e,\theta,\varphi)$ is a constant, we may write the wave function as a constant C times a function dependent on θ and ϕ. This gives

$$C\frac{\hat{L}^2}{2I}Y(\theta,\varphi) = CE_{\text{rot}}Y(\theta,\varphi)$$

$$-\frac{\hbar^2}{2I}\left[\frac{1}{\sin\theta}\frac{\partial}{\partial\theta}\left(\sin\theta\frac{\partial}{\partial\theta}\right)+\frac{1}{\sin^2\theta}\frac{\partial^2}{\partial\varphi^2}\right]Y(\theta,\varphi) = E_{\text{rot}}Y(\theta,\varphi) \tag{4.17}$$

as the differential equation to be solved. The eigenfunctions of the $\hat{L}^2$ operator are the spherical harmonics (see Chapter 3). The solution gives two quantum numbers, J and M, which are analogous to the l and m quantum numbers of atomic systems, respectively. Thus,

$$Y_{JM}(\theta,\varphi) = N_{J,|M|}P_J^{|M|}(\cos\theta)e^{iM\varphi} \tag{4.18}$$

Here, $P_J^{|M|}$ are the associated Legendre polynomials. As observed in Chapter 3, the Legendre functions are odd for odd J and even for even J. Hence, the rotational wave functions are alternately even and odd. The eigenvalues are $J(J+1)\hbar^2$, with $J = 0, 1, 2\ldots$ In the absence of an external field, each of these energy levels is $(2J + 1)$-fold degenerate because each of the states with different M has the same energy (i.e., the energy does not depend on the *orientation* of the rotating molecule). The degeneracy, g_J is $2J + 1$, corresponding to the allowed M values of 0, ±1, ±2, …, ±J.
Therefore,

$$E_J = \frac{\hbar^2}{2I}J(J+1) \quad J = 0, 1, 2\ldots \tag{4.19}$$

The quantity J, which can take integral values from zero upwards, is called the *rotational quantum number*; because of this restriction, only certain discrete rotational energy levels are allowed to the molecule. This J should not be confused with the total angular momentum quantum number of an atom, which has the same symbol.

For spectroscopic work, the expression for the energies of the rotational states allowed by the quantum restrictions is of prime importance. These allowed energies, given by equation (4.19), are shown schematically in Figure 4.5. Note that the ground state rotational energy is zero, but this is not in contradiction to the Uncertainty Principle since in the lowest quantum state all angles of rotation (0 to 2π) are equally probable. This is so because the spherical harmonic $Y_0^0(\theta,\varphi) = 1/\sqrt{4\pi}$ has no angular dependence. Since the angle (and hence the position) is completely uncertain, the momentum, and hence the energy, can be certainly known. Moreover, zero-point energy is only encountered for systems in which the particle is confined (particle-in-a-box, hydrogen atom, harmonic oscillator), but not for freely moving systems (free particle, rigid rotator). In free rotation, there is no confinement, and hence, there is no zero-point energy. However, if the rotation is *restricted*, so that some orientations become *preferred*, the zero-point energy becomes finite. The molecule may thus exist in its allowed state with a rotational energy of zero, and in this state, the molecule is not rotating.

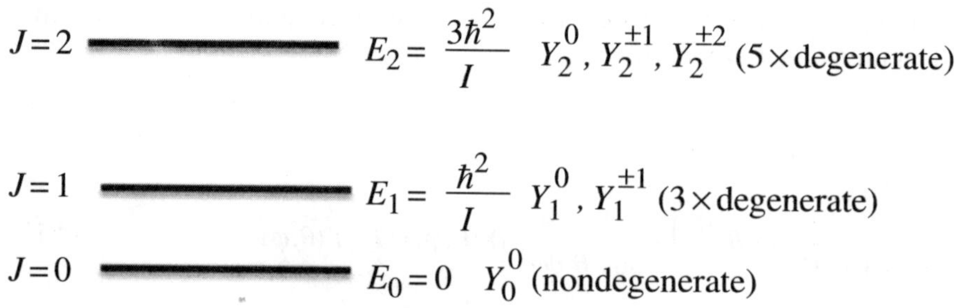

Figure 4.5. The allowed energies of a diatomic molecule

Example 4.2

(a) The moment of inertia of a solid sphere is given by ⅖mr^2. What is the moment of inertia of a cricket ball of mass 0.160 kg and radius 0.036 m? If it spins at 20 rad s^{-1}, what is its angular momentum and kinetic energy?

(b) What is the value of the orbital angular momentum quantum number, J, for this ball? How many different orientations may it take up with respect to the z-axis?

Solution

(a) For the cricket ball (sphere), the moment of inertia $I = \tfrac{2}{5}mr^2 = 2\times0.160\times0.036^2/5 = 8.3\times10^{-5}\,\text{kg m}^2$.

The angular momentum $L = I\omega_{\text{rot}} = 0.017\ \text{kg m}^2\ \text{s}^{-1}$.

The kinetic energy E_k is given by $E_k = \tfrac{1}{2}I\omega_{\text{rot}}^2 = 0.17\ \text{J}$.

(b) $E_J = E_{\text{rot}} = \dfrac{\hbar^2}{2I}J(J+1) \Rightarrow 0.17 = \dfrac{(6.626\times10^{-34})^2}{8\times3.142^2 I} = \dfrac{5.56\times10^{-69}}{8.3\times10^{-5}}J(J+1)$

$\Rightarrow J^2 + J - 2.5\times10^{63} = 0 \Rightarrow J \approx 5\times10^{31}$

Number of possible orientations $= 2J+1 \approx 10^{32}$

Hence, in the macroscopic world, quantization does not matter much, because J is very large, and the energy levels so closely spaced that they (and the angular momentum and its orientation) can be thought of as continuous, as in the classical model. Let us now look at a system of molecular dimensions.

Example 4.3

(a) What are the energies of the first four rotational energy levels of H_2 (r_e = 74.1 pm)? How do their values compare with k_BT at room temperature?

(b) What are the magnitudes of the angular momenta corresponding to these energies? Calculate the angular velocities (ω_{rot}) with which the molecule can rotate in those energy levels.

(c) What is the numerical value of the degeneracy of the energy levels?

Solution

(a) To calculate the moment of inertia of the hydrogen molecule, we require its reduced mass, which is given by

$$\mu_{red} = \frac{m_1 m_2}{m_1 + m_2} = \frac{m_H^2}{2m_H} = \frac{m_H}{2}$$

Note: The reduced mass of a homonuclear diatomic molecule is just half the atomic mass. The relative atomic mass of ^{1}H is 1.0078 *u*. Thus, $\mu_{red} = \frac{m_H}{2}$ $= 1.0078 / 2\,u = 0.5039\,u = 0.5039 \times 1.6605 \times 10^{-27} = 8.37 \times 10^{-28}$ kg.

The moment of inertia is thus $I = \mu_{red} r_e^2 = 4.59 \times 10^{-48}$ kg m^2.
The rotational energies are given by

$$E_J = \frac{\hbar^2}{2I} J(J+1) = (5.5606 \times 10^{-69} / 1.4945 \times 10^{-28}) J(J+1) = 1.2103 \times 10^{-21} J(J+1)$$

Thus, the first four energy levels are

$$E_0 = 0$$

$$E_1 = 1.21 \times 10^{-21} \times 1 \times 2 = 2.42 \times 10^{-21}\,\text{J}$$

$$E_2 = 1.21 \times 10^{-21} \times 2 \times 3 = 7.26 \times 10^{-21}\,\text{J}$$

$$E_3 = 1.21 \times 10^{-21} \times 3 \times 4 = 1.45 \times 10^{-20}\,\text{J}$$

At room temperature, 298 K, $k_B T = 1.38 \times 10^{-23} \times 298 = 4.12 \times 10^{-21}$ J. Hence, the rotational energies are comparable to the thermal energy, in contrast to the cricket ball for which the rotational energies are much smaller ~10^{-64} J and so a large number of rotational levels are occupied at room temperature.

(b) Since $E_{\text{rot}} = \frac{L^2}{2I}$, the rotational angular momentum is zero for $J = 0$.
For $J = 1$, applying $L^2 = 2IE_{\text{rot}}$ gives $|L| = 1.05 \times 10^{-34}$ kg m^2 s^{-1}. Since $L = I\omega_{\text{rot}}$, it follows that $\omega_{\text{rot}} = 1.05 \times 10^{-34}/4.59 \times 10^{-48} = 2.30 \times 10^{13}$ rad s^{-1}. Alternatively, the equation $E_{\text{rot}} = \frac{1}{2} I\omega_{\text{rot}}^2$ may be used to calculate ω_{rot}. Compared to the cricket ball, the molecule rotates much faster.

(c) The degeneracy of a level is $2J + 1$. Hence, the first four energy levels have degeneracies of 1, 3, 5 and 7, much smaller than ~10^{32} for the cricket ball.

Equation (4.19) expresses the allowed energies in Joule (the moment of inertia is expressed as kg m^{-2}); we are, however, interested in differences between these energies, or, more particularly, in the corresponding frequency, $\nu = \Delta E / h$ Hz, or wavenumber, $\tilde{\nu} = \Delta E / hc$ cm^{-1}, of the radiation emitted or adsorbed as a consequence of changes between energy levels. In the discussion of rotational spectra, transitions are usually expressed in terms of wavenumber, so it is convenient to consider energies expressed in these units. Hence, in the spectroscopic literature, the rotational term values $F(J) = E_J / hc$ are used instead of the energies.

We write

$$F(J) = \frac{E_J}{hc} = \frac{h}{8\pi^2 Ic} J(J+1) \qquad (4.20)$$

where c, the velocity of light, is expressed in cm s^{-1}, since the usual unit of wavenumber is the reciprocal centimetre. Equation (4.20) is usually abbreviated to

$$F(J) = \tilde{B}J(J+1)\ \text{cm}^{-1},\ J = 0, 1, 2... \tag{4.21}$$

where $\tilde{B}$, the *rotational constant*, is given by

$$\tilde{B} = \frac{h}{8\pi^2 Ic}\ \text{cm}^{-1} \tag{4.22}$$

Plainly, for $J = 0$, we have $F(0) = 0$ and the molecule is not rotating at all. For $J = 1$, the rotational term is $F(1) = 2\tilde{B}$ and a rotating molecule then has its lowest angular momentum.

4.2.1 Interaction of radiation with a rotating molecule

Equation (4.19) implies that the spacing between states increases as J increases, since

$$E_{J+1} - E_J = \frac{\hbar^2}{2I}[(J+1)(J+2) - J(J+1)] = \frac{\hbar^2}{I}(J+1) \tag{4.23}$$

As observed in Chapter 2, the intensity of a transition is proportional to the square of the transition dipole moment. For the transition $J'' \rightarrow J'$ (in spectroscopy, it is customary to designate the ground state with a double prime and the excited state by a single prime),

$$I_{J'J''} \propto \left| \int_{-\infty}^{+\infty} \psi_{J'}^{*} \hat{\mu}_0 \psi_{J''}\, d\tau \right|^2 \tag{4.24}$$

where μ_0 is the dipole moment of the rotor ($= \int \psi_{el} \hat{\mu} \psi_{el} d\tau_{el}$). We immediately see that the molecule must possess a *permanent dipole moment* for it to show pure rotational spectra, and that the intensity of a transition is proportional to the square of the dipole moment. The oscillating electric field grabs charges and torques the molecule (Figure 4.6).

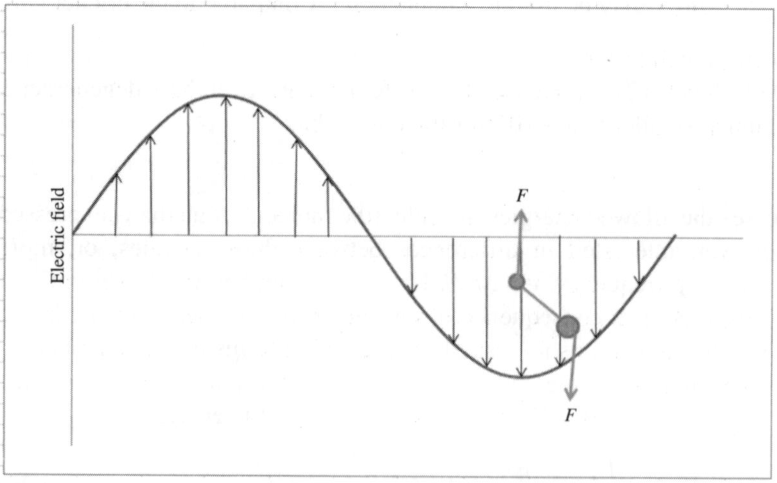

Figure 4.6. Interaction of a diatomic molecule with electromagnetic radiation

This selection rule implies that pure rotational transitions in *homonuclear* diatomic molecules (H_2, C_2, N_2, O_2), symmetric linear polyatomics like CO_2 and H-C≡C-H or molecules possessing tetrahedral or octahedral symmetry (e.g., CH_4, SF_6), etc. are electric dipole forbidden, but that they are allowed in *heteronuclear* diatomics, such as CO, NO, OH, SiO and other linear polyatomics like HCN and OCS. Typically, dipole moments are about ~3×10^{-30} C m – ~6×10^{-30} C m, although that of CO happens to be comparatively small (3×10^{-31} C m). The intensity of absorption in the pure rotation spectrum can be therefore used to calculate the permanent dipole moment of a molecule, but accurate intensity measurements are difficult, and the method is rarely applied.

The main gases in the atmosphere are N_2, O_2 and CO_2, none of which is microwave active. This allows radar, which operates with microwave frequencies, to be used as a detecting device without serious loss of signal due to absorption by the atmosphere through which the signals must pass. For the same reason, mobile phones operate on microwave radiation.

Example 4.4 Which of the following molecules exhibit pure rotational spectra: H_2, NH_3, CO_2, H_2O, benzene?

Solution

To exhibit a pure rotational spectrum, a molecule must possess a permanent dipole moment, and thus, only NH_3 and H_2O of those listed have a pure rotational spectrum. Note that our hydrogen molecule of Example 4.3 is transparent to microwave radiation. This implies that though hydrogen is rotating and has rotational energy levels, microwave spectroscopy does not allow us to study them. Other techniques, such as Raman spectroscopy, exist for studying the rotations of such molecules, as we shall see later.

A further restriction, besides that requiring the molecule to have a dipole, exists on the rotational energy changes that can occur. As we saw for atomic spectroscopy, conservation of angular momentum requires that the rotational quantum number can change by ±1 as a result of absorption or emission of radiation. When *absorption* of radiation is studied, only the change $\Delta J = +1$ is appropriate. Thus, this restriction allows the molecule to move up only one rotational energy step at a time. With this selection rule, the expected transitions can be indicated by the vertical arrows of Figure 4.7.

The selection rules for pure rotational transitions are therefore:

1. The molecule must possess a permanent dipole moment.
2. $\Delta J = \pm 1$, $\Delta M = 0, \pm 1$ [as in atomic spectroscopy, the latter is observable only in the presence of an applied electric or magnetic field (Stark or Zeeman Effect)].

When computing the *spectrum*, we need to consider *differences* between the levels. In general,

$$E_{\substack{\text{photon}\\ J\to J+1}} = h\nu_{\substack{\text{photon}\\ J\to J+1}} = \Delta E_{\text{rot}} = E_{J+1} - E_J = \frac{\hbar^2}{2I}\{(J+1)(J+2) - J(J+1)\}$$

$$= \frac{\hbar^2}{I}(J+1)$$

$$\therefore \nu_{\substack{\text{photon}\\ J\to J+1}} = \frac{h}{4\pi^2 I}(J+1) \quad (4.25)$$

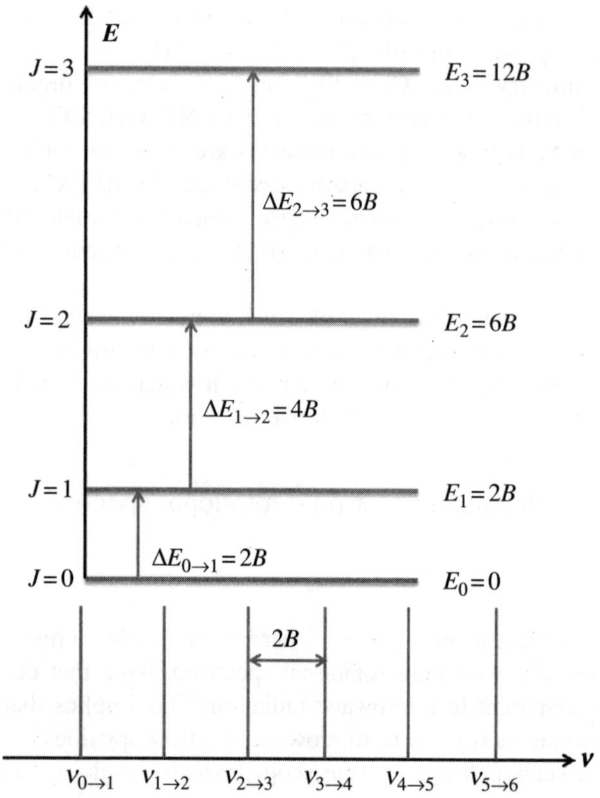

Figure 4.7. Allowed transitions between the energy levels of a rigid diatomic molecule and the rotational spectrum

In terms of the rotational constant,

$$B[\text{Hz}] \equiv \frac{h^2}{8\pi^2 I}, \nu_{\substack{\text{photon} \\ J\to J+1}}[\text{Hz}] = 2B(J+1) \tag{4.26}$$

or, in wavenumber,

$$\tilde{B}[\text{cm}^{-1}] \equiv \frac{h}{8\pi^2 Ic}, \tilde{\nu}_{\substack{\text{photon} \\ J\to J+1}}[\text{cm}^{-1}] = 2\tilde{B}(J+1) \tag{4.27}$$

Thus, a stepwise raising of the rotational energy results in an absorption spectrum consisting of lines at $2\tilde{B}$, $4\tilde{B}$, $6\tilde{B}$, ... cm^{-1}, while a similar lowering would result in an identical emission spectrum. This is shown at the bottom of Figure 4.7. In general, the spacing of adjacent lines is given by

$$\tilde{\nu}_{J+1\to J+2} - \tilde{\nu}_{J\to J+1} = 2\tilde{B}[(J+1)+1] - 2\tilde{B}(J+1) = 2\tilde{B} \tag{4.28}$$

This gives rise to a rigid rotor absorption spectrum with *evenly spaced* lines. The spacing between the transitions is $2B$ [Hz] or $2\tilde{B}\,[\text{cm}^{-1}]$ (Figure 4.7).

4.2.2 Rotational spectra and bond lengths of diatomic molecules

The next step is to see what molecular properties can be determined from rotational spectroscopy. Let us first calculate what radiation has quanta of energy that can cause a change in the rotational energy of a molecule from one allowed value to the next higher allowed value when absorbed by the molecule.

Example 4.5 Calculate the moment of inertia of the HF molecule (r_e = 92 pm), its rotational constant and hence the energy required to excite it from the $J = 0$ to the $J = 1$ level.

At what temperature does this energy equal the thermal energy, k_BT?

What wavelength of radiation is required for this excitation? Where in the electromagnetic spectrum does it fall?

Solution

The relative atomic masses of H and F are 1.00783 and 18.9984 *u*, respectively, to six significant places. Therefore, the reduced mass

$$\mu_{red} = \frac{m_H m_F}{m_H + m_F} = 0.957055\,u = 1.58923 \times 10^{-27}\ \text{kg}$$

since 1 u = 1.6605386 × 10^{-27} kg. The moment of inertia $I = \mu_{red} r_e^2 = 1.58923 \times 10^{-27}\text{kg} \times (92 \times 10^{-12}\ \text{m})^2 = 1.4 \times 10^{-47}\ \text{kg m}^2$.

The rotational constant is

$$B = \frac{\hbar^2}{2I} = 5.56062 \times 10^{-69} / 1.4 \times 10^{-47} = 4.1 \times 10^{-22}\ \text{J}$$

The energy require to excite the molecule from the $J = 0$ to $J = 1$ level is $2B = 8.2 \times 10^{-22}$ J.

Thermal energy $k_BT = 8.2 \times 10^{-22}$ J when $T = 8.2 \times 10^{-22} / 1.38066 \times 10^{-23} = 60$ K. Rotational energy differences are sometimes expressed in temperature units. The rotational temperature for the $J = 1$ level is thus 60 K (E_{rot} / k_B).

The corresponding wavelength $\lambda = hc / \Delta E = 2.4 \times 10^{-4}$ m = 2.4×10^{5} nm which falls between the microwave and infrared regions (Figure 1.7). Precisely, it is in the thermal infrared region.

Remember: The atomic mass unit is defined such that exactly one-twelfth of the mass of a gram-atom of carbon-12 is Avogadro's number times *u*. 1 atomic mass unit, $u = 1 / (1000\ N_A)$ = 1.6605386 × 10^{-27} kg. Since the energies pertain to a particular molecule, but because in a single molecule we have a particular isotope of each atom, we cannot use average molar masses. We use the *relative atomic mass of a single isotope* instead of the *standard atomic mass*, which is an average of the atomic masses of the different isotopes of an element in their natural abundance. The relative atomic mass of ^{1}H is 1.007825 *u* and that of ^{19}F is 18.99840 *u*. Note that the reduced mass of HF is almost 1 *u*, the atomic mass of a hydrogen atom. This is because

$$\frac{1}{\mu_{red}} = \frac{1}{m_H} + \frac{1}{m_F} \cong \frac{1}{m_H}$$

since $m_H << m_F$ and $1/m_H >> 1/m_F$.

We take up another example, this time that of a non-hydride.

Example 4.6 Assume that the carbon monoxide molecule is a rigid rotor with $r_e = 1.128 \times 10^{-10}$ m. Find the rotational energy (in Joule) for $J = 0$ and $J = 1$ and the wavelength of the transition between the two levels.

Solution

The masses of C and O (most prominent isotopes) are $m_C = 12.0000\ u$ (by definition) and $m_O = 15.9949\ u$, respectively. The reduced mass is therefore

$$\mu_{red} = \frac{12.0000u \times 15.9949u}{27.9949u} \times \frac{1.6605386 \times 10^{-27}\,\text{kg}}{u} = 1.13850 \times 10^{-26}\ \text{kg}$$

The moment of inertia is, therefore, $I = \mu_{red} r_e^2 = 1.4486 \times 10^{-46}$ kg m^2, and the rotational constant is $B = h^2 / 8\pi^2 I = 5.5606 \times 10^{-69} / I = 3.8386 \times 10^{-23}$ J

$E_0 = 0$

$E_1 = BJ(J + 1) = 7.6772 \times 10^{-23}$ J

The wavelength $\lambda = hc / \Delta E = 2.59 \times 10^{-3}$ m, which corresponds to the microwave range (10^{-3} to 1 m).

Notice the large rotational constant of the hydride, HF (Example 4.5). It is an order of magnitude higher than that of CO. Most of the heteronuclear diatomics are hydrides, which have very small moments of inertia and consequently large rotational constants and large rotational energy spacings. They generally absorb in the region between the microwave and infrared, for which very few instruments are available. Though large molecules absorb rotational energy in the microwave region, this is generally not true for most diatomic molecules.

Equation (4.27) shows that the wavenumber of the absorbed photon is $\tilde{\nu}_{\substack{\text{photon}\\ J\to J+1}} = 2\tilde{B}(J+1)$, and thus, an examination of the rotational spectrum can give a value of the rotational constant. The molecular quantity that can be related to this is the moment of inertia, since $\tilde{B} = h / 8\pi^2 Ic$. The moment of inertia is in turn related to the reduced mass and the bond length of the molecule. Thus, if one of these, usually the reduced mass, is known, the other can be easily determined. In the following example, we shall apply equation (4.27) to the observed spectrum of carbon monoxide in order to determine its moment of inertia and hence the CO bond length.

Example 4.7 Gilliam et al. (1950) measured the first line in the rotational spectrum of CO as 3.84235 cm^{-1}. Calculate $\tilde{B}$, I and r_{CO}.

Solution

$$\tilde{\nu}_{0-1} = 3.84235\ \text{cm}^{-1} = 2\tilde{B}(J+1) = 2\tilde{B}$$

Therefore, $\tilde{B} = 3.84235/2 = 1.92118\ \text{cm}^{-1}$. The moment of inertia is given by

$$I = \frac{h}{8\pi^2\tilde{B}c} = \frac{6.626068\times10^{-34}}{8\times(3.141592654)^2(2.99792458\times10^8)} = \frac{2.799274\times10^{-46}}{\tilde{B}}\ \text{kg m}^2$$
$$= 1.45706\times10^{-46}\ \text{kg m}^2$$

The reduced mass is given in Example 4.6 as 1.13850×10^{-26} kg. Therefore, $r_{CO}^2 = 1.45706 \times 10^{-46} / 1.13850 \times 10^{-26} = 1.27981 \times 10^{-20}\ \text{m}^2$ and $r_{CO} = 1.13129 \times 10^{-10}\ \text{m} = 113.129$ pm.

Note: The expression $I = \dfrac{2.799274\times10^{-46}}{\tilde{B}}\ \text{kg m}^2$ is a useful one to remember.

Example 4.7 illustrates how the microwave spectrum of a diatomic molecule allows us to accurately measure the internuclear distance, or bond length, of the molecule. Notice that we have determined the internuclear distance to six significant figures. This is because the *u* to kg conversion (1.6605386×10^{-27}) is known to 8 sig. figs., the speed of light (2.99792458×10^8) to 9 sig. figs, and Planck's constant (6.626068×10^{-34}) is also known to at least 7 sig. figs., so $\tilde{B}$ is the limiting sig. fig. Microwave spectroscopy has high resolving power and the number of significant figures to which $\tilde{B}$ has been determined is six. Hence, we report r_e = 113.129 pm. The application of quantum mechanics to find the allowed energy levels of a rotating diatomic molecule and microwave spectroscopic data has therefore allowed us to take a peek at molecular dimensions.

4.2.3 The effect of isotopic substitution

As seen above, the moment of inertia depends on both the reduced mass and the internuclear distance of the molecule. In cases where the latter is known, the reduced mass can be obtained, and this has led to the determination of precise atomic weights. When a particular atom in a molecule is replaced by its isotope, the resultant isotopologue is identical chemically with the original. In particular, there is no appreciable change in internuclear distance on isotopic substitution, since the bonding is produced by the electronic configuration and nuclear charges, which do not change on isotopic substitution. (We shall see in Chapter 5 that this applies only to the equilibrium internuclear distance at the bottom of the potential energy well.) There is, however, a change in the total mass and hence in the reduced mass and moment of inertia and $\tilde{B}$ value for the molecule.

Example 4.8 Calculate the reduced masses of $^{1}H^{35}Cl$, $^{2}H^{35}Cl$ and $^{1}H^{37}Cl$.

Solution

Reduced mass, $\mu_{red} = \dfrac{m_1 m_2}{m_1 + m_2}$.

Hence, for $^1H^{35}Cl$,

$$\mu_{red} = \frac{1.0078u.34.9689u}{35.9767u} = 0.9796u \times 1.6605386\ \times 10^{-27}\ \text{kg}\ u^{-1} = 1.6267 \times 10^{-27}\ \text{kg}$$

Likewise for $^2H^{35}Cl$,

$$\mu_{red} = \frac{2.0141u.34.9689u}{36.9830u} = 1.9044\ u = 3.1624 \times 10^{-27}\ \text{kg}$$

And for $^1H^{37}Cl$,

$$\mu_{red} = \frac{1.0078u.36.9659u}{37.9737u} = 0.9811\ u = 1.6291 \times 10^{-27}\ \text{kg}$$

Example 4.8 shows that, on going from $^1H^{35}Cl$ to $^2H^{35}Cl$, there is a mass increase (almost double) and therefore a decrease in the B value. The example also illustrates that isotopic substitution of the lighter atom has the greatest effect on the reduced mass, since substitution of ^{35}Cl by its heavier isotope ^{37}Cl has negligible effect on the reduced mass. If we designate the $^2H^{35}Cl$ molecule with a prime, we have $B' < B$. This change will be reflected in the rotational energy levels of the molecule, and the spectrum of the heavier species will show a smaller separation between the lines ($2B'$) than that of the lighter one ($2B$). Therefore,

$$\frac{\tilde{B}}{\tilde{B}'} = \frac{h}{8\pi^2 Ic} \cdot \frac{8\pi^2 I'c}{h} = \frac{I'}{I} = \frac{\mu_{red}'}{\mu_{red}} \tag{4.29}$$

where we have considered the internuclear distance unchanged by isotopic substitution.

Example 4.9 The rotational constant for $^1H^{35}Cl$ is 10.5909 cm^{-1}. Determine the rotational constants for $^2H^{35}Cl$ and $^1H^{37}Cl$.

Solution

From equation (4.29) and Example 4.8,

$$\frac{\tilde{B}_{^1H^{35}Cl}}{\tilde{B}_{^2H^{35}Cl}} = \frac{3.16235 \times 10^{-27}}{1.62666 \times 10^{-27}} = 1.94408;$$

hence, $\tilde{B}_{D^{35}Cl} = 5.44778\ \text{cm}^{-1}$

Likewise,

$$\frac{\tilde{B}_{^1H^{35}Cl}}{\tilde{B}_{^1H^{37}Cl}} = \frac{1.629125 \times 10^{-27}}{1.62666 \times 10^{-27}} = 1.00152\ \text{and}\ \tilde{B}_{^1H^{37}Cl} = 10.5749\ \text{cm}^{-1}$$

The above example illustrates how isotopic substitution, particularly of H by its D isotope, has a profound effect on the rotational constants. Isotopic substitution has been the basis of

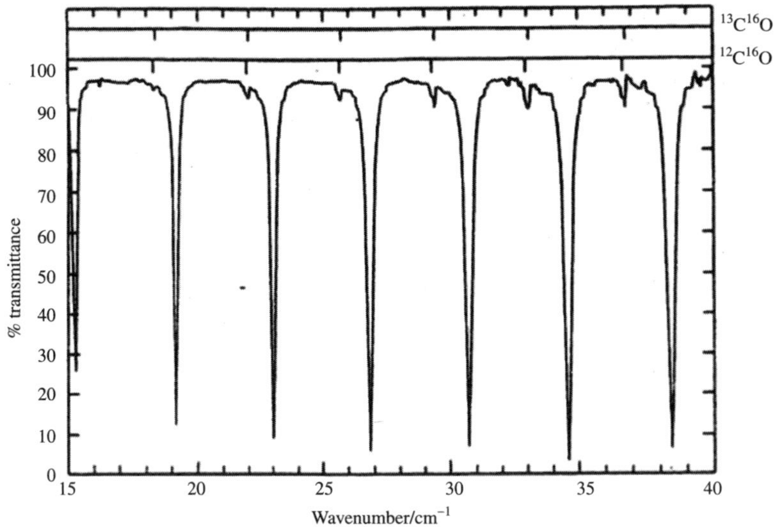

Figure 4.8. Part of the rotational (far infrared) spectrum of CO showing transitions with $J'' = 3$ to 9 [Fleming and Chamberlain (1974), reproduced with permission from Elsevier]

determination of accurate atomic masses. Figure 4.8 is part of the rotational spectrum of CO. One can see that, for every peak in the spectrum, there is another peak at lower wavenumber, corresponding to the $^{13}C^{16}O$ in its natural abundance.

Accurate measurements of the $J = 0 \rightarrow 1$ transition for $^{12}C^{16}O$ and $^{13}C^{16}O$ give separations of 3.84235 and 3.67337 cm^{-1}, respectively. Using the mass of oxygen (15.9949 u) and that of carbon-12 (12.0000 u by definition), we have from equation (4.29)

$$\frac{\mu_{red}'}{\mu_{red}} = \frac{3.84235}{3.67336} = 1.04600 = \frac{15.9949m'}{15.9949 + m'} \cdot \frac{12.0000 + 15.9949}{12.0000 \times 15.9949}$$

from which m', the atomic mass of carbon-13, is found to be 13.0007 u. This is within 0.02% of the currently accepted value (13.0034 u). It must be appreciated that the data were obtained from normal carbon monoxide without any additional isotopic substitution. Moreover, these data were published by Gilliam et al. (1950) at a time when Planck's constant was not accurately known. Equation (4.29) completely eliminates the Planck's constant. The relative absorption intensities of the two absorptions in the microwave also allow an estimate of the natural abundance of the ^{13}C carbon isotope (1.1% of ordinary carbon monoxide).

4.2.4 Rotational energy level populations: Intensities of spectral lines

Substitution of the various quantities in (4.24) shows that the transition probabilities show little variation with J, i.e., to a good approximation, all changes with $\Delta J = \pm 1$ are equally probable. However, this does not imply that all spectral lines will be equally intense. As we observed in

Chapter 1, the population of a level also plays a key role in deciding the intensity of a line emanating from that level. Since the intrinsic probabilities are identical, the relative intensities of the lines are proportional to the populations in the various levels. The rotational levels lie close in energy; hence, several levels can be populated significantly, even at very low temperatures (see Example 1.9). For example, for CO, the energy spacing between the $J = 0$ and 1 levels corresponds to only 5.5 K ($k_BT = 7.6 \times 10^{-23}$ J). Since the moment of inertia appears in the denominator of (4.25), the spacing between the allowed energies decreases as the moment of inertia increases.

The population of a level is governed by the Boltzmann distribution law (see Section 1.6)

$$N_J = g_J \frac{N}{q} \exp\left(\frac{-E_J}{k_BT}\right) = (2J+1)\frac{N}{q}\exp\left(-\frac{BJ(J+1)}{k_BT}\right) = (2J+1)\frac{N}{q}\exp\left(-\frac{\tilde{B}hcJ(J+1)}{k_BT}\right) \quad (4.30)$$

where N is the total number of molecules, g_J is the *degeneracy* of the Jth energy level $(= 2J + 1)$, q is the rotational partition function $\left(= \sum_{J=0}^{\infty}(2J+1)\exp\left(-\frac{BJ(J+1)}{k_BT}\right)\right)$ and the other symbols have their usual meaning. Since the degeneracy factor increases linearly with J, while the exponential factor decreases, the population rises to a maximum, and then diminishes, as shown in Figure 4.9.

The J_{max} value, i.e., the rotational level with the maximum population, can be obtained by differentiating N_J with respect to J and finding the nearest whole number for which the differential is zero.

Since $N_J = N(2J+1)e^{-E_J/kT}/q = (2J+1)Ne^{-\tilde{B}hcJ(J+1)/k_BT}/q$,

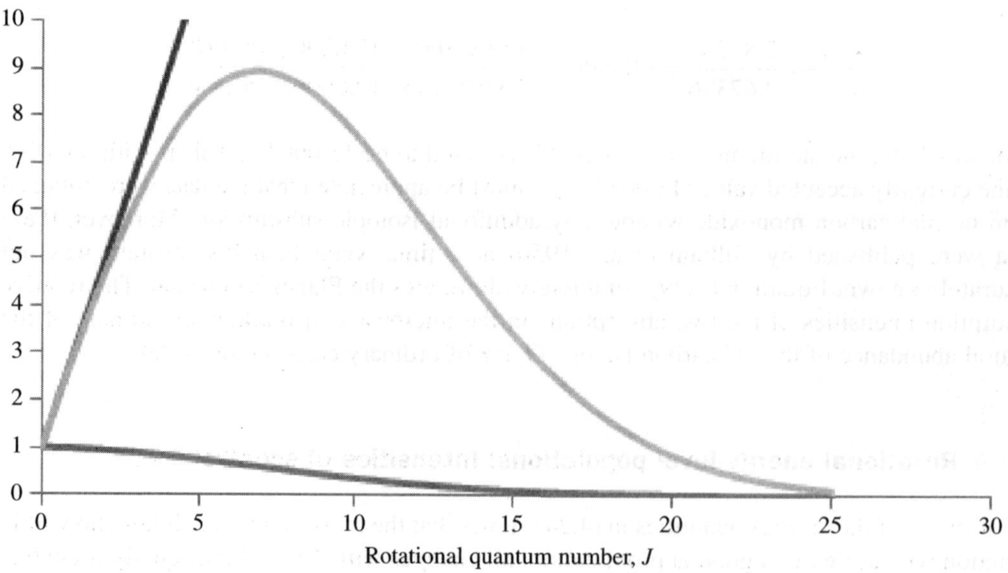

Figure 4.9. Variation of the population with rotational quantum number for CO at 300 K

$$\frac{dN_J}{dJ} = e^{-E_J/kT}\left[2-(2J+1)\frac{\tilde{B}hc}{k_BT}(2J+1)\right] = 0$$

$$\Rightarrow (2J+1)^2\frac{\tilde{B}hc}{k_BT} = 2$$

$$\therefore (2J+1)^2 = \frac{2k_BT}{\tilde{B}hc}$$

$$(2J+1) = \sqrt{\frac{2k_BT}{\tilde{B}hc}}$$

$$\therefore 2J = \sqrt{\frac{2k_BT}{\tilde{B}hc}} - 1$$

$$J_{\text{max}} = \frac{1}{2}\sqrt{\frac{2k_BT}{\tilde{B}hc}} - \frac{1}{2} = \sqrt{\frac{k_BT}{2\tilde{B}hc}} - \frac{1}{2} \tag{4.31}$$

The rotational energy of the level which has the highest population is, therefore,

$$E_{\text{rot}}(\text{max}) = J_{\text{max}}(J_{\text{max}}+1)\tilde{B}hc$$

$$= \left(\sqrt{\frac{k_BT}{2\tilde{B}hc}} - \frac{1}{2}\right)\left(\sqrt{\frac{k_BT}{2\tilde{B}hc}} + \frac{1}{2}\right)\tilde{B}hc$$

$$= \left(\frac{k_BT}{2\tilde{B}hc} - \frac{1}{4}\right)\tilde{B}hc$$

$$= \frac{k_BT}{2} - \frac{\tilde{B}hc}{4} \tag{4.32}$$

Except at very low temperatures, $\frac{k_BT}{2} >> \frac{\tilde{B}hc}{4}$; the energy of the most populated level is, therefore, $\approx \frac{k_BT}{2}$.

Substituting the constants in equation (4.31), we obtain $J_{\text{max}} = \sqrt{\frac{T}{2.87755\tilde{B}}} - \frac{1}{2}$, where $\tilde{B}$ is expressed in cm^{-1}. Except at very low temperatures, the first term is much larger than the second, and we may neglect the latter. The rotational quantum number corresponding to the most populated level J_{max} is then the integer nearest to $0.59\sqrt{\frac{T(K)}{\tilde{B}(\text{cm}^{-1})}}$ and the frequency of the band envelope maximum, measured relative to the band origin, is $\tilde{\nu}_{\text{max}} = 2\tilde{B}J_{\text{max}}(J_{\text{max}}+1) \approx 1.18\sqrt{\tilde{B}T}$, where we have assumed that $J_{\text{max}} >> 1$. This is a useful rule of thumb.

Example 4.10 Which rotational state of CO would be most populated at 300 K?

Solution

In Example 4.7, we saw that $\tilde{B}$ for CO is 1.92118 cm^{-1}. Inserting in equation (4.31), we obtain

$$J_{\text{max}} = \sqrt{\frac{k_BT}{2\tilde{B}hc}} - \frac{1}{2} = \sqrt{\frac{1.38066\times10^{-23}\times300}{2\times1.92118\times6.62608\times10^{-34}\times2.99792\times10^{10}}} - \frac{1}{2} = 6.87$$

The nearest whole number is 7.

The temperature dependence of the intensities of the microwave lines can be used to determine the temperature, as in flames or stellar temperatures. This is often referred to as *rotational thermometry*.

Example 4.11 For the ground state of CN, the maximum occurs at $J = 10$ at 300 K. In a flame, it occurs at $J = 27$. Estimate the temperature of the flame.

Solution

At 300 K,

$$J_{\max} = 10 = \sqrt{\frac{300k_B}{2\tilde{B}hc}} - \frac{1}{2} \Rightarrow \sqrt{\frac{300k_B}{2\tilde{B}hc}} = 10.5$$

At the flame temperature, T, this becomes

$$J_{\max} = 27 = \sqrt{\frac{k_BT}{2\tilde{B}hc}} - \frac{1}{2} \Rightarrow \sqrt{\frac{k_BT}{2\tilde{B}hc}} = 27.5$$

Dividing the second equation by the first, we obtain

$$\sqrt{\frac{T}{300}} = \frac{27.5}{10.5}$$

Hence, the flame temperature, $T = 2058$ K.

According to Example 4.10, we would expect the rotational transition originating from $J'' = 7$ to be the most intense in the rotational spectrum of CO. However, this is not the case, and the observed spectrum (dark lines) shows that the maximum intensity is for a transition corresponding to a higher J'' value (Figure 4.10). This is because we have ignored another

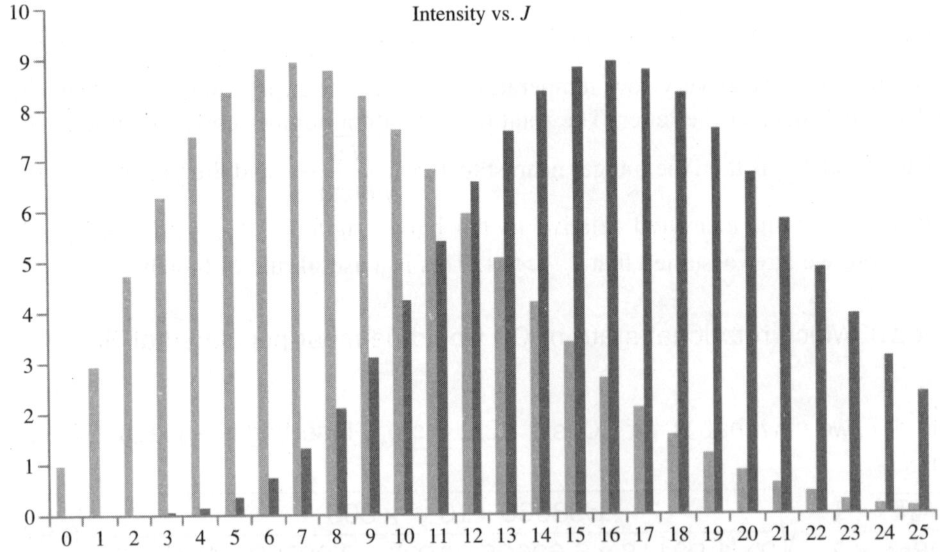

Figure 4.10. Schematic spectrum of CO at 300 K. Dark lines show the observed spectrum

factor that governs the intensities of rotational transitions. As discussed in Chapter 1, there are three processes by which radiation interacts with the sample, i.e., stimulated absorption, stimulated emission and spontaneous emission. In the microwave region, we need only be concerned with stimulated processes; spontaneous emission is negligibly slow, as we noted in Section 1.7. The rate of stimulated absorption depends on the population of the lower level, whereas that for stimulated emission depends on the population of the upper level. In the case of rotational energy levels, it is likely that *both* energy levels have significant populations, so we need to consider both processes.

Consider the allowed transitions between levels J and $J + 1$, with populations N_J and N_{J+1}, respectively, shown in the diagram below. Note that the degeneracy of the lower level is $(2J+1)$ and that of the upper level is $(2(J + 1) + 1) = (2J + 3)$.

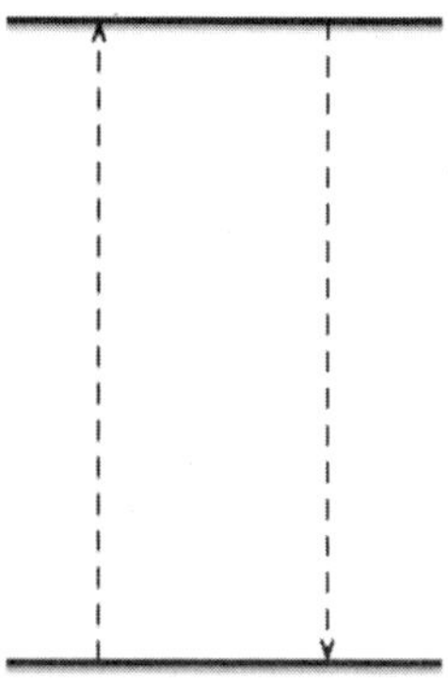

The net rate of absorption of photons is given by

$$\frac{dN_J}{dt} = -B_{J,\,J+1}\rho(\nu_J)N_J + B_{J+1,\,J}\rho(\nu_J)N_{J+1} \tag{4.33}$$

The first term is due to stimulated absorption, which decreases the population of level J, and is hence negative; the second term is due to stimulated emission and is thus positive. $\rho(\nu_J)$ is the energy density at the frequency of the transition, ν_J. The Einstein B coefficients are *not* the same for both processes, as the upper and lower levels have different degeneracies. According to equation (1.12), the relationship between the coefficients is

$$\frac{B_{lm}}{B_{ml}} = \frac{g_m}{g_l}$$

In this specific case, the relationship becomes

$$\frac{B_{J,\,J+1}}{B_{J+1,\,J}} = \frac{2J+3}{2J+1};$$

hence,

$$B_{J+1,\,J} = \frac{2J+1}{2J+3}B_{J,\,J+1} \tag{4.34}$$

The populations are given by the Boltzmann distribution (4.30). Taking the ratio of the populations of the upper and lower levels,

$$\frac{N_{J+1}}{N_J} = \frac{g_{J+1}}{g_J}\exp\left(-\frac{(E_{J+1}-E_J)}{k_BT}\right)$$

$$\Rightarrow N_{J+1} = N_J\frac{2J+3}{2J+1}\exp\left(\frac{-h\nu_J}{k_BT}\right) \tag{4.35}$$

where ν_J is the frequency of the transition from J to $J+1$, i.e., $h\nu_J = E_{J+1} - E_J$.

Substituting for N_{J+1} from equation (4.35) and for $B_{J+1,\,J}$ from equation (4.34) in equation (4.33) gives

$$\frac{dN_J}{dt} = -B_{J,\,J+1}\rho(\nu_J)N_J + B_{J+1,\,J}\rho(\nu_J)N_{J+1}$$

$$= -B_{J,\,J+1}\rho(\nu_J)N_J + \frac{2J+1}{2J+3}B_{J,\,J+1}\rho(\nu_J)N_J\frac{2J+3}{2J+1}\exp\left(-\frac{h\nu_J}{k_BT}\right)$$

$$= -B_{J,\,J+1}\rho(\nu_J)N_J\left[1-\exp\left(-\frac{h\nu_J}{k_BT}\right)\right]$$

where the degeneracy terms cancel out. For rotational transitions, $h\nu_J \ll k_BT$, so the exponential can be approximated by the first two terms in the Taylor expansion

$$\frac{dN_J}{dt} = -B_{J,\,J+1}\rho(\nu_J)N_J\left[1-\left(1-\frac{h\nu_J}{k_BT}\right)\right]$$

$$= -B_{J,\,J+1}\rho(\nu_J)N_J\frac{h\nu_J}{k_BT} \tag{4.36}$$

$$= -B_{J,\,J+1}\rho(\nu_J)(2J+1)\frac{N}{q}\exp\left(-\frac{E_J}{k_BT}\right)\frac{h\nu_J}{kT}$$

where we have used N_J from the Boltzmann equation (4.30). The rate is negative, meaning that, overall, stimulated absorption dominates over stimulated emission.

Detailed calculations show that the Einstein coefficients depend on rotational levels according to

$$B_{J,\,J+1} \propto \frac{J+1}{2J+1}; \tag{4.37}$$

The variation is rather slight and, for large J, the coefficients become independent of J. Substitution in equation (4.36) gives

$$\frac{dN_J}{dt} = -B_{J,\,J+1}\rho(\nu_J)(2J+1)\frac{N}{q}\exp\left(-\frac{E_J}{k_BT}\right)\frac{h\nu_J}{k_BT}$$

$$\propto \frac{J+1}{2J+1}\rho(\nu_J)(2J+1)\frac{N}{q}\exp\left(-\frac{E_J}{k_BT}\right)\left(\frac{h\nu_J}{k_BT}\right)$$

$$\propto (J+1)\rho(\nu_J)\frac{N}{q}\exp\left(-\frac{E_J}{k_BT}\right)\left(\frac{h\nu_J}{k_BT}\right) \tag{4.38}$$

Finally, we need to realize that what the spectrometer measures is the rate of absorption (or emission) of *energy* (i.e., the *power*). Each time the population changes by one, a photon of energy $h\nu_J$ is absorbed (or emitted). So, the rate of absorption of energy, I, is given by the rate of change of the population, times the energy of the photon:

$$I = \frac{dN_J}{dt} \times h\nu_J \tag{4.39}$$

Substituting equation (4.38), we find the following variation in the intensity, assuming that $\rho(\nu_J)$ is constant for the range of frequencies of interest:

$$I \propto (J+1)\nu_J^2 \exp\left(-\frac{E_J}{kT}\right) \tag{4.40}$$

This dependence is quite different from (4.31), where we had considered that the intensity is proportional only to the population of the lower energy level, and hence, the J value at which the intensity reaches a maximum, as predicted by these two equations, is quite different, as is shown in Figure 4.10.

4.2.5 Centrifugal distortion

In the previous sections, we indicated how internuclear distances could be determined from microwave spectra. The data for $\tilde{B}$ were taken for the first few rotational transitions. However, examination of higher spectral lines does not show the constant $2\tilde{B}$ separation produced by equation (4.28). In fact, the separation of spectral lines steadily decreases, ending in a continuum when the J value is sufficiently high. The decrease in line separations, shown by the spectrum of HCl in the far infrared, is given in Table 4.1; it is evident that the separation between successive lines (and hence the apparent $\tilde{B}$ value) decreases steadily with increasing J.

The last column of Table 4.1 gives the calculated values of the internuclear distance using the observed $\tilde{B}$ values, which show a steady increase with increasing rotational quantum number. As J increases, the rotational energy also increases, the molecule rotates faster and the increasing centrifugal distortion tends to move the atoms apart, thus increasing the moment of inertia. Thus, our assumption of rigidity is not correct. Real bonds are elastic and can be distorted, and finally, at large values of J, the rotational energy may be enough

Table 4.1. Rotational spectrum of HCl

$\tilde{\nu}_J$/cm^{-1}	$J \to J+1$	$\Delta\tilde{\nu}$/cm^{-1}	$\tilde{B}$/cm^{-1}	r_e / nm
84.32	3→4	-	-	0.1272
104.13	4→5	20.81	10.41	0.1273
124.73	5→6	20.60	10.30	0.1279
145.37	6→7	20.64	10.32	0.1278
165.89	7→8	20.52	10.26	0.1282
186.23	8→9	20.34	10.17	0.1287
206.60	9→10	20.37	10.19	0.1286
226.86	10→11	20.26	10.13	0.1290

to break the bond. This manifests itself in the appearance of a continuum at large wavenumbers.

The centrifugal distortion is taken into account by modifying equation (4.21) for the rotational energy levels by adding a term

$$F(J) = \tilde{B}J(J+1) - \tilde{D}J^2(J+1)^2 \tag{4.41}$$

where $\tilde{D}$, known as the *centrifugal distortion constant*, can be evaluated from spectral results and is positive in sign. The centrifugal distortion constant is expected to be somehow related to the strength of the bond, i.e., its force constant k, which can be determined from its vibrational frequency ($k = 4\pi^2\omega_{\text{vib}}^2 c^2 \mu_{red}$) (this is discussed in detail in Chapter 5). In fact, it can be shown that $\tilde{D} = \dfrac{4\tilde{B}^3}{\omega_{\text{vib}}^2}$ (see Appendix I). Note that the vibrational frequency occurs in the denominator: larger the vibrational frequency and hence the force constant, stronger is the bond and smaller the centrifugal distortion. We shall see that vibrational frequencies are usually of the order of 10^3 cm^{-1}, while $\tilde{B}$ we have found to be of the order of 10 cm^{-1}. Thus, we see that $\tilde{D}$, being of the order of 10^{-3} cm^{-1}, is very small compared with $\tilde{B}$. For small J, therefore, the correction term $\tilde{D}J^2(J+1)^2$ is almost negligible, while for J values of 10 or above, it may become appreciable. Since $\tilde{D}$ is positive, its effect is to decrease the spacing of rotational lines with increasing J, leading to a bunching of lines in the high wavenumber region of the spectrum.

The selection rule for the non-rigid rotor remains the same as that for the rigid rotator, $\Delta J = \pm 1$. We may easily write an analytical expression for the transitions

$$\begin{aligned} F(J+1) - F(J) = \tilde{\nu}_J &= \tilde{B}[(J+1)(J+2) - J(J+1)] - \tilde{D}\left[(J+1)^2(J+2)^2 - J^2(J+1)^2\right] \\ &= 2\tilde{B}(J+1) - 4\tilde{D}(J+1)^3 \text{ cm}^{-1} \end{aligned} \tag{4.42}$$

where $\tilde{\nu}_J$ represents equally the upward transition from J to $J + 1$, or the downward from $J + 1$ to J. Thus, we see analytically that the spectrum of the elastic rotor is similar to the rigid molecule except that each line is displaced slightly to low frequency, the displacement increasing with $(J + 1)^3$.

Example 4.12 The wavenumbers of the pure rotational spectrum of HF can be fit to an equation

$$\tilde{\nu}_J = 41.122(J+1) - 8.52 \times 10^{-3}(J+1)^3$$

Deduce the values of the rotational constant ($\tilde{B}$), centrifugal distortion constant ($\tilde{D}$), the vibrational frequency (ω_{vib}) and the force constant (k) of the H-F bond.

Solution

Comparison with (4.42) gives $2\tilde{B} = 41.122$ cm^{-1} and $4\tilde{D} = 8.52 \times 10^{-3}$ cm^{-1}. Therefore, $\tilde{B} = 20.561$ cm^{-1} and $\tilde{D} = 2.13 \times 10^{-3}$ cm^{-1}. Furthermore, $\tilde{D} = \dfrac{4\tilde{B}^3}{\omega_{\text{vib}}^2}$, which gives

$$\omega_{\text{vib}}^2 = \frac{4\tilde{B}^3}{\tilde{D}} = \frac{4 \times 20.561^3}{2.13 \times 10^{-3}} = 1.63235 \times 10^7 \text{cm}^{-2} \text{ and } \omega_{\text{vib}} = 4040 \text{ cm}^{-1}.$$

The force constant $k = 4\pi^2\omega_{\text{vib}}^2 c^2 \mu_{red}$. Substitution of the constants, the vibrational frequency and reduced mass of HF from Example 4.5 gives k = 960 N m^{-1}, which is a large value, indicating the strength of the HF bond.

Equation (4.42) can be rearranged to a linear form by dividing throughout by $(J + 1)$

$$\frac{\tilde{\nu}_J}{J+1} = 2\tilde{B} - 4\tilde{D}(J+1)^2 \text{ cm}^{-1} \tag{4.43}$$

Thus, a plot of $\frac{\tilde{\nu}_J}{J+1}$ versus $(J + 1)^2$ yields a straight line with intercept $2\tilde{B}$ and slope $-4\tilde{D}$. The data from Table 4.1 for HCl yields $2\tilde{B} = 20.87 \text{ cm}^{-1}$ and $4\tilde{D} = 0.002053 \text{ cm}^{-1}$. This gives $\tilde{B} = 10.435 \text{ cm}^{-1}$ and $\omega_{\text{vib}}^2 = \frac{4\tilde{B}^3}{\tilde{D}} = 8.8554043 \times 10^6 \ (\text{cm}^{-1})^2$, from which $\omega_{\text{vib}} \approx 2976$ cm^{-1}.

A more precise value obtained from infrared spectroscopy (to be studied in the next chapter) is 2880 cm^{-1}, a difference of ~3%, which is excusable, considering the small, and hence relatively inaccurate, value of $\tilde{D}$.

It is not always possible to assign J values to the transitions. In that case, if three consecutive lines are isolated, the three unknowns, J, $\tilde{B}$ and $\tilde{D}$ can be determined. Equation (4.42) is derived for harmonic vibrations only. If anharmonicity of the vibrations is taken into account, more terms need to be added

$$F(J) = \tilde{B}J(J+1) - \tilde{D}J^2(J+1)^2 + \tilde{H}J^3(J+1)^3 + \tilde{K}J^4(J+1)^4 + \ldots \tag{4.44}$$

where $\tilde{H}$, $\tilde{K}$, etc., are small constants dependent on the geometry of the molecule. They are, however, negligible compared with $\tilde{D}$, and most modern spectroscopic data are adequately fitted by equation (4.42).

4.3 ROTATION OF POLYATOMIC MOLECULES: CLASSIFICATION OF MOLECULES

Except linear molecules, all molecules possess three rotational degrees of freedom. In order to interpret their rotational spectra, the molecules are classified according to their *principal moments of inertia*. We initially ignore centrifugal distortion and assume that the relative nuclear positions are fixed, i.e., the molecule is rigid. In mechanics, the inertial properties of a rotating rigid body are fully described by its moment of inertia I, which is a measure of its resistance to changes in its rotation rate. The position of the centre of mass (COM) for a given axis α may be found from the equation $\vec{r}_\alpha(COM) = \frac{\sum_i m_i \vec{r}_i}{\sum_i m_i}$ and the moment of inertia of a rigid molecule about any axis passing through the centre of mass is defined by $I_\alpha = \sum_i m_i r_i^2$, where m_i is the mass of the ith nucleus and r_i its perpendicular (shortest) distance from the axis. The total mass of the molecule is $M = \sum_i m_i$.

We now draw lines of length $(1/\sqrt{I_\alpha})$, where I_α is the moment of inertia calculated along the line as an axis, radially from the centre of mass of the molecule. The envelope of all these lines is then a triaxial ellipsoid, known as the *momental* or *moment of inertia* ellipsoid. The three mutually perpendicular axes of the ellipsoid, about which the rotation is dynamically balanced, coincide with the *principal axes of inertia* of the molecule. The axes are labelled *a*, *b* and *c* in order of decreasing length, so that *a* is the major axis, *b* the intermediate axis and *c* the minor axis. Since the lengths of the axes are inversely proportional to the square root of the moment of inertia along that axis, it follows that $I_a \le I_b \le I_c$.

In practice, the moments of inertia for any rigid three-dimensional body can be obtained from its 3 × 3 moment of inertia tensor (i.e., matrix),

$$\begin{vmatrix} I_{xx} & -I_{xy} & -I_{xz} \\ -I_{yx} & I_{yy} & -I_{yz} \\ -I_{zx} & -I_{zy} & I_{zz} \end{vmatrix} \tag{4.45}$$

The molecule is put in a Cartesian frame of coordinates (*xyz*), and the various diagonal terms of the moment of inertia tensor are calculated as (using the *x*-axis, for example)

$$I_{xx} = \sum_i m_i(y_i^2 + z_i^2) - \frac{\left(\sum_i m_i y_i\right)^2}{\sum_i m_i} - \frac{\left(\sum_i m_i z_i\right)^2}{\sum_i m_i}, \tag{4.46}$$

and the off-diagonal elements are given by

$$I_{xy} = \sum_i m_i x_i y_i - \frac{\sum_i m_i x_i \sum_i m_i y_i}{\sum_i m_i}, \text{ etc.,} \tag{4.47}$$

where x_i, y_i and z_i are the coordinates of the *i*th atom of mass m_i and the sum is taken over all the atoms present in the molecule. If the origin of the coordinate system is taken at the centre of mass, the various terms simplify since, by definition of the centre of mass, terms such as $\sum_i m_i x_i$ vanish, leaving only the first term in each expression.

The moment of inertia tensor is clearly symmetric, and hence Hermitian, and so can be diagonalized, yielding three eigenvalues, I_a, I_b and I_c, labelled such that $I_a \le I_b \le I_c$. The product of the three eigenvalues equals the determinant of the matrix. These eigenvalues are called the *principal moments of inertia* and the eigenvectors corresponding to the diagonalized coordinate system are referred to as the *principal axes* of the molecule. The matrix containing the eigenvectors is the rotation matrix that rotates the coordinate system so as to coincide with the principal axes of the molecule. In the new coordinate system, the principal axis system (*a,b,c*), $I_a^2 a^2 + I_b b^2 + I_c c^2 = 1$. Comparison with the equation of an ellipse $\frac{x^2}{a^2} + \frac{y^2}{b^2} + \frac{z^2}{c^2} = 1$ shows that the lengths of the axes of the momental ellipsoid in this rotated coordinate system are $1/\sqrt{I_a}$, $1/\sqrt{I_b}$ and $1/\sqrt{I_c}$, respectively. The principal moments of inertia are then

$$I_a = \sum_{i=1} m_i(b_i^2 + c_i^2)$$

$$I_b = \sum_{i=1} m_i(a_i^2 + c_i^2)$$

$$I_c = \sum_{i=1} m_i(a_i^2 + b_i^2) \tag{4.48}$$

where the coordinates of atom i in the principal axis system are given by (a_i, b_i, c_i). The calculation of the moments and principal axes is straightforward, but tedious! However, computer routines that calculate the various terms in the moment of inertia tensor and diagonalize it to obtain the principal axes and rotational constants exist, making it possible to evaluate these for even complex molecules.

Fortunately, even without the aid of these computer routines, some simplification is possible. In symmetrical molecules, the direction of the principal axes of inertia can often be determined by inspection, since the axes of the momental ellipsoid must coincide with the symmetry axes of the molecule. Similarly, every plane of symmetry of the molecule must contain the principal section of the ellipsoid, i.e., it must contain two of the principal axes of inertia and be perpendicular to the third. We may therefore choose our Cartesian frame in such a way that the Cartesian axes coincide with the rotation axes of the molecule or are perpendicular to the planes of symmetry of the molecule.

Let us use these ideas to find the orientation of the principal axes of inertia in the C_{2v} molecule formaldehyde. The elements of symmetry in the molecule are the two-fold rotation axis C_2 and two mutually perpendicular planes of symmetry σ_v and σ_v' containing the C_2 axis. One of the axes of inertia must then coincide with the C_2 axis, and the other two must lie in each of the planes σ_v. The centre of mass of the molecule lies on the C = O bond axis. We choose our z-axis to coincide with the C_2 axis. Then, according to the group theoretical convention, the molecular plane is yz and the x-axis is normal to the molecular plane. As all the x_i's are zero since the molecule is in the yz plane, equation (4.47) implies that I_{xy} and I_{xz} are also zero, and the determinant factorizes into block-diagonal form, i.e., one linear and one quadratic expression. The rotation matrix then entails only a rotation about the x-axis to bring the y and z axes to also coincide with the principal axes of the ellipse.

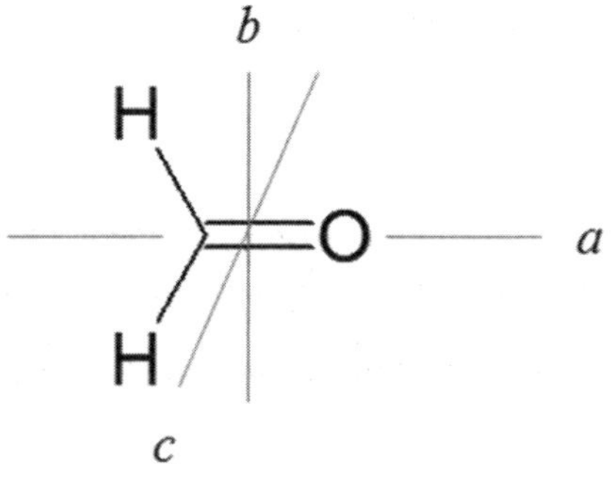

Having decided that the three principal axes coincide with our Cartesian axes, it remains to be seen which moment of inertia is smallest and which the largest. Since the C = O group lies along the z-axis, the only atoms that contribute to I_{zz} are the light hydrogen atoms, and hence, the moment of inertia is least along this axis. By convention $I_a \leq I_b \leq I_c$, and consequently this coordinate must be *a*. For all planar molecules, the moment of inertia is largest along the axis normal to the molecular plane and, by convention, it is labelled *c*. I_b then lies in the molecular plane (*y*-axis). Thus, all the principal axes are uniquely determined for this molecule by inspection alone.

Moreover, in the case of a *planar* rigid molecule such as formaldehyde, only two of the three moments are independent. When the molecule is planar, the values of the c_i coordinates are zero for all atoms. From the *perpendicular axis theorem*, one can straightforwardly derive the constraint as follows:

$$I_c - I_b - I_a = 0$$

This establishes the fact that only two moments of inertia for a planar rigid rotator are independent. This condition is also used to test for planarity of a molecule.

In case the *x*, *y* and *z* coincide with the principal axes of inertia, all the off-diagonal elements of the inertia tensor are zero, and the matrix is already in diagonal form, the diagonal elements being the three principal moments of inertia. However, if the molecule lacks symmetry, so that the positions of the axes are not obvious, the moments have to be obtained from the corresponding determinantal equation (4.45).

Why do we bother at all to diagonalize the moment of inertia tensor? Diagonalization allows us to separate the rotational energy into three independent components since the three principal axes are orthogonal. If we use these principal axes, then the components of the rotational angular momentum $\vec{L}$ along these axes can be shown to be $L_a = I_a\omega_a$, $L_b = I_b\omega_b$, $L_c = I_c\omega_c$ and the kinetic energy (= total energy) for a rigid-rotor simplifies to

$$E_{\text{rot}} = \frac{L_a^2}{2I_a} + \frac{L_b^2}{2I_b} + \frac{L_c^2}{2I_c} \tag{4.49}$$

In all other coordinate systems, things are much more complicated! (There will be cross terms).

We first describe the classification of the molecules based on the symmetry of the momental ellipsoid, and the possible simplification of the classical expression (4.49) for molecules that possess some degree of symmetry. Real molecules are divided into four separate classes: linear, spherical tops, symmetric tops and asymmetric tops. After giving examples of molecules in each category, we discuss their rotational spectra.

4.3.1 Linear molecules

For a *linear molecule*, $I_a = 0$ (corresponding to zero moment about the internuclear axis since all atoms lie along the axis and all r_i in $I = \sum_i m_i r_i^2$ are zero) and $I_b = I_c$, where the *b*- and *c*-axes may be in any direction perpendicular to the internuclear *a*-axis.

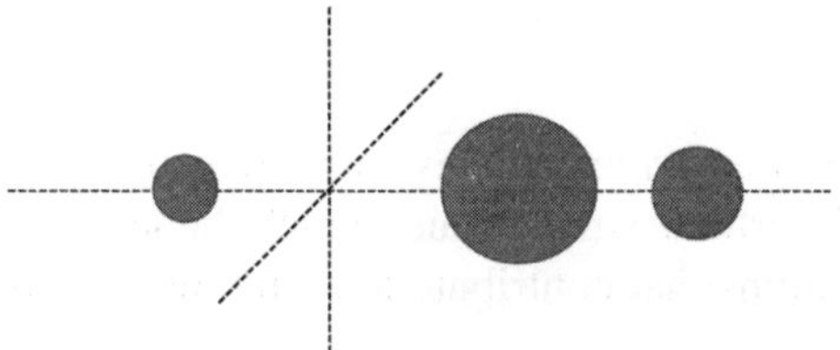

The linear molecule is a special case of a planar one. Here, $I_a = 0$, and hence, by the perpendicular axis theorem, $I_b = I_c = I$.

Therefore, the principal axes of inertia are mutually perpendicular and perpendicular to the internuclear axis: the momental ellipsoid is a circular cylinder of infinite length. Because I_a is zero, the component of angular momentum L_a ($= I_a\omega_a$) along a is also zero. Thus,

$$L^2 = L_b^2 + L_c^2$$

and

$$E_{\text{rot}} = \frac{L_b^2}{2I_b} + \frac{L_c^2}{2I_c} = \frac{L^2}{2I_b} \tag{4.50}$$

Thus, linear molecules have only two rotational degrees of freedom. Applying quantization, we find that the rotational energies are

$$\frac{J(J+1)\hbar^2}{2I_b} = BJ(J+1)$$

where $B = \frac{\hbar^2}{2I_b}$ as already discussed.

Diatomic molecules, discussed above, belong to this class of molecules. As we saw, the selection rules are $\Delta J = \pm 1$. Since we have already discussed the rotational spectra of diatomic molecules in detail, we need only consider any new aspects in the spectra of linear polyatomics. The first is that, since polyatomics have larger number of atoms, their moments of inertia are generally larger than those of diatomics. Consequently, their rotational constants are much smaller and may be of the order of ~1 cm^{-1} for linear triatomics, whereas they are of the order of ~10 cm^{-1} for diatomics. Plainly, in case of linear molecules with larger number of atoms, the rotational constants are still smaller. The second concerns the requirement of a permanent dipole moment in the molecule. Thus, molecules like carbon dioxide (OCO) and acetylene (HCCH) are microwave inactive, but OCS, HCN and NNO are microwave active. It should also be noted that isotopic substitution does not affect the dipole moment, and $^{18}OC^{16}O$ is also microwave inactive though HD does have a small dipole moment. The third difference is a problem encountered when applying microwave spectroscopy for structure determination of polyatomic linear molecules. Linear molecules have only one rotational constant that can be determined from rotational spectroscopy. However, there may be two or more bond lengths to be determined. For example, for OCS, there are two bond lengths (C=O and C=S) to be determined, but only one rotational constant. This problem is solved by isotopic substitution, giving another equation with a new rotational constant corresponding to the changed masses. Again, we assume that the bond lengths remain unchanged on isotopic substitution. In this case, the normal isotope of O, C or S may be replaced by another isotope, e.g., ^{18}O, ^{13}C or ^{34}S. If there are more unknowns, more than one isotopic substitution may be necessary, so that there are as many equations as the number of unknowns.

We consider the specific case of a linear triatomic molecule **XYZ** with internuclear distances r_{XY} and r_{YZ}, as shown below:

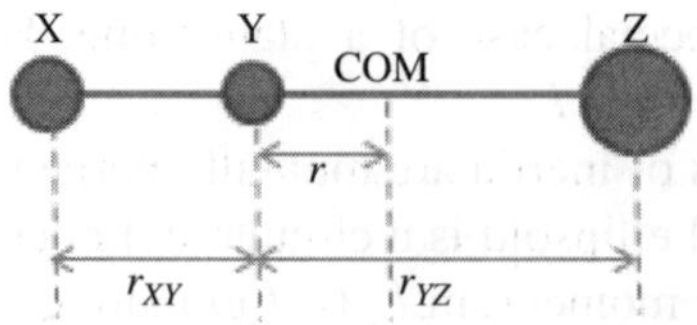

The masses of the three atoms are m_X, m_Y and m_Z. Let the centre of mass lie along the Y–Z bond at a distance r from atom Y. Then, from the relation $r_\alpha(COM) = \sum_i m_i r_i / \sum_i m_i$, we get

$$r = \frac{m_Z r_{YZ} - m_X r_{XY}}{M} \tag{4.51}$$

where $M = m_X + m_Y + m_Z$ is the total mass of the molecule.

We now determine the moment of inertia of the molecule. By the parallel axis theorem, the moment of inertia about the central atom (Y) is related to that about the centre of mass by $I_Y = I + Mr^2$. Substituting for r from equation (4.51) gives

$$I_b = m_X r_{XY}^2 + m_Z r_{YZ}^2 - M\left(\frac{m_Z r_{YZ} - m_X r_{XY}}{M}\right)^2 = m_X r_{XY}^2 + m_Z r_{YZ}^2 - \frac{(m_Z r_{YZ} - m_X r_{XY})^2}{M} \tag{4.52}$$

A general, easy to remember, expression for the moment of inertia of a linear triatomic molecule, with the atoms labelled 1, 2, 3 in sequence, is

$$I_b = \frac{m_1 m_2 r_{12}^2 + m_2 m_3 r_{23}^2 + m_1 m_3 r_{13}^2}{m_1 + m_2 + m_3} = (m_1 m_2 r_{12}^2 + m_2 m_3 r_{23}^2 + m_1 m_3 r_{13}^2)/M \tag{4.53}$$

where r_{12}, r_{23} and r_{13} are the individual bond distances. Since the molecule is linear, $r_{13} = r_{12} + r_{23}$. For XYZ, this equation becomes

$$\begin{aligned} I_b &= \frac{m_X m_Y r_{XY}^2 + m_Y m_Z r_{YZ}^2 + m_X m_Z r_{XZ}^2}{M} \\ &= (m_X m_Y r_{XY}^2 + m_Y m_Z r_{YZ}^2 + m_X m_Z (r_{XY} + r_{YZ})^2)/M \end{aligned} \tag{4.54}$$

The two bond lengths can be determined by determining the moments of inertia of two isotopologues of the molecule. We then have two moments of inertia and two equations, and hence, the two unknown bond lengths can be determined.

For example, if the atom Z in the molecule XYZ is replaced with an isotope of mass m_Z' and the bond lengths are unchanged, the new moment of inertia is I_b'. From equation (4.54)

$$I_b' = (m_X m_Y r_{XY}^2 + m_Y m_Z' r_{YZ}^2 + m_X m_Z'(r_{XY} + r_{YZ})^2)/M' \tag{4.55}$$

Multiplication and division of equation (4.54) by m_Z / m_Z' gives

$$I_b' M' m_Z / m_Z' = m_X m_Y m_Z r_{XY}^2 / m_Z' + m_Y m_Z r_{YZ}^2 + m_X m_Z (r_{XY} + r_{YZ})^2$$

Subtraction from the expression for $I_b M$ from equation (4.55) gives

$$I_b M - I_b' M' \frac{m_Z}{m_Z'} = m_X m_Y \left(1 - \frac{m_Z}{m_Z'}\right) r_{XY}^2 \tag{4.56}$$

Rearrangement of equation (4.56) gives

$$r_{XY}^2 = \frac{I_b M m_Z' - I_b' M' m_Z}{m_X m_Y (m_Z' - m_Z)} \tag{4.57}$$

The other bond length r_{YZ} can be determined by substitution in equation (4.54).

Example 4.13 Determine the CO and CS bond lengths in OCS from the rotational constants $B(^{16}O^{12}C^{32}S) = 6081.5$ MHz and $B(^{16}O^{12}C^{34}S) = 5932.8$ MHz.

Solution

Here, isotopic substitution of sulphur has been carried out, and hence, r_{CO} can be determined by substitution of the appropriate masses in equation (4.57).

From the given rotational constants, the moments of inertia can be obtained from the expression

$$I_b = \frac{h}{8\pi^2 B} = \frac{8.39202 \times 10^{-36}}{B}.$$

Thus, for the normal isotope, one obtains

$$I_b = \frac{8.39202 \times 10^{-36}}{6081.5 \times 10^6} 1.3799 \times 10^{-45} \text{ kg m}^2.$$

For this molecule, $m_O + m_C + m_S = (15.9949 + 12.0000 + 31.9721)u = 59.9670u = 9.95775 \times 10^{-26}$ kg. Therefore, $I_b M = 1.3733 \times 10^{-70}$ kg^2 m^2.

Similarly, for the isotopologue,

$$I_b' = \frac{h}{8\pi^2 B'} = \frac{8.39202 \times 10^{-36}}{B'} = \frac{8.39202 \times 10^{-36}}{5932.8 \times 10^6} = 1.4145 \times 10^{-45} \text{ kg m}^2$$

$$M' = m_O + m_C + m_S = (15.9949 + 12.0000 + 33.9679)u = 61.9628u = 1.02892 \times 10^{-25} \text{ kg}$$

$$I_b' M' = 1.4544 \times 10^{-70} \text{ kg}^2 \text{ m}^2$$

Equation (4.57) becomes

$$r_{CO}^2 = \frac{I_b M m_S' - I_b' M' m_S}{m_C m_O (m_S' - m_S)} = \frac{1.3799 \times 10^{-70} \times 33.9679 - 1.4544 \times 10^{-70} \times 31.9721}{12 \times 15.9949 \times 1.9958 \times (1.6605 \times 10^{-27})^2}$$
$$= 1.3450 \times 10^{-20} \text{ m}^2$$

whence $r_{CO} = 1.1600 \times 10^{-10}$ m = 116 pm.

Once one of the bond lengths is determined, the other, r_{CS}, can be obtained by substitution in equation (4.52), which, for OCS, becomes

$$I_b = m_O r_{CO}^2 + m_S r_{CS}^2 - (m_S r_{CS} - m_O r_{CO})^2 / M \tag{4.58}$$

Substituting the various values, we obtain r_{CS} = 156 pm. Alternatively, the Solver utility of Excel may be employed for this purpose or a further isotopic substitution of ^{16}O by ^{18}O can yield a value for r_{CS}.

The above examples show the utility of rotational spectroscopy for determining molecular structure. In general, as many equations are required as there are unknown parameters in the structure, and the additional equations can be obtained by measuring the spectra of the same molecule with different isotopes substituted.

In the case of OCS, a single spectrum of an ordinary sample can provide all the required information, since the rotational spectra of this molecule are sufficiently intense for the rotational constant of each of the four isotopic species in their natural isotopic abundance (^{18}O: 0.20%; ^{13}C: 1.11%; ^{34}S: 4.22%) to be determined. Thus, in this case, no isotopic labelling chemistry is required. Though only two equations are necessary, the extra equations can be used to test the assumption of constancy of bond lengths on isotopic substitution.

4.3.2 Spherical tops

The most symmetric momental ellipsoid is a sphere, i.e., when $a = b = c$, and this happens when the three principal moments of inertia are equal. Molecules in this class are known as *spherical tops* and belong to I_h, O_h and T_d symmetries. Some examples are methane, carbon tetrachloride, sulphur hexafluoride and fullerene. Putting $I = I_a = I_b = I_c$ (and hence $L = L_a = L_b = L_c$) gives

$$E_{\text{rot}} = \frac{L_a^2 + L_b^2 + L_c^2}{2I} = \frac{L^2}{2I} \tag{4.59}$$

Applying quantization to the angular momentum, we have

$$E_{\text{rot}} = BJ(J+1) \tag{4.60}$$

The selection rule is $\Delta J = \pm 1$, as for linear molecules, and the spectra of spherical top molecules are expected to display the same regular $2B$ spacing. However, since these molecules do not possess a permanent dipole moment, they are microwave inactive. Centrifugal distortion may, however, distort the molecule and impart a small dipole moment, resulting in a weak rotational spectrum. The spherical top molecule of prime astrophysical and atmospheric importance is CH_4. Rotation about any of the four axes containing a C-H bond results in a centrifugal distortion in which the other three hydrogen atoms are thrown outwards slightly from the axis, converting the molecule into a symmetric rotor (Section 4.3.3) with a small dipole moment, 10^{-6} D. Another tetrahedral molecule, SiH_4, shows similar behaviour. Part of its far-infrared spectrum is shown in Figure 4.11. It can be seen that it has the same regular spacing as the spectrum of a linear rotor.

4.3.3 Symmetric tops

Next in symmetry are molecules in which two of the principal moments of inertia are equal and the third is non-zero, and these are known as *symmetric tops*. Since the unique moment of inertia may be either smaller or larger than the other two, symmetric tops are further subdivided into two categories.

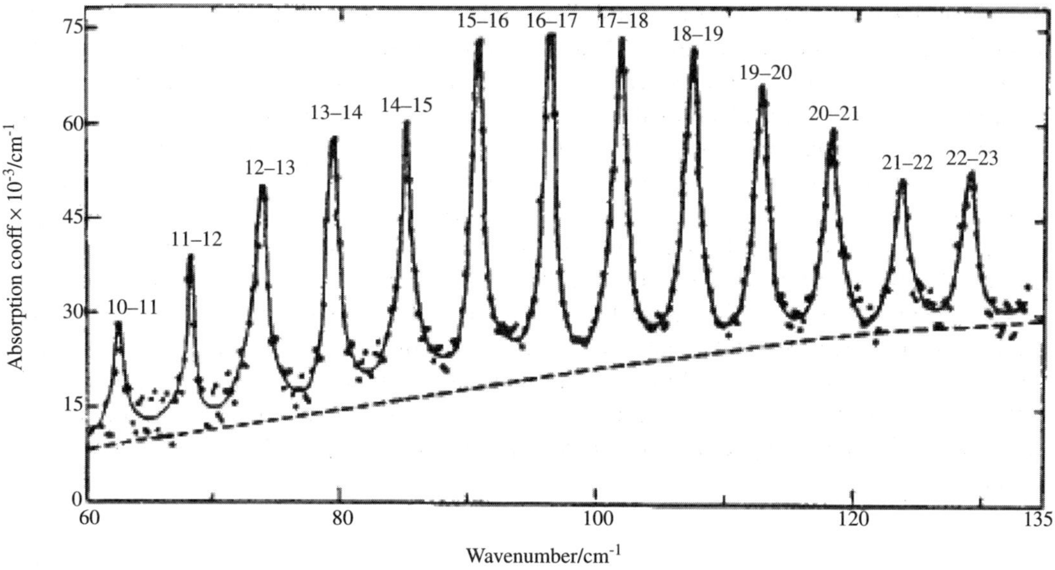

Figure 4.11. Far-infrared pure rotational spectrum of the silane spherical top [Rosenberg and Ozier (1974); reprinted with permission]

In case the unique moment of inertia is smaller than the other two, $I_a \leq I_b = I_c$, the molecule is said to be a *prolate* symmetric top. These are generally elongated molecules (Figure 4.12) and resemble in general shape a cigar or a shuttlecock. Examples of molecules belonging to this class are methyl iodide and CH_3CN, both of which are C_{3v} molecules. The unique axis is the C_3 axis through which all the atoms except the hydrogens pass. Thus, only the hydrogen

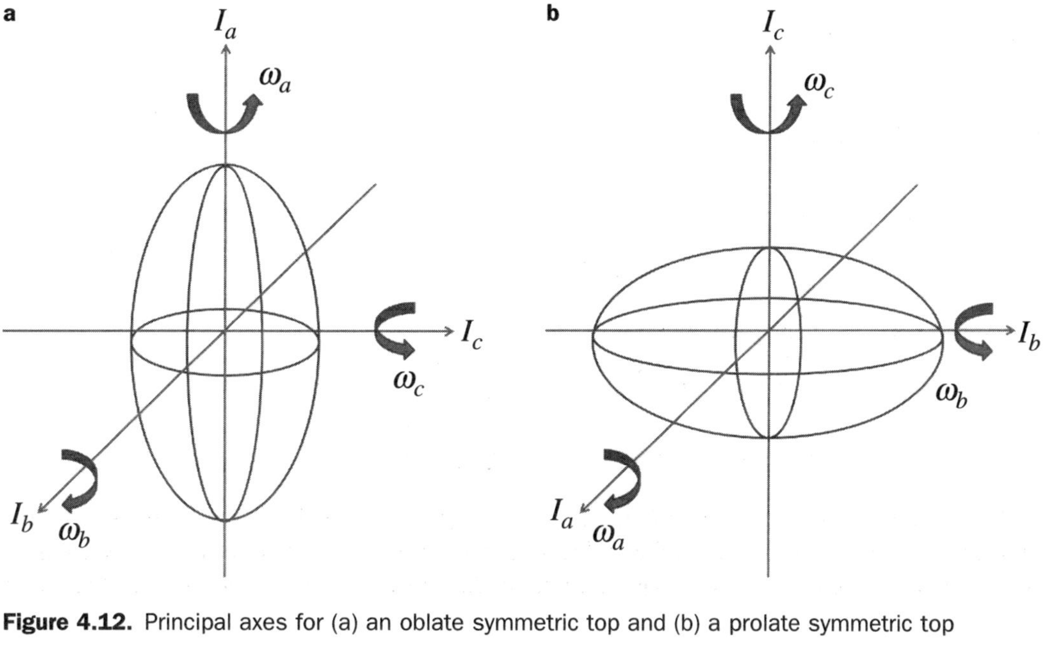

Figure 4.12. Principal axes for (a) an oblate symmetric top and (b) a prolate symmetric top

atoms contribute to the moment of inertia along this axis, and the principal moment of inertia along this axis is small (but non-zero). Thus, linear molecules, such as HCN, are special cases of symmetric rotors in which I_a is zero. As in the case of linear molecules, the end-over-end rotation in and out of the plane of the paper are still identical, and we have $I_b = I_c$. Such a molecule can be imagined to be spinning about its main axis (I_a through which its centre of mass passes) like a top, and hence the name of this class of molecules.

The other possibility is $I_a = I_b \leq I_c$, and we have an *oblate* symmetric top. Such molecules are flat and saucer shaped (like a frisbee) and example molecules are planar ones like benzene and BF_3. Let us consider the latter molecule (D_{3h}). The central atom is boron, and the centre of mass of the molecule lies on it. The highest order axis is the C_3 axis passing through boron and perpendicular to the plane of the molecule; hence, one of the principal moments of inertia must lie along this axis. Since all the fluorines contribute to the moment of inertia along this axis, it is the largest and is plainly *c*. The second axis will be perpendicular to this, passing through one of the C_2 axes along a B-F bond, and the third is perpendicular to the other two (Figure 4.13).

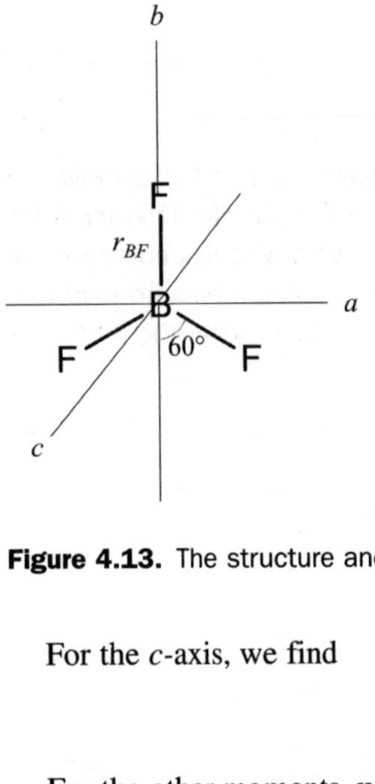

Figure 4.13. The structure and principal axes of boron trifluoride (BF_3) with r_{BF} = 0.13 nm

For the *c*-axis, we find

$$I_c = m_B \times 0.0 + 3m_F r_{BF}^2 = 3m_F r_{BF}^2$$

For the other moments, we obtain

$$I_a = m_B \times 0.0 + m_F r_{BF^2} + 2m_F (r_{BF} \sin 30)^2 = \tfrac{3}{2} m_F r_{BF^2}$$

$$I_b = m_B \times 0.0 + m_F \times 0.0 + 2m_F (r_{BF} \sin 60)^2 = \tfrac{3}{2} m_F r_{BF^2}$$

Thus, we find that $I_a = I_b = \frac{1}{2} I_c$ for BF_3. This nice relationship is a consequence of the numerous symmetry properties of the molecule. As the above example indicates, for any planar molecule, the *c*-axis is always perpendicular to the plane containing the nuclei. The three moments of inertia also satisfy the perpendicular axis theorem, i.e., $I_c = I_a + I_b$. As shown for the also planar molecule, formaldehyde, the off-diagonal matrix elements, I_{zx} and I_{zy} vanish. The moment of inertia tensor can be diagonalized simply by diagonalizing a 2×2 matrix. The direction cosines then collapse to a single rotation about the *c*-axis that brings the arbitrary axes to coincide with the principal *a*,*b*-axes.

Let us consider the kind of molecules which belong to the symmetric top class. All molecules having a rotation axis C_n with $n > 2$, which coincides with one of the principal axes of inertia, with two of the moments equal, belong to this class. Hence, any molecule with a single three-fold or higher principal axis of symmetry is necessarily a symmetric top, the principal axis of the molecule being coincident with the unique axis of the top. Molecules that possess an S_4 axis also belong to this class of molecules. An example is allene.

4.3.3.1 Quantization of the rotational energy of a symmetric top

For a prolate symmetric top ($I_b = I_c$), the expression (4.49) for the energy can be rearranged to

$$E_{\text{rot}} = \frac{L_a^2}{2I_a} + \frac{L_b^2}{2I_b} + \frac{L_c^2}{2I_c} = \frac{L_a^2}{2I_a} + \frac{L_b^2 + L_c^2}{2I_b} = \frac{L_a^2}{2I_a} + \frac{L^2 - L_a^2}{2I_b} = \frac{L^2}{2I_b} + \frac{L_a^2}{2}\left(\frac{1}{I_a} - \frac{1}{I_b}\right) \tag{4.61}$$

If the top is oblate, the only effect is that the subscript *c* replaces *a*, i.e.,

$$E_{\text{rot}} = \frac{L^2}{2I_b} + \frac{L_c^2}{2}\left(\frac{1}{I_c} - \frac{1}{I_b}\right) \tag{4.62}$$

Let us first consider a prolate top. The energy in expression (4.61) can be translated to the corresponding quantum mechanical operator

$$\hat{H}_{\text{rot}} = \frac{\hat{L}^2}{2I_B} + \frac{\hat{L}_a^2}{2}\left(\frac{1}{I_a} - \frac{1}{I_b}\right) \tag{4.63}$$

We know from equation (4.17) that the rotational wave functions are eigenfunctions of the $\hat{L}^2$ operator. We also know the commutation rules

$$[\hat{L}_x, \hat{L}_y] = i\hbar \hat{L}_z, \text{ etc.}$$
$$\left[\hat{L}^2, \hat{L}_z\right] = 0 \tag{4.64}$$

Thus, the total angular momentum and one of its components (usually along the *z*-axis) are quantized:

$$\hat{L}^2\psi = J(J+1)\hbar^2\psi$$
$$\hat{L}_z\psi = M\hbar\psi \tag{4.65}$$

The laboratory *xyz*-axes are fixed in space, while the *abc*-axes are tied to the rotating molecule. The former are thus called *space-fixed* coordinates, while the latter are called *body-fixed* coordinates. The angular momenta follow the same commutation rules in any coordinate system. Thus, for prolate symmetric tops, we also have

$$[\hat{L}_b, \hat{L}_c] = i\hbar \hat{L}_a$$
$$\left[\hat{L}^2, \hat{L}_a\right] = 0 \tag{4.66}$$

Hence, the total angular momentum and one of its components corresponding to one of the principal axes is also quantized. Since *a* is the unique axis for a prolate top, we let this component be $\hat{L}_a$ and introduce a new quantum number, *K*, which refers to the component L_a of the total angular momentum $\vec{L}$ along the axis of a symmetric top. For example, for a prolate symmetric top like CH_3I, $\hat{L}_a$ represents the angular momentum due to motion about the *a*-axis (Figure 4.14).

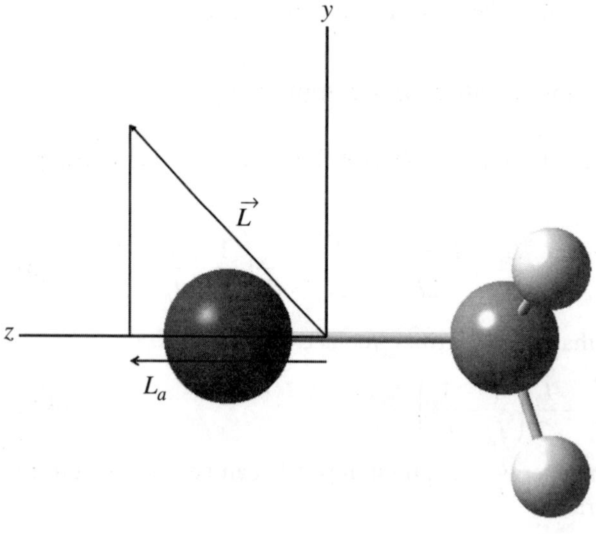

Figure 4.14. $\vec{L}$, L_a (*K*) definitions for a prolate top (CH_3I)

For every value of *J*, there are $2J + 1$ values of *K* given by

$$K = 0, \pm 1, ..., \pm J \tag{4.67}$$

$$\left|\vec{K}\right| = K\hbar \tag{4.68}$$

$\vec{K}$ thus gives the *component* of the total angular momentum along the unique axis, i.e., $\vec{K}$ is the projection of $\vec{L}$ onto the *a*-axis. The diagram below illustrates this for $J = 2$ and $K = 0, 1, 2$.

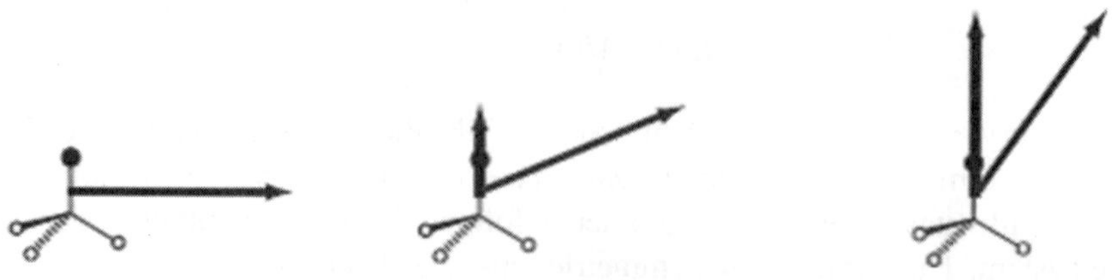

The molecule itself rotates about the $\vec{L}$ vector; at the same time, unless $K = 0$, it is also rotating about the a-axis. When $K = 0$, there is no rotation about the a-axis and so the molecule is just tumbling end over end. As K increases, the molecule has greater angular momentum about the a-axis, thus bringing the $\vec{L}$ vector towards the a-axis. Negative values of K would correspond to the $\vec{K}$ vector pointing downwards.

Let us determine the quantized rotational energy levels of a prolate symmetric top. Substituting the quantum numbers for the classical expression (4.61), we obtain

$$\begin{aligned} E_{\text{rot}} &= \frac{L^2}{2I_b} + \frac{L_a^2}{2}\left(\frac{1}{I_a} - \frac{1}{I_b}\right) \\ &= J(J+1)\frac{\hbar^2}{2I_b} + \frac{K^2\hbar^2}{2}\left(\frac{1}{I_a} - \frac{1}{I_b}\right) \\ &= J(J+1)B + K^2(A - B) \end{aligned} \tag{4.69}$$

If both sides are divided by hc, the equation simplifies in form to

$$F(J,K) = J(J+1)\tilde{B} + K^2(\tilde{A} - \tilde{B}) \tag{4.70}$$

in which $\tilde{B} = \dfrac{h}{8\pi^2 I_b c}$ and $\tilde{A} = \dfrac{h}{8\pi^2 I_a c}$.

The axial component K of the symmetric top may be positive or negative. However, from (4.70), $F(J, K)$ depends on the magnitude of K but not upon its sign; thus, every energy level for which $K > 0$ is doubly degenerate. The degeneracy corresponds physically to opposite directions of rotation about the top axis, because it does not matter if the rotation about the symmetry axis is 'clockwise' or counter-clockwise (as viewed down the symmetry axis). The total number of sublevels belonging to any value of J is therefore $J + 1$, of which $2J$ sublevels occur as degenerate pairs and one is non-degenerate.

When the top is oblate, the subscript C replaces A, and therefore, $\tilde{C} = h/8\pi^2 I_c c$ replaces $\tilde{A}$. The formula for the energy levels of an oblate top, then, is

$$F(J,K) = J(J+1)\tilde{B} + K^2(\tilde{C} - \tilde{B}) \tag{4.71}$$

Figure 4.15 illustrates the energy levels for a prolate and an oblate top. On the top of the diagram is shown, to scale, the energy levels for the prolate symmetric top CH_3I which has $\tilde{A} = 5.173\ \text{cm}^{-1}$ and $\tilde{B} = 0.250\ \text{cm}^{-1}$; J values up to 5 and K values up to 2 are shown. On the bottom is the corresponding set of energy levels for NH_3, an oblate symmetric top with $\tilde{C} = 6.449\ \text{cm}^{-1}$ and $\tilde{B} = 10.001\ \text{cm}^{-1}$. As the energy depends on K^2, only positive values of K are shown. Since $\tilde{A} \geq \tilde{B} \geq \tilde{C}$, the energy of each J level increases with K for a prolate top, but decreases with K for an oblate top.

It may be mentioned here that NH_3 is not a good example of an oblate type molecule because of the 'umbrella inversion' of the molecule, which destroys the $\pm K$ degeneracy and

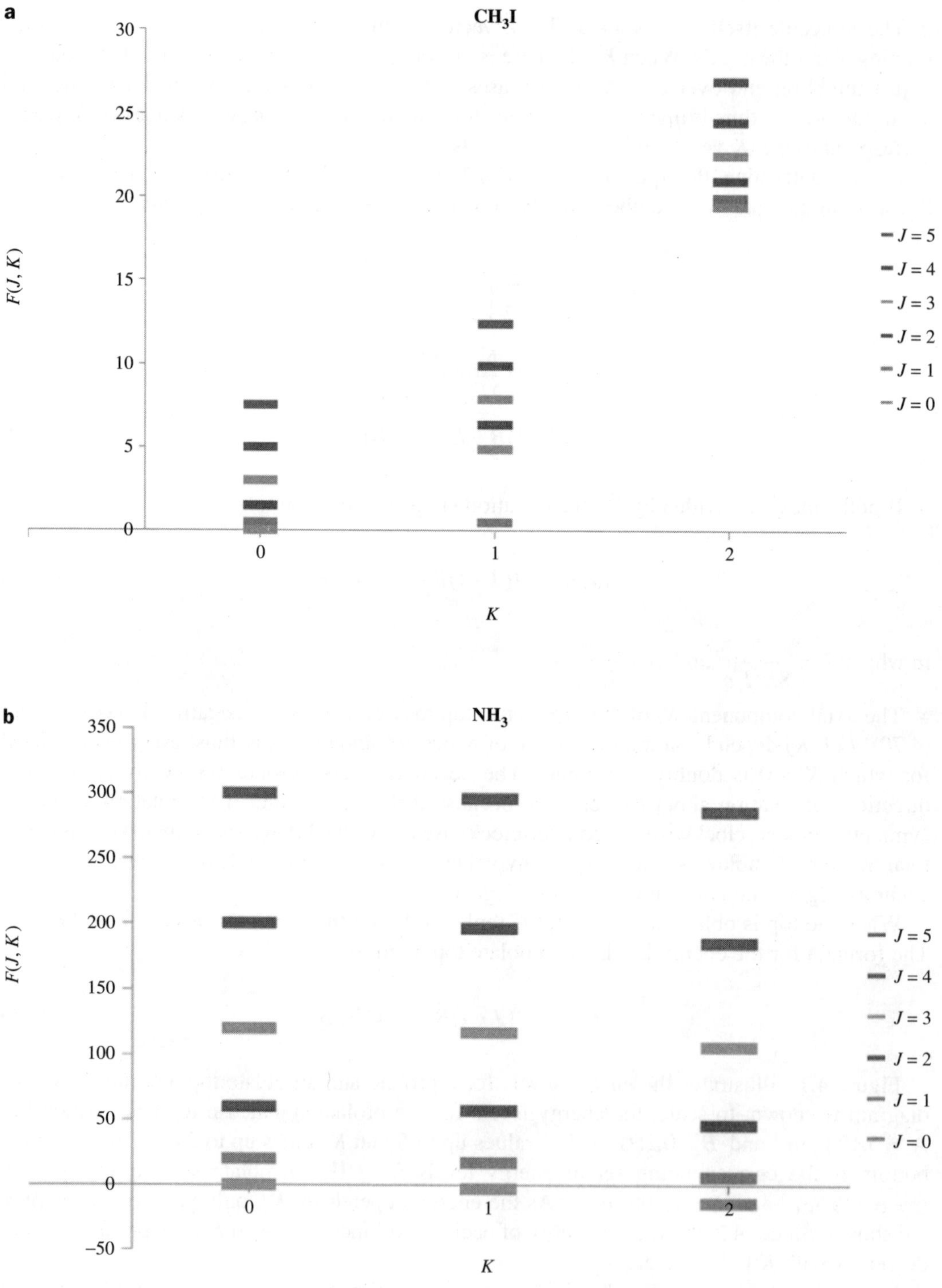

Figure 4.15. Rotational energy levels (cm^{-1}) for (a) a prolate (CH_3I) and (b) an oblate (NH_3) symmetric top

leads to an inversion splitting that depends slightly on J and K. Because of the umbrella inversion, the average symmetry of the molecule is D_{3h}.

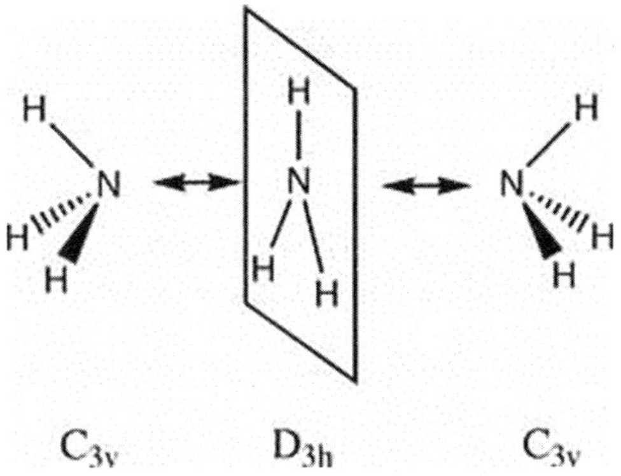

For a closed shell linear rotor, which is a special case of a symmetric top, $K = 0$. Thus, each J level then has the usual $2J + 1$ different M states, corresponding to the different possible projections of $\vec{L}$ on a 'space-fixed' axis. For symmetric top molecules, each (J, K) level also has $2J + 1$ different K states corresponding to the different possible projections of $\vec{L}$ on a 'molecule-fixed' axis. Thus, the statistical weight (degeneracy) for each J level is $(2J + 1)^2$, rather than the $2J + 1$ that is true for a linear (or diatomic) rotor.

The general selection rules are $\Delta J = 0, \pm 1$, $\Delta K = 0$. However, the $\Delta J = 0$ selection rule is meaningless, since there is no change in the rotational energy or its component in this transition, as ΔK is also zero. The $\Delta K = 0$ rule also results in a simple expression for the transition frequencies. The rotational energy levels of a rigid symmetric top are

$$F(J,K) = J(J+1)\tilde{B} + K^2(\tilde{A} - \tilde{B}) \qquad \text{(Prolate)}$$

$$F(J,K) = J(J+1)\tilde{B} + K^2(\tilde{C} - \tilde{B}) \qquad \text{(Oblate)}$$

We obtain for the wavenumber of the absorption lines for the $\Delta J = 1$ transition

$$\tilde{\nu} = F(J+1,K) - F(J,K) = 2\tilde{B}(J+1) \qquad (4.72)$$

which is the same as that for a linear molecule. The pure rotation spectrum therefore consists of a set of evenly spaced lines which obey the same formula as a linear molecule. It is important to note, however, that each line contains $2J + 1$ components, corresponding to the possible K values for each J value. Centrifugal distortion frequently causes a partial separation of the K components, and this allows one to distinguish the symmetric top spectrum from a linear molecule. Nevertheless, only one rotational constant $\tilde{B} = h/8\pi^2 I_b c$ can be evaluated from the wavenumbers. For a complete structure determination, therefore, we need at least as many isotopic moments as there are independent bond distances and interbond angles in the molecule; e.g., in PCl_3, there are two unknowns, the P-Cl bond length and the Cl-P-Cl bond angle, and data for two isotopologues, $P^{35}Cl_3$ and $P^{37}Cl_3$, are sufficient to determine both.

Example 4.14 Ammonia has two defined inertial moments: $I = 4.413 \times 10^{-47}$ kg m^2 and $I = 2.806 \times 10^{-47}$ kg m^2.

(a) Label these as I_a, I_b and I_c.
(b) Calculate the rotational constants A, B and C.
(c) What is the value of the lowest non-zero rotational energy?

Solution

(a) Since the molecule is oblate, the c-axis is unique. Therefore, the largest of the moments of inertia is I_c, and the other two are equal to each other, i.e., $I_c = 4.413 \times 10^{-47}$ kg m^2 and $I_a = I_b = 2.806 \times 10^{-47}$ kg m^2.

(b) $A = B = h/8\pi^2 I_a = 8.392 \times 10^{-36}/2.806 \times 10^{-47} = 2.991 \times 10^{11}$ Hz $= 299.1$ GHz
$C = 1.902 \times 10^{11}$ Hz $= 190.2$ GHz
In wavenumber, $\tilde{A} = \tilde{B} = A/c = 9.976\ \text{cm}^{-1}, \tilde{C} = C/c = 6.343\ \text{cm}^{-1}$

(c) For an oblate top, $F(J,K) = J(J+1)\tilde{B} + K^2(\tilde{C} - \tilde{B})$ and $\tilde{C} < \tilde{B}$. Therefore, the lowest non-zero energy level is when $J = 1$ and $K = \pm 1$. The energy is then

$$F(J, K) = 1(1 + 1)9.976 + (6.343 - 9.976) = 15.319\ \text{cm}^{-1}$$

When centrifugal stretching is taken into account, the energy levels become

$$F(J,K) = \tilde{B}J(J+1) + (\tilde{A} - \tilde{B})K^2$$
$$-\tilde{D}_J J^2(J+1)^2 - \tilde{D}_{JK} J(J+1)K^2 - \tilde{D}_K K^4$$

The three distortion constants deal with the stretching of the pseudo-diatomic bond due to end-over-end rotation of the prolate top ($\tilde{D}_J$), the change in the symmetric top bond angle due to rotational excitation about the symmetry axis ($\tilde{D}_K$) and the coupling of these two motions ($\tilde{D}_{JK}$). A similar expression holds for an oblate symmetric rotor.

Using the $\Delta J = 0, \pm 1$ and $\Delta K = 0$ electric dipole selection rules, the wavenumber expression for the rotational transitions in a prolate top becomes

$$\tilde{\nu}_{J,K} = F(J+1,K) - F(J,K) = 2(J+1)\left[\tilde{B} - 2\tilde{D}_J(J+1)^2 - \tilde{D}_{JK}K^2\right]\text{cm}^{-1} \quad (4.73)$$

We see that the spectrum will be basically that of a linear molecule (including centrifugal stretching). The centrifugal distortion constants $\tilde{D}_J$ and $\tilde{D}_{JK}$ can be considered as correction terms to the rotational constant $\tilde{B}$, and hence as perturbing the moment of inertia I_b.

Since each value of J is associated with $2J + 1$ values of K, we see that each line characterized by a certain J value must have $2J + 1$ components. However, since K only appears as K^2, there will be only $J + 1$ *different* frequencies, all those with $K > 0$ being doubly degenerate. We may tabulate a few lines for methyl fluoride as in Table 4.2.

Fitting the data for the observed lines to equation (4.73) leads directly to

$$\tilde{B} = 0.851204\ \text{cm}^{-1}$$
$$\tilde{D}_J = 2.00 \times 10^{-6}\ \text{cm}^{-1}$$
$$\tilde{D}_{JK} = 1.47 \times 10^{-5}\ \text{cm}^{-1}$$

Table 4.2. Some lines in the rotational spectrum of methyl fluoride

J	K	$\tilde{\nu}_{JK}$	$\tilde{\nu}_{obs}$	$\tilde{\nu}_{calc}$
0	0	$2\tilde{B}-4\tilde{D}_J$		
1	0	$4\tilde{B}-32\tilde{D}_J$	4.40475	4.40475
	±1	$4\tilde{B}-32\tilde{D}_J-4\tilde{D}_{JK}$	4.40470	4.40469
2	0	$6\tilde{B}-108\tilde{D}_J$	5.10701	5.10701
	±1	$6\tilde{B}-108\tilde{D}_J-6\tilde{D}_{JK}$	5.10692	5.10692
	±2	$4\tilde{B}-32\tilde{D}_J-24\tilde{D}_{JK}$	5.10665	5.10666

The calculated wavenumbers in the last column using these values show how precisely such measurements may now be made. Once again, each spectrum examined yields only one value of $\tilde{B}$, but the spectra of the isotopologues may be used to calculate all the bond lengths and bond angles.

4.3.4 Asymmetric top molecules

This is the least symmetrical class of molecules, having all three principal moments unequal. Therefore, the momental ellipsoid has three unequal axes, and the classical expression cannot be reduced to a simpler form. *A molecule is an asymmetric top unless it possesses at least one three-fold or higher symmetry axis, or an S_4 axis*. A majority of molecules found in nature are asymmetric tops. Simple examples are H_2O and vinyl chloride, CH_2=CHCl. In asymmetric top molecules, the K-degeneracy of the symmetric top is lifted and there are $2J + 1$ levels of different energy for each value of J. This is a familiar pattern: reduction in symmetry leads to loss of degeneracy.

Although only a small fraction of all known molecules are true symmetric or spherical tops, there is a more sizeable group that falls in the category of 'near prolate' tops with $I_a < I_b \approx I_c$, or 'near oblate' tops with $I_a \approx I_b < I_c$. Examples of near prolate tops are C_{2v} molecules such as H_2O and H_2CO and a D_{2h} molecule such as ethylene, C_2H_4. Classifying a molecule as a near symmetric top helps in interpreting its rotational spectra.

It is useful to have a quantitative measure of the degree of asymmetry of a molecule. Several possible relationships between the rotational constants exist for the purpose, but Ray's asymmetry parameter, κ (kappa), is one of the most frequently used. It is defined as

$$\kappa = \frac{2B - A - C}{A - C} \tag{4.74}$$

For a prolate symmetric top, $B = C$, and $\kappa = -1$, while for an oblate symmetric top, $A = B$, and $\kappa = +1$. For any other situation, κ lies between these limits and the molecule is an asymmetric top. If κ is close to either of the two limits -1 or $+1$, we describe the rotor as *near prolate* or *near oblate*, respectively, while a situation with $\kappa = 0$ represents the most asymmetric case. For the case of formaldehyde, the three rotational constants are $A = 281.98$ GHz, $B = 38.83$ GHz and $C = 34.00$ GHz. Substitution in equation (4.74) leads to a Ray's asymmetry parameter of $\kappa = -0.96$, and hence, formaldehyde is a near prolate top.

Before we end this section, we would like to emphasize the importance of symmetry in the classification of molecules. Group theory greatly simplifies the assignment of classes to the different rotors. Thus, we have the following criteria for classification:

1. Linear rotors must, obviously, belong to $C_{\infty v}$ or $D_{\infty h}$ molecular point groups.
2. A symmetric rotor must have either a single C_n axis with $n > 2$ or an S_4 axis. Point groups include S_n ($n \geq 4$), C_n, C_{nv}, C_{nh}, D_n, D_{nh} (all with $n \geq 3$) and all D_{nd} (note that D_{2d} has an S_4 axis).
3. Molecules in the higher symmetry groups, T_d, O_h, I_h, are spherical rotors. These molecules have two or more non-coincident C_n or S_n axes with $n \geq 3$.

Figure 4.16 illustrates the principal inertial axes for a number of molecules of different symmetry types. The expressions for moments of inertia of the different categories of molecules are also given.

Molecule		Point group	Classification	Moment of inertia
CO	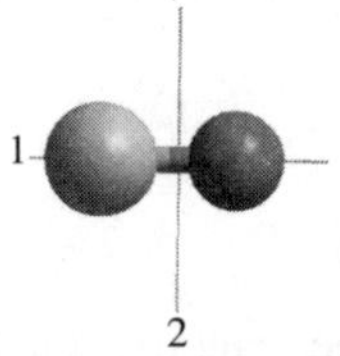	$C_{\infty v}$	Linear	$I = \mu_{red} r_e^2$ $\mu_{red} = \frac{m_C m_O}{m_C + m_O}$
HCN	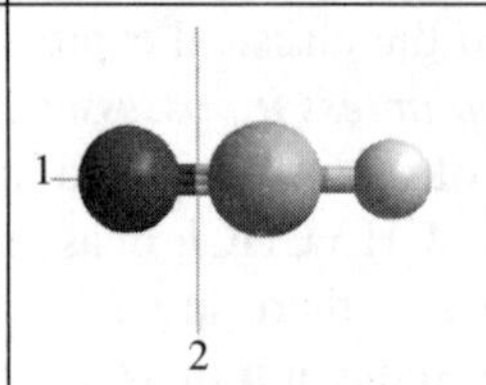	$C_{\infty v}$	Linear	$I_b = m_H r_{HC}^2 + m_N r_{CN}^2 - \frac{(m_N r_{CN} - m_H r_{HC})^2}{M}$
SF_6	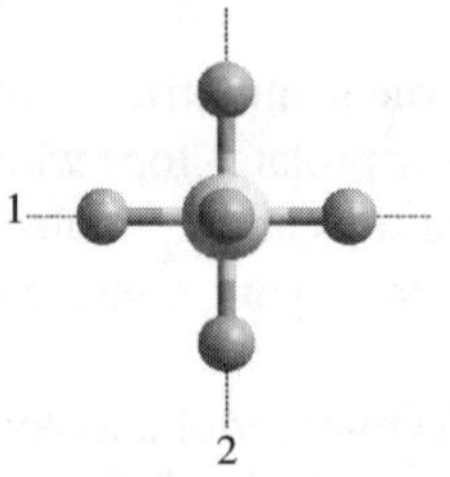	O_h	Spherical top	$I = 4m_F r_{SF}^2$
CH_4	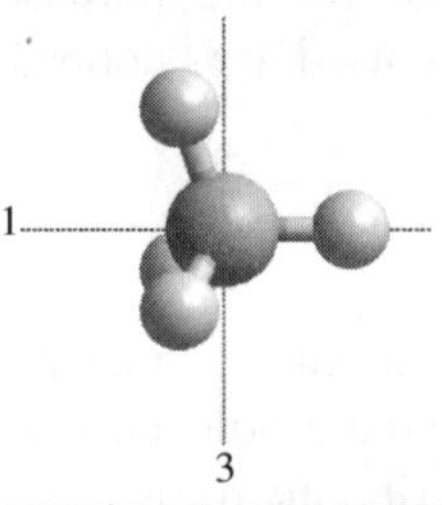	T_d	Spherical top	$I = \frac{8}{5} m_H r_{CH}^2$

Molecule		**Point group**	**Classification**	**Moment of inertia**
CH_3Cl	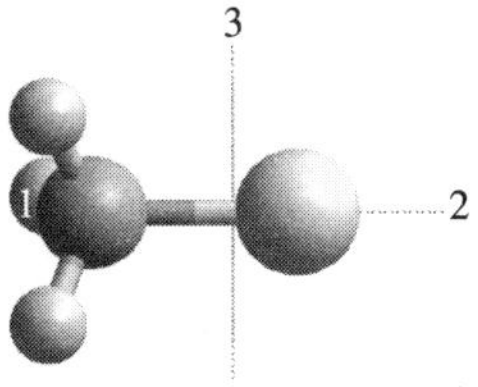	C_{3v}	Symmetric top - prolate	$I_{//} = 2m_H(1-\cos\theta)r_{CH}^2$ $I_\perp = m_H(1-\cos\theta)r_{CH}^2 + \frac{m_H}{M}(m_C + m_{Cl})(1+2\cos\theta)r_{CH}^2 + \frac{m_{Cl}}{M}\left\{ \begin{matrix} (3m_H + m_C)r_{CCl} \\ +6m_H r_{CH}\left[\frac{1}{3}(1+2\cos\theta)\right]^{1/2} \end{matrix} \right\} r_{CCl}$ $\theta = \angle HCH$
NH_3	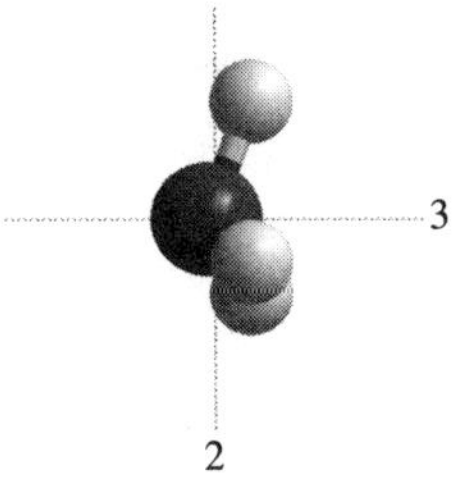	C_{3v}	Symmetric top - oblate	$I_{//} = 2m_H(1-\cos\theta)r_{NH}^2$ $I_\perp = m_H(1-\cos\theta)r_{CH}^2 + \frac{m_H m_N}{M}(1+2\cos\theta)r_{NH}^2$ $\theta = \angle HNH$
BF_3	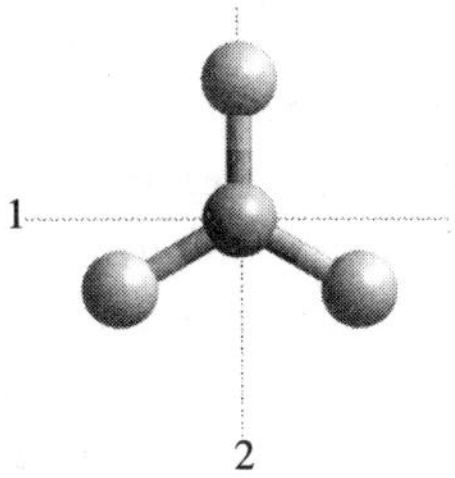	D_{3h}	Symmetric top - oblate	$I_{//} = 3m_F r_{BF}^2$ $I_\perp = \frac{3}{2}m_F r_{BF}^2$
C_6H_6	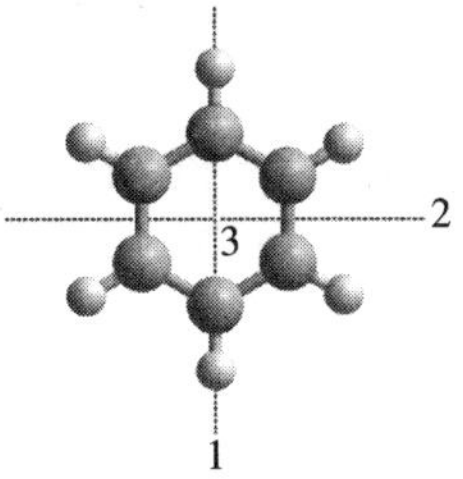	D_{6h}	Symmetric top - oblate	$I_{//} = 3(m_C r_{CC}^2 + m_H(r_{CC} + r_{CH})^2)$ $I_\perp = 6(m_C r_{CC}^2 + m_H(r_{CC} + r_{CH})^2)$

Figure 4.16. Principal inertial axes and expressions for moments of inertia of the different categories of molecules

---

Example 4.15 Classify the following molecules as linear, spherical, symmetric or asymmetric tops, and state which will give pure rotational spectra: SF_6, BrF_5, NH_3, NO_2, CO_2, N_2O, CF_3I, BF_3, BeH_2, SO_2, C_6H_6.

Solution

Molecule	Point group	Classification	Microwave activity
SF_6	O_h	Spherical top	Inactive
BrF_5	C_{4v}	Oblate symmetric top	Active
NH_3	C_{3v}	Oblate symmetric top	Active
NO_2	C_{2v}	Asymmetric top	Active
CO_2	$D_{\infty h}$	Linear	Inactive
N_2O	$C_{\infty v}$	Linear	Active
CF_3I	C_{3v}	Prolate symmetric top	Active
BF_3	D_{3h}	Oblate symmetric top	Inactive
BeH_2	$D_{\infty h}$	Linear	Inactive
SO_2	C_{2v}	Asymmetric top	Active
C_6H_6	D_{6h}	Oblate symmetric top	Inactive

It is not usually difficult to decide whether a symmetric top molecule is prolate or oblate, but there are some molecules for which the assignment is difficult. For example, if ammonia were to adopt a planar geometry, similar to BF_3, it would be an oblate top molecule, as it is in one of its excited states. On the other hand, if the H-N-H bond angle is extremely small, as it is in some other excited state, it would be classified as prolate. Clearly, at some intermediate geometry, there would be a transition from a prolate to an oblate top, and at angles close to this geometry, it is not obvious without doing calculations whether to classify it as an oblate or a prolate top. In the ground electronic state, ammonia is an oblate symmetric top.

Example 4.16 The moment of inertia about an axis perpendicular to the principal axis ($I_\perp$) for NH_3 is 2.82×10^{-47} kg m². Which type of rotor is NH_3? Calculate the separation of the pure rotational spectrum lines for NH_3.

Solution

The moment of inertia about the principal axis ($I_{//}$) is given by $2m_H r^2(1-\cos\theta)$, where the mass of a hydrogen atom = m_H; the N–H bond length = $r = 1.014 \times 10^{-10}$ m and the bond angle is 106.78°. Therefore,

$$I_{//} = 2 \times 1.0078 \times 1.6605 \times 10^{-27}(1.014 \times 10^{-10})^2(1-\cos(106.78°)) = 4.43 \times 10^{-47}\ \text{kg m}^2$$

Since $I_{//}$ is larger than $I_\perp$, it must be I_c and the molecule is an oblate top.

For symmetric top molecules, the spacing of the pure rotational lines is $2\tilde{B} = 2h/8\pi^2 I_b c = 6.60\times10^{-46}/2.82\times10^{-47} = 19.9\ \text{cm}^{-1}$.

4.4 THE EFFECT OF ELECTRIC FIELDS: THE STARK EFFECT

Although the rotational spectra of linear molecules are characterized by two quantum numbers, the energies of these states depend only on the total angular momentum quantum number J. The $2J + 1$ degeneracy of these rotational levels, due to the various values of M for a given value of J, can be lifted in the presence of an external field, which then defines a direction along which the M values are quantized. When an electric field is used, the shifting and splitting of rotational lines that results is known as the *Stark effect*.

In classical physics, an electric dipole, $\vec{\mu}$, is a vector quantity whose energy of interaction with an electric field, $\vec{E}$ is given by the dot product $-\vec{\mu}.\vec{E}$. If θ is the angle between the two vectors, the energy of interaction can be written as $-\mu E\cos\theta$, where μ and E are the lengths of the vectors which represent the dipole moment and the field, respectively. Another way of thinking about this is that the interaction depends on the *projection* of $\vec{\mu}$ along the direction of the electric field.

4.4.1 Effect on the energy levels of a symmetric top

Consider a prolate symmetric top, such as CH_3F, in which the molecular dipole lies along the C_3 axis, which is also the direction of the unique axis, a. Recall that the molecule is rotating about the vector $\vec{L}$ (which represents the total angular momentum) and that in general this vector is pointing at some angle to the a-axis. The dipole is thus precessing on a cone at this angle to the direction of $\vec{L}$, as shown in Figure 4.17(a) below. As a result of this precession, the dipole is averaged so that only its projection onto $\vec{L}$ survives; the components perpendicular to $\vec{L}$ are averaged to zero.

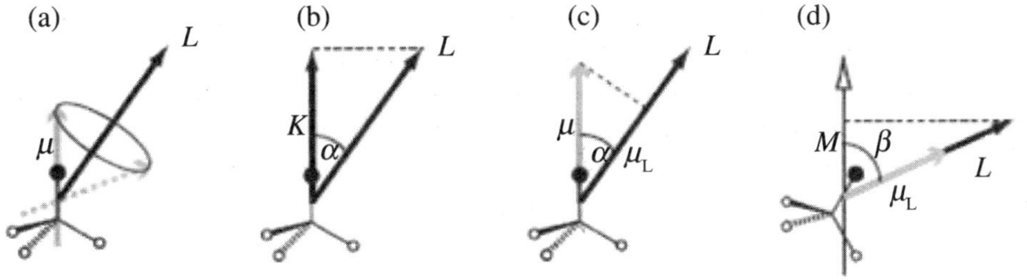

Figure 4.17. Precession of the dipole

The projection of the dipole moment on $\vec{L}$, represented by μ_L, is given by $\mu\cos\alpha$, where α is the angle between the dipole moment vector (a-axis) and the angular momentum vector $\vec{L}$ (Figure 4.17(b)). Since the dipole points along the direction of the vector $\vec{K}$, which is the projection of $\vec{L}$ onto the a-axis, the cosine of the angle α is easily found and is equal to the magnitude of $\vec{K}(K\hbar)$ divided by that of $\vec{J}$ $(\sqrt{J(J+1)}\hbar)$

$$\cos\alpha = \frac{K}{\sqrt{J(J+1)}} \tag{4.75}$$

Hence, the projection of the dipole moment on $\vec{L}$ is given by

$$\mu_L = \mu \frac{K}{\sqrt{J(J+1)}} \tag{4.76}$$

Now suppose that an electric field is applied and that the molecule is oriented such that $\vec{L}$ makes an angle β to the electric field direction, as shown in Figure 4.17(d). The angle β cannot take on arbitrary values but is determined by the allowed values of the M quantum number, such that

$$\cos\beta = \frac{M}{\sqrt{J(J+1)}} \tag{4.77}$$

since M represents the projection of $\vec{L}$ in the field direction. Hence, the energy of interaction is

$$\begin{aligned} E_{\text{Stark}}(J,K,M) &= -\mu_L E \cos\beta \\ &= -\mu E \frac{K}{\sqrt{J(J+1)}} \frac{M}{\sqrt{J(J+1)}} \\ &= -\mu E \frac{KM}{J(J+1)} \end{aligned} \tag{4.78}$$

Note that it is the projection of the dipole moment onto $\vec{L}$ which determines the interaction with the field. The energy correction E_{Stark} depends on the three quantum numbers J, K and M and is zero whenever K or M is zero; otherwise the correction varies linearly with the product μE of the electric field intensity and the permanent dipole moment, and is described as a *first-order* Stark correction. The energy shift caused by the electric field is thus proportional to the field and the dipole moment, and also depends on the three quantum numbers J, K and M.

From equation (4.78), we can see immediately that only energy levels with $K > 0$ will be affected by an electric field. Let us consider, as an example, the transition between $J = 1$, $K = 1$ and $J = 2$, $K = 1$. In the presence of an electric field, the $J = 1$ level will split into three sublevels, with $M = -1$, 0 and 1; using equation (4.78), the energy shifts will be

$$E_{\text{Stark}}(1,1,-1) = \frac{\mu E}{2} \quad E_{\text{Stark}}(1,1,0) = 0 \quad E_{\text{Stark}}(1,1,+1) = \frac{-\mu E}{2}. \tag{4.79}$$

The $J = 2$ level will split into five sublevels with M running from -2 to $+2$; the energy shifts are

$$E_{\text{Stark}}(2,1,\pm 2) = \frac{\mp \mu E}{3} \quad E_{\text{Stark}}(2,1,\pm 1) = \frac{\mp \mu E}{6} \quad E_{\text{Stark}}(2,1,0) = 0. \tag{4.80}$$

The splitting of the energy levels is shown in Figure 4.18.

The frequency of the displaced lines that appears in the Stark effect depends on the selection rule for the quantum number M. It is usual to arrange that the external electric field is parallel to the electric field of the incident microwave radiation, and then the selection rule is $\Delta M = 0$

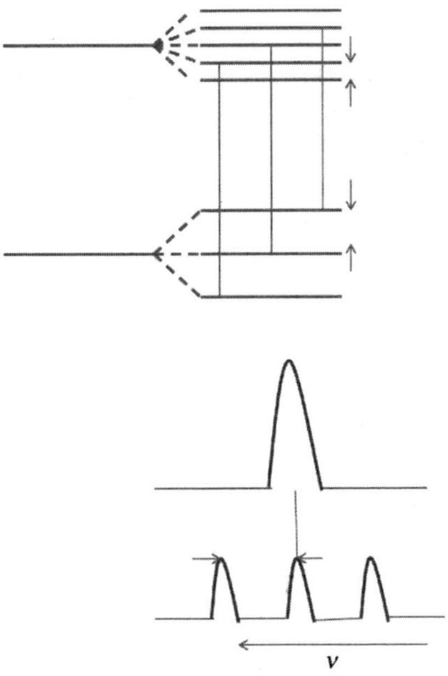

Figure 4.18. The Stark effect

(this selection rule is derived in Chapter 3). This condition, together with the usual selection rules for a symmetric top, $\Delta J = 1$ and $\Delta K = 0$, can be used to calculate the frequency of the displaced lines in the Stark effect.

$$\begin{aligned}\Delta E &= E_{\text{Stark}}(J+1,K,M) - E_{\text{Stark}}(J,K,M)\\ &= -\mu E\left\{\frac{KM}{(J+1)(J+2)} - \frac{KM}{J(J+1)}\right\}\\ &= 2\mu E\frac{KM}{J(J+1)(J+2)}\end{aligned} \tag{4.81}$$

In general, a transition from a level with quantum numbers J and K will split into $2J + 1$ lines, where J is the lower energy state, so there are three allowed transitions for the $J = 1 \to 2$, $K = 1$ transition of methyl fluoride considered above. The single line, observed in the absence of the field, splits symmetrically into three and, from the shifts in the energy levels, it can be seen that the splitting of the lines in the spectrum is $\mu E/3$.

Example 4.17 An electric field of magnitude 2×10^4 V m^{-1} is applied parallel to the electric field vector of the microwave radiation, and the rotational transition $(J = 2, K = 1) \leftarrow (J = 1, K = 1)$ of CH_3Cl at 1.67360 cm^{-1} is split into three components, with lines at 1.67569, 1.67360 and 1.67151 cm^{-1}. Determine the rotational constant, $\tilde{B}$, of CH_3Cl and its dipole moment.

Solution

The splitting of the lines is $\mu E/3$. The separation of the lines is 0.00209 cm^{-1}. This energy is changed to the SI units of Joule by multiplication by hc, giving 4.15173×10^{-26} J = $\mu E / 3$. Therefore, $\mu = 3 \times 4.15173 \times 10^{-26}$ J / 2×10^4 V m^{-1} = 6.2276×10^{-30} C m.

The central line corresponds to a pure $J = 2 \leftarrow J = 1$ transition with no Stark splitting. Thus, $4\tilde{B} = 1.67360\ \text{cm}^{-1}$, or $\tilde{B} = 0.41840\ \text{cm}^{-1}$.

Note the units. In SI units, the electric field is expressed in V m^{-1} and the dipole moment in C m to give the energy in Joule.

The above example shows that the Stark effect can be used to determine accurate values for the dipole moment. Second, the observed splittings can help in the assignment of spectra, i.e., identification of the quantum numbers of the energy levels involved in each transition.

4.4.2 Stark effect for linear molecules

As we have seen, linear molecules have no rotation about the internuclear axis, and so effectively $K = 0$ at all times. From equation (4.78), we see that the energy levels will not be split by the application of an electric field; thus, there is no first-order Stark effect.

One way of thinking about this is that the $\vec{L}$ vector is always perpendicular to the internuclear axis and hence to the dipole moment. Thus, rotation averages the dipole completely to zero, i.e., $\mu_L = \mu\cos\alpha = \mu\cos(\pi/2) = 0$.

However, in strong electric fields, linear molecules do show a *second-order* Stark effect, where the splitting of the energy levels goes as the *square* of the electric field strength. The effect described for the symmetric top is *first order*, as it is linear in the field. Thus, the Stark correction to the energy can be expressed as

$$\begin{aligned} E_{\text{Stark}} &= E^{(1)} + E^{(2)} + \ldots \\ E_{J,M} &= E_J + a_{J,M}E + \frac{1}{2}b_{J,M}^2E^2 + \ldots \\ E^{(2)} &= \frac{1}{2}b_{J,M}^2E^2 \end{aligned} \tag{4.82}$$

The exact quantum mechanical expression for the second-order Stark effect of linear molecules is

$$\begin{aligned} E_{\text{Stark}}^{(2)} &= \frac{\mu^2E^2\left[J(J+1) - 3M^2\right]}{2hc\tilde{B}J(J+1)(2J+1)(2J+3)} \quad \text{for } J \neq 0 \\ &= -\frac{\mu^2E^2}{6h\tilde{B}c} \quad \text{for } J = 0 \end{aligned} \tag{4.83}$$

4.5 APPLICATIONS OF MICROWAVE SPECTROSCOPY

An example of an application of microwave spectroscopy is the domestic microwave oven, whose frequency is tuned to the rotational frequency of water. The principle of the microwave oven involves heating the molecules of water in food through high-speed rotations induced by microwaves. The glass container, however, remains cold since it does not contain rotating dipoles.

Microwave spectroscopy is commonly used to determine the structure of small molecules (such as ozone, methanol or water), planarity and non-planarity of molecules, and accurate determination of geometric parameters such as bond lengths and bond angles with high precision. This is because it is highly sensitive, gives high resolution and is non-destructive. Other common techniques for determining molecular structure, such as X-ray crystallography, do not work very well for some of these molecules (especially the gases) and are not as precise. However, microwave spectroscopy is not useful for determining the structures of large molecules such as proteins.

Modern microwave spectrometers have very high resolution, allowing hyperfine structure to be observed. This technique can also provide information on the electronic structures of molecules, number and energy difference of rotational isomers, and is useful for the determination of electric properties of the molecules, such as dipolar and quadrupolar moments.

Microwave spectroscopy is one of the principal means by which the constituents of the universe are determined from the earth. It is particularly useful for analysis of the chemical composition in the interstellar medium (ISM). One early surprise in interstellar chemistry was the discovery of long-chain carbon molecules in the ISM. This prompted Harry Kroto to collaborate with Rick Smalley and Robert Curl, in whose laboratory it was possible to vapourize carbon under enormous energy conditions and study these molecules. The rest is history: the collaboration resulted in the discovery of C_{60}, buckminsterfullerene and the 1996 Nobel Prize in chemistry to the trio.

4.6 SUMMARY

- The moment of inertia about an axis is computed from the relation $I = \sum_i m_i r_i^2$, where r_i is the perpendicular distance of the ith atom from the axis.
- The three principal moments of inertia are labelled such that $I_a \le I_b \le I_c$. These axes pass through the centre of mass and are perpendicular to one another.
- A spherical top has $I_a = I_b = I_c$; a prolate symmetric top has $I_a < I_b = I_c$; an oblate symmetric top has $I_a = I_b < I_c$. A symmetric top must possess a single C_n axis with $n \ge 3$ or an S_4 axis.
- The energy levels of a symmetric top (ignoring centrifugal distortion) are
  - *prolate*: $E_{J,K} = BJ(J+1) + (A-B)K^2 \quad J = 0, 1, \ldots \quad K = -J, \ldots, +J$
  - *oblate*: $E_{J,K} = BJ(J+1) + (C-B)K^2 \quad J = 0, 1, \ldots \quad K = -J, \ldots, +J$

  where

  $$A = \frac{\hbar^2}{2I_a} \qquad \text{or} \qquad \tilde{A} = \frac{h}{8\pi^2 I_a c}, \text{ etc.}$$
- The selection rules are (1) the molecule must have a permanent dipole moment and (2) $\Delta J = \pm 1$, $\Delta K = 0$. As a result of the $\Delta K = 0$ selection rule, the spectrum of a symmetric top resembles that of a diatomic, with peaks at $2B$, $4B$, …
- In the presence of centrifugal distortion, the energies become
  - prolate: $$F(J,K) = \tilde{B}J(J+1) + (\tilde{A} - \tilde{B})K^2 - \tilde{D}_J J^2(J+1)^2 - \tilde{D}_{JK} J(J+1)K^2 - \tilde{D}_K K^4$$

Each transition splits into $(J + 1)$ closely spaced components.

- For linear molecules, centrifugal distortion does not result in extra lines, but simply causes the spacing of the lines to decrease as J increases.
- Stark effect: an electric field shifts the energy levels of a symmetric top according to

$$E_{\text{Stark}}(J,K,M) = -\mu E \frac{KM}{J(J+1)}.$$

- Microwave spectroscopy finds several uses, some of which are as follows: It is used for structure determination using the calculated moments of inertia and for calculating the atomic weights of isotopes. The approximate force constant (k) can be determined by calculating D_J and $D_{J,K}$. Line breadth measurements give indications about intermolecular forces. The Stark effect can be used to determine accurate dipole moments of molecules.

4.7 EXERCISES

1. Why does the allowed value $J = 0$ for the rotational quantum number not violate the Heisenberg's Uncertainty Principle?
2. What is the gross selection rule for microwave (rotational) spectroscopy? What is the selection rule for J? Which of the following molecules may show a pure rotational microwave absorption spectrum: H_2O, H_2, N_2O, NH_3, IF, O_2, KCl, Cl_2, HF, CH_4, CH_3F, BF_3, $C_2F_2H_2$, O_3, CO_2, toluene, Argon…HCl?
3. Given the recursion relation for associated Legendre polynomials

$$zP_J^{|M|}(z) = \frac{J+|M|}{2J+1} P_{J-1}^{|M|}(z) + \frac{J-|M|+1}{2J+1} P_{J+1}^{|M|}(z)$$

where $z = \cos\theta$, show that, for a diatomic molecule, the selection rule for rotational transitions in the rigid rotator model is $\Delta J = \pm 1$ (see Chapter 3).

4. Calculate the wavenumbers and wavelengths for pure rotational lines in the spectrum of HCl corresponding to the following changes in rotational quantum numbers $0 \rightarrow 1$, $1 \rightarrow 2$ and $8 \rightarrow 9$. The rotational constant of HCl is 10.5909 cm^{-1} (*Ans.* 21.1818 cm^{-1}, 42.3636 cm^{-1}, 190.636 cm^{-1}; 4.72×10^{-4} m, 2.36×10^{-4} m, 5.25×10^{-5} m).
5. Assuming that a $D^{35}Cl$ molecule can be approximated by a rigid rotor with internuclear distance $r_e = 127.5$ pm, calculate the expected positions in cm^{-1} for the absorption peaks corresponding to the rotational transitions $J = 1 \leftarrow 0$, $2 \leftarrow 1$ and $3 \leftarrow 2$.
6. The rotational spectrum of $^1H^{131}I$ consists of a series of lines separated by 12.85274 cm^{-1}. Determine the bond length of the molecule (*Ans.* 161.952 pm).
7. The table below compares the reduced masses of $^1H^{35}Cl$, $^1H^1H$ and $^{35}Cl^{35}Cl$. Explain both mathematically and in terms of the rotational motion of the molecule, why the reduced mass of Cl_2 is so different from that of HCl, and why the reduced mass of H_2 is so much smaller than that of HCl.

$^1H^{35}Cl$	1.62661×10^{-27} kg
$^1H^1H$	8.36745×10^{-28} kg
$^{35}Cl^{35}Cl$	3.0691×10^{-26} kg

8. The rotational constant of $^1H^{35}Cl$ is 10.5909 cm^{-1}. Calculate the rotational constants of $^1H^{37}Cl$ and $^2H^{35}Cl$.
9. The following absorptions were observed in the microwave spectrum of HI: 64.275, 77.130 and 89.985 cm^{-1}. Determine B, I and r_e for the molecule. Is there any evidence of centrifugal distortion?
   When H is isotopically substituted with D, the following lines are observed: 65.070, 71.577, 78.094 and 84.591 cm^{-1}. Determine B and I for DI. Usually it is assumed that bond lengths are unchanged by isotopic substitution (why?) – do the data support this assumption?
10. Consider the linear molecule, H-C≡C-Cl. There are two isotopes of chlorine, ^{35}Cl (~75%) and ^{37}Cl (~25%). Therefore, one will observe two series of lines in the rotational spectrum, resulting from transitions of H-C≡C-^{35}Cl and H-C≡C-^{37}Cl.
    Can the structure of H-C≡C-Cl be determined from these two series? If not, what additional information could be used to determine all three bond distances?
11. The intensity of a transition is proportional, to a first approximation, to the population of the originating energy level. For a diatomic molecule treated as a rigid rotor, the population of the Jth level,

$$N_J \propto (2J+1)\exp\left(-\frac{hc\tilde{B}J(J+1)}{k_B T}\right)$$

This is just the Boltzmann formula. The pre-exponential factor is the degeneracy which increases linearly with J and the exponential energy factor decreases with J.

(a) Sketch N_J against J and show that the most populated J-level is

$$J_{\max} = \sqrt{\frac{kT}{2\tilde{B}hc}} - \frac{1}{2}$$

(b) Show that the rotational quantum number of the rotational energy level with the maximum population can be approximated as

$$J_{\max} = \sqrt{\frac{kT}{2\tilde{B}hc}} - \frac{1}{2} \approx 0.59\sqrt{\frac{T(\mathrm{K})}{\tilde{B}(\mathrm{cm}^{-1})}}$$

(c) The wavenumber corresponding to this level is $\tilde{\nu}_{\max} \approx 1.18\sqrt{\tilde{B}T}$.

(d) The rotational energy of the most populated state $\tilde{E}_{\mathrm{rot}} \approx \dfrac{kT}{2hc}$ and $E_{\mathrm{rot,\,max}} \approx \dfrac{1}{2}kT$.

12. (a) Show that, for a linear molecule subject to centrifugal distortion, the frequencies of the allowed lines in the pure rotation spectrum are given by

$$\tilde{\nu}(J) = 2\tilde{B}(J+1) - 4\tilde{D}(J+1)^3.$$

Explain how a graphical method can be used to determine the values of $\tilde{B}$ and $\tilde{D}$. Is it necessary for the lines to be assigned correctly for this method to be successful?

(b) The following wavenumbers were observed for the rotational transitions of $^1H^{35}Cl$:

104.38 124.87 145.36 165.85 186.34 206.59 226.84

Assign these lines, explaining how you arrived at your assignments. Use a graphical method to determine $\tilde{B}$ and $\tilde{D}$; hence, determine the moment of inertia and the corresponding bond length.

Explain why the bond length you have determined is *not* the equilibrium bond length.

13. In a high-resolution microwave study of $^{2}H^{19}F$, the *first* four lines in the spectrum are observed at 22.0180, 44.0218, 65.9970 and 87.9295 cm^{-1}. By drawing a suitable straight-line graph, deduce the values of $\tilde{B}$ and $\tilde{D}$ for $^{2}H^{19}F$. Hence, determine the $^{2}H^{19}F$ bond length.
14. The first four lines in the rotational absorption spectrum of CO in its ground vibrational level are observed at the following wavenumbers: 4.862536 cm^{-1}, 7.724924 cm^{-1}, 11.587019 cm^{-1} and 15.448673 cm^{-1}. Draw a suitable straight line graph and determine $\tilde{B}$ and $\tilde{D}$ for the CO molecule. Hence, determine the equilibrium bond length.
15. Calculate the bond length of $^{12}C^{16}O$ using $\tilde{B} = 1.9302\ \text{cm}^{-1}$ and the reduced mass $\mu_{red} = 1.14 \times 10^{-26}$ kg.
16. The rotational terms of a diatomic molecule showing centrifugal distortion are given by

$$F(J) = \tilde{B}J(J+1) - \tilde{D}J^2(J+1)^2$$

where $\tilde{D}$ is the *centrifugal distortion constant*, a small positive quantity.

(a) Explain the meaning of J, $\tilde{B}$ and $\tilde{D}$ in this expression.

(b) Show that this formula predicts lines in the microwave spectrum at

$$\tilde{\nu}_{J+1\leftarrow J} = 2\tilde{B}(J+1) - 4\tilde{D}(J+1)^3$$

Since $\tilde{D}$ is positive, the lines get closer together as J increases, as observed.

(c) In the harmonic approximation, the centrifugal distortion constant can be calculated from the formula

$$\tilde{D} = \frac{h^3}{32\pi^4 I^2 r^2 kc} = \frac{4\tilde{B}^3}{\omega_{\text{vib}}^2}$$

where k is the force constant (in N m^{-1}) and c is expressed in cm s^{-1}. Show that $\tilde{D}$ has the correct units (cm^{-1}) and find its value for the N_2 molecule using $k(N_2) =$ 2294 N m^{-1} and $r(N_2) = 109.8$ pm.

17. The molecule $^{1}H^{35}Cl$ exhibits rotational absorption lines in the far infrared at the following wavenumbers (cm^{-1}): 83.32, 104.13, 124.73, 145.37, 165.89, 186.23, 206.60 and 226.86 (note that there may be other lines in the microwave region too).

(a) Identify the transitions and use a graphical procedure to determine the rotational and centrifugal distortion constants. Calculate the bond length of HCl.

(b) Predict the rotational constant for DCl.

(c) Determine the most populated rotational level in HCl at 300 K.

18. The bond length of $^{79}Br^{81}Br$ is 228 pm. Calculate the moment of inertia and the rotational constant, B.
19. The rotational constant of $^{127}I^{35}Cl$ is 0.1142 cm^{-1}. Calculate the equilibrium bond length.
20. When a hydrogen molecule is chemisorbed on a crystalline surface, its rotation can be approximated as that of a rigid rotor. The hydrogen bond length is 74 pm. Calculate the wavenumber of the lowest energy rotational transition of chemisorbed H_2 (*Ans.* 60.9 cm^{-1}).

21. Pure microwave absorptions are observed at 84.421, 90.449 and 96.477 GHz on flowing dibromine gas over hot copper metal at 1100 K. What transitions do these frequencies represent? What is the bond length of the species formed? Assume a linear rotor (*Ans.* 218.6 pm).
22. A space probe was designed to observe $^{12}C^{16}O$ in the solar atmosphere using a microwave technique. Given that the bond length of $^{12}C^{16}O$ is 112.82 pm, at which wavenumbers will the first four transitions lie? At which wavenumbers would the same transitions lie in $^{13}C^{16}O$?
23. The free radical $^{12}C^{1}H$ shows a microwave spectrum consisting of a series of lines equally spaced by 28.92 cm^{-1}.
    (a) Calculate the moment of inertia of the $^{12}C^{1}H$ radical.
    (b) Use this value to calculate the bond length of $^{12}C^{1}H$.
    (c) Calculate the line separation in the microwave spectrum of $^{12}C^{1}H$.
24. (a) Derive from classical dynamics the formula for the moment of inertia of two mass points of masses m_C and m_O separated by a distance r_e.
    (b) Thence determine r_e for carbon monoxide.
    (c) At what approximate frequency would you expect to find the J_{1-0} absorption of $^{13}C^{16}O$?
    (d) Rosenblum et al. (1958) found that the frequencies for the J_{1-0} transition are 112.359276 and 109.782182 GHz for $^{12}C^{17}O$ and $^{12}C^{18}O$, respectively. Evaluate the masses of ^{17}O and ^{18}O to six significant figures. Ignore the effects of bond elasticity.
25. (a) Starting with the angular momentum of the *i*th element of mass

$$\vec{L}_i = \vec{r}_i \times \vec{p}_i = m_i \vec{r}_i \times (\omega \times \vec{r}_i)$$

show that

$$L_{x_i} = m_i \left[\omega_x (r_i^2 - x_i^2) - x_i y_i \omega_y - x_i z_i \omega_z \right]$$

or

$$L_x = \sum_i L_{x_i} = \sum_i m_i \left[\omega_x \left(y_i^2 + z_i^2 \right) - x_i y_i \omega_y - x_i z_i \omega_z \right]$$

Derive the inertia matrix such that $|L\rangle = I|\omega\rangle$.
(Use the relation $\vec{A} \times \vec{B} \times \vec{C} = \vec{B}(\vec{A}.\vec{C}) - \vec{C}(\vec{A}.\vec{B})$)

    (b) A hypothetical linear rigid molecule, AB_2, may be represented by three point masses
    B: $m_1 = ½$ at (−1, −1, −1)
    A: $m_2 = 1$ at (0,0,0)
    B: $m_3 = ½$ at (1,1,1)
    (i) Find the moment of inertia matrix.
    (ii) Diagonalize the inertia matrix, obtaining the eigenvalues and the principal axes (as orthonormal eigenvectors).
    (iii) Display the matrix of rotation and the new coordinates of the three atoms.
    (iv) Explain the physical significance of the $\lambda = 0$ eigenvalue. What is the significance of the corresponding eigenvector?
26. Show that the moment of inertia of a linear triatomic molecule XYZ can be written:

$$I = m_X r_{XY}^2 + m_Z r_{YZ}^2 - \frac{1}{M}(m_X r_{XY} - m_Z r_{YZ})^2$$

where $M = m_X + m_Y + m_Z$ is the total mass of the molecule. If the atom Z is replaced with an isotope of mass m_Z' such that the bond lengths are unchanged and the new moment of inertia is I', show that the XY bond length can be obtained from the equation

$$r_{XY}^2 = \frac{IMm_Z' - I'M'm_Z}{m_X m_Y (m_Z' - m_Z)}$$

Determine the CO and CS bond lengths in OCS (which is linear) from the rotational constants $B(^{16}O^{12}C^{32}S) = 6081.5$ MHz, $B(^{16}O^{12}C^{34}S) = 5932.8$ MHz.

27. The rotational constant B is given by DeLucia and Gordy (1969) for HCN as 44,315.97 MHz and for DCN as 36,207.40 MHz. Deduce the moments of inertia of the two molecules. Assuming that the bond lengths are independent of isotopic substitution, calculate the H–C and C–N bond lengths.
28. Given that the CO bond length in the molecule OCS is 0.1165 nm and the CS bond length is 0.1558 nm, determine its moment of inertia. At which frequencies do the $J = 1 \leftarrow 0$ and $2 \leftarrow 1$ transitions occur in the rotational spectrum of OCS?
29. What information about the molecular geometry of N_2O can be determined from knowing that a pure rotational absorption spectrum is observed for this molecule?
30. (a) Describe, with examples, the classification of symmetric top molecules as *oblate* or *prolate*.
    (b) Classify the following molecules as
    (i) spherical top;
    (ii) symmetric top (prolate or oblate); or
    (iii) asymmetric top.
    CH_4, CH_3Cl, CH_2Cl_2, $CH_2{=}CH_2$, benzene, SF_6, SO_3, SF_4, NF_3, ethylene, ethane, allene.
    Which of these molecules will show a pure rotational (microwave) spectrum?
    For those which are symmetric tops, make a sketch showing where the principal axes lie and indicate which is the unique axis.
31. From the general shape of the following molecules, define them as either prolate or oblate symmetric tops, or neither.
    (a) Tetrafluoroethylene, CF_2CF_2
    (b) Boron trifluoride, BF_3
    (c) Trimethylamine, $N(CH_3)_3$
    (d) Dimethyldiacetylene, $CH_3\text{-}C{\equiv}C\text{-}C{\equiv}C\text{-}CH_3$
32. For the planar D_{3h} molecule boron trifluoride (BF_3), show that $I_a = I_b = I_c / 2$. Given that the B–F bond length is 0.13 nm, calculate the value of I_a.
33. Trifluoro methyl iodide (CF_3I) is a prolate symmetric top, with rotational constants $\tilde{A} = 0.1891$ cm^{-1} and $\tilde{B} = 0.0498$ cm^{-1}. Calculate all the rotational energy levels for the $J = 2$ level.
    Make a similar calculation for the oblate symmetric top molecule NH_3 for which $\tilde{B} = 9.443$ cm^{-1} and $\tilde{C} = 6.196$ cm^{-1}, and make a qualitative comparison with the energy levels for CF_3I.
34. The rotational constants of the asymmetric top molecule, formaldehyde, are $A = 282$ GHz, $B = 39$ GHz and $C = 4$ GHz. Calculate the degree of asymmetry, $\kappa = \dfrac{2B - A - C}{A - C}$ for the molecule. Is the molecule near prolate or near oblate?

35. Consider the molecules CCl_4, $CHCl_3$ and CH_2Cl_2.
    (a) What kind of rotor are they (spherical top, symmetric top or asymmetric top)?
    (b) Will they show pure rotational spectra?
    (c) Assume that ammonia shows a pure rotational spectrum. If the rotational constants are 9.44 and 6.20 cm^{-1}, use the energy expression:

$$E = (\tilde{A} - \tilde{B})K^2 + \tilde{B}J(J+1)$$

    to calculate the energies of the first three lines (i.e., those with lowest K, J quantum number for the absorbing level) in the absorption spectrum (ignoring higher order terms in the energy expression).
36. The molecule $CClF_3$ is a prolate symmetrical top with $\tilde{A} = 0.1908\ \text{cm}^{-1}$ and $\tilde{B} = 0.1111\ \text{cm}^{-1}$. Calculate the energy corresponding to $J = 2$ and $K = \pm 1$.
37. Ammonia, NH_3, is an oblate symmetric top.
    (a) State the selection rules which apply to the molecule undergoing changes in rotational energy. Give any rationalization you can of these rules.
    Hence, determine an expression for the frequencies of the allowed transitions (ignore centrifugal distortion); draw a labelled sketch of the spectrum you expect.
    (b) The idealized microwave spectrum of ammonia shows absorptions at the following frequencies, in GHz

    1798.9 2398.6 2998.2

    Assign the transitions (giving your reasons), and determine what you can about the molecule.
38. When centrifugal distortion is taken into account, the energy levels of a prolate symmetric top are given by

$$F(J,K) = \tilde{B}J(J+1) + (\tilde{A} - \tilde{B})K^2 - \tilde{D}_J J^2(J+1)^2 - \tilde{D}_{JK} J(J+1)K^2 - \tilde{D}_K K^4$$

    (a) Show that the frequencies of the allowed transitions are given by

$$\tilde{\nu}(J,K) = 2(\tilde{B} - \tilde{D}_{JK}K^2)(J+1) - 4\tilde{D}_J(J+1)^3$$

    Explain how it is that centrifugal distortion splits what was a single line into $J + 1$ components.
    (b) In the rotational spectrum of CH_3F, absorption maxima were found at *wavelengths* of 1.958 and 2.937 mm. At higher resolution, these absorption peaks split into three and two components, respectively.
    Explain the origin of these observations, assign the transitions and determine as many rotational and centrifugal distortion constants of CH_3F as you can. Give your constants in cm^{-1} and in MHz.
39. (a) The energy levels of a symmetric top molecule including centrifugal distortion are given by the expression

$$F(J,K) = \tilde{B}J(J+1) - (\tilde{A} - \tilde{B})K^2 - D_J J^2(J+1)^2 - D_{JK} J(J+1)K^2 - D_K K^4$$

where D_J, D_{JK} and D_K are centrifugal distortion constants. Derive an expression for the wavenumbers of the allowed transitions in the rotational spectrum ($\Delta J = \pm 1$, $\Delta K = 0$).

(b) In the rotational spectrum of CH_3F, lines were observed at the frequencies 51071.8, 102142.6, 102140.8, 15210.3, 153207.6 and 153199.6 MHz. Assign quantum numbers to the transitions and determine $\tilde{B}$, D_J and D_{JK} for CH_3F.

40. In the microwave spectrum of CH_3F, the following absorption lines (in cm^{-1}) were observed: 5.098 6.797 8.496 10.195. Assign the transitions and determine what you can from the spectrum.

41. (a) Identify the molecules that will exhibit a pure rotational absorption microwave spectrum: N_2O, NO_2, $CClF_3$, NF_3, SF_6, CH_4, CO_2, H_2, HCl, CH_3Cl, CH_2Cl_2, H_2O, NH_3, benzene, toluene.

(b) Suggest a structure for XY_3 so that it (i) gives a rotational spectrum and (ii) does not give a rotational spectrum (give reasons).

42. Consider the bent AB_2 type molecule (C_{2v}), shown below; the bond angle is θ and the bond length is R. Compute the moments of inertia about axes **1** and **2**, where axis **2** passes through the centre of mass, located a distance x from atom A.

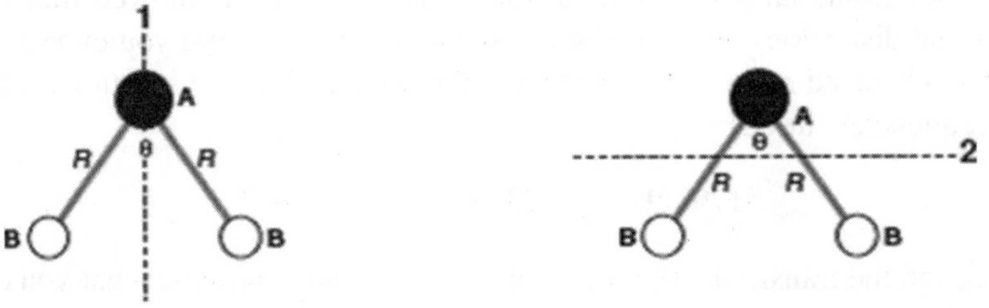

OF_2 has C_{2v} symmetry; the bond angle is 103° 18′ and the O-F bond length is 140.9 pm.

(a) Determine the moment of inertia about axis **1** as shown in the diagram; hence, determine the corresponding rotational constant in cm^{-1}. You may take the masses of O and F as 16 and 19 u, respectively.

(b) Determine the value of the distance x between axis **2** and the centre of mass; hence, determine the moment of inertia about axis **2** (which passes through the centre of mass) and the corresponding rotational constant, in cm^{-1}.

(c) Is OF_2 a symmetric top? Give reasons for your answer.

43. Both linear and asymmetric top molecules have a small quadratic Stark effect. For a linear molecule (only second order effect) with a dipole moment μ, the additional energy in an electric field E is given by

$$\frac{\left[J(J+1)-3M^2\right]}{J(J+1)(2J-1)(2J+3)}\frac{\mu^2E^2}{2hB}$$

apart from the special case of $J = 0$ when the additional energy is $-\mu^2E^2 / 6hB$.

(a) Tabulate the states and their additional energies as multiples of $\mu^2E^2 / 2hB$ for $J =$ 0, 1, 2 and 3.

(b) Use the results to tabulate the contribution to the transition frequencies for the transitions allowed between these states by the selection rules $\Delta J = 1$, $\Delta M = 0$, which apply when the static and microwave electric fields are parallel.

(c) For the linear molecule OCS, $B_0 = 6.081$ GHz and the displacement of the two lines from their natural field-free positions for the J_{21} transition near 24.3 GHz are −0.406 MHz and +0.330 MHz with a field of 50 kV m^{-1} and −1.623 and +1.319 MHz for a field of 100 kV m^{-1}. Confirm that the effect is second order and evaluate the dipole moment, μ.

44. (a) Draw up a diagram showing the effect of an electric field on the energy levels ($J = 2$, $K = 1$) and ($J = 3$, $K = 1$) of a prolate symmetric top. Indicate the splittings (in terms of μ and E) of the M levels.
Identify the allowed transitions and hence predict the expected form of the spectrum.
(b) For CH_3I, the rotational constant, B, is 1523 MHz and the dipole moment is 1.0 Debye. Using your results from part (a), predict the frequencies (in MHz) of the lines you would expect to see from the ($J = 2$, $K = 1$) $\rightarrow$ ($J = 3$, $K = 1$) transition (i) when no electric field is applied, (ii) when an electric field of 10^4 V m^{-1} is applied parallel to the electric field vector of the radiation.

45. How many lines will appear in the microwave spectrum of a diatomic molecule for a transition from $J = 1$ to $J = 2$ in the presence of an electric field?

46. Calculate the energies of the $J = 2$ levels of CH_3I ignoring centrifugal distortion, given $\tilde{A} = 0.191\ \text{cm}^{-1}$ and $\tilde{B} = 0.0507\ \text{cm}^{-1}$. What would be the magnitudes of the Stark splittings of these levels in a static electric field of 10 V m^{-1} if the dipole moment of CH_3I is 3.34×10^{-30} C m (1 Debye)? Draw an energy level diagram illustrating the splittings.

47. For the $J = 1 \rightarrow 2$, $K = 1$ transition, compute the expected splitting in the spectrum of CH_3F, which has a dipole moment of 1.857 Debye, when a field of 20,000 V m^{-1} is applied.

48. The literature value of r_e for the ground electronic state of iodine molecule is 266.6 pm. Calculate the moment of inertia for iodine molecule, rotational constant $\tilde{B}$, and the difference in energy (in cm^{-1}) for transitions from $J' = 128$ to $J'' = 127$ and $J'' = 129$.

49. The moment of inertia of $^{12}C^{16}O$ is 14.49×10^{-47} kg m^2. Calculate the energies of the first three rotational energy levels and the angular velocities (ω_{rot}) with which the molecule can rotate in those energy levels. Show that the number of molecules in the $J = 0$ and $J = 1$ states is almost equal. Make a population versus J diagram for CO at 25 °C and 1500 °C. What would be the nearest integral J value corresponding to the maximum population for each temperature?

50. The rotational constants B (in units of MHz) for several alkali metal halides are as in the accompanying table. Calculate the bond distances in each. Do you observe a pattern?

Molecule	B/MHz	Molecule	B/MHz	Molecule	B/MHz
$Na^{35}Cl$	6336.9	$Na^{79}Br$	4534.1	NaI	3531.8
$^{39}K^{35}Cl$	3856.4	$^{39}K^{79}Br$	2434.9	^{39}KI	1826.0
$^{85}Rb^{35}Cl$	2627.4	$^{85}Rb^{79}Br$	1424.8	^{85}RbI	984.3
$Cs^{35}Cl$	2161.2	$Ca^{79}Br$	1081.3	CsI	708.4

5

Vibrational Spectroscopy of Diatomics

Do not Bodies and Light act mutually upon one another;
that is to say, Bodies upon Light in emitting, reflecting,
refracting and inflecting it,
and Light upon Bodies for heating them,
and putting their parts into a vibrating motion wherein heat consists?
—Sir Isaac Newton

5.1 INTRODUCTION

We now move on to the next higher energy range, the infrared region, where vibrational motions of molecules are studied. As we shall see, vibrational spectra provide information about the stiffness of bonds, i.e., how easy it is to distort bond lengths and bond angles. However, when these studies are made in the gas phase, an additional complexity arises in the form of simultaneous changes in the rotational energy. This problem is resolved by first considering the vibrational spectra in liquid or solution phase, in which the close neighbours prevent rotational motion. Having considered pure vibrations, we shall turn our attention to the accompanying rotational energy changes in later sections.

As in the previous study of the rotation of molecules, we shall first consider the classical treatment of systems with one and two degrees of freedom and then extend our treatment to systems that do not behave according to classical (Newtonian) laws of motion, i.e., those that behave as quantum mechanical systems.

5.2 OSCILLATIONS OF SYSTEMS WITH ONE DEGREE OF FREEDOM

Consider a mass m attached to a rigid wall by a spring.

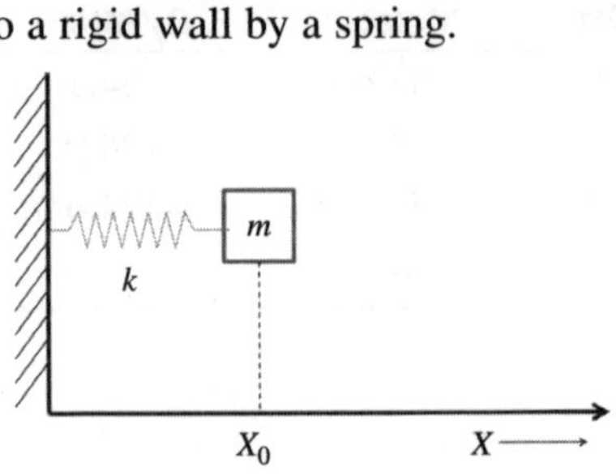

When the spring is stretched by an amount $x \equiv (X - X_0)$ from its equilibrium position, X_0, the force on the mass is given from Newton's law as $F = ma = m\dfrac{d^2x}{dt^2} = m\ddot{x}$, where $\ddot{x}$ represents, as usual, the second derivative of x with respect to time. This is balanced by an equal and opposite restoring force, which, according to Hooke's law, is proportional to the extension of the spring, $F = -k(X - X_0) \equiv -kx$, where the proportionality constant, k, has a positive value and is known as the *force constant* of the spring. The rectilinear motion of a mass m attached to a spring of constant k is, therefore, governed by the differential equation

$$m\ddot{x} = -kx \tag{5.1}$$

The general solutions of this second-order differential equation are in terms of the trigonometric sine and cosine functions. Therefore,

$$x(t) = A\cos(\alpha t) + B\sin(\alpha t) \tag{5.2}$$

or $x(t) = A'\cos(\alpha t + \phi)$, or, alternatively, $x(t) = Ce^{-i\alpha t}$, where C is a complex constant.

The constants A and B, or A' and φ, or C, are determined by the initial conditions. Let the mass be released with non-trivial initial conditions, i.e., $x(0)$ and $\dot{x}(0) = \text{v}(0)$ (the velocity at time zero) are not simultaneously zero. Then, if $x(0) = x_0$ and $\text{v}(0) = 0$ are the initial position and velocity, respectively, then we have from equation (5.2),

$$x(0) = A\cos(0) + B\sin(0) = x_0 \Rightarrow A = x_0 \tag{5.3}$$

since $\sin(0) = 0$ and $\cos(0) = 1$.

Similarly, for the initial velocity, we have

$$\text{v}(0) = \left(\frac{dx}{dt}\right)_{x=0} = -A\alpha\sin(0) + \alpha B\cos(0) = 0 \Rightarrow B = 0 \tag{5.4}$$

So the time dependence of the displacement x is given by the equation

$$x(t) = x_0\cos(\alpha t) \tag{5.5}$$

Substituting in the differential equation (5.1), we have $-mx_0\alpha^2\cos(\omega t) = -kx_0\cos(\alpha t)$, or $\alpha^2 = k/m$. Therefore, the spring performs harmonic oscillations with angular frequency $\omega_{\text{osc}} = \alpha = \sqrt{k/m}$ and maximum displacement x_0 (the amplitude) from equilibrium, since the maximum magnitude of the cos term is unity. The displacement is shown as a function of time in Figure 5.1.

This is an example of oscillations of a system with *one degree of freedom*. The state of the system can be described by a single function $x = x(t)$ representing the coordinate of the mass m. The total energy of this harmonic oscillator is the sum of its kinetic (T) and potential (V) energies. The kinetic energy is given, as usual, by

$$\begin{aligned} T &= \frac{1}{2}m\text{v}^2 = \frac{1}{2}m\left(\frac{dx}{dt}\right)^2 = \frac{1}{2}m[-\alpha x_0\sin(\alpha t)]^2 = \frac{1}{2}m\alpha^2x_0^2\sin^2\alpha t \\ &= \frac{1}{2}kx_0^2\sin^2(\alpha t) \end{aligned} \tag{5.6}$$

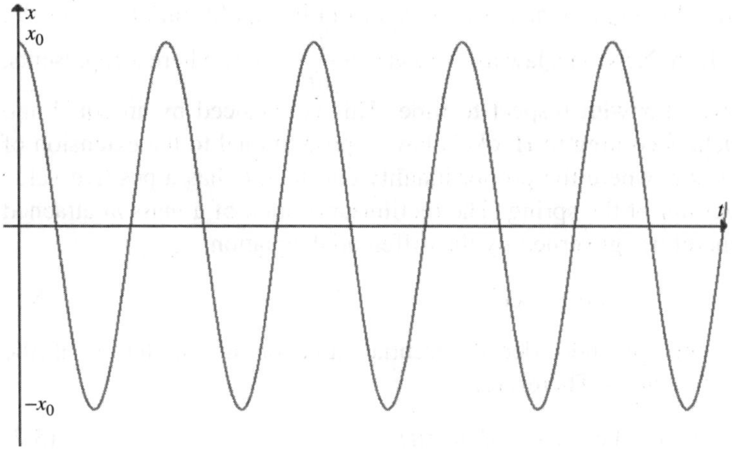

Figure 5.1. Displacement of the spring as a function of time

since $\alpha^2 = \frac{k}{m}$. The potential energy can be treated similarly. The force on a particle, $F(x)$, is the negative gradient of the potential energy with respect to x. Therefore, the potential energy may be obtained as follows:

$$F(x) = -\frac{dV}{dx} \Rightarrow V = -\int F(x)\,dx = \int (kx)\,dx = \frac{1}{2}kx^2$$
$$= \frac{1}{2}kx_0^2\cos^2(\alpha t) \tag{5.7}$$

The total vibrational energy, E, is therefore

$$E = T + V = \frac{1}{2}kx_0^2\left[\sin^2(\alpha t) + \cos^2(\alpha t)\right] = \frac{1}{2}kx_0^2 \tag{5.8}$$

The energy therefore oscillates between the kinetic and potential energies in such a manner that its total value is constant, as shown in Figure 5.2.

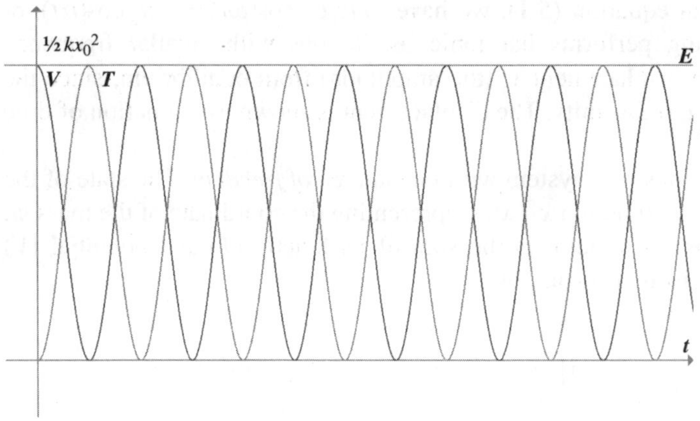

Figure 5.2. Variation of potential and kinetic energies with time

Similar oscillatory motion is also observed for systems with many degrees of freedom. In fact, most real systems can be approximated as harmonic oscillators. Let us first take an example from the real world, that of a diatomic molecular bond.

5.3 THE DIATOMIC MOLECULE

For simplicity's sake, let us first take the case of a homonuclear diatomic molecule, i.e., the case where the atoms in the diagram below are identical, e.g., what might be a H_2 or N_2 molecule constrained to move in a fixed line. A reasonable model for this system is one in which the whole of the mass (m) resides at the nuclei and the covalent bond behaves as a spring.

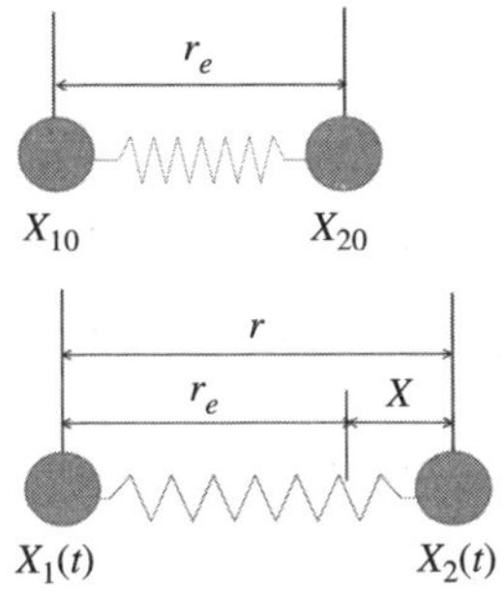

Let the two atoms (of mass m) be moving with negligible friction. The instantaneous positions of the masses are denoted by X_1 and X_2 and their equilibrium positions by X_{10} and X_{20}. It is convenient to describe the motion in terms of displacement

$$x_1(t) = X_1(t) - X_{10}$$
$$x_2(t) = X_2(t) - X_{20} \quad (5.9)$$

Evidently, the net elongation of the bond is ($X = x_2 - x_1$) and hence the restoring force is $k(x_2 - x_1)$. Therefore, the equation of motion, $F = ma$, for the first atom reads

$$m\ddot{x}_1 = k(x_2 - x_1) \quad (5.10a)$$

Similarly, for the second atom,

$$m\ddot{x}_2 = -k(x_2 - x_1) \quad (5.10b)$$

Note the opposite sign in this case. The restoring force is opposite in sign to the displacement x_2 of the second atom. As atom **2** moves, so will atom **1** in the same direction to keep the bond length constant. The two force constants are the same, since moving either atom **1** or atom **2** has the same effect on the bond length and, therefore, the force on atom **1**.

We are now dealing with a system with two degrees of freedom. The state of the system is described by two functions $x_1(t)$ and $x_2(t)$, which satisfy a system of *coupled* differential equations (5.10a,b). Experiments indicate that this system is capable of exercising harmonic oscillations with certain characteristic frequencies.

As before, we can look for solutions of the type

$$x_1(t) = A_1 \cos(\alpha t)$$
$$x_2(t) = A_2 \cos(\alpha t) \tag{5.11}$$

where the A terms give the amplitude of vibration and α is the angular frequency ($= 2\pi\nu$, where ν is the frequency of vibration). The equation implies that the amplitude terms for atoms 1 and 2 may be different, but the angular frequency α must be the same for both atoms. Substituting equation (5.11) in (5.10), we get

$$-\alpha^2 m A_1 \cos(\alpha t) = k\cos(\alpha t)(A_2 - A_1)$$
$$-\alpha^2 m A_2 \cos(\alpha t) = k\cos(\alpha t)(A_1 - A_2) \tag{5.12}$$

or

$$(k - m\alpha^2)A_1 - kA_2 = 0$$
$$-kA_1 + (k - m\alpha^2)A_2 = 0 \tag{5.13}$$

which can be written in determinantal form as

$$\begin{pmatrix} k - m\alpha^2 & -k \\ -k & k - m\alpha^2 \end{pmatrix}\begin{pmatrix} A_1 \\ A_2 \end{pmatrix} = 0 \tag{5.14}$$

This is an eigenvalue problem for the matrix. The eigenvalues α^2 give the allowed vibration frequencies

$$(k - m\alpha^2)^2 - k^2 = 0$$
$$(k - m\alpha^2) = \pm k$$
$$\Rightarrow \alpha_1^2 = 0;\ \alpha_2^2 = \frac{2k}{m} \tag{5.15}$$

Thus,

$$\nu_1 = 0;\quad \nu_2 = \frac{1}{2\pi}\sqrt{\frac{2k}{m}}\ \text{Hz} \tag{5.16}$$

Note that $\nu_1 = 0$ implies that the motion is not a vibration; ν_2 is the only *characteristic frequency* of the system. The masses can vibrate (in simple harmonic motion) only with this frequency and no other. We reach the conclusion that the system has a natural frequency of vibration and that this frequency is related to the force constant and the masses of the system. To convert the frequency to wavenumber, the unit most commonly used in vibrational spectroscopy, we divide by c, the speed of light (expressed in cm s^{-1}) to get $\omega_{\text{osc}} = \frac{1}{2\pi c}\sqrt{\frac{2k}{m}}\ \text{cm}^{-1}$.

5.3.1 Lagrange's equations of motion

It is instructive to derive these equations of motion by a more sophisticated method, using the Lagrangian equations of motion. We shall show that the Lagrangian and Newton methods are equivalent, but the Lagrangian method is easier to interpret than Newton's method. If $q_i (i = 1, 2, \ldots, n)$ are the generalized coordinates describing the configuration of a system with n degrees of freedom, then its motion is given by n equations

$$\frac{\partial}{\partial t}\left(\frac{\partial L}{\partial \dot{q}_i}\right) - \left(\frac{\partial L}{\partial q_i}\right) = 0 \quad (i = 1, 2, 3, \ldots, n) \tag{5.17}$$

where L denotes the Lagrangian of the system. For conservative systems, the Lagrangian can be taken as the difference between the kinetic and potential energies

$$L = T - V \tag{5.18}$$

In our example ($n = 2$), the generalized coordinates can be chosen as x_1 and x_2. The kinetic energy is then

$$T = \frac{1}{2}m\dot{x}_1^2 + \frac{1}{2}m\dot{x}_2^2 \tag{5.19}$$

and, if the motion is harmonic, the potential energy is

$$V = \frac{1}{2}k(x_2 - x_1)^2 = \frac{k}{2}\left(x_2^2 - 2x_1x_2 + x_1^2\right) \tag{5.20}$$

since the net extension of the bond is $(x_2 - x_1)$. This expression is an example of the *quadratic form*. A general quadratic form in n variables q_i is an expression of the type $\sum_{i=1}^{n}\sum_{j=1}^{n} a_{ij}q_iq_j$, where a_{ij} are coefficients independent of q_i. If a quadratic form contains only terms with squares of q_i, namely, $a_{11}q_1^2$ and $a_{22}q_2^2$, but no cross terms like $a_{12}q_1q_2$, then it is called *diagonal*. Diagonal quadratic forms correspond to diagonal matrices.

Observe that the kinetic energy T is also represented by a quadratic form in the variables $\dot{x}_1$ and $\dot{x}_2$ and it happens to be diagonal. It can be written as

$$\begin{pmatrix}\dot{x}_1 & \dot{x}_2\end{pmatrix}\begin{pmatrix}\frac{m}{2} & 0 \\ 0 & \frac{m}{2}\end{pmatrix}\begin{pmatrix}\dot{x}_1 \\ \dot{x}_2\end{pmatrix} = \dot{x}^T T\dot{x} = \langle \dot{x}|T|\dot{x}\rangle \tag{5.21}$$

The potential energy V is not a diagonal form because of the presence of the term $-2kx_1x_2$. This term is responsible for the *coupling* between the two motions because if it were not present, the resulting system would reduce to two independent differential equations. Thus, we have the case of the so-called static coupling where the coupling terms appear in V but not in T. A real quadratic form in variables q_i, $\sum_{i=1}^{n}\sum_{j=1}^{n} a_{ij}q_iq_j$, is completely defined by the matrix of its coefficients. The matrix is essentially a *symmetric matrix* (all quantum mechanical matrices must be Hermitian, and hence symmetric) because the term containing the product $q_i q_j$ can always be split into two equal parts, one of them denoted by $a_{ij}q_iq_j$ and the other by $a_{ji}q_jq_i$.

$$\left(x_1\, x_2\right)\begin{pmatrix} \frac{k}{2} & -\frac{k}{2} \\ -\frac{k}{2} & \frac{k}{2} \end{pmatrix}\begin{pmatrix} x_1 \\ x_2 \end{pmatrix} = x^T V x = \langle x|V|x\rangle \tag{5.22}$$

Let us deduce the equations of motion by substituting in equation (5.17). We have

$$\begin{aligned}
\frac{\partial L}{\partial \dot{x}_1} &= \frac{\partial T}{\partial \dot{x}_1} = m\dot{x}_1 \\
\frac{\partial L}{\partial \dot{x}_2} &= m\dot{x}_2 \\
\frac{\partial L}{\partial x_1} &= -\frac{\partial V}{\partial x_1} = k(x_2 - x_1) \\
\frac{\partial L}{\partial x_2} &= -\frac{\partial V}{\partial x_2} = -k(x_2 - x_1)
\end{aligned} \tag{5.23}$$

Substituting in Lagrange's equations, we obtain the equations of motion in the same form as before [equation (5.10)].

$$\begin{aligned}
m\ddot{x}_1 + kx_1 - kx_2 &= 0 \\
m\ddot{x}_2 - kx_1 + kx_2 &= 0
\end{aligned} \tag{5.24}$$

It is not difficult to trace the effect of the quadratic form for V on the coupling of the two differential equations.

5.3.2 Normal coordinates and linear transformations

When we talk about the motion of a diatomic molecule, we imply the physical fact that, in general, one atom is bound to influence the motion of the other. This is reflected in the mathematical formulations (5.24). The differential equation satisfied by x_1 depends on x_2 and vice versa. On a more sophisticated level, the quadratic expression for the potential energy contains cross terms.

By adding and subtracting the first equation from the second one in (5.24), we obtain

$$\begin{aligned}
m(\ddot{x}_1 + \ddot{x}_2) &= 0 \\
m(\ddot{x}_2 - \ddot{x}_1) + 2k(x_2 - x_1) &= 0
\end{aligned} \tag{5.25}$$

If we now define two new coordinates, $Q_1 = (x_1 + x_2)$ and $Q_2 = (x_2 - x_1)$, it appears from equation (5.25) that variations in the quantity Q_1 are independent of variations in Q_2. This suggests that we can change to generalized coordinates and write the equations of motion in terms of Q_1 and Q_2:

$$\begin{aligned}
m\ddot{Q}_1 &= 0 \\
m\ddot{Q}_2 + 2kQ_2 &= 0
\end{aligned} \tag{5.26}$$

We see that the equations of motion are *decoupled*, and we may solve both equations independently. Putting $Q_1 = A_1 \cos\alpha_1 t$ in equation (5.26), we get $-mA_1\alpha_1^2 \cos\alpha_1^2 t = 0 \Rightarrow \alpha_1 = 0$. Similarly, insertion of $Q_2 = A_2 \cos\alpha_2 t$ in equation (5.26) results in $-mA_2\alpha_2^2 \cos\alpha_2^2 t + 2k\cos\alpha_2 t = 0 \Rightarrow \alpha_2^2 = \frac{2k}{m}$. Thus, the characteristic frequencies are

$$\alpha_1 = 0; \quad \alpha_2 = \sqrt{\frac{2k}{m}} \tag{5.27}$$

The generalized coordinates Q_1 and Q_2 are known as *normal* coordinates. In this case, the two normal coordinates are

$$\begin{aligned} Q_1 &= A_1 \cos\alpha_1 t = A_1 \\ Q_2 &= A_2 \cos\alpha_2 t = A_2 \cos\sqrt{\frac{2k}{m}}t \end{aligned} \tag{5.28}$$

with A_1 and A_2 as the respective amplitudes. While it is true that if one atom is set in motion, the other will start moving as well, the coupling property does not hold for the normal modes; we can *excite* one normal mode *without affecting the other.* We are led towards the statement that normal modes are not coupled to each other. If we had known how to select them from the very beginning and expressed the Lagrangian in terms of them, then the solution to the problem would have been greatly simplified. We now approach the finding of the normal coordinates of any arbitrary system in a more sophisticated way.

5.3.3 Simultaneous diagonalization

The eigenvalue problem for molecules can be approached from the point of view of *diagonalizing the Lagrangian.* Indeed, the equations of motion arise from the Lagrangian

$$L = \frac{1}{2}\left\{m\left(\dot{x}_1^2 + \dot{x}_2^2\right) - k\left(x_1^2 + x_2^2 - 2x_1x_2\right)\right\} \tag{5.29}$$

in which the kinetic energy is a diagonal matrix, $T = \frac{1}{2}\begin{pmatrix} m & 0 \\ 0 & m \end{pmatrix}$, but the potential energy matrix, V, is not diagonal $V = \frac{1}{2}\begin{pmatrix} k & -k \\ -k & k \end{pmatrix}$. In terms of these two matrices, the secular equations in (5.14) may be written as

$$\left(V - \alpha^2 T\right)(A) = 0 \tag{5.30}$$

where A is the column of eigenvectors. This equation is similar to that used for diagonalization of matrices (with $\lambda = \alpha^2$), but, instead of the identity matrix, we have the kinetic energy matrix, which is a multiple of the unit matrix (the factor being ½m).

We approach the problem from another angle. We note that T is already in diagonal form, and hence it is essentially the V matrix which needs to be diagonalized by a translation to normal coordinates. We diagonalize the V matrix as follows:

$$|V-\lambda I| = \frac{1}{2}\begin{pmatrix} k-\lambda & -k \\ -k & k-\lambda \end{pmatrix} = 0 \\ \Rightarrow \lambda = 0, 2k \tag{5.31}$$

The eigenvectors are determined by substituting the respective eigenvalue in the secular determinant:

For $\lambda = 0$,

$$\begin{vmatrix} k & -k \\ -k & k \end{vmatrix} \begin{vmatrix} A_{11} \\ A_{12} \end{vmatrix} = \begin{vmatrix} 0 \\ 0 \end{vmatrix}$$

$$\Rightarrow A_{11} = A_{12} = \frac{1}{\sqrt{2}}$$

where we have used the normalization condition. Similarly, for $\lambda = 2k$, we find that $A_{21} = -A_{22} = 1/\sqrt{2}$. Here, in A_{ik}, i represents the root and k the nucleus. Thus, A_{12} is the coefficient of the second atom for the first root.

The diagonalization is performed by a linear transformation involving the orthogonal matrix A, which is the matrix of eigenvectors written as column vectors.

$$\begin{pmatrix} x_1 \\ x_2 \end{pmatrix} = \begin{pmatrix} \frac{1}{\sqrt{2}} & \frac{1}{\sqrt{2}} \\ \frac{1}{\sqrt{2}} & -\frac{1}{\sqrt{2}} \end{pmatrix} \begin{pmatrix} Q_1 \\ Q_2 \end{pmatrix} \quad \text{or } x = AQ \tag{5.32}$$

In terms of the new coordinates, the potential energy matrix, $x^T V x = Q^T A^T V A Q$, where we have used the relation, $x^T = Q^T A^T$. Since A is the matrix of eigenvectors of V, it diagonalizes the potential energy matrix to $V' = A^T V A$, where V' is a diagonal matrix containing the eigenvalues of V, i.e.,

$$V' = \begin{vmatrix} 0 & 0 \\ 0 & 2k \end{vmatrix} \tag{5.33}$$

The kinetic energy is a quadratic form in $\dot{x}_1$ and $\dot{x}_2$. Since we have transformed the coordinates in which the potential energy matrix is expressed, we also have to express the kinetic energy in the same coordinate system. This mandatory transformation of coordinates

$$\begin{pmatrix} \dot{x}_1 \\ \dot{x}_2 \end{pmatrix} = \begin{pmatrix} \frac{1}{\sqrt{2}} & \frac{1}{\sqrt{2}} \\ \frac{1}{\sqrt{2}} & -\frac{1}{\sqrt{2}} \end{pmatrix} \begin{pmatrix} \dot{Q}_1 \\ \dot{Q}_2 \end{pmatrix} \quad \text{or } \dot{x} = A\dot{Q} \tag{5.34}$$

does not disrupt its diagonal form. The unit matrix I commutes with any matrix, and so does any multiple of I. We have $A^TcIA = cIA^TA$, where c is a constant (here ½m). Moreover, in our case, $A^TA = I$, since A is an orthogonal matrix, so the kinetic energy elements remain unchanged. We see that the entire Lagrangian is diagonalized by the linear transformation, $x = AQ$. Indeed, if we use T and V to denote the matrices corresponding to quadratic forms for kinetic or potential energy, then in matrix notation, $L = \dot{x}^T T\dot{x} - x^T Vx$, where $T = (½m)I$. After the transformation, the Lagrangian is expressed in the form $L = \dot{Q}^T A^T TA\dot{Q} - Q^T A^T VAQ$. Now, $T' = A^TTA$ and $V' = A^TVA$ are diagonal matrices, and hence $L = \dot{Q}^T T'\dot{Q} - Q^T V'Q$. The statement follows that the total energy

$$E = T' + V' = \frac{1}{2}\sum_{i=1}^{n}\left(\dot{Q}_i^2 + \lambda_i Q_i^2\right) \tag{5.35}$$

Equation (5.35) is an important one, for it shows that just as the principal axes diagonalized the moment of inertia tensor (Chapter 4), normal coordinates 'diagonalize' the kinetic energy and potential energy expressions to yield a set of independent differential equations. The principal axes are fairly straightforward to set up in the general case, but normal coordinates are highly variable and depend sensitively on both the symmetry of the molecule and the potential. It is also worth remembering that normal mode theory depends on the harmonic approximation.

5.3.4 Heteronuclear diatomic molecules

We now move on to a slightly more complicated case. Consider a heteronuclear diatomic molecule AB, i.e., one in which the two atoms have different masses.

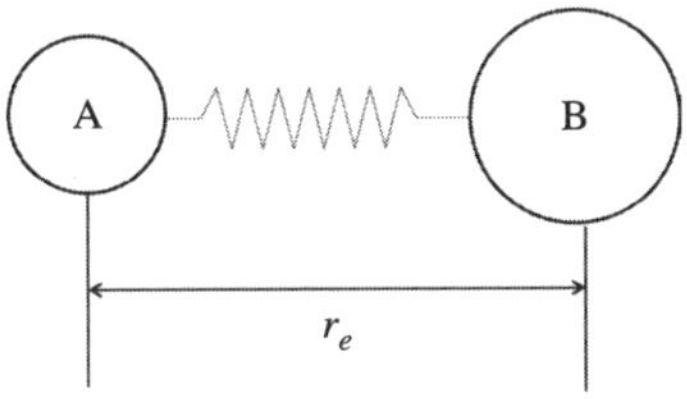

For this case, the matrix that diagonalizes V is the same as that for the homonuclear system, since the potential energy form is the same, but here T is different because the two masses are different.

$$T = \begin{pmatrix} \frac{1}{2}m_1 & 0 \\ 0 & \frac{1}{2}m_2 \end{pmatrix}$$

We find that, while the same transformation (5.32) diagonalizes the potential energy matrix, it disrupts the diagonal character of the kinetic energy matrix, since

$$\begin{pmatrix} \frac{1}{\sqrt{2}} & \frac{1}{\sqrt{2}} \\ \frac{1}{\sqrt{2}} & -\frac{1}{\sqrt{2}} \end{pmatrix} \begin{pmatrix} \frac{1}{2}m_1 & 0 \\ 0 & \frac{1}{2}m_2 \end{pmatrix} \begin{pmatrix} \frac{1}{\sqrt{2}} & \frac{1}{\sqrt{2}} \\ \frac{1}{\sqrt{2}} & -\frac{1}{\sqrt{2}} \end{pmatrix} = \frac{1}{4}\begin{pmatrix} m_1+m_2 & m_1-m_2 \\ m_1-m_2 & m_1+m_2 \end{pmatrix} \tag{5.36}$$

On diagonalizing the potential energy matrix, we have ended up with the kinetic energy matrix losing its diagonal character.

We therefore look for a transformation that simultaneously diagonalizes the kinetic and potential energy matrices. Such methods do exist, for example, by solving $|V - \lambda T| = 0$ [*c.f.* (5.30)] to obtain the simultaneous eigenvalues and eigenvectors of the two matrices V and T [two matrices can be simultaneously diagonalized, provided one of the matrices (here T) is positive definite, i.e., is a diagonal matrix with the diagonal elements all positive], but in normal coordinate analysis, it is usual to adopt another procedure that involves mass-weighted coordinates.

5.3.5 Mass-weighted coordinates

The problem in (5.36), i.e., the kinetic energy matrix losing its diagonal character on diagonalizing the potential energy matrix, arose because here the kinetic energy matrix is not a multiple of the unit matrix as in the case of the homonuclear system. If this can be converted to a unit matrix, the procedure $\det|V - \lambda T| = 0$ [equation (5.30)] would be replaced by the more familiar $\det|V' - \lambda I| = 0$, where I is, as usual, the identity matrix and V' is the modified potential energy in the new coordinate system.

We repeat the procedure of diagonalization of V, but this time converting the kinetic energy matrix to a unit matrix. This can be done conveniently by replacing the linear displacements x_i by mass-weighted displacements q_i given by $q_i = \sqrt{m_i}x_i$. In terms of these coordinates, the kinetic energy is

$$2T = m_1\dot{x}_1^2 + m_2\dot{x}_2^2 = m_1\left(\frac{\dot{q}_1}{\sqrt{m_1}}\right)^2 + m_2\left(\frac{\dot{q}_2}{\sqrt{m_2}}\right)^2 = \dot{q}_1^2 + \dot{q}_2^2 \tag{5.37}$$

or

$$2T = \begin{pmatrix} \dot{q}_1 & \dot{q}_2 \end{pmatrix} \begin{pmatrix} 1 & 0 \\ 0 & 1 \end{pmatrix} \begin{pmatrix} \dot{q}_1 \\ \dot{q}_2 \end{pmatrix} \tag{5.38}$$

The potential energy is given similarly by

$$2V = k(x_2 - x_1)^2 = k\left(\frac{q_2}{\sqrt{m_2}} - \frac{q_1}{\sqrt{m_1}}\right)^2 = k\left(\frac{q_2^2}{m_2} - \frac{2q_1q_2}{\sqrt{m_1m_2}} + \frac{q_1^2}{m_2}\right) \tag{5.39}$$

or

$$2V = \begin{pmatrix} \frac{k}{m_1} & -\frac{k}{\sqrt{m_1m_2}} \\ -\frac{k}{\sqrt{m_1m_2}} & \frac{k}{m_2} \end{pmatrix} \tag{5.40}$$

Therefore, $|V - \lambda I| = 0$ becomes

$$\begin{vmatrix} \frac{k}{m_1} - \lambda & -\frac{k}{\sqrt{m_1 m_2}} \\ -\frac{k}{\sqrt{m_1 m_2}} & \frac{k}{m_2} - \lambda \end{vmatrix} = 0$$

$$\Rightarrow \left(\frac{k}{m_1} - \lambda\right)\left(\frac{k}{m_2} - \lambda\right) - \frac{k^2}{m_1 m_2} = 0$$

$$\Rightarrow \left(\frac{k^2}{m_1 m_2} - \lambda k\left(\frac{1}{m_1} + \frac{1}{m_2}\right) + \lambda^2\right) - \frac{k^2}{m_1 m_2} = 0$$

$$\Rightarrow \left(\lambda^2 - \lambda \frac{k}{\mu_{red}}\right) = 0 \tag{5.41}$$

yielding the eigenvalues $\lambda = 0, k/\mu_{red}$. Therefore, conversion to mass-weighted coordinates reduces the problem to that of diagonalization of the potential energy matrix alone.

We note that, as in the previous case, one of the eigenvalues is zero. We shall see that this corresponds to translational motion. The other eigenvalue is $\lambda = \alpha^2 = k/\mu_{red}$, or the vibrational frequency $\nu = \frac{1}{2\pi}\sqrt{\frac{k}{\mu_{red}}}$. As in the treatment of rotational spectroscopy, separation out of translation replaces m by μ_{red}, reducing the system effectively to that with one degree of freedom, as in Section 5.2.

The A matrix (of eigenvectors) is obtained in the usual way. For $\lambda = 0$, the equation is

$$\frac{k}{m_1} A_{11} - \frac{k}{\sqrt{m_1 m_2}} A_{12} = 0$$

$$\Rightarrow A_{12} = \sqrt{\frac{m_2}{m_1}} A_{11}$$

$$\therefore A_{11} = \sqrt{\frac{m_1}{m_1 + m_2}}; \; A_{12} = \sqrt{\frac{m_2}{m_1 + m_2}} \tag{5.42}$$

Similarly, for $\lambda = k/\mu_{red}$, the eigenvectors are

$$A_{21} = \sqrt{\frac{m_2}{m_1 + m_2}}; \; A_{22} = -\sqrt{\frac{m_1}{m_1 + m_2}} \tag{5.43}$$

The linear transformation A is

$$\begin{pmatrix} \sqrt{\frac{m_1}{m_1 + m_2}} & \sqrt{\frac{m_2}{m_1 + m_2}} \\ \sqrt{\frac{m_2}{m_1 + m_2}} & -\sqrt{\frac{m_1}{m_1 + m_2}} \end{pmatrix} \tag{5.44}$$

and $q = AQ$ or $Q = A^T q$. Thus,

$$\begin{pmatrix} Q_1 \\ Q_2 \end{pmatrix} = \begin{pmatrix} \sqrt{\dfrac{m_1}{m_1+m_2}} & \sqrt{\dfrac{m_2}{m_1+m_2}} \\ \sqrt{\dfrac{m_2}{m_1+m_2}} & -\sqrt{\dfrac{m_1}{m_1+m_2}} \end{pmatrix} \begin{pmatrix} q_1 \\ q_2 \end{pmatrix} \tag{5.45}$$

or

$$\begin{aligned} Q_1 &= \frac{1}{\sqrt{m_1+m_2}}\left(\sqrt{m_1}q_1 + \sqrt{m_2}q_2\right) \\ Q_2 &= \frac{1}{\sqrt{m_1+m_2}}\left(\sqrt{m_2}q_1 - \sqrt{m_1}q_2\right) \end{aligned} \tag{5.46}$$

If we substitute the relations $q_i = \sqrt{m_i}x_i$ into equation (5.46), we obtain the (unnormalized) normal coordinates in terms of the displacement coordinates:

$$\begin{aligned} Q_1 &= \frac{1}{\sqrt{m_1+m_2}}\left(m_1x_1 + m_2x_2\right) = \sqrt{m_1+m_2}\,x_{\text{COM}} \\ Q_2 &= \frac{1}{\sqrt{m_1+m_2}}\left(\sqrt{m_1m_2}x_1 - \sqrt{m_2m_1}x_2\right) = \sqrt{\frac{m_1m_2}{m_1+m_2}}\left(x_1-x_2\right) = -\sqrt{\mu_{red}}\,(x_2-x_1) \end{aligned} \tag{5.47}$$

Therefore, Q_1 is proportional to the centre of mass coordinates, $x_{\text{COM}} = \dfrac{m_1x_1+m_2x_2}{m_1+m_2}$, while Q_2 is proportional to the net displacement $(x_2 - x_1)$. Indeed, if we had expressed our differential equations in terms of the displacement coordinates, x_1 and x_2,

$$\begin{aligned} m_1\ddot{x}_1 - k(x_2-x_1) &= 0 \\ m_2\ddot{x}_2 + k(x_2-x_1) &= 0 \end{aligned}$$

addition of the two equations would have given $m_1\ddot{x}_1 + m_2\ddot{x}_2 = 0$, or

$$M\ddot{x}_{\text{COM}} = 0 \tag{5.48}$$

which is the translational motion of a mass M. However, subtraction of the first equation from the second gives $m_2\ddot{x}_2 - m_1\ddot{x}_1 + 2k(x_2-x_1) = 0$ and the equations still remain coupled. Multiplication of the first equation by m_2 and the second by m_1, followed by subtraction of the first equation from the second, yields $m_1m_2\left(\ddot{x}_2-\ddot{x}_1\right) + k(m_1+m_2)(x_2-x_1) = 0$, and division by m_1+m_2 results in $\mu_{red}\left(\ddot{x}_2-\ddot{x}_1\right) + k(x_2-x_1) = 0$. In terms of the displacement coordinate, $x = x_2 - x_1$, this equation reduces to

$$\mu_{red}\,\ddot{x} + kx = 0 \tag{5.49}$$

In terms of these two coordinates, the two equations are exactly soluble. The first (5.48) pertains to the *translation* energy of the molecule, since changes in the centre of mass coordinates reflect translation of the molecule as a whole. The second equation (5.49) directly gives us the square of the angular frequency of vibration. It is clear that, in order to reduce a two-body problem (which cannot be solved quantum mechanically) to a one-body soluble problem, we switch over to a new set of coordinates – one in terms of the centre of mass of the system and the second in terms of relative, or molecule-fixed, coordinates, in which the mass is replaced by the reduced mass. Changes in the latter reflect changes in the potential energy of the system resulting from vibrational motion.

However, it is not always possible to find the normal coordinates by simple substitutions, particularly for polyatomic molecules with a large number of vibrational degrees of freedom, and we have to resort to the normal coordinate method.

Substitution of the eigenvalues into equation (5.35) gives for the total vibrational energy of a diatomic molecule

$$E = \frac{1}{2}\dot{Q}^2 + \frac{k}{2\mu_{red}}Q^2 \tag{5.50}$$

where we have dropped the subscript 2, i.e., $Q = Q_2$ since only Q_2 pertains to a genuine vibration (Q_1 is the coordinate for translational motion). Note that the definition of the normal coordinates adjusts the effective mass to unity, so μ_{red} drops out from the kinetic energy term. The first term refers to the kinetic energy and the second to the potential energy. In terms of the displacement coordinates, the substitution (5.47) in (5.50) gives

$$E = \frac{1}{2}\mu_{red}\,\dot{x}^2 + \frac{1}{2}kx^2 \tag{5.51}$$

Insertion of $x = A\cos\sqrt{\frac{k}{\mu_{red}}}t$ in equation (5.51) gives

$$E = \frac{1}{2}kA^2 \tag{5.52}$$

The total energy is thus a constant as in equation (5.8) for the oscillator with a single degree of freedom. Consider the energy changes during vibration. At maximum compression, there is no motion (i.e., no *kinetic* energy) since the mass changes direction and hence the velocity is zero at the point, and all energy is stored as *potential* energy in the spring. When it passes through r_e ($Q = 0$), all the energy is now kinetic energy (maximum speed) i.e., potential energy is zero. At full extension, again there is no motion instantaneously, so the kinetic energy is zero, i.e., all the energy is stored as potential energy. There is therefore continual interconversion of potential and kinetic energies during a period of vibration. The potential energy is a maximum at the extremes of extension and zero at equilibrium.

Remember that the normalized amplitudes A_{ij} are expressed in terms of the mass-weighted coordinates q_i; therefore, to visualize the actual displacements (x_i), these must be divided by the square root of the appropriate mass. In general, to display the motion of the atoms during vibration, we use equation (5.9)

$$x_j(t) = X_j(t) - X_{j0} = \frac{A_{ij}}{\sqrt{m_j}}\cos(\sqrt{\lambda_i}t) \tag{5.53}$$

For example, the coefficients $A_{ij}/\sqrt{m_j}$ from equation (5.44) for the two atoms are

$$X_1(t) = X_{10} + \frac{1}{\sqrt{m_1 + m_2}}$$
$$X_2(t) = X_{20} + \frac{1}{\sqrt{m_1 + m_2}} \tag{5.53a}$$

for $\lambda = 0$, and

$$X_1(t) = X_{10} + \sqrt{\frac{m_2}{m_1(m_1 + m_2)}} \cos\left(\sqrt{\frac{k}{\mu_{red}}} t\right)$$
$$X_2(t) = X_{20} - \sqrt{\frac{m_1}{m_2(m_1 + m_2)}} \cos\left(\sqrt{\frac{k}{\mu_{red}}} t\right) \tag{5.53b}$$

for $\lambda = \sqrt{k/\mu_{red}}$. The eigenvectors A_{ij} are the normal modes of vibration. For each normal mode, all the atoms move with the same frequency and phase, but the amplitudes are different.

Two points need mention. The first is that translational motion [equation (5.53a)] requires both atoms to move in the same direction by the same amount, but vibrational motion [equation (5.53b)] requires them to move in opposite directions, as observed from the opposite signs for the displacements of the two atoms. Furthermore, the amount of displacement is proportional to the relative masses of the two atoms. For example, in HCl, the lighter hydrogen atom moves approximately six times ($\sqrt{m_{Cl}/m_H}$) as much as chlorine in the normal mode.

In the above example, we had constrained the atoms to move along the bond axis only. If we remove this constraint, the system will have six degrees of freedom (three per atom). However, we shall still obtain only one non-zero value for the vibrational frequency. The remaining five eigenvalues are zero corresponding to translational motion in each of the three directions and rotational motion in the two directions perpendicular to the bond axis. Our treatment of rotational motion in the previous chapter had shown that a linear molecule has only two rotational degrees of freedom.

5.4 QUANTUM MECHANICAL TREATMENT

Let us now take up the quantum mechanical treatment of the diatomic molecule. The classical expression for the total energy (5.50) can be translated to the Hamiltonian operator for the motion of the atoms in a diatomic molecule in the form (in normal coordinates Q):

$$\hat{H} = -\frac{\hbar^2}{2}\frac{d^2}{dQ^2} + \frac{k}{2\mu_{red}} Q^2 \tag{5.54}$$

Thus,

$$\left(-\frac{\hbar^2}{2}\frac{d^2}{dQ^2} + \frac{k}{2\mu_{red}} Q^2\right)\psi(Q) = E\psi(Q) \tag{5.55}$$

together with a further set of five equations that correspond to zero roots of the secular determinant and lead to the wave equations for translation and rotation, which need not concern us here.

This equation is simply the Schrödinger amplitude equation of a harmonic oscillator expressed in the normal coordinates Q. You may check for yourself that the substitution (5.47) converts (5.50) to the more familiar form

$$E = \frac{1}{2}\mu_{red}\dot{x}^2 + \frac{1}{2}kx^2 \tag{5.56}$$

and the Hamiltonian to

$$\hat{H} = -\frac{\hbar^2}{2\mu_{red}}\frac{d^2}{dx^2} + \frac{1}{2}kx^2 \tag{5.57}$$

The solutions to (5.57) are

$$\psi_\upsilon(x) = N_\upsilon H_\upsilon(y)e^{-y^2/2} \tag{5.58}$$

with $y = \sqrt{\alpha}x$, $\alpha = \sqrt{\mu_{red}k}/\hbar$ and the normalization constant $N_\upsilon = \left(\frac{\sqrt{\alpha}}{2^\upsilon \upsilon!\sqrt{\pi}}\right)^{1/2}$. This α should not be confused with the angular frequency symbol used up to now in this chapter.

Here $\nu = \frac{\sqrt{\lambda}}{2\pi} = \frac{1}{2\pi}\sqrt{\frac{k}{\mu_{red}}}$ and $H_v(y)$ is the Hermite (pronounced 'air-MEET') polynomial of degree υ in x. The first five Hermite polynomials are

$$\begin{aligned} H_0 &= 1 \\ H_1(y) &= 2y \\ H_2(y) &= 4y^2 - 2 \\ H_3(y) &= 8y^3 - 12y \\ H_4(y) &= 16y^4 - 48y^2 + 12 \end{aligned} \tag{5.59}$$

Thus, for the wave functions of the first two states, we find

$$\begin{aligned} \psi_0(x) &= \left(\frac{\alpha}{\pi}\right)^{1/4} e^{-\alpha x^2/2} \\ \psi_1(x) &= \left(\frac{4\alpha^3}{\pi}\right)^{1/4} xe^{-\alpha x^2/2} \end{aligned} \tag{5.60}$$

The energy levels are the eigenvalues of equation (5.57).

$$E_\upsilon = \left(\upsilon + \frac{1}{2}\right)h\nu \tag{5.61}$$

Therefore, the vibrational energy levels are uniformly spaced. The vibrational terms are the energy levels expressed in wavenumber.

$$G(\upsilon) = \left(\upsilon + \frac{1}{2}\right)\omega; \quad \omega = \frac{1}{2\pi c}\sqrt{\frac{k}{\mu_{red}}}$$

Example 5.1 Calculate the *frequency* of oscillation and hence the *period* of oscillation of the CO molecule, given that the force constant is 1901 N m^{-1}. How does this compare with the period of rotation of CO in its most populated rotational state at 300 K?

Solution

$$\nu_{\text{vib}} = \frac{1}{2\pi}\sqrt{\frac{1901}{\left(\frac{12.0000 \times 15.9949}{(12.0000 + 15.9949)}\right) \times 1.6605386 \times 10^{-27}}} \text{ s}^{-1} = 6.503 \times 10^{13} \text{s}^{-1}$$

Period of oscillation $\tau = 1/\nu_{\text{vib}} = 1.538 \times 10^{-14}$ s

As given in Chapter 4, the rotational constant for CO is 1.92118 cm^{-1}. From this, the calculated value of J_{max} is 7 (Example 4.10). Putting $E_{\text{rot}} = \frac{1}{2} I\omega_{\text{rot}}^2 = \tilde{B}hcJ(J+1)$ and $\tilde{B} = \frac{h}{8\pi^2 Ic}$, we have $\omega_{\text{rot}} = 2\pi\nu_{\text{rot}} = 4\pi\tilde{B}c\sqrt{J(J+1)}$, which gives $\nu_{\text{rot}} = 2\tilde{B}c\sqrt{J(J+1)} = 8.6 \times 10^{11}$ s^{-1} and $\tau = 1.2 \times 10^{-12}$ s. Thus, hundreds of vibrations take place during a single rotation.

Notice the better resolution of the microwave spectrum.

Example 5.2 Calculate the difference in wavenumber (cm^{-1}) of:

(a) The lowest two vibrational levels of CO.

(b) The lowest two rotational levels of CO.

Solution

$F(J) = \tilde{B}J(J+1)$ cm^{-1}

$F(0)$ (the ground state rotational term) = 0 cm^{-1}

$F(1)$ (the first excited rotational term) = $2\tilde{B}$ cm^{-1}

Therefore, the gap between them $= 2\tilde{B} = 3.84235$ cm^{-1}

$G(\upsilon) = \omega(\upsilon + ½)$ cm^{-1}

$G(0)$ (the ground state vibrational term) = 0.5ω cm^{-1}

$G(1)$ (the first excited vibrational term) = 1.5ω cm^{-1}

Therefore, the gap between them $= \omega = \nu / c = 65.03 \times 10^{12} / 2.998 \times 10^{10} = 2169$ cm^{-1}, which lies in the infrared region.

The allowed vibrational energies of a diatomic molecule may now be shown schematically as in Figure 5.3.

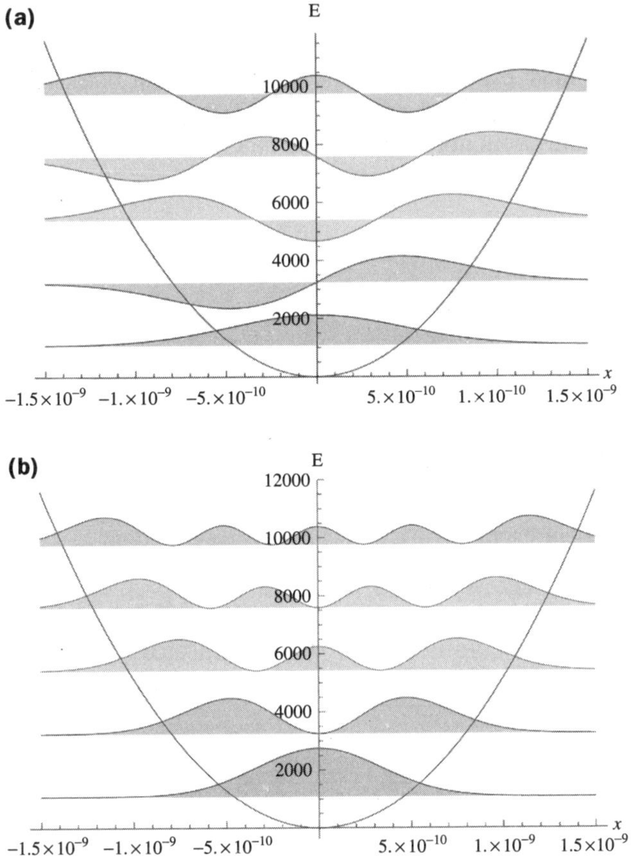

Figure 5.3. Plot of harmonic oscillator (a) wave functions and (b) probability densities and energy levels for CO ($\tilde{\nu} = 2168 \text{ cm}^{-1}$; bond extension in m, energies in cm^{-1})

The potential energy diagram (Figure 5.3) shows that

- The energy increases linearly with υ and the energy levels are equally spaced (by $h\nu$)
- All vibrational levels have a degeneracy of 1, i.e., there is only one quantum state per level, or, in other words, $g_\upsilon = 1$ for all υ.
- The lowest energy possible, for $\upsilon = 0$, is *not* zero!

$$G(0) = (0 + \tfrac{1}{2})\omega = 0.5\omega$$

This is called *zero-point energy (ZPE)*. Even with a vibrational quantum number $\upsilon = 0$, the energy in a bond is non-zero. A bond is *always* vibrating – even at the absolute zero of temperature. This is a manifestation of the Heisenberg Uncertainty Principle:

$$\Delta x \Delta p \geq \hbar / 2$$

We cannot simultaneously know both the momentum (velocity) and the position of a particle to arbitrary precision. The atoms are confined by the potential – if they were to lie completely at rest, there would be zero uncertainty in the position (defined by the bond length) and also zero uncertainty in the momentum, which would be certainly zero.

Figure 5.3 also depicts the wave functions and probability densities of a quantum mechanical harmonic oscillator. If we wish to compare the probability densities of the quantum and classical oscillator, we must examine the same energy level for both, since the ground state is different for the two – the classical oscillator can exist with zero energy (and hence zero motion), but the quantum oscillator cannot. Since the ground state of the quantum oscillator has an energy $\frac{1}{2}h\nu$, we compare it with a classical oscillator of the same energy:

$$E_\upsilon = (\upsilon + \frac{1}{2})h\nu = \frac{1}{2}kA^2 \tag{5.62}$$

Solving for A, we get

$$A = \left[\frac{h\nu}{k}(2\upsilon + 1)\right]^{\frac{1}{2}} \tag{5.63}$$

This is the amplitude of a classical oscillator with the same energy as the ground state quantum oscillator. The length of a level in Figure 5.3 is equal to *twice the amplitude* of the motion, since the classical oscillator oscillates between $-A$ and $+A$.

Example 5.3 Calculate the amplitude of vibration of CO in its ground vibrational level, assuming it to be a classical harmonic oscillator.

Solution

For CO, putting $\upsilon = 0$ and $k = 1901$ N m^{-1} (see Exercise 5.1), we find

$$A = \left[\frac{6.626\times10^{-34}\,\text{J s}\times6.503\times10^{13}\text{s}^{-1}}{1901\ \text{N m}^2}\right]^{\frac{1}{2}} = 4.761\ \text{pm}$$

The equilibrium bond length is 112.8 pm. Therefore, the amplitude of vibration is ~4% of the bond length for this strong double bond.

Example 5.4 Repeat for HCl for which the force constant is 480 N m^{-1}.

Solution

For HCl, the reduced mass is $\mu_{red} = 1.6267 \times 10^{-27}$ kg (Example 4.8). Therefore $\nu = \frac{1}{2\pi}\sqrt{\frac{k}{\mu_{red}}} = 8.6\times10^{13}\,\text{Hz}$ and A = 11 pm. The equilibrium bond length in HCl is 127 pm, and so the percent amplitude is ~9%.

Note the dependence of A on the force constant and hence the strength of the bond.

Examples 5.3 and 5.4 imply that, in the ground vibrational level, the displacement is no more than 10% of the bond length. Equation (5.63) shows that the amplitude increases with increasing vibrational energy. This is a consequence of the parabolic shape of the potential energy curve (Figure 5.3), which widens up with increasing energy.

We first examine the probability distribution of a classical oscillator. Classically, the probability of finding a particle at a given point is inversely proportional to its velocity at that point. This is so because the more rapidly a particle moves, the less likely are we to observe it. Therefore, the particle is more likely to be found at the classical turning points ($\pm A$) where its kinetic energy, and hence its velocity, is zero. At $x = 0$, the probability will be minimum. Furthermore, the classical probability is the same for all energy levels; only the amplitude of vibration increases with the energy.

We now compare this with the probability density of a quantum oscillator. First and foremost, the probability density now varies with the energy through the vibrational quantum number, since it is given by $\psi_\upsilon(x)^2$. We begin with the $\upsilon = 0$ energy level. Figure 5.3 shows that the probability is maximum at the equilibrium position ($x = 0$), in opposite behaviour to the classical oscillator. More importantly, the wave function extends beyond the classical limits. This is similar to the quantum mechanical 'tunnelling' through a finite barrier. At the classical turning points, the potential energy is equal to the total energy, since the kinetic energy is zero. Any extension beyond this limit increases the potential energy, so that we have the strange situation of negative kinetic energy!

Moving to the next energy level, $\upsilon = 1$, two peaks are observed and the regions of maximum probability move away from the centre, where there is a point of zero probability density, called a *node*. This localization of probabilities (restriction) causes this state to have higher energy than the ground state. A general principle of quantum mechanics is that, the more confined a particle, higher is its energy. Recall that the 1s orbital has no node, but the higher energy orbital, 2s, has a node in its wave function. The number of nodes in the *n*s wave function increases with the principal quantum number, *n*, and so does the energy.

When we look at the $\upsilon = 2$ level (Figure 5.3), we find that the number of nodes has risen to two. We observe from Figure 5.3 that the number of nodes is equal to the vibrational quantum number. An additional observation is that the central peak of the $\upsilon = 2$ probability distribution is smaller than the other two. We see the beginnings of approach to classical behaviour. This trend continues with increasing quantum numbers, so that for higher quantum numbers, the probability density maxima become sharper and move to the classical turning points, in agreement with classical behaviour. Moreover, the number of nodes increases, so that the distance between them decreases beyond the limits of measurement set by the Heisenberg Uncertainty Principle. Beyond a certain point, we cannot distinguish between nodes and antinodes, and the probability distribution gets smeared out. Also, for higher energy levels, the wave functions outside the classical limit die down faster, and the quantum oscillator also shows classical behaviour. This crossing over from quantum to classical behaviour was called the 'correspondence principle' by Bohr: "in the limit of large quantum numbers, quantum behaviour merges with classical behaviour."

The probability distributions in Figure 5.3 have an important bearing on electronic spectra (Chapter 8) and are the basis of the Franck-Condon principle.

5.4.1 Mechanism of infrared radiation absorption: Qualitative ideas

The mechanism of transfer of energy between radiation and a molecule is very similar to that discussed in connection with rotational transitions. We again ask if the electric field of radiation can interact with a vibrating molecule to cause it to jump from its initial $\upsilon = 0$ level to a higher υ level. Such interactions can be accounted for if it is supposed that the molecule has positively and negatively charged ends (i.e., a dipole) and that the amount of charge in the ends varies with the internuclear distance. If such is the case, the molecule might have a larger (or smaller) dipole moment in the stretched configuration than it has in the compressed configuration. It follows that the magnitude of the charges of the ends of the molecule will oscillate as the molecule vibrates. One might expect, as for rotational transitions, that if the oscillating charges are in phase with the electric field of the radiation and the radiation field has the same frequency as that of the molecular vibration, radiation can push the molecule to vibrate more. This is based on the classical description of a light wave as an oscillating field.

According to Figure 1.7, however, infrared radiation has wavelength between 10^2 and 10^5 nm, which is much larger than the size of small molecules, typically less than 1 nm. For example, consider the wavelength of infrared light, which is typically absorbed by a carbonyl group near 1700 cm^{-1}. The corresponding wavelength, $\lambda = 1/\tilde{\nu} = 5.882 \times 10^{-4}$ cm, which is much longer than the size of formaldehyde, a prototype carbonyl compound. Thus, during the absorption process, what the molecule sees is a constant electric field, $\vec{E}$. If the molecular dipole couples with the electric field, the bond may be lengthened or compressed slightly by the light wave, leading to larger or smaller vibrational amplitudes, depending on the magnitude of the interaction term $\vec{E}.\vec{\mu}$. Here, $\vec{\mu} = q\vec{r}$ is the dipole moment for some charge separation q along a distance $\vec{r}$.

We see that the same type of mechanism acts in the vibrating molecule as in the rotating molecule to allow the radiation to transfer energy to the molecule. For this to occur for vibrational energies, the molecule must have an *oscillating dipole moment*. The dipole moments of heteronuclear molecules such as CO and HCl can generally be expected to be a function of the internuclear distance, and such molecules can therefore absorb radiation and go from the $\upsilon = 0$ to a higher υ state. On the other hand, homonuclear diatomic molecules such as H_2 and N_2 have no dipole moment, no matter what the internuclear distance. It follows that, as for rotation, they cannot interact with radiation to change their vibrational energy. Such molecules are in fact found to have no absorption band in the infrared spectral region.

As for rotational energy changes, a further restriction limits the vibrational change that a molecule makes as a result of interaction with radiation to jump to a higher or lower allowed vibrational energy. The vibrational wave functions given in equation (5.58) are products of even functions (the exponential function and the normalization constant) and the Hermite polynomials. Equation (5.59) shows that Hermite polynomials $H_n(y)$ are odd when n is odd and even otherwise. It follows that the vibrational wave functions ψ_υ are odd for odd υ and even for even υ. Some of these functions are plotted in Figure 5.3(a). Since we know that products of odd and even functions are odd and that the dipole moment operator is odd, transitions between odd and odd υ values and even and even υ values are forbidden. However, symmetry alone cannot decide which odd–even transitions are allowed. We need to do a more quantitative treatment to determine the precise selection rules.

5.4.2 Quantitative treatment of selection rules

The selection rule can be determined by evaluating the transition moment integral $\left|\vec{M}_{\upsilon''\upsilon'}\right|$ for the υ'' to υ' level transition. The dipole moment, $\vec{\mu}$, necessary for the evaluation of $\left|\vec{M}_{\upsilon''\upsilon'}\right|$ for a heteronuclear diatomic molecule, might be expected to be some function of the internuclear distance. For small displacements from the equilibrium configuration, the dipole moment may be represented as a Taylor series expansion

$$\vec{\mu} = \vec{\mu}_{r=r_e} + \left(\frac{d\vec{\mu}}{dr}\right)_{r=r_e}(r - r_e) + \frac{1}{2!}\left(\frac{d^2\vec{\mu}}{dr^2}\right)_{r=r_e}(r - r_e)^2 + \cdots \tag{5.64}$$

where $\vec{\mu}_{r=r_e}$ is essentially the quantity usually referred to as the permanent dipole moment, $\vec{\mu}_0$.

In terms of the displacement coordinate $x = r - r_e$, the dipole moment operator is

$$\hat{\mu} = \hat{\mu}_{x=0} + \left(\frac{d\hat{\mu}}{dx}\right)_{x=0} x + \frac{1}{2!}\left(\frac{d^2\hat{\mu}}{dx^2}\right)_{x=0} x^2 + \cdots$$

Since x is small, every higher order term becomes smaller and smaller. The first term is a constant and its contribution $\left|\vec{M}_{\upsilon''\upsilon'}\right| = \int_{-\infty}^{+\infty} \psi^*_{\upsilon''}\hat{\mu}_{x=0}\psi_{\upsilon'}\, dx = \mu_0 \int_{-\infty}^{+\infty} \psi^*_{\upsilon''}\psi_{\upsilon'}\, dx$ is zero because of the orthogonality of $\psi_{\upsilon''}$ and $\psi_{\upsilon'}$. This shows that a constant dipole moment is not necessary for a vibrational transition to be induced by electromagnetic radiation. Thus, the first term does not tell us anything apart from the fact that, whether the molecule possesses a permanent dipole moment or not, this term does not contribute to the transition dipole moment.

Let us now examine the next term,

$$\left|\vec{M}_{\upsilon''\upsilon'}\right| = \left(\frac{d\vec{\mu}}{dx}\right)_{x=0} \int_{-\infty}^{\infty} \psi_{\upsilon''}\hat{x}\psi_{\upsilon'}dx \tag{5.65}$$

In this expression, we have omitted the asterisk (*) signifying the complex conjugate of $\psi_{\upsilon''}$, since harmonic oscillator wave functions are all real. We immediately have the condition that $(d\vec{\mu}/dx)_{x=0}$ should be non-zero for a vibration to be infrared active. In other words, the vibration should induce an oscillating dipole moment. This is the *gross* selection rule. The second condition arises from the symmetry properties of the integral, which does not allow transitions between two odd or two even functions. In fact, closer examination of the wave functions discloses that transitions between even and odd quantum states are also forbidden unless the two states combining in this way are adjacent states.

The normalized harmonic oscillator wave functions are given in equation (5.58). With these expressions for ψ_n and $\hat{\mu}$, the transition moment can be set up:

$$\begin{aligned}\left|\vec{M}_{\upsilon''\upsilon'}\right| &= \left(\frac{d\vec{\mu}}{dx}\right)_{x=0} \int_{-\infty}^{\infty} \psi_{\upsilon''}\hat{x}\psi_{\upsilon}dx \\ &= N_{\upsilon''}N_{\upsilon'}\left(\frac{d\vec{\mu}}{dx}\right)_{x=0} \int_{-\infty}^{+\infty} H_{\upsilon''}(x)\hat{x}H_{\upsilon'}(x)e^{-\alpha x^2}\, dx\end{aligned} \tag{5.66}$$

The recursion formula of Hermite polynomials is

$$yH_n(y) = nH_{n-1}(y) + \frac{1}{2}H_n(y)$$

$$\therefore xH_\upsilon(y) = \frac{\upsilon}{\sqrt{\alpha}}H_{\upsilon-1}(y) + \frac{1}{2\sqrt{\alpha}}H_{\upsilon+1}(y)$$

$$\left|\vec{M}_{\upsilon''\upsilon'}\right| = N_{\upsilon''}N_{\upsilon'}\left(\frac{d\vec{\mu}}{dx}\right)_{x=0} \int_{-\infty}^{+\infty} e^{-y^2}H_{\upsilon''}(y)\left[\frac{\upsilon'}{\sqrt{\alpha}}H_{\upsilon'-1}(y) + \frac{1}{2\sqrt{\alpha}}H_{\upsilon'+1}(y)\right]dx$$

$$= \frac{N_{\upsilon'}N_{\upsilon'}}{\sqrt{\alpha}}\left(\frac{d\vec{\mu}}{dx}\right)_{x=0}\left[\int_{-\infty}^{+\infty} \upsilon' e^{-y^2}H_{\upsilon''}(y)H_{\upsilon'-1}(y)\,dx + \frac{1}{2}\int_{-\infty}^{+\infty} e^{-y^2}H_{\upsilon''}(y)H_{\upsilon'+1}(y)\,dx\right]$$

The first integral on the right is zero unless $\upsilon'' = \upsilon' - 1$ and the second unless $\upsilon'' = \upsilon' + 1$ because of orthogonality of Hermite polynomials. The condition for $\left|\vec{M}_{\upsilon''\upsilon'}\right|$ to be different from zero is therefore

$$\upsilon'' = \upsilon' \pm 1 \text{ or } \Delta\upsilon = \pm 1$$

i.e., the only transitions that take place in absorption or induced emission are between adjacent states of the oscillator.

For absorption, $\upsilon' = \upsilon'' + 1$, and

$$\left|\vec{M}_{\upsilon''\upsilon'}\right| = \frac{N_{\upsilon''}N_{\upsilon''+1}}{2\sqrt{\alpha}}\left(\frac{d\mu}{dx}\right)_{q=0}\int_{-\infty}^{\infty} e^{-y^2}H_{\upsilon''+1}^2(y)\,dx = \frac{N_{\upsilon''}}{N_{\upsilon''+1}}\frac{1}{2\sqrt{\alpha}}\left(\frac{d\mu}{dx}\right)_{x=0}$$

$$= \sqrt{\frac{\upsilon''+1}{2\alpha}}\left(\frac{d\mu}{dx}\right)_{x=0} \tag{5.67}$$

Having established the conditions for observation of vibrational spectra, we now examine the intensities of vibrational lines. Equation (5.67) tells us that the intensities depend on the vibrational quantum number. The other factor that governs the intensities is the population. Let us calculate the population of the first excited vibrational state relative to the ground state for the carbon monoxide molecule.

Example 5.5 Calculate the room temperature population of the CO $\upsilon = 1$ level relative to the $\upsilon = 0$ level, given that the fundamental wavenumber is 2169 cm^{-1}.

Solution

We have

$$\frac{N_{\upsilon=1}}{N_{\upsilon=0}} = \frac{g_1}{g_0}e^{-\Delta E/k_BT}$$

Recall that all vibrations are singly degenerate.

Always put the *units* into the equation as it helps to avoid errors. In this case, we need a dimensionless number, so all the units in the exponential term must cancel to 1.

$$\frac{N_{\upsilon=1}}{N_{\upsilon=0}} = \frac{1}{1}e^{-hc\omega/k_BT} = \exp\left(\frac{-6.626\times10^{-34}\,\text{J s}\times2.998\times10^{10}\,\text{cm s}^{-1}\times2169\,\text{cm}^{-1}}{1.381\times10^{-23}\,\text{J K}^{-1}\times298\,\text{K}}\right) = 0.000028$$

i.e., most CO molecules are in the ground vibrational state at room temperature.

Thus, less than 28 out of a million carbon monoxide molecules are in the $\upsilon = 1$ state, and negligibly small numbers will be in still higher states. It follows that in experiments not above room temperature, the transitions that begin with the $\upsilon = 0$ state will be of major importance.

We must also note what factors influence the relative populations.

Example 5.6 Calculate the ratio of molecules in the excited vibration energy level to that in the lowest energy level at 25 °C and 1000 °C. Assume that the excited energy level is at 1000 cm^{-1} above the lowest one.

Solution

Given: $T_1 = 298$ K, $T_2 = 1273$ K

At T_1,

$$\frac{N_{\upsilon=1}}{N_{\upsilon=0}} = e^{-hc\omega/k_BT_1} = \exp\left(\frac{-6.626\times10^{-34}\,\text{J s}\times2.998\times10^{10}\,\text{cm s}^{-1}\times1000\,\text{cm}^{-1}}{1.381\times10^{-23}\,\text{J K}^{-1}\times298\text{K}}\right) = 0.008$$

and at T_2, $\dfrac{N_{\upsilon=1}}{N_{\upsilon=0}} = e^{-hc\omega/k_BT_2} = 0.323$

A 40-fold increase in the population of the higher level is caused by an increase in temperature to 1000 °C. Moreover, comparison with Example 5.5 reveals that a two-fold (approx.) decrease in the vibrational frequency increases the population 300 times!!! Thus, either a decrease in the vibrational frequency or increase in temperature raises the population of the excited vibrational level.

Since at room temperature most of the molecules are in the ground $\upsilon = 0$ state, the only transition of interest in absorption is $\upsilon = 0 \rightarrow \upsilon = 1$, i.e., $\upsilon'' = 0$, $\upsilon' = 1$. For this transition, one may obtain the transition moment integral directly by putting $\upsilon'' = 0$ in equation (5.67):

$$\left|\vec{M}_{01}\right| = \frac{1}{\sqrt{2\alpha}}\left(\frac{d\vec{\mu}}{dx}\right)_{x=0} \tag{5.68}$$

An alternate derivation is insertion of the wave functions (5.60) in the expression for the transition dipole moment (5.65) as follows:

$$\left|\vec{M}_{01}\right| = \int_{-\infty}^{+\infty}\left(\frac{\alpha}{\pi}\right)^{1/4} e^{-\alpha x^2/2}\left(\frac{d\mu}{dx}\right)_{x=0} x\left(\frac{4\alpha^3}{\pi}\right)^{1/4} xe^{-\alpha x^2/2}dx = \left(\frac{d\mu}{dx}\right)_{x=0}\left(\frac{2}{\pi}\right)^{1/2}\alpha\int_{-\infty}^{+\infty} x^2e^{-\alpha x^2}dx$$

$$= \left(\frac{d\mu}{dx}\right)_{x=0}\left(\frac{2}{\pi}\right)^{1/2}\alpha\times\frac{2}{4\alpha}\left(\frac{\pi}{\alpha}\right)^{1/2}$$

$$\left[\because \int_0^{\infty} x^2e^{-\alpha x^2}dx = \frac{1}{4\alpha}\sqrt{\frac{\pi}{\alpha}}\right]$$

$$= \frac{1}{\sqrt{2\alpha}}\left(\frac{d\mu}{dx}\right)_{x=0}$$

As stated in Chapter 2, the experimental integrated absorption coefficient ($\bar{A}$) can be used for determining the transition dipole moment. In practice, this is the method used for the determination of $\left(\frac{d\mu}{dx}\right)_{x=0}$ from infrared measurements. The integrated absorption intensity is related to molecular properties [equation (2.36)] according to

$$\begin{aligned}\bar{A} &= 9.784\times10^{60}\left|\vec{M}_{01}\right|^{2}\omega_{01}\\ &= 9.784\times10^{60}\frac{\hbar}{2\sqrt{\mu_{red}k}}\left(\frac{d\mu}{dx}\right)_{x=0}^{2}\frac{1}{2\pi c}\sqrt{\frac{k}{\mu_{red}}}\\ &= 2.739\times10^{15}\frac{1}{\mu_{red}}\left(\frac{d\mu}{dx}\right)_{x=0}^{2}\end{aligned} \tag{5.69}$$

Equation (5.69) is an important result that shows how the absorption intensity of a fundamental vibration transition is related to the properties μ_{red} and $(d\vec{\mu}/dx)_{x=0}$ of the molecule under study. This method is used to obtain the dipole moment derivative from the molecular absorption intensity. Rearrangement of equation (5.69) gives

$$\left(\frac{d\mu}{dx}\right)_{x=0} = \pm\sqrt{\frac{\mu_{red}\bar{A}}{2.739\times10^{15}}} = \pm1.911\times10^{-8}\sqrt{\mu_{red}\bar{A}} \tag{5.70}$$

The measured value of $\bar{A}$ for the fundamental transition of BrCl at 439 cm^{-1}, for example, is

$$\bar{A} = 105 \pm 14\ \text{cm}^{-2}\ \text{mol}^{-1}\ \text{dm}^{3}$$

From this value and the reduced mass of 4.091×10^{-26} kg, one calculates

$$\left(\frac{d\mu}{dx}\right)_{x=0} = \pm3.961\times10^{-20}\ \text{C}$$

Although distortion of a chemical bond from its equilibrium distance is undoubtedly generally accompanied by a complicated redistribution of the electrons of the bonded atoms, the order of magnitude that is to be expected for values of $(d\vec{\mu}/dr)_{r=0}$ can be obtained by assuming a constant charge of $+\delta e$ on one atom of the bond and $-\delta e$ on the other atom. With this model, the bond dipole moment is

$$\vec{\mu} = (\delta e)\vec{r}$$

and, since $x = r - r_e$,

$$\left(\frac{d\vec{\mu}}{dr}\right)_{r=r_e} = \left(\frac{d\vec{\mu}}{dx}\right)_{x=0} = \delta e$$

This result suggests that the value of $(d\vec{\mu}/dx)_{x=0}$ should be of the order of magnitude of the electronic charge 1.602×10^{-19} C. The result obtained for BrCl is consistent with this expectation and suggests that $\delta e = 3.961\times10^{-20}$ C, or ~0.25e.

Thus, the strength of a vibrational band in the infrared depends on the magnitude of the derivative of the dipole moment with respect to the displacement x and hence the internuclear distance, since $r = r_e + x$. Figure 5.4 shows how the dipole moment μ varies with r in a typical heteronuclear diatomic molecule. Obviously, $\mu \to 0$ when $r \to 0$, since the nuclei coalesce to atoms. For neutral diatomics, $\mu \to 0$ when $r \to \infty$ because the molecule dissociates into neutral atoms. Therefore, between $r = 0$ and ∞, there must be a maximum value of μ. In Figure 5.4, this maximum occurs at $r < r_e$, giving a negative slope $(d\mu/dr)$ at $\vec{r}_e$. Conversely, if the maximum were to occur at $r > r_e$, the slope would have been positive at r_e.

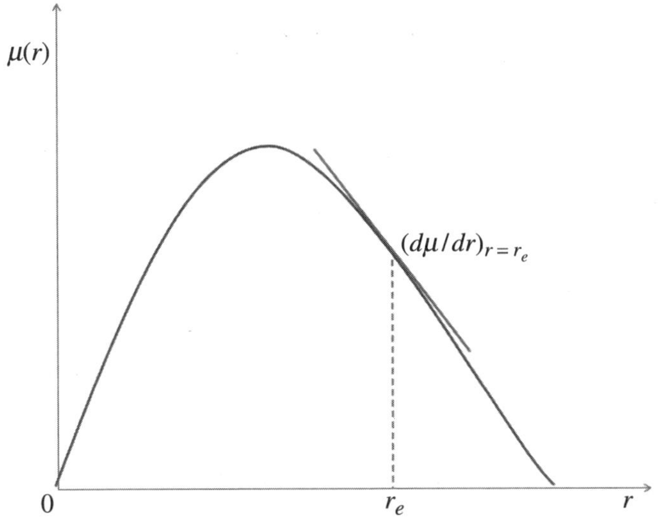

Figure 5.4. Variation of dipole moment with internuclear distance in a heteronuclear diatomic molecule

A molecule with a relatively small dipole moment may have a large dipole derivative, and, conversely, a molecule with a very large dipole moment may have a small dipole derivative if the dipole moment is near its maximum value at $r = r_e$, so that $(d\mu/dr) = 0$ at r_e. For example, CO, which has a permanent moment of only 0.11 D, possesses a large dipole derivative and thus one of the strongest known infrared absorptions. A homonuclear molecule, however, for which $\mu = 0$ at all internuclear separations, has a dipole derivative that is zero everywhere and thus no electric dipole vibrational absorption at all. Thus, homonuclear molecules such as H_2, O_2 and N_2 have neither vibrationally nor rotationally electric dipole allowed transitions.

It is at present a difficult matter to decide whether the curve with a positive value of $(d\mu/dr)_{r=0}$ or a negative value is correct for a given molecule. Nevertheless, the measurement of values of $\bar{A}$ and the deduction of values for $\pm(d\mu/dr)_{r=0}$ leads to the accumulation of data that are of considerable potential value in considerations of the electronic structure of molecules.

Example 5.7 The fundamental transition of HCl at 2886 cm^{-1} has an integrated absorption coefficient of 33.2 km mol^{-1}. Compute the dipole moment derivative.

Solution

From the measured value of the intensity, 33.2 km mol^{-1}, which is equivalent to 3320 dm^3 mol^{-1} cm^{-2}, and the reduced mass of HCl, 1.627×10^{-27} kg (Example 4.8), the calculated value of the dipole moment derivative

$$\left(\frac{d\mu}{dx}\right)_{x=0} = \pm4.441\times10^{-20}\ \text{C}$$

5.5 THE POTENTIAL ENERGY FUNCTION FOR A CHEMICAL BOND

The above discussion shows that if the molecule is modelled as a simple harmonic oscillator, only one absorption frequency should be observed, since all vibration energy levels are equally spaced and the selection rule is $\Delta\upsilon = \pm1$. However, examination of observed spectra shows that, besides the absorption frequency at ω, considerably weaker absorption lines are also observed at ~2ω, ~3ω, etc., with the intensity falling off with increasing frequency. Obviously, the $\Delta\upsilon = \pm1$ selection rule is not valid, and $\Delta\upsilon = \pm2, \pm3$, etc. transitions are also allowed, albeit with lower intensities.

To understand the breakdown of the harmonic oscillator selection rule, let us re-examine our model of the behaviour of bonds as ideal springs, i.e., harmonic oscillators. Real molecules do not behave as harmonic oscillators – at some point molecules suffer bond dissociation – i.e., the atoms are pulled apart. So what does the potential energy curve for a real molecule look like? The 'spring' in real molecules is provided by the electrons which form the bond holding the two atoms together. The form of the potential energy curve is dependent on the electronic structure of the bond.

Let us examine the simplest of all molecules, the hydrogen molecule. The ground state of the hydrogen atom has an electron in a 1s orbital, with spherical probability density. As we bring two hydrogen atoms together, they can form either a σ bonding orbital or a σ^* anti-bonding orbital. At large internuclear separations, we have just two atoms, each with the energy of the 1s orbital. At smaller internuclear separations, the energy must go down because a chemical bond forms. Energy has to be supplied if we want to break the bond. *The bonding state is a low energy (stable) state.* However the coulombic repulsion between the two positively charged nuclei prevents them from getting too close, so at small separations the energy must rise again above that of the isolated atoms. Therefore, the potential energy function goes through a minimum, corresponding to the equilibrium internuclear distance.

In other words, as the bond is stretched, it becomes weaker and the restoring force decreases, until at some value of the internuclear distance, the restoring force becomes zero, and the bond breaks. On the other hand, when the bond is compressed, the strong coulombic repulsion between the oppositely charged nuclei increases the resistance of the bond to compression and the restoring force increases. The potential energy curve is therefore more steep at internuclear distances less than r_e, and wider at $r > r_e$, as shown in Figure 5.5. Such an oscillator is known as an *anharmonic* oscillator.

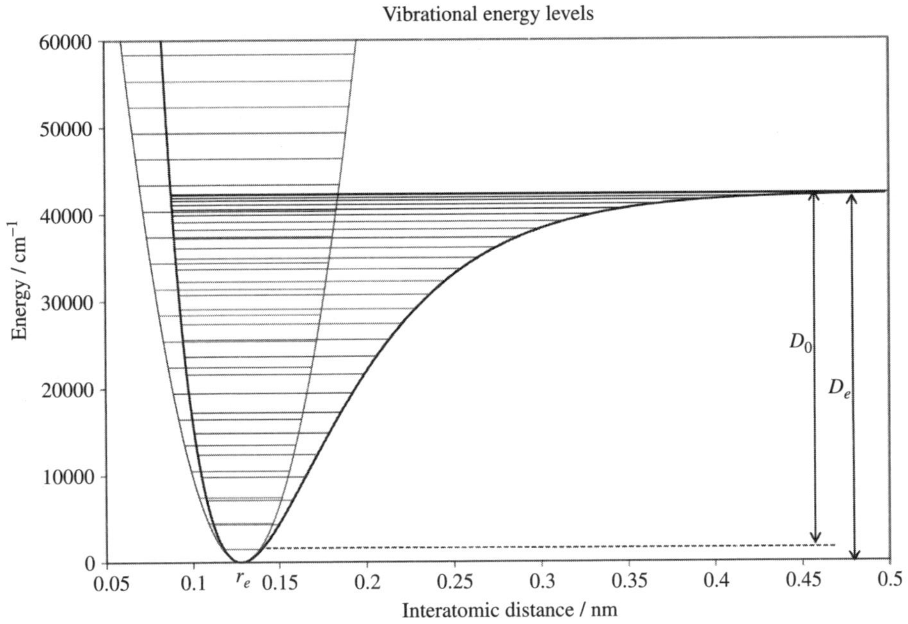

Figure 5.5 Energy levels of a harmonic oscillator (grey line), compared with those for an anharmonic potential (black line) for HCl

The experimental observation of absorption lines at ~2ω, ~3ω, etc., indicates that the transitions $\Delta\upsilon = \pm2, \pm3$, etc. are also allowed for an anharmonic oscillator. Indeed, for the harmonic oscillator, the $\Delta\upsilon = \pm2$ transition is forbidden on the basis of symmetry because this transition is between two even (symmetric) levels. As seen in Figure 5.5, the reflection symmetry is lost in the anharmonic oscillator and symmetry does not forbid the transition any longer. The $\Delta\upsilon = \pm2, \pm3$, etc. transitions are known as the *first* and *second overtone* (also called first and second harmonic), and so on, and the $\Delta\upsilon = \pm1$ transition is called the *fundamental*. The overtones are much weaker (<10% intensity) than the fundamental. Third and higher overtones are rarely observed. Further, the positions of the first and second overtones are not quite double or triple of the fundamental frequencies, but are at slightly lower frequencies, indicating that the spacing between the energy levels decreases for higher vibrational quantum numbers. As Figure 5.5 shows, the anharmonic potential is wider than the harmonic one, and, as the limits of confinement become larger, the energy levels come closer to each other, until at some high value of the quantum number, the energy levels become continuous, and the molecule dissociates. This is another example of the Bohr correspondence principle mentioned earlier.

The dissociation energy, D_e, is the energy required to break the bond and is the difference in energies between the separated atoms and the molecule at its equilibrium internuclear separation. However, since the lowest energy accessible to the molecule is the zero-point energy, the energy it has when it is in the $\upsilon = 0$ level, experimentally determined dissociation energies are with respect to this level, and are known as *thermochemical dissociation energies*, D_0, in contrast to the theoretical values, D_e, known as *spectroscopic dissociation energies*. Obviously, the two are related by the expression $D_0 = D_e - \frac{1}{2}\, h\nu$ (see Figure 5.5).

Example 5.8 Calculate the ratios of ZPEs of H-H to those of H-D and D-D, assuming that the force constants are the same and that the mass of deuterium is twice that of hydrogen.

Solution

The ZPE is given by $E_0 = \frac{1}{2}h\nu = \frac{h}{4\pi}\sqrt{\frac{k}{\mu_{red}}}$.

Therefore, $\frac{E_0(\text{HH})}{E_0(\text{HD})} = \sqrt{\frac{\mu_{red}(HD)}{\mu_{red}(HH)}} = \sqrt{\frac{m_H m_D \times 2m_H}{(m_H + m_D) \times m_H m_H}} = \sqrt{\frac{1\times 2\times 2}{3\times 1\times 1}} = \sqrt{\frac{4}{3}} = 1.15$

The zero-point energy of H_2 is therefore higher than that of HD.

In a similar fashion, the ratio $\frac{E_0(\text{HH})}{E_0(\text{DD})} = \sqrt{\frac{\mu_{red}(DD)}{\mu_{red}(HH)}} = \sqrt{\frac{2\times 2\times 2}{4\times 1\times 1}} = \sqrt{2} = 1.42$

The higher zero-point energy of X-H bonds makes them easier to cleave than X-D bonds. This explains why the electrolysis of water leads to a gradual enrichment of D_2O over H_2O. Although it would be quite expensive to use the electrolysis deliberately for the sole purpose of producing heavy water, the process occurs as a side reaction in all electrolytic processes that are carried out in water as a solvent. Using water that has been in the process for long enough is the starting point for the industrial manufacturing of heavy water.

A potential function that is often used to describe the potential energy curve of the anharmonic oscillator shown in Figure 5.5 was developed by Morse. This simple and useful potential, which provides considerable insights, is known as the *Morse potential*, and has the form

$$V_{\text{Morse}} = D_e[1 - \exp(-\beta(r - r_e))]^2 = D_e[1 - \exp(-\beta x)]^2 \tag{5.71}$$

At $x = 0$, $V_{\text{Morse}} = 0$ and at $x = \infty$, $V_{\text{Morse}} = D_e$. The constant β is related to the force constant by the expression $k = 2\beta^2 D_e$.

The next step would be to use the Morse potential in the expression for the potential energy of the Hamiltonian, and solve for the allowed wave functions and energies, as was done for the harmonic potential. Such a procedure results in the energy levels shown in Figure 5.5. However, the mathematics involved in this procedure is rather complicated and we shall follow a simplified procedure. Just as for the dipole moment, the potential can be expanded as a Taylor series in powers of $(r - r_e)$ about r_e:

$$V(r) = V_{r=r_e} + \left(\frac{dV}{dr}\right)_{r=r_e}(r - r_e) + \frac{1}{2!}\left(\frac{d^2V}{dr^2}\right)_{r=r_e}(r - r_e)^2 + \frac{1}{3!}\left(\frac{d^3V}{dr^3}\right)_{r=r_e}(r - r_e)^3 + \cdots$$

or, in terms of the displacement coordinate, x,

$$V(x) = V_{x=0} + \left(\frac{dV}{dx}\right)_{x=0} x + \frac{1}{2!}\left(\frac{d^2V}{dx^2}\right)_{x=0} x^2 + \frac{1}{3!}\left(\frac{d^3V}{dx^3}\right)_{x=0} x^3 \cdots \tag{5.72}$$

Since we can start our potential energy scale at any point, it is convenient to put $V = 0$ at $r = r_e$ or $x = 0$. Moreover, at $r = r_e$, the first derivative vanishes (since we are at a minimum with respect to the potential energy), and the first non-zero term is the quadratic term, $V(x) = \frac{1}{2}\left(\frac{d^2V}{dx^2}\right)_{x=0} x^2$. On comparison with the harmonic oscillator potential, we find that $k = \left(\frac{d^2V}{dx^2}\right)_{x=0} = \left(\frac{d^2V}{dr^2}\right)_{r=r_e}$. The force constant is therefore the second derivative of the potential energy at the minimum and is hence positive. It is also the *curvature*, defined as the amount by which the potential energy curve deviates from being *flat* at the minimum.

The shape of the parabola therefore depends on the force constant. The stronger the bond, the narrower is the parabola, and hence smaller the *amplitude* of the motion. Strong bonds, such as the C=O bond, correspond to stiff springs in the classical model and lead to steep potentials with large spacing between the energy levels [Figure 5.6(a)], while weak bonds give shallow potentials with smaller separations.

(a)

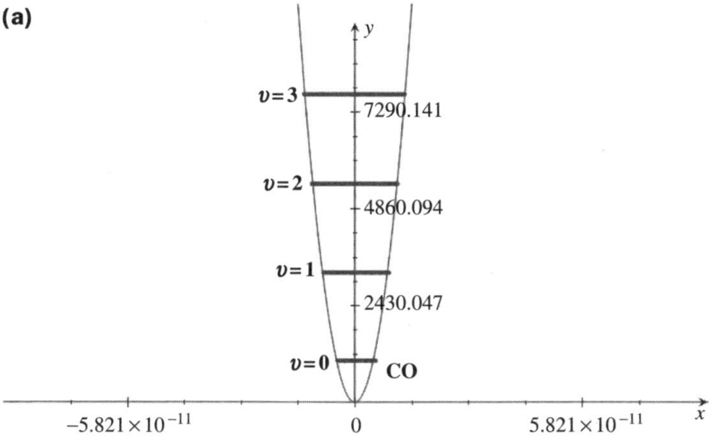

(b)

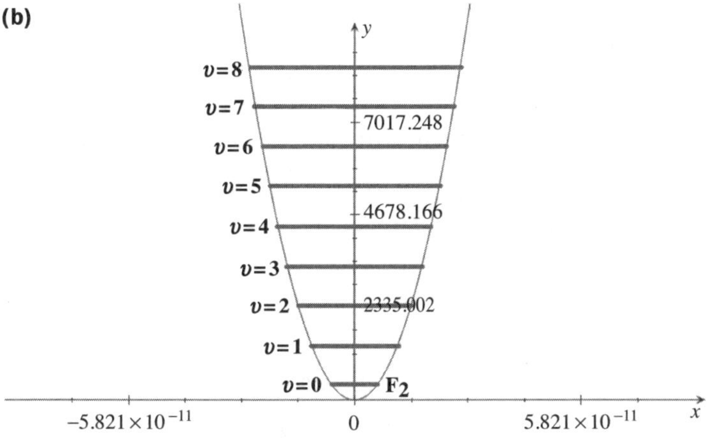

Figure 5.6. Effect of k on the energy levels of CO (k = 1870 N m^{-1}) and F_2 (k = 450 N m^{-1})

We can now improve the model of the harmonic oscillator by taking more terms, i.e., cubic and quartic terms in equation (5.72) to model our anharmonic oscillator. It must be realized, however, that the terms become smaller and smaller because of the presence of the x^n terms, which become smaller with increasing n because, as mentioned previously, x is small, not more than 10% of the bond length.

Including the cubic term in our potential leads to the following expression for the vibrational energy terms of an anharmonic oscillator

$$G(\upsilon) = \left(\upsilon + \frac{1}{2}\right)\omega_e - x_e\omega_e\left(\upsilon + \frac{1}{2}\right)^2 \tag{5.73}$$

In this expression, x_e is a small positive dimensionless constant of the order of ~0.01 and $x_e\omega_e$ is called the *anharmonicity*. It measures the extent of anharmonic behaviour of the oscillator. Obviously, when x_e is zero, we have a harmonic oscillator. The quantity ω_e is the hypothetical frequency the oscillator would have had if it were at the bottom of the potential energy curve corresponding to the hypothetical state $\upsilon = -½$. The extra term in equation (5.73) is negative to reflect the decrease in the energy separation of adjacent levels for higher vibrational quantum numbers. Equation (5.73) can be rearranged to

$$G(\upsilon) = \left(\upsilon + \frac{1}{2}\right)\omega_e\left[1 - x_e\left(\upsilon + \frac{1}{2}\right)\right] \tag{5.74}$$

from which it is clear that the vibrational level spacing decreases with increasing υ.

Referring to the Morse potential (5.71), the anharmonicity constant, x_e, is related to the depth of the potential well by the relation $x_e = \dfrac{h\nu_e}{4D_e} = \dfrac{\omega_e}{4\tilde{D}_e}$, where the tilde over D_e indicates that it is expressed in cm^{-1}. In a similar fashion, we can show that $\omega_e = \dfrac{\beta}{2\pi c}\sqrt{\dfrac{2D_e}{\mu_{red}}}$.

The harmonic oscillator selection rules were derived by taking the first two terms in the expansion of the dipole moment as a function of the internuclear distance. Now that we have added another term in the potential energy expansion, we can add another term to the dipole moment expansion also, leading to an additional term like

$$|\vec{M}_{02}| = N_0 N_2 \frac{1}{2}\left(\frac{d^2\mu}{dx^2}\right)_{x=0} \int_{-\infty}^{\infty} H_2(\sqrt{\alpha}x)\hat{x}^2 H_0(\sqrt{\alpha}x)dx$$

All the terms in the integrand are even and so it is non-zero, but $\left(\dfrac{d^2\mu}{dx^2}\right)_{x=0} << \left(\dfrac{d\mu}{dx}\right)_{x=0}$ and so this transition is much weaker than the $\upsilon = 0 \rightarrow \upsilon = 1$ transition, which remains the most intense. We have assumed that the effect of anharmonicity is small, and the wave functions can still be described by the harmonic oscillator wave functions.

Considering the fundamental $\upsilon = 0 \rightarrow 1$ transition and inserting the υ values in equation (5.71), we get

$$\omega_0 = G(1) - G(0) = \frac{3}{2}\omega_e - \frac{9}{4}x_e\omega_e - \left(\frac{1}{2}\omega_e - \frac{1}{4}x_e\omega_e\right) = \omega_e(1 - 2x_e) \tag{5.75a}$$

$$\omega_1 = G(2) - G(0) = \frac{5}{2}\omega_e - \frac{25}{4}x_e\omega_e - \left(\frac{1}{2}\omega_e - \frac{1}{4}x_e\omega_e\right) = 2\omega_e(1 - 3x_e) \tag{5.75b}$$

$$\omega_2 = G(3) - G(0) = \frac{7}{2}\omega_e - \frac{49}{4}x_e\omega_e - \left(\frac{1}{2}\omega_e - \frac{1}{4}x_e\omega_e\right) = 3\omega_e(1-4x_e) \quad (5.75c)$$

ω_0, ω_1 and ω_2 are the wavenumbers for the fundamental, first and second overtone, respectively. Equations (5.75a–c) involve two unknowns and any two of the equations can be used to determine both x_e and ω_e.

Example 5.9 The fundamental, first and second overtones of HCl are observed at 2886, 5668 and 8347 cm^{-1}, respectively. Determine ω_e, k_e and x_e.

Solution

Substitution in (5.75) gives

$$\omega_0 = \omega_e(1-2x_e) = 2886 \text{ cm}^{-1}$$
$$\omega_1 = 2\omega_e(1-3x_e) = 5668 \text{ cm}^{-1}$$

Multiplication of the first equation by 3 and subtraction of the second equation gives $\omega_e = 2990$ cm^{-1} and substitution of this value in any of the two equations yields $x_e = 0.0174$.
From the expression $k_e = 4\pi^2c^2\omega_e^2\mu_{red}$, we get $k_e = 516$ N m^{-1}.

Besides the fundamental and overtones, another band is sometimes observed, particularly for vibrations having small ω_e. This arises due to $\upsilon = 1 \rightarrow 2$ transitions and is called a *hot band.* Hot bands are characterized by an increase in intensity as the temperature is raised, since increasing the temperature raises the population of the $\upsilon = 1$ level. Substitution in equation (5.74) gives for the hot band

$$\omega_{\text{hot}} = G(2) - G(1) = \frac{5}{2}\omega_e - \frac{25}{4}x_e\omega_e - \left(\frac{3}{2}\omega_e - \frac{9}{4}x_e\omega_e\right) = \omega_e(1-4x_e)$$

Vibrational data can also be used to get an estimate of the dissociation energy of the molecule. From equation (5.74), the separation between adjacent levels is

$$\Delta\omega = G(\upsilon+1) - G(\upsilon) = \left(\upsilon+\frac{3}{2}\right)\omega_e - \left(\upsilon+\frac{3}{2}\right)^2 x_e\omega_e - \left[\left(\upsilon+\frac{1}{2}\right)\omega_e - \left(\upsilon+\frac{1}{2}\right)^2 x_e\omega_e\right]$$
$$= \omega_e(1-2x_e(\upsilon+1)) \quad (5.76)$$

Equation (5.76) shows that the vibrational level spacing decreases with increasing vibrational quantum number. Now, dissociation occurs when the energy levels become continuous, or $\Delta\omega$ becomes zero. Putting this condition in (5.76) leads to the value of the vibrational quantum number, $\upsilon_{\max}$, at which this occurs. Thus,

$$\Delta\omega = \omega_e(1-2x_e(\upsilon_{\max}+1)) = 0$$
$$\Rightarrow \upsilon_{\max} = \frac{1}{2x_e} - 1 \quad (5.77)$$

Returning to our HCl example, for which $x_e = 0.0174$, we find that $\upsilon_{max} = 27.74$. Since the quantum number can have only integer values, the maximum value of the vibrational quantum number, υ_{max} is 27 (since 27.74 is the upper bound, so 28 is ruled out). With this value for υ_{max}, we can now calculate the dissociation energy, D_e. This is the energy of the $\upsilon = 27$ level. Substituting in (5.74), we get

$$G(27) = \left(27 + \frac{1}{2}\right)2990 - 0.0174 \times 2990\left(27 + \frac{1}{2}\right)^2 = 42890 \text{ cm}^{-1}$$

This is equal to 513 kJ mol^{-1}. However, the experimentally determined value (thermochemical) is much lower (427.2 kJ mol^{-1}). Note that the calculated value is D_e and the experimental value is D_0 and $D_e = D_0 + ½ h\nu$. However, the zero-point energy is not too large to account for the difference between the experimental and theoretical results. The large difference is because, at higher values of the vibrational quantum number, the energy gap decreases much faster than equation (5.76) would predict. We need to take more terms in the expansion of the potential energy expression as a function of the internuclear distance to account for this effect, and cubic and quartic terms, along with other anharmonicity terms, y_e and z_e, need to be introduced in (5.73). However, infrared spectroscopy can at most provide three equations – the fundamental, first and second overtones, and these are not enough to solve for all the unknowns. We shall see later that electronic spectroscopy provides this information.

5.5.1 Experimental accuracy of the Morse potential

Table 5.1 gives the measured infrared wavenumbers of gaseous HCl and those calculated with the harmonic and anharmonic potentials. The harmonic potential is a good approximation for the fundamental vibration, but deviates substantially from the experimental data for the higher vibration levels. The anharmonic potential on the other hand gives excellent accuracy (> 0.1%) for all levels.

Table 5.1. Comparison of the harmonic and anharmonic oscillators

$\Delta\upsilon$	Vibration	ω_{obs} / cm^{-1}	Oscillator	
			Harmonic	Anharmonic
$1 \leftarrow 0$	Fundamental	2885.9	2885.9	2885.7
$2 \leftarrow 0$	First harmonic	5668.0	5771.8	5668.2
$3 \leftarrow 0$	Second harmonic	8347.0	8657.7	8347.5
$4 \leftarrow 0$	Third harmonic	10923.1	11543.6	10923.6
$5 \leftarrow 0$	Fourth harmonic	13396.5	14429.5	13396.5

The increasing inaccuracy of the harmonic approximation is readily understood if one recalls that the anharmonic correction term $-x_e\omega_e\left(\upsilon + \frac{1}{2}\right)^2$ is proportional to υ^2 and thus rapidly gains importance as υ increases.

5.5.2 Force constants

The quantity that is determined from vibrational spectroscopy is ω, which is related to the force constant of the bond, k, and the inverse of the reduced mass, μ_{red} through the expression

$\omega = \frac{1}{2\pi c}\sqrt{\frac{k}{\mu_{red}}}$. As the name suggests, force constants describe the strength of the bond. Some typical values for observed wavenumbers (ω_0 for the $\upsilon = 0$ level) and the force constants calculated from them, as well as the dissociation energies for some diatomics, are listed in Table 5.2. Some of these are homonuclear diatomics that are 'infrared inactive'. The vibrational data for these molecules has been obtained from the complementary technique, Raman spectroscopy, to be discussed in Chapter 7.

Table 5.2. Force constants and dissociation energies of some bonds

Molecule	ω_0 / cm^{-1}	k / N m^{-1}	D_0 / kJ mol^{-1}
H_2	4159.2	520	435
D_2	2990.3	530	436
HF	3958.4	880	565
HCl	2885.6	480	431
HBr	2559.3	380	364
HI	2230.0	290	297
CO	2143.3	1870	1075
NO	1876.0	1550	628
F_2	892.0	450	159
Cl_2	556.9	320	243
Br_2	321.0	240	192
I_2	231.4	170	151
O_2	1556.3	1140	498
N_2	2330.7	2260	950
Li_2	246.3	130	109
Na_2	157.8	170	75
NaCl	378.0	120	410
KCl	278.0	80	423

Example 5.10 Use the force constant given in Table 5.2 for N_2 with $v = \frac{1}{2\pi}\sqrt{\frac{k}{\mu_{red}}}$ to calculate the vibrational frequency.

Solution

Given $k = 2260$ N m^{-1}, $\mu_{red} = (14.003/2) \times 1.6605 \times 10^{-27} = 1.163 \times 10^{-26}$ kg.
Therefore, $\nu = 7.0 \times 10^{13}$ s^{-1}

Table 5.2 shows that stronger bonds (higher force constants) show higher wavenumber transitions. The observed dissociation energies also show the expected trend. Thus, double bonds (O_2) and triple bonds (CO, N_2) have higher force constants and hence wavenumbers and dissociation energies. In fact, the calculated force constants of the C-C, C=C and C≡C bonds from the observed data (Table 5.3) are ~ 500, ~ 1000 and ~ 1500 N m^{-1}, roughly in the ratio 1:2:3.

Table 5.3. Relationship between vibrational frequency and strength of carbon–carbon bonds for hydrocarbons

Bond type	ω/ cm^{-1}	C_2H_x ΔH_0 / kJ mol^{-1}
C-C	1100	368
C=C	1650	699
C≡C	2200	962

5.5.3 Isotopic substitution

The other quantity on which ω depends is the reduced mass. As we observed in Chapter 4, if $m_1 \gg m_2$, then $\mu_{red} \approx m_2$. Due to this reason, bonds with hydrogen atoms have very high wavenumber vibrations (Table 5.2). Table 5.4 illustrates the effect of atomic mass on the vibrational wavenumbers of C-X bonds. As the atom masses change, so do the transition wavenumbers and the heats of formation (measures of stabilities of bonds).

Table 5.4. Effect of atomic masses on the vibrational frequencies of C-X bonds

Bond type	ω/ cm^{-1}	CH_3-X ΔH_0 / kJ mol^{-1}
C-Cl	700	452
C-Br	600	293
C-I	500	243

The dependence of the transition wavenumber on the reduced mass indicates that isotopic substitution can cause change in the wavenumber of the band. As for rotational spectroscopy, we assume that isotopic substitution merely changes the nuclear mass without any effect on the electronic distribution, and hence the strength and force constant of the bond. Assuming harmonic behaviour then, the wavenumber is inversely related to the square root of the reduced mass.

Example 5.11 Assuming that the bond force constant k = 484 N m^{-1} is the same for ^{1}H-^{35}Cl, ^{1}H-^{37}Cl and ^{2}H-^{35}Cl, calculate the isotope shift in the fundamental vibrational frequencies of ^{1}H-^{37}Cl and ^{2}H-^{35}Cl with respect to ^{1}H-^{35}Cl.

Solution

From Example 4.8, we know that the reduced mass of ^{1}H^{35}Cl is 1.6267 × 10^{-27} kg and that of ^{1}H^{37}Cl is 1.6291 × 10^{-27} kg. Therefore, their respective vibrational wavenumbers are

$$(\omega_0)_{35} = \frac{1}{2\pi c}\sqrt{\frac{k}{\mu_{red}}} = 1.168\times 10^{-10} / \sqrt{\mu_{red}} = 2896 \text{ cm}^{-1}$$

and

$$(\omega_0)_{37} = 1.168\times 10^{-10} / \sqrt{\mu_{red}} = 2894 \text{ cm}^{-1}$$

Hence, the isotopic shift $(\omega_0)_{37} - (\omega_0)_{35}$ is –2 cm^{-1}.
Substitution of the reduced mass for $^2H^{35}Cl$ (3.1624×10^{-27} kg, Example 4.8) gives

$$(\omega_0)_D = 1.168 \times 10^{-10} / \sqrt{\mu_{red}} = 2080 \text{ cm}^{-1}$$

The isotopic shift $(\omega_0)_D - (\omega_0)_{35}$ is –816 cm^{-1}.

Hence, substitution of the ^{35}Cl isotope of chlorine by the heavier ^{37}Cl isotope in HCl lowers the wavenumber of the transition by a factor of 0.99924. Substitution of hydrogen by its isotope deuterium has a larger effect owing to the nearly double mass of the substituted atom. The reduced mass consequently almost doubles (since $1 / m_{Cl} << 1 / m_H$) and hydrogen atoms have a profound effect on the reduced mass (the factor is precisely 1.944). Thus, the vibrational wavenumbers are approximately 0.7 ($\sqrt{2}$) times those of the non-deuterated sample. The same sort of isotope shift is observed in polyatomic molecules as well. The hydrogen/deuterium isotope shift is the largest observed in vibrational spectroscopy and is easily seen when comparing the infrared spectra of H_2O and D_2O. This fact is used in organic chemistry to study the mechanisms of reactions using kinetic isotope effects.

Up till now we have considered solution phase spectra, where the molecules are not far enough for rotational fine structure to be observed. However, in the gas phase, an additional complication of simultaneous rotational transitions is observed. We shall find that the rotation–vibration transitions provide not only information about the rotational constants and hence the structure of molecules, but their contours also provide valuable information. We next study the effect of rotational transitions on vibration spectra.

Transitions between vibrational levels require photons with wavenumbers in the region of 1000 to 4000 cm^{-1} ($\lambda = 10 - 2.5$ mm), i.e., the infrared region of the spectrum. The photon has sufficient energy to also cause a simultaneous rotational transition since $E_{\text{rot}} << E_{\text{vib}}$. The combined rotation–vibrational term $S(\upsilon, J)$ of a vibrating rotor is therefore given by

$$S(\upsilon, J) = G(\upsilon) + F(J) = \omega_e(\upsilon + \frac{1}{2}) - x_e\omega_e\left(\upsilon + \frac{1}{2}\right)^2 + \tilde{B}J(J+1) - \tilde{D}J^2(J+1)^2 \tag{5.78}$$

according to the Born-Oppenheimer approximation where we treat the two energies as independent of each other. Since the orders of magnitude of $\tilde{D}$, $\tilde{B}$ and ω_e are 0.001 cm^{-1}, 1 cm^{-1} and 1000 cm^{-1}, respectively, the effect of centrifugal distortion will be observed only in spectrophotometers of high resolution, but the effect of anharmonicity certainly cannot be ignored. The selection rules are also the same as before, i.e., $\Delta\upsilon = \pm1, \pm2, \ldots$; $\Delta J = \pm1$. This implies that a vibrational transition of a diatomic molecule must be accompanied by a rotational transition as well, since the $\Delta J = 0$ transition is not permitted.

Each vibrational level has its own set of rotational levels. At room temperature, many of the J levels of the $\upsilon'' = 0$ level are occupied. From $J'' = 0$, the only transition possible is to the $J' = 1$ level, since the $J'' = 0 \rightarrow J' = 0$ transition is forbidden. That is to say, the *band centre*, corresponding to a pure vibrational transition is missing. Some of the allowed transitions are shown in Figure 5.7.

The individual lines are separated by $2\tilde{B}$, except at the band centre, where the spacing is $4\tilde{B}$. In spectroscopic notation, the $\Delta J = -1$ transitions are named the P branch, and the individual lines are labelled P(1), P(2), etc., where the number in parenthesis refers to the lower J level. The $\Delta J = +1$ transitions form the R branch. Note that there is no P(0) transition, but there is an R(0)

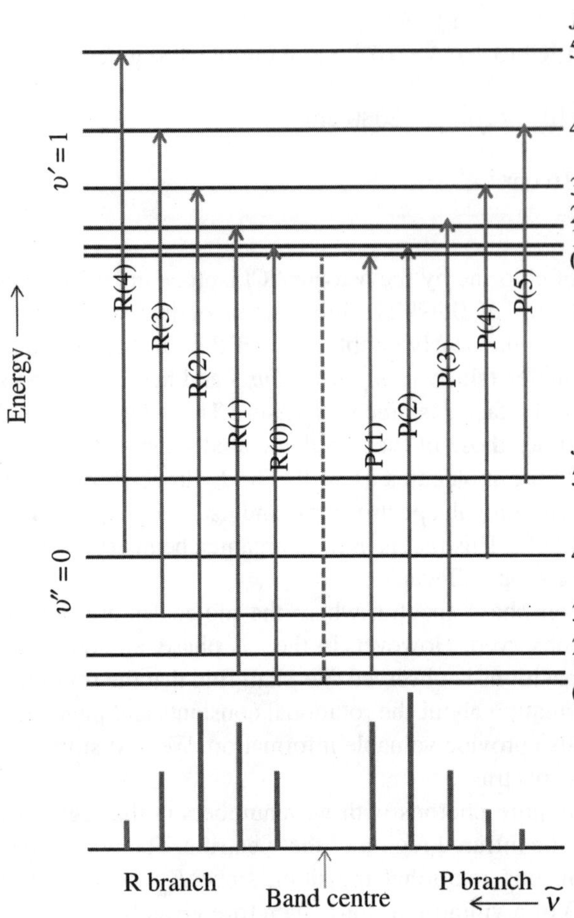

Figure 5.7. Vibration-rotation transitions of a diatomic molecule

transition corresponding to the $J'' = 0 \rightarrow J' = 1$ transition. In fact, the branches are labelled O, P, Q, R or S according to the following notations:

ΔJ	−2	−1	0	1	2
Label	O	P	Q	R	S

We shall encounter Q branches in our study of polyatomic molecules and the O and S branches in Raman spectroscopy.

We may now state that heteronuclear diatomic molecules exhibit PR contours in their infrared spectra, as the Q branch corresponding to pure vibrational transitions is missing in their spectra. This PR structure is characteristic of diatomic and other linear molecules. No non-linear molecule can display this simple structure. The observation of a PR contour in the vibrational spectrum of a molecule is evidence that the molecule is linear. There are exceptions, however. Linear molecules that have a component of the angular momentum perpendicular to the bond axis, such as the odd-electron molecules, NO and OH, display Q branches. In these cases, the additional spin angular momentum allows a Q transition.

Example 5.12 For carbon monoxide, the band centre corresponding to ω_0 (the dotted line in Figure 5.7) is located at 2143 cm^{-1} and the spacing $2\tilde{B}$ is equal to 3.83 cm^{-1}. From the data, determine the force constant and the length of the bond in CO.

Solution

For CO, using the values $m_C = 12.0000\ u$ and $m_C = 15.9949\ u$, we obtain

$$\mu_{CO} = \frac{m_C m_O}{m_C + m_O} 1.6605386 \times 10^{-27} = 1.1385 \times 10^{-26}\ \text{kg}$$

From the given data, $\tilde{B} = 3.83/2 = 1.915\ \text{cm}^{-1}$
The moment of inertia is given by

$$I_{CO} = \frac{h}{8\pi^2 \tilde{B}c} = \frac{2.799274 \times 10^{-46}}{\tilde{B}} = 1.462 \times 10^{-46}\ \text{kg m}^{-2}$$

From the relation $I = \mu_{CO} r_{CO}^2$, we obtain $r_{CO} = 1.133 \times 10^{-10}$ m = 113.3 pm.
This compares well with the best value of 112.83 pm obtained from microwave spectroscopy.
Further, using the relation $\omega_0 = \frac{1}{2\pi c}\sqrt{\frac{k}{\mu_{red}}}$, we obtain k = 1857 N m^{-1}, showing that it is a very strong bond, as expected for a triple bond.

Thus, rotation–vibration spectra can provide information about bond lengths and strengths, even though the band centre is not observed. The transition energy for the $\upsilon'' = 0 \rightarrow \upsilon' = 1$ transition is given by

$$\begin{aligned}\tilde{\nu} &= \omega_e(1-2x_e) + \tilde{B}J'(J'+1) - \tilde{B}J''(J''+1) \\ &= \omega_0 + \tilde{B}\left(J'^2 + J' - J''^2 - J''\right) \\ &= \omega_0 + \tilde{B}\left[\left(J' - J''\right) + (J' - J'')(J' + J'')\right] \\ &= \omega_0 + \tilde{B}(J' - J'')(1 + J' + J'')\end{aligned} \tag{5.79}$$

where we have written $\omega_0 = \omega_e(1-2x_e)$ as before.
For the P branch, $\Delta J = -1$, or $J' = J'' - 1$. Equation (5.79) becomes

$$\tilde{\nu}_P = \omega_0 + \tilde{B}(-1)(1 + J' + J' + 1) = \omega_0 - 2\tilde{B}(J'+1)$$

Similarly, substituting, $\Delta J = +1$ and $J'' = J' - 1$ for the R branch gives

$$\tilde{\nu}_R = \omega_0 + 2\tilde{B}(J''+1)$$

We may now combine the two expressions into a single equation,

$$\tilde{\nu}_{P,R} = \omega_0 + 2\tilde{B}m \tag{5.80}$$

where m can take on the integer values ±1, ±2, etc. and is positive and equal to $J'' + 1$ for the R branch, and is negative and equal to $J' + 1$ for the P branch. In general, none of the three quantities in the expression, *viz.*, ω_0, $\tilde{B}$ and m are known. Identification of three consecutive lines in the IR spectrum can serve to determine all these quantities using three simultaneous equations in the three unknowns.

In the above derivation, we have ignored the effect of centrifugal distortion. Had we included it, an additional term would have entered into equation (5.80), giving

$$\tilde{\nu}_{P,R} = \omega_0 + 2\tilde{B}m - 4\tilde{D}m^3 \tag{5.81}$$

However, as stated earlier, this term is generally ignored for lower J values, as it is much smaller than ω_0.

As discussed in the previous chapter, one J level corresponding to J_{max}, dependent on temperature, is the most highly populated. The P and R branches emanating from this level will then have the maximum intensity. The R branch thus has an appearance similar to pure rotational spectra, and the P branch is its mirror image. Sometimes, under low resolution, the individual lines are not resolved, and the PR structure is recognized only by its contour. Even then, the position of the maxima can be used to get a rough idea about the rotational constants. Since the P and R branch lines emanating from J_{max} are the most intense, and $m = J + 1$ in equation (5.80), the separation x between the P and R branch maxima is given by $4\tilde{B}\,m$, where $m = J_{max} + 1$ and $J_{max} = \sqrt{\dfrac{k_BT}{2\tilde{B}hc}} - \dfrac{1}{2}$. Thus,

$$x = 4\tilde{B}\left(\sqrt{\frac{k_BT}{2\tilde{B}hc}} + \frac{1}{2}\right)$$

The factor ½ is much smaller than the other term, and we may ignore it, so that we have finally

$$x \approx 4\tilde{B}\left(\sqrt{\frac{k_BT}{2\tilde{B}hc}}\right) = \sqrt{\frac{8\tilde{B}k_BT}{hc}} \tag{5.82}$$

Example 5.13 The infrared spectrum of gaseous carbon monoxide was run on a spectrophotometer of resolution insufficient to show the rotational fine structure. The separation between the maxima of the humps representing the P and R branches was 55 cm^{-1}. Estimate $\tilde{B}$, assuming that the temperature is 298 K.

Solution

Rearrangement of equation (5.82) gives

$$\tilde{B} = \frac{hcx^2}{8k_BT} = \frac{6.626\times10^{-34}\,\text{J s}\times2.998\times10^{10}\,\text{cm s}^{-1}\times(55\ \text{cm}^{-1})^2}{8\times1.381\times10^{-23}\,\text{J K}^{-1}\times298\ \text{K}} = 1.8\ \text{cm}^{-1}$$

Though this value is smaller than the more precise value 1.92 cm^{-1} obtained from microwave spectroscopy, it must be remembered that it was obtained with a spectrophotometer of very low resolution.

The first overtone of CO is observed at 4260 cm^{-1}. Using ω_0 = 2143.3 cm^{-1} for the fundamental (Table 5.2) and insertion in equation (5.73) gives $\omega_e = 2169.9$ cm^{-1} and $x_e = 0.0061$. If we compare this value for x_e with that obtained for the HCl molecule (0.0174), we find that the CO molecule can be approximated as a harmonic oscillator, as it has very little anharmonicity.

5.5.4 Vibrational dependence of rotational constants

If the vibration–rotation spectrum of HCl is examined under high resolution, we find that the spacing between the rotational lines is not quite constant. In fact, the spacing increases on the red side (P branch) and decreases on the blue side (R branch). The contours are not quite symmetrical as expected from a mirror relationship. This implies that some of our assumptions are incorrect, and the rotational constant of the upper vibrational level differs from that of the lower level. In Example 5.1, we saw that hundreds of vibrations take place during a single rotation. What the molecule sees during a rotation is an average bond length. If the vibrations are simple harmonic, the amount of extension and compression from the equilibrium position is equal, and the average bond length is the same as the equilibrium value. However, the rotational constant is proportional to $<1/r^2>$. Consider a typical bond having an equilibrium bond length of 0.10 nm executing simple harmonic motion in the ground vibrational level. We saw that the amplitude of vibration in this state is ~10%, which means that the bond length varies between 0.09 nm and 0.11 nm in a vibration, but the average bond length remains 0.10 nm. The average value of $<1/r^2>$ is, however, given by $0.5(1/0.11^2 + 1/0.09^2)$, which is equal to $1/0.103^2$. In the upper vibrational levels, the parabola shape of the harmonic oscillator implies that the amplitude is even larger, and the measured value of the bond length is larger. If the vibration is anharmonic, the amplitudes are even larger, and so the average bond length increases even faster. Thus, the average bond length, and hence the rotational constant, is not independent of the vibrational quantum number, and the Born-Oppenheimer approximation breaks down. The rotational constant is thus some function of the vibrational quantum number, and increases with increasing υ. We expand the dependence in a Taylor series, retaining only the linear term,

$$\tilde{B}_\upsilon = \tilde{B}_e - \alpha_e\left(\upsilon + \frac{1}{2}\right) \tag{5.83}$$

where $\tilde{B}_\upsilon$ and $\tilde{B}_e$ are, respectively, the rotational constants for the vibrational level corresponding to a quantum number υ and the hypothetical level $\upsilon = -\frac{1}{2}$ corresponding to the bottom of the potential energy curve, and α_e is a small positive constant (since $\tilde{B}_\upsilon < \tilde{B}_e$). Once the values of $\tilde{B}_e$ and α_e are known from the values of any two rotational constants, the $\tilde{B}_\upsilon$ value for any vibrational state can be determined.

This means that we must derive equation (5.79) for the fundamental transition again, this time taking different rotational constants $\tilde{B}_0$ and $\tilde{B}_1$ for the ground and excited vibrational levels.

$$\tilde{\nu} = \omega_0 + \tilde{B}_1 J'(J'+1) - \tilde{B}_0 J''(J''+1) \tag{5.84}$$

For the P-branch transitions, $\Delta J = -1$, or $J'' = J' + 1$. Substituting in equation (5.84), we get

$$\begin{aligned}
\tilde{\nu}_P &= \omega_0 + \tilde{B}_1 J'(J'+1) - \tilde{B}_0 (J'+1)(J'+2) \\
&= \omega_0 + (J'+1)(\tilde{B}_1 J' - \tilde{B}_0 J' - 2\tilde{B}_0) \\
&= \omega_0 + J'(J'+1)(\tilde{B}_1 - \tilde{B}_0) - 2(J'+1)\tilde{B}_0 \\
&= \omega_0 + J'(J'+1)(\tilde{B}_1 - \tilde{B}_0) + (\tilde{B}_1 - \tilde{B}_0)(J'+1) - 2(J'+1)\tilde{B}_0 - (\tilde{B}_1 - \tilde{B}_0)(J'+1) \\
&= \omega_0 + (\tilde{B}_1 - \tilde{B}_0)(J'+1)^2 - (\tilde{B}_1 + \tilde{B}_0)(J'+1)
\end{aligned} \tag{5.85a}$$

For the R branch, $\Delta J = +1$, or $J' = J'' + 1$. Substituting this in equation (5.84) gives

$$\Delta \tilde{\nu}_R = \omega_0 + (\tilde{B}_1 - \tilde{B}_0)(J''+1)^2 + (\tilde{B}_1 + \tilde{B}_0)(J''+1) \tag{5.85b}$$

The only difference is the sign of the last term and the appearance of J'' instead of J' in the expression for the R branch. As before, we combine the two expressions in a single expression

$$\tilde{\nu}_{P,R} = \omega_0 + (\tilde{B}_1 + \tilde{B}_0)m + (\tilde{B}_1 - \tilde{B}_0)m^2 \tag{5.86}$$

where $m = -(J' + 1)$ for the P branch and $J'' + 1$ for the R branch.

Let us look at the spacing of the two branches. For the P branch, m is negative, and hence the second term is also negative in sign. The sign of the third term depends on the quantity in parenthesis, since m^2 is always positive. We have seen that the internuclear distance in the excited vibrational state is larger than that in the ground state, and hence $\tilde{B}_1 < \tilde{B}_0$ and the last term is also negative. Since both terms are of the same sign, the spacing increases with increasing m. For the R branch, m is positive and hence the second term is positive and the last negative. Since the two terms are of opposite signs, the spacing decreases with increasing m. In certain cases, the negative term will dominate at higher values of m, so that the overall term changes sign from positive to negative, and the band may return at some value of m. This is termed the *band head*. We shall discuss this phenomenon in Chapter 8 in the context of electronic spectroscopy, where the differences in rotational constants are larger, as they pertain to different electronic states.

For CO, for example, the spectral lines were fit to a quadratic equation in m using a spreadsheet program, giving an equation

$$\tilde{\nu}_{\text{spectral}} = 2143.28 + 3.813m - 0.0176m^2$$

Comparison with equation (5.86) gives

$$\omega_0 = 2143.28 \text{ cm}^{-1};\ \tilde{B}_1 + \tilde{B}_0 = 3.813 \text{ cm}^{-1};\ \tilde{B}_1 - \tilde{B}_0 = -0.0176 \text{ cm}^{-1},$$

from which $\tilde{B}_1 = 1.898 \text{ cm}^{-1}$ and $\tilde{B}_0 = 1.915 \text{ cm}^{-1}$. Substitution in equation (5.83) gives

$$\tilde{B}_1 = \tilde{B}_e - \frac{3}{2}\alpha_e$$

$$\tilde{B}_0 = \tilde{B}_e - \frac{1}{2}\alpha_e$$

giving $\alpha_e = 0.018$ and $\tilde{B}_e = 1.924 \text{ cm}^{-1}$. The small value of α_e indicates very little rotation–vibration interaction in CO.

Using the three rotational constants, the bond lengths r_e, r_0 and r_1 are obtained as 0.1130 nm, 0.1133 nm and 0.1136 nm, respectively, clearly showing that the average bond distance increases with increasing vibrational quantum number (Figure 5.8).

The bunching of lines in the R branch can be seen in the fundamental band of HCl (Figure 5.9). Closer examination reveals the offset of $H^{35}Cl$ and $H^{37}Cl$ lines due to different reduced mass (and hence different ω_e and B constants) and the smaller intensities of the heavier isotopologue because of its smaller natural abundance (~25%).

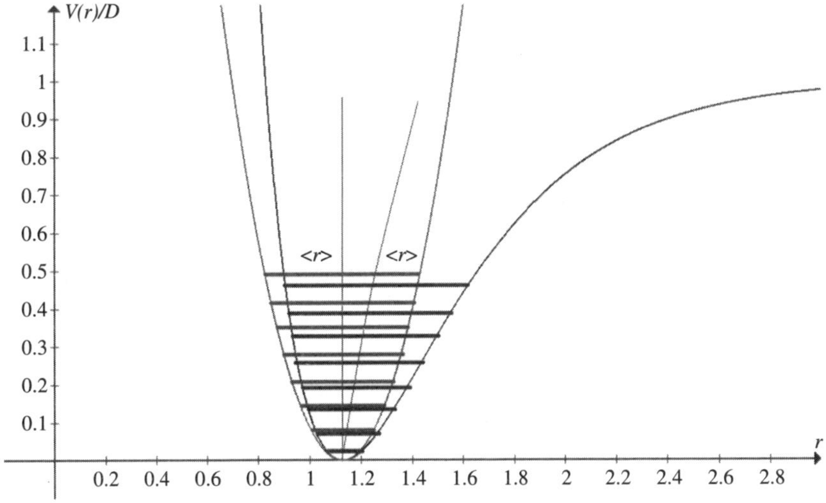

Figure 5.8. Mean internuclear distance <r> for the harmonic and anharmonic potentials, showing the first few vibrational energy levels

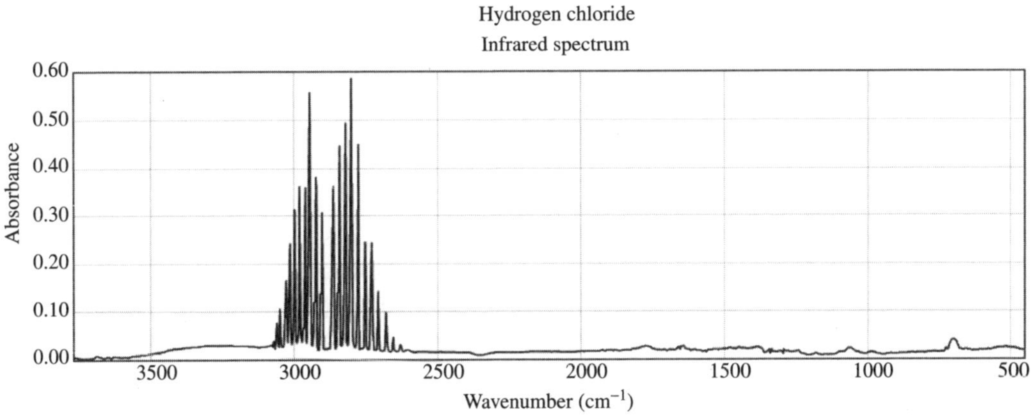

Figure 5.9. Fundamental band of HCl (*NIST Chemistry WebBook*)

5.5.5 Combination differences

Another method that can be used to determine the upper or lower vibrational level rotational constant is the method of *combination differences*. In this method, we identify transitions with either common upper or lower levels to determine the rotational constant in the other state (Figure 5.10). For example, consider the two lines P(J'') and R(J''), both of which originate from $\upsilon = 0$, J'' but end at $J' = J'' - 1$ and $J'' + 1$, respectively.
Therefore,

$$\tilde{\nu}_R(J'') - \tilde{\nu}_P(J'') = 2\tilde{B}_1(2J'' + 1)$$

from which the upper state rotational constant can be determined.

Similarly P(J''+1) and R(J''−1) originate from different ground vibrational levels, but terminate at the same upper rotational level. For this pair of lines

$$\tilde{\nu}_R(J''-1)-\tilde{\nu}_P(J''+1)=2\tilde{B}_0(2J''+1)$$

and the lower level rotational constant can be determined.

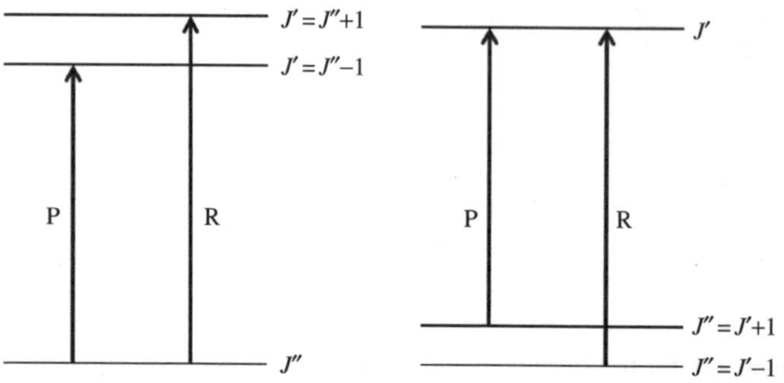

Figure 5.10. The method of combination differences (a) common J'' and (b) common J'

For example, for the third overtone of CO, $\tilde{\nu}_R(1)=8421.728$ cm^{-1} and $\tilde{\nu}_P(1)=8412.612$ cm^{-1}. This gives $6\tilde{B}_4=8421.728-8410.613$ cm^{-1} and $\tilde{B}_4=1.8525$ cm^{-1}, which gives $r_4=0.115$ nm. Compared to the equilibrium bond length of 0.113 nm, the bond is considerably elongated in the $\upsilon=4$ level.

Finally, we consider the effect of isotopic substitution on the rotational constants. In the previous chapter, we had assumed that isotopic substitution does not affect bond lengths. This is not strictly true. It is true only for the equilibrium bond length r_e, which is unaffected by isotopic substitution. We had stated earlier that substitution of hydrogen by deuterium has the greatest effect on the isotopic masses, as the reduced mass nearly doubles. This leads to a decrease in the vibrational frequency by a factor of ~0.7. This means that the $\upsilon=0$ level of DCl, for example, is lower than that of HCl, and hence its amplitude of vibration is also smaller. This implies that r_0(DCl) < r_0(HCl). Similarly, its bond distance in all other vibrational states is smaller than that of HCl. Only the equilibrium bond length is common for the two species. Hence, the assumption that bond lengths are independent of isotopic substitution is not strictly true.

5.6 SUMMARY

In this chapter, we have studied vibrational spectra of diatomic molecules.

- As a first approximation, bonds may be considered idealized springs obeying Hooke's law, i.e., as harmonic oscillators.
- Quantum mechanical solution of the harmonic oscillator model leads to equally spaced vibrational levels. Even at the absolute zero of temperature, the molecules are vibrating and possess a zero-point energy.

- In the harmonic oscillator model, only transitions between adjacent levels are permitted. The vibration should also lead to changes in the dipole moment, and so homonuclear diatomic molecules are excluded from pure vibration spectroscopy.
- Investigation of vibrational spectroscopy leads to elucidation of force constants of bonds, which are related to bond strengths.
- Real molecules are not harmonic oscillators, as they eventually break.
- The selection rules break down, allowing larger vibrational jumps, but with lower intensities.
- Investigations of these bands give approximate values of the dissociation energies of bonds.
- Isotopic substitution affects the spectra. In particular, substitution of hydrogen by deuterium has the greatest effect.
- Rotational fine structure provides information about the rotational constants and bond lengths.

5.7 EXERCISES

1. Show that the bond-stretching vibration frequency ν_{osc} of a diatomic molecule AB is given by

$$\nu_{osc} = \frac{1}{2\pi}\sqrt{\frac{k}{\mu_{red}}}\ \mathrm{s}^{-1}$$

where k is the force constant and μ_{red} is the reduced mass. Given that the stretching frequency of a C-C bond is approximately 1000 cm^{-1},

(a) Calculate the stretching frequencies of C=C and C≡C bonds on the assumption that double and triple bonds have force constants which are, respectively, two or three times those of a single bond;

(b) Calculate the frequencies of C-H and S-S bonds on the assumption that all single bond force constants are equal. Compare the frequencies calculated on the above assumptions with the following observed values: wavenumbers C=C, 2200; C-H, 3000; S-S, 500 cm^{-1}.

(c) Which of the following vibrational transitions will be observed for a diatomic molecule (treated as a harmonic oscillator): $\upsilon = 1$ to $\upsilon = 3$; $\upsilon = 2$ to $\upsilon = 3$; $\upsilon = 5$ to $\upsilon = 4$?

2. Given the recursion formula for Hermite polynomials $yH_n(y) = nH_{n-1}(y) + \frac{1}{2}H_{n+1}(y)$, where $y = \sqrt{\alpha}x$, $\alpha = \frac{\sqrt{\mu_{red}k}}{\hbar}$, and x is the displacement coordinate, show that, in the harmonic oscillator model, transitions are allowed only between adjacent levels, i.e., $\Delta\upsilon = \pm 1$.

3. (a) The vibrational wave functions of a diatomic molecule are given by

$$\Phi_n = \left(\frac{\sqrt{\alpha}}{2^n n!\sqrt{\pi}}\right)^{1/2} H_n e^{-y^2/2}$$

where $y = \sqrt{\alpha}x$, α is a constant and x is the displacement coordinate. Evaluate the transition moment integrals $|\vec{M}_{01}|$ and $|\vec{M}_{12}|$ for the $0 \rightarrow 1$ and $1 \rightarrow 2$ transitions.

(b) For BrCl, $(d\mu / dx)_{x=0} = 3.96 \times 10^{-20}$ C, where μ is the dipole moment. Show that this magnitude is of the correct order.

4. (a) Explain why the overtone of a particular vibrational band of an anharmonic oscillator will not be found at exactly 2, 3, 4, … times the frequency of the fundamental.
   (b) Determine ω_e and x_e from the data given below for HCl.
   (c) Use Boltzmann's distribution law to calculate the number of molecules in the $\upsilon = 1$ state compared with the number in the $\upsilon = 0$ state at 25 °C.
   (d) Calculate the spectroscopic and chemical dissociation energies, $\tilde{D}_e$ and $\tilde{D}_0$.

Absorptions due to vibrational transitions

υ	$H^{35}Cl$ (cm^{-1})
0–1	2886.2
0–2	5668.5
0–3	8347.6
0–4	10923.6

5. The infrared spectrum of gaseous carbon monoxide was run on a spectrophotometer of resolution insufficient to show the rotational fine structure. The separation between the maxima of the humps representing the P and R branches was 56 cm^{-1}. Estimate $\tilde{B}$ ignoring the difference between $\tilde{B}_0$ and $\tilde{B}_1$. What resolution would be required to show the fine structure of the P and R branches?
6. The rotational constants of HCl in different vibrational levels of its electronic ground state are given below. Extrapolate the values to $\upsilon = -½$, and hence calculate the moment of inertia and equilibrium internuclear distance (r_e) of this molecule. Why is r_e less than all r values?

υ	$\tilde{B}$ / cm^{-1}
0	10.44
1	10.14
2	9.83
3	9.53
4	9.23
5	8.93

7. (a) Three absorption lines of $^{12}C^{16}O$ are at 115.271201, 230.537974 and 345.795900 GHz. Assuming these to be the lowest frequencies and that the centrifugal distortion D is small, suggest the initial and final levels for each transition.
   (b) Calculate B and D as accurately as possible.
   (c) Infrared measurements on the transition to the $\upsilon = 1$ level give $B_1 = 57.11073$ GHz. Using this value and B_0 from (b), calculate α_e and thence B_e and I_e.

8. What is the gross selection rule for infrared (vibrational) spectroscopy? What is the selection rule for υ? Which of the following molecules may show an infrared vibrational absorption spectrum?
   (a) Cl_2 (b) HBr (c) NH_3 (d) CO_2
9. The wavenumber of the fundamental vibrational transition of $^{79}Br^{81}Br$ is 323.2 cm^{-1}. Calculate the force constant of the bond. Would the force constant of $^{81}Br^{81}Br$ be different? Estimate the wavenumber of the fundamental vibrational transition of $^{81}Br^{81}Br$.
10. Calculate the relative numbers of Cl_2 molecules (ω = 559.7 cm^{-1}) in the ground and first excited vibrational states at (a) 298 K (b) 500 K.
11. Explain why the C=O stretching vibration of an aldehyde gives rise to a strong absorption in the infrared, but the absorption due to the C=C vibration in an alkene is normally weak.
12. The vibrational wavenumber of $^1H^{35}Cl$ is 2991 cm^{-1}. Without calculating the bond force constant, estimate the wavenumber for (a) $D^{35}Cl$ (b) $H^{37}Cl$ (c) $D^{37}Cl$.
13. The Morse oscillator is often used as a model for a real, anharmonic molecule because it is simple and its Schrödinger equation can be solved exactly. The energy levels that emerge from this solution are

$$G(\upsilon) = \omega_e\left(\upsilon + \frac{1}{2}\right) - x_e\omega_e\left(\upsilon + \frac{1}{2}\right)^2, \qquad \omega_e = \frac{1}{2\pi c}\sqrt{\frac{k}{\mu_{red}}}$$

   where ω_e is the *equilibrium vibrational wavenumber* (the hypothetical frequency for infinitely small oscillations about the equilibrium geometry ($r = r_e$, the bottom of the potential well), and x_e is the *anharmonicity constant*, a small, positive number.
   (a) Derive expressions for the wavenumbers of the following transitions:
      (i) the fundamental ($\upsilon = 1 \leftarrow \upsilon = 0$)
      (ii) the first two overtone bands ($\upsilon = 2 \leftarrow \upsilon = 0$) and ($\upsilon = 3 \leftarrow \upsilon = 0$)
      (iii) the first two hot bands ($\upsilon = 2 \leftarrow \upsilon = 1$) and ($\upsilon = 3 \leftarrow \upsilon = 2$)
   (b) Show that the energy levels get closer together linearly, according to the relation:

$$\Delta E_\upsilon = G(\upsilon + 1) - G(\upsilon) = \omega_e - 2x_e\omega_e(\upsilon + 1)$$

   and hence find an expression for υ_{max}, the largest value of υ before dissociation (i.e., where the energy levels have converged).
   (c) The values of ω_e and x_e in the $^3\Pi_u$ and $^1\Pi_g$ states of the C_2 molecule are (you need not worry about this notation – it will become clear in Chapter 8 and need not bother us here):

State	ω_e / cm^{-1}	x_e
$^3\Pi_u$	1641.4	7.11×10^{-3}
$^1\Pi_g$	1788.2	9.19×10^{-3}

   Find the number of vibrational energy levels below the dissociation limit and hence the dissociation energies, D_e of C_2 for both states in cm^{-1} and kJ mol^{-1}.
14. The vibrational frequencies of H_2^+, D_2 and H_2 are approximately 2322, 3118 and 4400 cm^{-1}, respectively. Calculate the force constants for these molecules and comment on the relative magnitudes of the values you obtain.

15. Calculate the difference between the zero-point energy, in kJ mol^{-1}, of the C-H and C-D bond stretches, given a typical C-H vibrational stretching frequency of 2900 cm^{-1}. Hence explain why C-H bonds are observed to react more rapidly than C-D bonds in many organic reactions (an example of the *kinetic isotope effect*).
16. Calculate the isotope effect for the $R(5)$ transition of ^{7}Li^{1}H and ^{7}Li^{2}H between the ground state X$^1\Sigma^+$ $\upsilon = 0$ and the excited state A$^1\Sigma^+$ $\upsilon = 0$ using the following molecular constants of ^{7}Li^{1}H:
    X$^1\Sigma^+$ $\upsilon=0$: $\omega_e = 1405.65$ cm^{-1}, $\omega_e x_e = 23.20$ cm^{-1}, $\tilde{B}_e = 7.5131$ cm^{-1}, $\bar{D}_e = 8.617\times10^{-4}$ cm^{-1}
    A$^1\Sigma^+$ $\upsilon=0$: $\omega_e = 280.96$ cm^{-1}, $\omega_e x_e = -28.0$ cm^{-1}, $\tilde{B}_e = 2.8536$ cm^{-1}, $\bar{D}_e = 1.187\times10^{-3}$ cm^{-1}
17. The Morse potential can be written [equation (5.71)]

$$V_{\text{Morse}} = D_e[1-\exp(-\beta x)]^2,$$

where x, the displacement, is $(r - r_e)$.
Expand the exponential up to the first term in x (i.e., $e^x \approx 1 + x$) and show that in this limit the potential is harmonic with force constant, $k_{\text{Morse}} = 2D_e\beta^2$. The vibrational frequency, ν, is defined in terms of the force constant in the usual way $\nu = \frac{1}{2\pi}\sqrt{\frac{k}{\mu_{red}}}$,
where μ_{red} is the reduced mass. Express β in terms of μ_{red}, D_e and ν.
Hence, given that the anharmonicity parameter for a Morse oscillator is $x_e = \frac{\hbar\beta^2}{4\pi\mu_{red}\nu}$, show that $D_e = \frac{h\nu}{4x_e}$.
18. For HBr, $\bar{D}_e = 31{,}590$ cm^{-1}, $\beta = 1.811 \times 10^{-10}$ m, $r_e = 1.414 \times 10^{-10}$ m.
    (a) Sketch the Morse potential for this molecule schematically and comment on the value of V at $r = 0$, $r = r_e$ and $r = \infty$.
    (b) Show that for small displacements from the equilibrium position, the above function is approximated by a simple harmonic potential and calculate the values of the force constant, the ground state vibrational frequency, the zero-point energy and the dissociation energy of HBr.
19. (a) A typical bond has $\omega_e = 1000$ cm^{-1} and $x_e = 0.0025$. What is its dissociation energy in kJ mol^{-1}?
    (b) In the (low resolution) infrared spectrum of ^{1}H^{79}Br, a strong absorption is observed at 2558.5 cm^{-1} and a weaker absorption at 5026.5 cm^{-1}. Use the Morse oscillator energy levels to account for these observations; determine all the parameters (ω_e, x_e, D_e and β), stating the units of your answers.
20. The fundamental infrared band of H^{35}Cl has lines at the following wavenumbers: 2775.77, 2799.00, 2821.59, 2843.63, 2865.14, 2906.25, 2925.92, 2944.99, 2963.35, 2981.05, 2998.05 cm^{-1}.
    Assign a pair of rotational quantum numbers to each transition, and label them P(J) or R(J). Illustrate your assignment with a diagram showing the observed transitions.
21. The following lines are observed in the fundamental (infrared) vibration–rotation band of H^{35}Cl:
    2998.05, 2981.05, 2963.35, 2944.99, 2925.92, 2906.25, 2865.14, 2843.63, 2821.59, 2799.00, 2775.77 cm^{-1}.

(a) Assign the rotational quantum numbers for $\upsilon = 0$ and $\upsilon = 1$ associated with each transition.
(b) Graphically or otherwise, using the method of combination differences, determine the rotational constants in both levels ($\tilde{B}_0$ and $\tilde{B}_1$); ignore centrifugal distortion.
(c) Hence, determine the equilibrium internuclear distance, r_e.

22. (a) Derive an expression for the separation between adjacent lines in the P branch of the fundamental vibration–rotation band of a diatomic molecule in terms of the rotational constants $\tilde{B}_0$ and $\tilde{B}_1$. Ignore centrifugal distortion.
(b) Lines are observed in the P branch of the fundamental band of carbon monoxide with the following wavenumbers (cm^{-1}):

P(9)	*P*(10)	*P*(14)	*P*(15)
2107.425	2103.271	2086.323	2082.003

Determine the rotational constants $\tilde{B}_0$ and $\tilde{B}_1$ and comment on their relative values.

23. The force constant of $^{79}Br_2$ is 240 N m^{-1}. Calculate the fundamental vibrational frequency and the zero-point energy of $^{79}Br_2$.

24. What is meant by zero-point energy? Calculate the energy change for the reaction

$$HD + HCl \rightarrow H_2 + DCl$$

assuming that each molecule is in its ground vibrational state. Ignore anharmonicity effects.

$$[\omega_e(H_2) = 4395 \text{ cm}^{-1};\ \omega_e(HCl) = 2990 \text{ cm}^{-1}]$$

25. Vibrational absorption lines for $H^{35}Cl$ lie at the following wavenumbers:
2885.9, 5668.0, 8347.0, 10922.9 cm^{-1}
Show that these values are in agreement with those expected for a Morse oscillator. Derive values for the force constant, zero-point energy and dissociation energy of the molecule.

26. The four central lines in the high resolution $\upsilon' = 1 \leftarrow \upsilon'' = 0$ infrared spectrum of $H^{37}Cl$ occur at
2837.6, 2858.8, 2899.2 and 2918.6 cm^{-1}.
Deduce as much as possible about the molecule.
Would the corresponding lines in $H^{35}Cl$ lie at the same positions?

27. Starting with a Morse oscillator, which has a vibrational constant (ω_e) of 500 cm^{-1} and taking the anharmonicity to be 1% of the vibrational constant, determine the bond dissociation energy in wavenumber for this oscillator.
What are the units of the anharmonicity constant?

28. (a) Account for the appearance of the infrared ro-vibrational spectrum of the CO molecule. At high resolution, lines are observed in the fundamental band for this molecule at the following wavenumbers: 2131.4, 2135.3, 2139.2, 2146.9, 2150.7, 2154.1 cm^{-1}. Assign these transitions and indicate them on a diagram of the energy levels involved. Is there any evidence of anharmonicity? Use the method of combination differences to obtain the rotational constants in the $\upsilon = 0$ and $\upsilon = 1$ vibrational levels, $\tilde{B}_0$ and $\tilde{B}_1$, and derive a value for the vibration–rotation interaction constant, α, and the equilibrium rotational constant, $\tilde{B}_e$. Hence, derive a value for the equilibrium bond length of CO.

(b) Show that the combination difference $\tilde{\nu}_R(J-1)-\tilde{\nu}_P(J+1)$ is equal to $4\tilde{B}_0(J+\frac{1}{2})$, where $\tilde{B}_0$ is the rotational constant in the $\upsilon = 0$ level. Find the corresponding expression for the combination difference $\tilde{\nu}_R(J)-\tilde{\nu}_P(J)$.

(c) Calculate these combination differences for different values of J and use the results to determine the rotational constants in the $\upsilon=0$ and $\upsilon=1$ levels, $\tilde{B}_0$ and $\tilde{B}_1$. Is there any evidence of centrifugal distortion? Is there any evidence of anharmonicity?

(d) Calculate the equilibrium rotational constant, $\tilde{B}_e$, and rotation–vibration interaction constant, α. Hence, deduce the equilibrium bond length, r_e.

(e) Closer examination reveals a further set of much weaker lines with a slightly different band origin but spacing very similar to those in the main band. On heating, the second set of lines becomes more intense, whereas the lines in the stronger band become slightly weaker. Explain these observations.

29. The fundamental stretching vibration for $^1H^{35}Cl$ is observed at 2886 cm^{-1}.
   (a) Draw a schematic energy diagram indicating the vibrational energy levels.
   (b) Calculate approximately the positions of the first and second overtones for HCl.
   (c) Calculate the zero-point energy of HCl.
   (d) Calculate the force that would be required to stretch the HCl bond by 0.1 nm assuming it follows Hooke's law.
   (e) Assuming that the internuclear distance is constant at 0.1273 nm, calculate the wavenumber of each of the two first lines in the P and R branches of HCl.

30. Because the vibrational spacing of diatomic molecules generally diminish with increasing energy, it is possible to extrapolate the vibrational spacing as a function of vibrational energy to zero, and thereby obtain a moderately accurate estimate of the dissociation energy of the molecule. Such graphs are known as Birge-Sponer plots named after R. Birge and H. Sponer.

   Using values of ω_e, $\omega_e x_e$ and $\omega_e y_e$ given below, construct such plots for Na_2, CH and HCl, then evaluate D_e and D_0, the dissociation energies from the bottom of the potential and from the ground vibrational state. Compare these values of D_0 with those given below:

Molecule	ω_e / cm^{-1}	$\omega_e x_e$ / cm^{-1}	$\omega_e y_e$ / cm^{-1}	D_0 / eV
Na_2	159.23	0.726	−0.0027	0.75
CH	2859.1	63.3	–	3.47
HCl	2991.09	52.82	0.2244	4.4361

31. The dissociation energy D_0 of H_2 is 4.46 eV and its zero-point energy $\frac{1}{2}h\nu_e$ is 0.26 eV. Estimate D_0 and $\frac{1}{2}h\nu_e$ for D_2 and T_2.

32. Calculate the vibrational wavenumber for HI, given that the force constant is 320 N m^{-1}, the reduced mass is 1.64×10^{-27} kg, and the equilibrium internuclear distance is 160 pm. Compare this figure with that for HF, for which the force constant, reduced mass and equilibrium internuclear distance are 970 N m^{-1}, 1.59×10^{-27} kg and 92 pm, respectively, and see that the comparison illustrates the general rule that when the motion of a system is more restricted, the quantum restrictions are more important, i.e., the steps between the allowed energies are greater.

6

Vibrational Spectroscopy of Polyatomic Molecules

If you want to find the secrets of the universe,
think in terms of energy,
frequency and vibration.
—Nikola Tesla

6.1 INTRODUCTION

We now discuss the complexities that arise for polyatomic molecules. Most of our discussion for diatomic molecules will apply to polyatomic molecules as well, but with two additions. One is the larger number of vibrational degrees of freedom, and the other is the existence of other modes of vibration. For diatomic molecules, the only possible mode of vibration is bond stretching, but the presence of additional internal coordinates (bond angles) makes other kinds of motion, such as bond bending and torsional motion, possible for polyatomic molecules. As in the previous chapters, we will first do an analysis using classical mechanics.

6.2 NORMAL MODES OF VIBRATION OF CARBON DIOXIDE

We begin by taking up the normal mode analysis of the linear symmetrical molecule, carbon dioxide, which is easily amenable to both classical and quantum mechanical treatment. From our normal mode analysis for heteronuclear diatomic molecules, we had concluded that the problem is reduced to the diagonalization of the Lagrangian. While the kinetic energy matrix is already in diagonal form, the potential energy matrix is not diagonal because of coupling of motions of the atoms. In order to simultaneously diagonalize both matrices, we had transformed the displacement coordinates to mass-weighted coordinates and then diagonalized the potential energy matrix in this coordinate system to obtain the eigenvalues and eigenvectors. The eigenvalues are related to the natural vibrational frequencies (one in the diatomic case) and the corresponding eigenvectors to the motions in these normal modes. We now apply the method of normal coordinate analysis to another linear, but polyatomic, molecule, carbon dioxide (O=C=O). We again constrain the atoms to move in the bond direction only, which we shall call the z direction.

We label the atoms **1**, **2** and **3**. Since both bonds are of the same type, i.e., C=O, we would expect both to have identical force constants, say k. The potential energy is then given by

$$V = k\left(z_2 - z_1\right)^2 + k\left(z_3 - z_2\right)^2 = k\left(z_1^2 + 2z_2^2 + z_3^2 - 2z_1z_2 - 2z_2z_3\right) \tag{6.1}$$

where z_1, z_2 and z_3 represent the Cartesian displacements of the three atoms. Writing the symmetrical force constant matrix, we obtain

$$\begin{pmatrix} k & -k & 0 \\ -k & 2k & -k \\ 0 & -k & k \end{pmatrix} \tag{6.2}$$

Since we now have three atoms, the force constant matrix is 3 × 3. We observe that the diagonal elements of the force constant matrix are all positive, and the off-diagonal elements are negative. If a bond exists between two atoms, the matrix element is filled with the corresponding force constant (with a negative sign). If there is no bond, the corresponding element is zero. After filling up all the off-diagonal elements, the diagonal elements are filled up with the sum of force constants of the bonds to which this atom is connected. The force constant matrix therefore is a kind of topology matrix – the matrix elements depend on the connectivity of atoms.

Looking at the force constant f_{ij} matrix in equation (6.2), we see that the f_{12} element is $-k$ because there exists a bond between atoms **1** and **2** with a force constant k. There is no bond between atoms **1** and **3**, and so the element f_{13} is zero. Since atom **1** is connected to only one atom with a force constant of k, f_{11} is set equal to k. Consider now the second row corresponding to the carbon atom. By symmetry, $f_{21} = f_{12} = -k$. The second atom is also connected to atom **3**, so the f_{23} element is $-k$. Since the second atom is connected to two atoms, each with a force constant k, the f_{22} element is $2k$. The third row is similarly explained. Verify that the sums of the elements of all rows and columns equal zero.

The next step is to transform the coordinates to mass-weighted ones and diagonalize the resultant force constant matrix. By definition, the mass-weighted coordinates are $q_i = \sqrt{m_i}z_i$. Substituting $z_i = q_i / \sqrt{m_i}$ in equation (6.1), we obtain for the potential energy matrix in mass-weighted coordinates

$$V = k\left(\frac{q_2}{\sqrt{m_2}} - \frac{q_1}{\sqrt{m_1}}\right)^2 + k\left(\frac{q_3}{\sqrt{m_3}} - \frac{q_2}{\sqrt{m_2}}\right)^2$$

$$= k\left(\frac{q_1^2}{m_1} + \frac{2q_2^2}{m_2} + \frac{q_3^2}{m_3} - 2\frac{q_1q_2}{\sqrt{m_1m_2}} - 2\frac{q_2q_3}{\sqrt{m_2m_3}}\right) \tag{6.3}$$

The force constant matrix in mass-weighted coordinates is thus

$$\begin{pmatrix} \frac{k}{m_O} & -\frac{k}{\sqrt{m_C m_O}} & 0 \\ -\frac{k}{\sqrt{m_C m_O}} & \frac{2k}{m_C} & -\frac{k}{\sqrt{m_C m_O}} \\ 0 & -\frac{k}{\sqrt{m_C m_O}} & \frac{k}{m_O} \end{pmatrix} \tag{6.4}$$

where we have substituted $m_1 = m_3 = m_O$ and $m_2 = m_C$. This matrix is simply obtained from (6.2) by dividing each element of the matrix by the square root of the product of the masses of the two atoms involved.

Example 6.1 Write down the potential energy matrix in displacement and mass-weighted coordinates for acetylene.

Solution

In displacement coordinates, the potential energy matrix is

$$\begin{pmatrix} k_{CH} & -k_{CH} & 0 & 0 \\ -k_{CH} & k_{CH}+k_{C\equiv C} & -k_{C\equiv C} & 0 \\ 0 & -k_{C\equiv C} & k_{CH}+k_{C\equiv C} & -k_{CH} \\ 0 & 0 & -k_{CH} & k_{CH} \end{pmatrix}$$

In mass-weighted coordinates, the matrix becomes

$$\begin{pmatrix} \frac{k_{CH}}{m_H} & -\frac{k_{CH}}{\sqrt{m_H m_C}} & 0 & 0 \\ -\frac{k_{CH}}{\sqrt{m_H m_C}} & \frac{k_{CH}+k_{C\equiv C}}{m_C} & -\frac{k_{C\equiv C}}{m_C} & 0 \\ 0 & -\frac{k_{C\equiv C}}{m_C} & \frac{k_{CH}+k_{C\equiv C}}{m_C} & -\frac{k_{CH}}{\sqrt{m_H m_C}} \\ 0 & 0 & -\frac{k_{CH}}{\sqrt{m_H m_C}} & \frac{k_{CH}}{m_H} \end{pmatrix}$$

Diagonalization of the matrix in (6.4) gives

$$\begin{pmatrix} \dfrac{k}{m_O} - \lambda & -\dfrac{k}{\sqrt{m_C m_O}} & 0 \\ -\dfrac{k}{\sqrt{m_C m_O}} & \dfrac{2k}{m_C} - \lambda & -\dfrac{k}{\sqrt{m_C m_O}} \\ 0 & -\dfrac{k}{\sqrt{m_C m_O}} & \dfrac{k}{m_O} - \lambda \end{pmatrix} = 0 \qquad (6.5)$$

which, when expanded, gives

$$\left(\frac{k}{m_O} - \lambda\right)\left[\left(\frac{2k}{m_C} - \lambda\right)\left(\frac{k}{m_O} - \lambda\right) - \frac{k^2}{m_C m_O}\right] + \frac{k}{\sqrt{m_C m_O}}\left[-\frac{k}{\sqrt{m_C m_O}}\left(\frac{k}{m_O} - \lambda\right)\right] = 0$$

$$\Rightarrow \left(\frac{k}{m_O} - \lambda\right)\left[\frac{2k^2}{m_C m_O} - \lambda\left(\frac{2k}{m_C} + \frac{k}{m_O}\right) + \lambda^2 - \frac{k^2}{m_C m_O}\right] - \frac{k^2}{m_C m_O}\left(\frac{k}{m_O} - \lambda\right) = 0$$

$$\Rightarrow \left(\frac{k}{m_O} - \lambda\right)\left[\frac{2k^2}{m_C m_O} - \lambda\left(\frac{2k}{m_C} + \frac{k}{m_O}\right) + \lambda^2 - \frac{2k^2}{m_C m_O}\right] = 0$$

$$\Rightarrow \lambda\left(\frac{k}{m_O} - \lambda\right)\left(\frac{2k}{m_C} + \frac{k}{m_O} - \lambda\right) = 0$$

The three eigenvalues are thus 0, $\dfrac{k}{m_O}$ and $\dfrac{k}{m_O} + \dfrac{2k}{m_C}$. As before, one of the eigenvalues is zero, corresponding to translational motion in the z (bond) direction, and there are only two genuine vibrations corresponding to bond stretching motions.

We now find the eigenvectors for each of the eigenvalues in order to visualize the motions. For $\lambda = 0$, we have, from the first row of equation (6.5)

$$\frac{k}{m_O} A_{11} - \frac{k}{\sqrt{m_C m_O}} A_{12} = 0$$

$$\Rightarrow A_{12} = \sqrt{\frac{m_C}{m_O}} A_{11}$$

Similarly, from the third row,

$$A_{12} = \sqrt{\frac{m_C}{m_O}} A_{13}$$

and hence $A_{11} = A_{13}$. The coefficients can be uniquely determined by applying the normalization condition

$$A_{11}^2 + A_{12}^2 + A_{13}^2 = 1$$

$$\Rightarrow A_{11}^2 + \frac{m_C}{m_O} A_{11}^2 + A_{11}^2 = 1$$

$$\Rightarrow A_{11} = A_{13} = \sqrt{\frac{m_O}{m_C + 2m_O}} = \sqrt{\frac{m_O}{M}}$$

$$A_{12} = \sqrt{\frac{m_C}{M}} \tag{6.6}$$

In this equation, M refers to the total mass of the molecule (= $m_C + 2m_O$).

For the second root, substitution of $\lambda = k/m_O$ in equation (6.5) gives

$$\begin{pmatrix} \frac{k}{m_O} - \frac{k}{m_O} & -\frac{k}{\sqrt{m_C m_O}} & 0 \\ -\frac{k}{\sqrt{m_C m_O}} & \frac{2k}{m_C} - \frac{k}{m_O} & -\frac{k}{\sqrt{m_C m_O}} \\ 0 & -\frac{k}{\sqrt{m_C m_O}} & \frac{k}{m_O} - \frac{k}{m_O} \end{pmatrix} \begin{pmatrix} A_{21} \\ A_{22} \\ A_{23} \end{pmatrix} = 0 \tag{6.7}$$

The first row gives

$$0A_{21} - \frac{k}{\sqrt{m_C m_O}} A_{22} + 0A_{23} = 0$$

$$\Rightarrow A_{22} = 0$$

The second row condition, after normalization, is $A_{21} = -A_{23} = 1/\sqrt{2}$. The third set of eigenvectors is similarly obtained. The normalized eigenvectors are summarized in Table 6.1.

As stated in the previous chapter, these coefficients represent the amplitudes in mass-weighted coordinates. To obtain the amplitudes in displacement coordinates, we divide by the square root of the mass of the concerned atom, i.e., $z_{ij} = \frac{A_{ij}}{\sqrt{m_j}} \cos\left(\sqrt{\lambda_i} t\right)$. If we do this for the first root, i.e., $\lambda_1 = 0$, the coefficients for all three atoms become $1/\sqrt{M}$. Thus, in this mode, the molecule

Table 6.1. Normalized coefficients (A_{ij}) for the three normal modes of carbon dioxide

Atom number	1	2	3
$\lambda_1 = 0$	$\sqrt{\frac{m_O}{M}}$	$\sqrt{\frac{m_C}{M}}$	$\sqrt{\frac{m_O}{M}}$
$\lambda_2 = \frac{k}{m_O}$	$\frac{1}{\sqrt{2}}$	0	$-\frac{1}{\sqrt{2}}$
$\lambda_3 = \frac{k}{m_O} + \frac{2k}{m_C}$	$\sqrt{\frac{m_C}{2M}}$	$-2\sqrt{\frac{m_O}{2M}}$	$\sqrt{\frac{m_C}{2M}}$

translates as a whole and there is no motion of the atoms with respect to one another. Moreover, λ, which represents the oscillation frequency, is zero, implying that this is not a vibration. For the second mode, we find the two oxygens moving in opposite directions with the same amplitude $(1\sqrt{(2m_O)}$. The central carbon atom is stationary. This kind of motion is called a *symmetric stretch*. Both the bonds stretch and compress in phase. In the third mode, both oxygens move in the same direction with the same amplitude, while the carbon moves in the opposite direction to maintain the centre of mass at the same position. Thus, one bond stretches while the other compresses. Such a motion is called *asymmetric* (or *antisymmetric*) *stretch.* Both these vibrational modes are depicted in Figure 6.1.

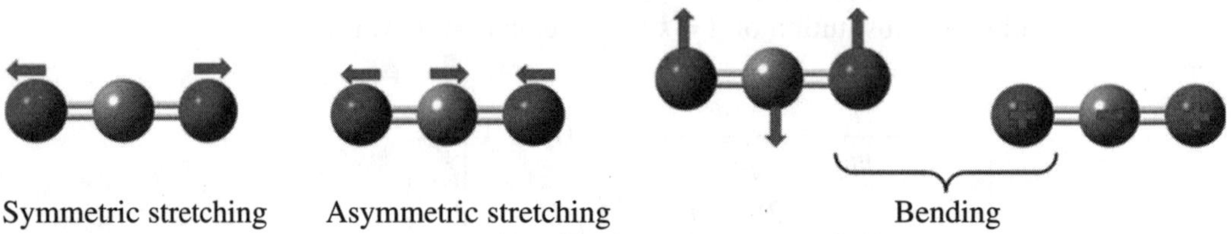

Figure 6.1. Normal modes of vibration of carbon dioxide

In the asymmetric stretch mode, all the atoms move, and this is reflected in the expression for the eigenvalue, which contains the masses of all atoms, in contrast to λ_2, in which only the masses of the oxygen atoms are involved. Moreover, $\nu_{as} > \nu_s$, since the former contains an additional term. This is a general observation and can be helpful in assigning vibrations. *The asymmetric stretch is observed at a higher wavenumber than the symmetric stretch.*

In Figure 6.1, we have also shown a third vibration, called a *bending mode*. Since we had constrained our atoms to move only along the internuclear (z) axis, this model provided only the symmetric and asymmetric stretching modes. In general, each atom in a molecule has three *translational degrees of freedom* – one each for translation along the x, y and z Cartesian axes of the molecule-based coordinate system. Thus, the molecule possesses a total of $3N$ degrees of freedom. Linear molecules have three unique translations, but only two unique rotations (Chapter 4). Thus, from the total of $3N$ degrees of freedom, we subtract three translations and two rotations, leaving $3N - 5$ vibrational degrees of freedom. For a diatomic molecule, this is consistent with the single vibration along the bond axis. Non-linear molecules possess $3N - 6$ vibrational degrees of freedom because of the additional rotational degree of freedom.

Returning to carbon dioxide, the $3N - 5$ rule for vibrational degrees of freedom predicts four vibrations. Since the internuclear axis is the z-axis, the symmetric stretch and the antisymmetric stretch are in the z direction. There are also two symmetric bending modes in each of the xz (perpendicular to the plane of the paper) and yz (in-plane) planes (Figure 6.1). In the figure, the '+' and '–' signify, respectively, movement above and below the plane of the paper. Since the two planes are equivalent, the two bending modes are *degenerate.* Hence, for carbon dioxide, there are two stretching modes, symmetric and asymmetric stretch, and a pair of degenerate bending modes, making a total of four modes, in agreement with the $3N - 5$ rule. However, there are only three fundamental bands, since the two bending modes are degenerate. It is easier to bend a molecule than stretch a bond, and therefore, the bending modes appear at the smallest wavenumber. The observed fundamental wavenumbers for carbon dioxide are 667 cm^{-1}, 1334 cm^{-1} and 2450 cm^{-1}, corresponding to bending, symmetric stretch and asymmetric stretch vibrations, respectively.

The experimental fundamental wavenumbers can be used to estimate the force constant of the C=O bonds in carbon dioxide. The eigenvalues represent the squares of the angular oscillation frequencies. Hence, $\lambda = 4\pi^2\nu_{\text{osc}}^2 = 4\pi^2c^2\omega^2$. For carbon dioxide, λ_2 corresponds to the symmetric stretch at 1334 cm^{-1}. Hence, $\lambda_2 = \dfrac{k}{m_{\text{O}}} = 4\pi^2c^2\omega^2$. Inserting the values, we obtain

$$k = 4\pi^2c^2\omega^2m_{\text{O}} = 4\pi^2(2.998\times10^{10}\ \text{cm s}^{-1})^2(1334\text{cm}^{-1})^2(15.9949\ u)(1.6605\times10^{-27}\,\text{kg}\ u^{-1})$$
$$= 1693\ \text{N m}^{-1}$$

Another value of k can be obtained from the expression for λ_3. Inserting $m_{\text{C}} = 12.0000\ u$, we obtain a new value for k, i.e., 1316 N m^{-1}. The two values are so different that we are forced to conclude that the assumption of equal force constants of the two bonds is in error. Below are shown two resonance structures of carbon dioxide, which clearly show that if one bond is elongated to a single bond, the other gets strengthened to a triple bond, and hence the two force constants do not remain equal.

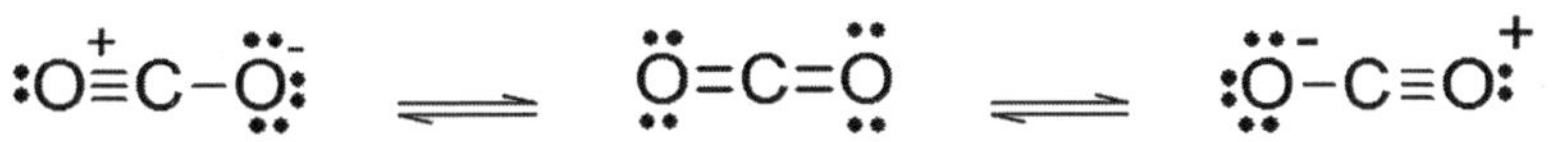

If we recognize that the two CO bonds have different force constants, we can rewrite the force constant matrix as

$$\begin{pmatrix} \dfrac{k}{m_{\text{O}}} & -\dfrac{k}{\sqrt{m_{\text{C}}m_{\text{O}}}} & 0 \\ -\dfrac{k}{\sqrt{m_{\text{C}}m_{\text{O}}}} & \dfrac{k+k'}{m_{\text{C}}} & -\dfrac{k'}{\sqrt{m_{\text{C}}m_{\text{O}}}} \\ 0 & -\dfrac{k'}{\sqrt{m_{\text{C}}m_{\text{O}}}} & \dfrac{k'}{m_{\text{O}}} \end{pmatrix}$$

and solve for the two force constants using the wavenumbers for the symmetric and asymmetric stretch vibrations.

6.2.1 Properties of normal modes

We may now extend equation (5.35) to polyatomic molecules and write $E_{\text{vib}} = \dfrac{1}{2}\sum_{i=1}^{3N-5/3N-6}\left(\dot{Q}_i^2 + \lambda_i Q_i^2\right)$, where the summation extends over $3N-5$ or $3N-6$ coordinates depending on whether the molecule is linear or nonlinear. Thus, the system may be treated as one consisting of $3N-5$ or $3N-6$ independent harmonic oscillators.

The quantum mechanical Hamiltonian operator for the ith normal coordinate can similarly be written from the classical expression as

$$\hat{H}_i = -\frac{\hbar^2}{2}\frac{\partial^2}{\partial Q_i^2} + \frac{1}{2}\lambda_i Q_i^2 \tag{6.8}$$

Notice that the mass drops out from the expression since the masses have been normalized to unity in mass-weighted coordinates. Expression (6.8) is similar to the Hamiltonian of a harmonic oscillator with an effective mass of unity, and the solutions are also similar.

The Schrödinger equation for each Q_i can be solved independently using the expression

$$\hat{H}_i\psi_i = E_i\psi_i \tag{6.9}$$

where E_i is the energy of the ith harmonic oscillator. The total energy is then the sum of individual energies

$$E = \sum_{i=1}^{3N-5/3N-6} E_i \tag{6.10}$$

and the total wave function is the product of the individual wave functions

$$\psi = \psi_1\psi_2 \ldots \psi_{3N-5/3N-6} = \Pi\psi_i \tag{6.11}$$

Each of the $3N - 5/3N - 6$ harmonic oscillators is like a diatomic molecule, and all that we have learnt for a diatomic molecule applies here too.

To summarize, normal modes have the following characteristics:

1. Each normal mode acts like a simple harmonic oscillator.
2. A normal mode is a concerted motion of many atoms.
3. The centre of mass does not move.
4. All atoms pass through their equilibrium positions at the same time.
5. Normal modes are independent; they do not interact.

In the asymmetric stretch and the two bending vibrations for CO_2, all the atoms move. The concerted motion of many of the atoms is a common characteristic of normal modes. In small molecules, all or most of the atoms move in a given normal mode; however, symmetry may require that a few atoms remain stationary for some normal modes. Thus, in the symmetric stretch, to keep the centre of mass in position, the central atom does not move.

The last characteristic, that normal modes are independent, means that normal modes do not exchange energy. While it is true that elongation of one bond will automatically set the other bond into motion, this is not true for the normal modes. You can excite one without exciting the other.

6.2.2 Selection rules for carbon dioxide

We may thus treat each normal mode as an independent harmonic oscillator. The same selection rules apply as for the diatomic case, i.e., the vibrational quantum number for each normal mode can change by only one unit, and secondly there should be a change in the dipole moment during a vibration for it to be infrared active. In the previous chapter, we had seen that a homonuclear diatomic molecule does not possess a dipole moment and during its stretching motion (the only mode available for a diatomic molecule), the dipole moment remains zero. Carbon dioxide is also a symmetrical molecule and has zero dipole moment (Figure 6.1). However, unlike the diatomic case, it has four modes of vibration and some of these are asymmetric and produce change in the dipole moment.

Mode	Symmetric stretch 	Asymmetric stretch 	Bend 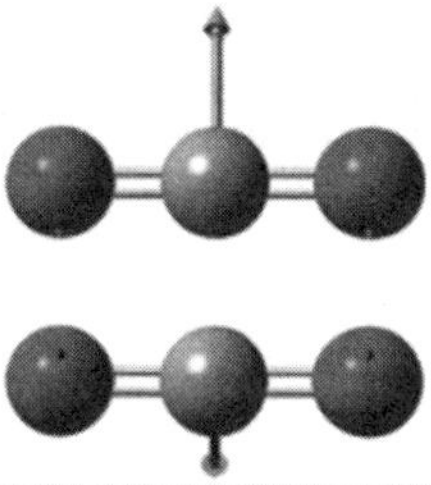
$(\partial\mu/\partial Q)$	0	⇦	⇧
IR activity	Inactive	Active	Active
Symmetry	Σ_g^+	Σ_u^+	Πu

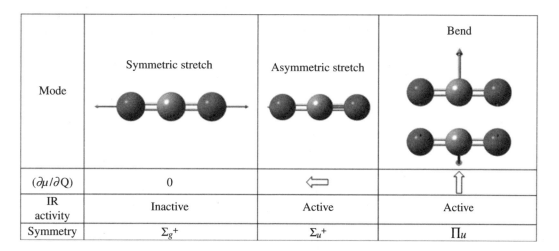

Figure 6.2. Dipole moment change in the normal modes of vibration of carbon dioxide

During the symmetric stretch mode of vibration, the two bonds elongate and compress in phase, and hence the dipole moment remains zero. No infrared absorption can take place for this vibrational mode, and we say that this mode is infrared inactive. On the other hand, in the asymmetric stretch mode, one bond compresses while the other elongates, leading to an oscillatory change in the dipole moment. This mode is therefore infrared active. In general, symmetric modes are infrared inactive and asymmetric modes are active. For the bending mode, too, we observe a change in the dipole moment (Figure 6.2) and this mode is also infrared active.

We also notice that the dipole moment change is along the internuclear axis for the asymmetric stretch vibration and perpendicular to the internuclear axis for the bending modes. It is for this reason that the former is called a *parallel* (//) *mode* and the latter a *perpendicular* (⊥) *mode* of vibration.

The selection rules for the accompanying rotational transitions are also different for the two kinds of vibrations. For parallel modes, the selection rules are the same as those for diatomic molecules

$$\Delta \upsilon = \pm 1, \pm 2, \ldots; \Delta J = \pm 1 \tag{6.12}$$

and we have the familiar PR contours observed for diatomic molecules. For the perpendicular mode, the $\Delta J = 0$ transition also becomes allowed (as there is a change in angular momentum perpendicular to the bond axis), so that the selection rules are

$$\Delta \upsilon = \pm 1, \pm 2, \ldots; \Delta J = 0, \pm 1 \tag{6.13}$$

Note that the $\Delta J = 0$ transition corresponds to a Q branch and we see a PQR structure for the first time in the perpendicular mode.

The Q branch therefore refers to a pure vibrational transition without any accompanying change in the rotational energy. For the fundamental $\upsilon = 0 \rightarrow 1$ transition, the transition wavenumber is given by

$$\tilde{\nu}_{\text{spectral}} = S(\upsilon') - S(\upsilon'') = F(J') + G(\upsilon') - F(J'') - G(\upsilon'') = \omega_e(1 - 2x_e) = \omega_0 \qquad (6.14)$$

where we have substituted $J' = J''$, $\upsilon' = 1$ and $\upsilon'' = 0$ in the expressions for the rotational and vibrational terms. This equation is valid for all J. Thus, the Q branch consists of coincident lines at the band centre, and hence, we expect a very intense line at the band centre. The actual reality is slightly different, as shown in Figure 6.3, where the Q branch appears as a somewhat broad absorption centred around ω_0.

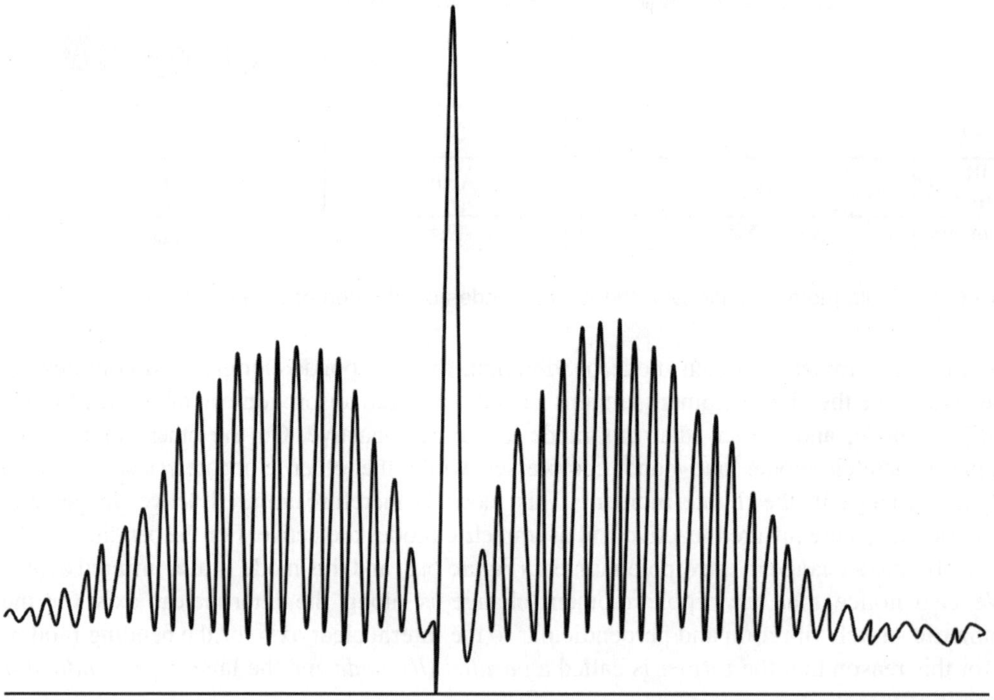

Figure 6.3. PQR band contour

The appearance of this band can be traced to the slight difference in the rotational constants for the ground and excited vibrational levels. On writing $\tilde{B}''$ and $\tilde{B}'$, respectively, for the two rotational constants, equation (6.14) gets modified to

$$\tilde{\nu}_{\text{spectral}} = \omega_0 + (\tilde{B}' - \tilde{B}'')J''(J'' + 1) \qquad (6.15)$$

The small negative value of the second term (since $\tilde{B}'' > \tilde{B}'$), which increases with increasing J'', explains the slight shift to the red with increasing J''. This difference is too small for individual lines to be resolved, but lends a distinctive shape to the Q branch, which tapers off to the red because of decreasing population of the higher levels.

In summary, the infrared spectrum of carbon dioxide shows two bands, one bending mode at 667 cm^{-1} with a PQR contour and another at 2350 cm^{-1} showing a PR contour, corresponding to the asymmetric stretch motion.

6.2.3 Vibration–rotation spectra of symmetric top molecules

We briefly discuss the vibration–rotation spectra of this important class of molecules, such as BF_3 (oblate) and CH_3Cl (prolate). The selection rules are again different for the parallel and perpendicular modes. Here, parallel refers to the modes in which the dipole moment change is parallel to the unique (C_3) axis, and perpendicular refers to those modes in which the change is perpendicular.

For the parallel modes, the selection rules are $\Delta J = 0, \pm 1, \Delta K = 0$ for $K \neq 0$, but if $K = 0$, the selection rule changes to $\Delta J = \pm 1, \Delta K = 0$. In other words, except when $K = 0$, there is a PQR structure, similar to that of the perpendicular band of a linear molecule, where the Q branch consists of lines superimposed on each other. For the perpendicular modes of symmetric tops, $\Delta J = 0, \pm 1, \Delta K = \pm 1$. Because $\Delta K \neq 0$, the individual components of the Q branch are no longer coincident and get spread out (Figure 6.4). The selection rules for the vibrational transitions are the same, i.e., $\Delta \upsilon = \pm 1$ (harmonic approximation) and $\Delta \upsilon = \pm 1, \pm 2, \pm 3, \ldots$ (anharmonic oscillator), as for a linear molecule.

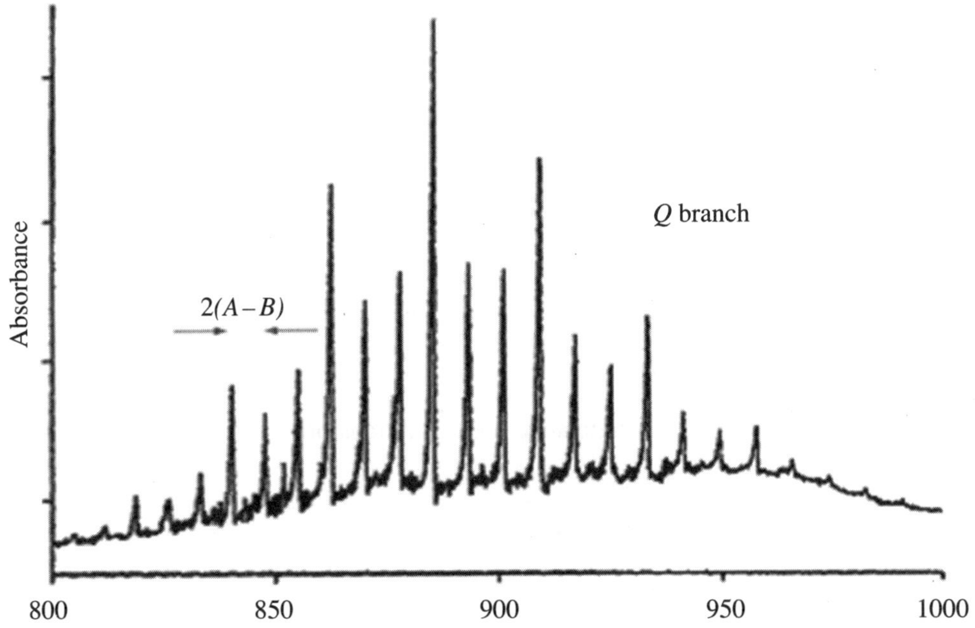

Figure 6.4. Q branch of perpendicular band of a symmetric top molecule

6.2.4 Symmetries of normal modes

Returning to carbon dioxide, we had found that solution of the force constant matrix yields three eigenvalues, one of which is for translational motion, and the remaining two are the symmetric and asymmetric stretches. We had also stated that the remaining degrees of freedom are two more translational, two rotational and a doubly degenerate bending mode.

While it was relatively easy to obtain expressions for the vibrational frequencies and the displacement equations when we had constrained the molecule to move in one dimension, you can imagine the amount of labour involved in solving the 9×9 matrix if these constraints are removed. This is the case for one of the simplest molecules. Fortunately, Group Theory comes to our aid by simplifying the computations, and qualitatively predicting the number and kind of vibrational motions in a molecule, and which of these are infrared (IR) active. We shall introduce its applications in this chapter, but discuss these in greater detail in the next chapter after Raman spectroscopy is introduced.

We first recognize that each of the four normal modes of vibration of carbon dioxide transforms as one of the irreducible representations of the point group to which the molecule belongs ($D_{\infty h}$). Picture the behaviour of the arrows in Figure 6.1 under the various operations of the point group of the molecule.

$D_{\infty h}$	E	$2C_\infty^\phi$	...	$\infty\sigma_v$	i	$2S_\infty^\phi$	...	∞C_2		
Σ_g^+	1	1	...	1	1	1	...	1		x^2+y^2, z^2
Σ_g^-	1	1	...	−1	1	1	...	−1	R_z	
Π_g	2	$2\cos\phi$	...	0	2	$-2\cos\phi$	...	0	(R_x, R_y)	(xz, yz)
Δ_g	2	$2\cos 2\phi$	...	0	2	$2\cos 2\phi$	...	0		$(x^2-y^2, 2xy)$
...	...	...	...	...	...	...	...	...		
Σ_u^+	1	1	...	1	−1	−1	...	−1	z	
Σ_u^-	1	1	...	−1	−1	−1	...	1		
Π_u	2	$2\cos\phi$	...	0	−2	$2\cos\phi$	...	0	(x, y)	
Δ_u	2	$2\cos 2\phi$	...	0	−2	$-2\cos 2\phi$	...	0		
...	...	...	...	...	...	...	...	...		

On examination of the symmetric stretch, we see that it transforms as

$D_{\infty h}$	E	$2C_\infty^\phi$	...	$\infty\sigma_v$	i	$2S_\infty^\phi$	...	∞C_2
Σ_g^+	1	1	...	1	1	1	...	1

under the various operations and hence forms a basis for the Σ_g^+ representation. Similarly, the asymmetric stretch transforms as Σ_u^+. The +/− superscript applies only to Σ states and labels the symmetry of the wave function with respect to reflection in a plane containing the nuclei. Both the singly degenerate representations for CO_2 (Σ_g^+ and Σ_u^+) are '+'. It is worth remembering that no vibrational mode has '–' symmetry.

$D_{\infty h}$	E	$2C_\infty^\phi$	...	$\infty\sigma_v$	i	$2S_\infty^\phi$	...	∞C_2
Σ_u^+	1	1	...	1	–1	–1	...	–1

It is not so readily apparent, but the bending modes transform as the doubly degenerate Π_u representation (Figure 6.2). Although it is fairly easy to assign the irreducible representation once we have the picture of the displacement vectors, obtaining these is itself not so straightforward for complex molecules. Identifying the point group of the molecule is the first crucial step. It is assumed that the reader is familiar with elementary Group Theory and can perform this step.

To the definition of normal modes, we may now add: normal modes are *independent, harmonic vibrations* which:

1. leave the centre of mass unmoved;
2. involve all atoms moving in phase (coherent motion);
3. transform as an irreducible representation of the molecular point group.

Normal modes of some other linear molecules are shown in Figure 6.5.

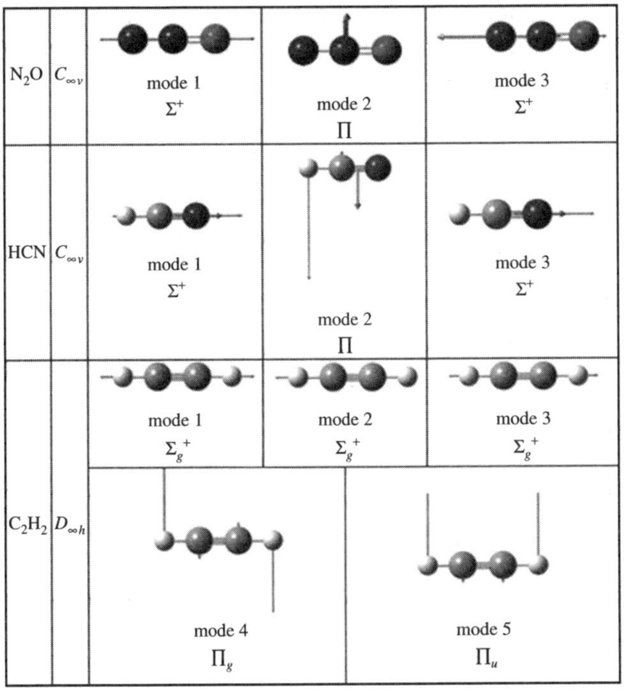

Figure 6.5. Normal modes of some linear molecules

Carbon dioxide belongs to an infinite group and is hence not amenable to the usual kind of manipulation possible for the finite groups. We first extend our results qualitatively to a bent triatomic molecule like water and work out a procedure for identifying the irreducible representations for the normal modes of a molecule and determining their infrared activity. Water belongs to the C_{2v} point group.

C_{2v}	E	$C_2(z)$	$\sigma_v(xz)$	$\sigma_v(yz)$		
A_1	1	1	−1	−1	z	x^2, y^2, z^2
A_2	1	1	−1	−1	R_z	xy
B_1	1	−1	1	−1	x, R_y	xz
B_2	1	−1	−1	1	y, R_x	yz

By analogy with carbon dioxide, we can picture its vibrational modes as under

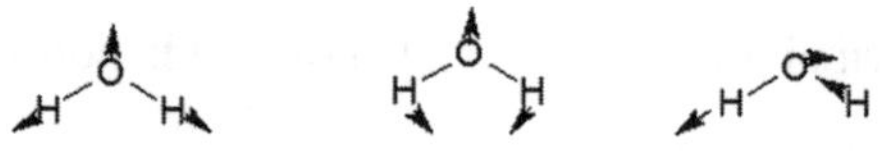

ν_1 ν_2 ν_3

The first is the symmetric stretch mode, the second is the bending mode and the third is the asymmetric stretch. Note that there is only one bending mode since the molecule is now non-linear and possesses only three vibrational degrees of freedom.

By convention, the molecule is put in a Cartesian frame in which the highest order axis is labelled z and the y axis is the next important one, in this case the direction perpendicular to z in the molecular plane. Finally, x is perpendicular to the molecular plane. The symmetry elements of water are as follows:

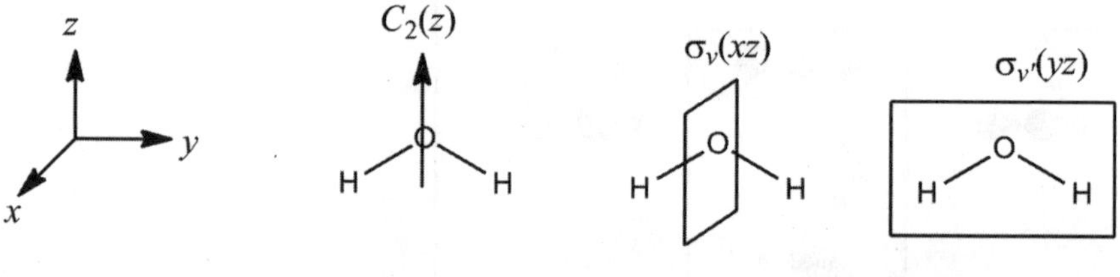

With this convention, we find that the first two vibrations transform as A_1 and the third as B_2. For example, the result of a C_2 operation on ν_2 and ν_3 is as under:

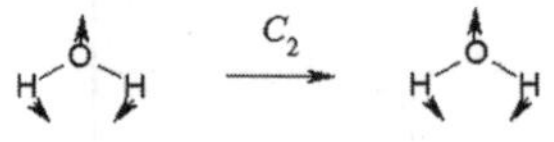

ν_2 ν_2

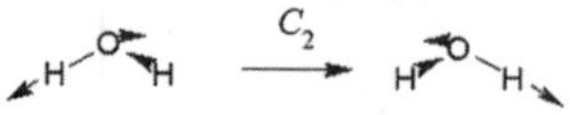

ν_3 ν_3'

and the character for C_2 is +1 for ν_2 and −1 for ν_3 since the arrows change direction for ν_3, i.e., $\nu_3' = -\nu_3$.

Coming to the labels, the vibrations are labelled with the highest symmetry first, or in the order they appear in the character table. Here, since two vibrations transform as A_1 (most symmetrical), the one with the higher frequency is labelled ν_1. We already know from our study of the carbon dioxide molecule that bond stretching requires higher energy than bending, and so the symmetric stretch is labelled ν_1 and the bending mode ν_2, leaving the asymmetric stretch, which is labelled ν_3.

Let us now start from scratch and assume that we cannot picture the vibrations of the water molecule. We label the atoms as under and place Cartesian coordinates on each atom.

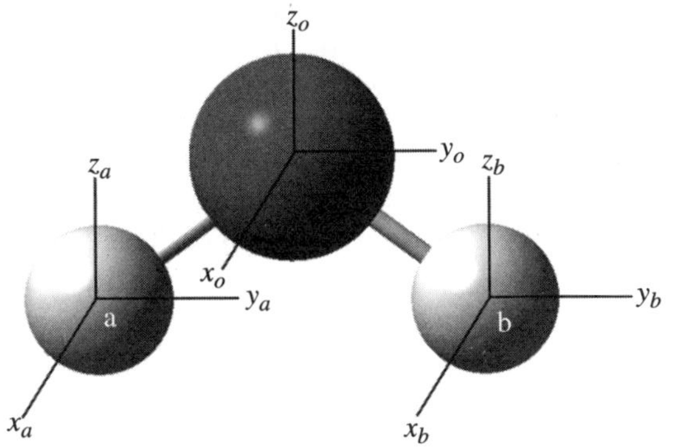

Let us now see the effect of each symmetry element on a nine-dimensional basis vector consisting of the Cartesian coordinates of each atom. For the identity element, the resulting matrix is

$$E\begin{pmatrix} x_a \\ y_a \\ z_a \\ x_O \\ y_O \\ z_O \\ x_b \\ y_b \\ z_b \end{pmatrix} = \begin{bmatrix} 1 & 0 & 0 & 0 & 0 & 0 & 0 & 0 & 0 \\ 0 & 1 & 0 & 0 & 0 & 0 & 0 & 0 & 0 \\ 0 & 0 & 1 & 0 & 0 & 0 & 0 & 0 & 0 \\ 0 & 0 & 0 & 1 & 0 & 0 & 0 & 0 & 0 \\ 0 & 0 & 0 & 0 & 1 & 0 & 0 & 0 & 0 \\ 0 & 0 & 0 & 0 & 0 & 1 & 0 & 0 & 0 \\ 0 & 0 & 0 & 0 & 0 & 0 & 1 & 0 & 0 \\ 0 & 0 & 0 & 0 & 0 & 0 & 0 & 1 & 0 \\ 0 & 0 & 0 & 0 & 0 & 0 & 0 & 0 & 1 \end{bmatrix} \begin{pmatrix} x_a \\ y_a \\ z_a \\ x_O \\ y_O \\ z_O \\ x_b \\ y_b \\ z_b \end{pmatrix} \quad (6.16)$$

The identity ('do nothing') operation leaves all vectors unchanged, i.e., $x_a \to x_a$, $y_a \to y_a$, and so on, so that all diagonal elements are equal to one, and the resulting matrix is a 9×9 identity matrix with a trace of 9.

We repeat for the C_2 operation. This is a rotation about the z-axis that interchanges the two hydrogens and also changes the sign of the x and y coordinates. Thus, we have $x_a \to -x_b$, $y_a \to -y_b$, $z_a \to z_b$, $x_O \to -x_O$, $y_O \to -y_O$, $z_O \to z_O$, $x_b \to -x_a$, $y_b \to -y_a$, and $z_b \to z_a$. Writing in matrix form,

$$C_2\begin{pmatrix} x_a \\ y_a \\ z_a \\ x_O \\ y_O \\ z_O \\ x_b \\ y_b \\ z_b \end{pmatrix} = \begin{bmatrix} 0 & 0 & 0 & 0 & 0 & 0 & -1 & 0 & 0 \\ 0 & 0 & 0 & 0 & 0 & 0 & 0 & -1 & 0 \\ 0 & 0 & 0 & 0 & 0 & 0 & 0 & 0 & 1 \\ 0 & 0 & 0 & -1 & 0 & 0 & 0 & 0 & 0 \\ 0 & 0 & 0 & 0 & -1 & 0 & 0 & 0 & 0 \\ 0 & 0 & 0 & 0 & 0 & 1 & 0 & 0 & 0 \\ -1 & 0 & 0 & 0 & 0 & 0 & 0 & 0 & 0 \\ 0 & -1 & 0 & 0 & 0 & 0 & 0 & 0 & 0 \\ 0 & 0 & 1 & 0 & 0 & 0 & 0 & 0 & 0 \end{bmatrix} \begin{pmatrix} x_a \\ y_a \\ z_a \\ x_O \\ y_O \\ z_O \\ x_b \\ y_b \\ z_b \end{pmatrix} \tag{6.17}$$

We see that the elements of the first three rows move off-diagonal, as do the last three rows. Only the middle three rows (corresponding to the central oxygen) remain on the diagonal and contribute to the trace, but the signs of x and y are changed. Thus, the trace of this matrix is -1.

For $\sigma_v(xz)$, again the hydrogens exchange position, but this time reflection is in the xz plane which changes the sign of the y coordinate, leaving x and z unchanged. The resulting matrix is

$$\sigma_v(xz)\begin{pmatrix} x_a \\ y_a \\ z_a \\ x_O \\ y_O \\ z_O \\ x_b \\ y_b \\ z_b \end{pmatrix} = \begin{bmatrix} 0 & 0 & 0 & 0 & 0 & 0 & 1 & 0 & 0 \\ 0 & 0 & 0 & 0 & 0 & 0 & 0 & -1 & 0 \\ 0 & 0 & 0 & 0 & 0 & 0 & 0 & 0 & 1 \\ 0 & 0 & 0 & 1 & 0 & 0 & 0 & 0 & 0 \\ 0 & 0 & 0 & 0 & -1 & 0 & 0 & 0 & 0 \\ 0 & 0 & 0 & 0 & 0 & 1 & 0 & 0 & 0 \\ 1 & 0 & 0 & 0 & 0 & 0 & 0 & 0 & 0 \\ 0 & -1 & 0 & 0 & 0 & 0 & 0 & 0 & 0 \\ 0 & 0 & 1 & 0 & 0 & 0 & 0 & 0 & 0 \end{bmatrix} \begin{pmatrix} x_a \\ y_a \\ z_a \\ x_O \\ y_O \\ z_O \\ x_b \\ y_b \\ z_b \end{pmatrix} \tag{6.18}$$

and the trace of the matrix is 1.

For rotation in plane, all atoms remain unshifted, but x changes sign.

$$\sigma_{v'}(yz)\begin{pmatrix} x_a \\ y_a \\ z_a \\ x_O \\ y_O \\ z_O \\ x_b \\ y_b \\ z_b \end{pmatrix} = \begin{bmatrix} -1 & 0 & 0 & 0 & 0 & 0 & 0 & 0 & 0 \\ 0 & 1 & 0 & 0 & 0 & 0 & 0 & 0 & 0 \\ 0 & 0 & 1 & 0 & 0 & 0 & 0 & 0 & 0 \\ 0 & 0 & 0 & -1 & 0 & 0 & 0 & 0 & 0 \\ 0 & 0 & 0 & 0 & 1 & 0 & 0 & 0 & 0 \\ 0 & 0 & 0 & 0 & 0 & 1 & 0 & 0 & 0 \\ 0 & 0 & 0 & 0 & 0 & 0 & -1 & 0 & 0 \\ 0 & 0 & 0 & 0 & 0 & 0 & 0 & 1 & 0 \\ 0 & 0 & 0 & 0 & 0 & 0 & 0 & 0 & 1 \end{bmatrix} \begin{pmatrix} x_a \\ y_a \\ z_a \\ x_O \\ y_O \\ z_O \\ x_b \\ y_b \\ z_b \end{pmatrix} \tag{6.19}$$

and the trace of the matrix is 3.

We begin to see a pattern. Only the atoms that remain unshifted during a symmetry operation contribute to the trace of the matrix. Second, every symmetry element has a characteristic 3×3 matrix. Thus, for example, the E matrix is

$$E = \begin{bmatrix} 1 & 0 & 0 \\ 0 & 1 & 0 \\ 0 & 0 & 1 \end{bmatrix}$$

and those for the other operations are likewise

$$C_2(z) = \begin{bmatrix} -1 & 0 & 0 \\ 0 & -1 & 0 \\ 0 & 0 & 1 \end{bmatrix}$$

$$\sigma_v(xz) = \begin{bmatrix} 1 & 0 & 0 \\ 0 & -1 & 0 \\ 0 & 0 & 1 \end{bmatrix}$$

and

$$\sigma_{v'}(yz) = \begin{bmatrix} -1 & 0 & 0 \\ 0 & 1 & 0 \\ 0 & 0 & 1 \end{bmatrix}$$

We see that when an atom shifts position, its matrix also shifts off-diagonal and does not contribute to the trace.

Therefore, in order to write the $3N$ reducible representation, we need only determine the number of unshifted atoms for each operation and then multiply by the trace of the characteristic matrix for that operation. What is needed now is to find a method to find the trace of the characteristic matrix of a symmetry operation.

For this purpose, we may conveniently divide the symmetry operations into two categories: *proper* and *improper* rotations. In the former category, we place the two operations, E and C_n, which are, respectively, rotated by 0° and $360°/n$, where n is the order of the rotation axis. The matrix of rotation about the z axis is

$$\begin{pmatrix} x' \\ y' \\ z' \end{pmatrix} = \begin{bmatrix} \cos\theta & \sin\theta & 0 \\ -\sin\theta & \cos\theta & 0 \\ 0 & 0 & 1 \end{bmatrix} \begin{pmatrix} x \\ y \\ z \end{pmatrix}$$

and its trace is $1 + 2\cos\theta$, where θ is the angle of rotation. For $E(\theta = 0°)$, the trace $\chi(E) = 3$. Similarly, $\chi(C_2) = -1$, as seen in the above example.

The other category of operations is that of improper rotations, which involve a proper rotation, followed by reflection in a plane perpendicular to the axis of rotation. If rotation is about the z-axis, the reflection is in the xy plane, which reverses the sign of z. Thus, the matrix for improper rotation of any arbitrary x-, y- and z-axis is

$$\begin{pmatrix} x' \\ y' \\ z' \end{pmatrix} = \begin{bmatrix} \cos\theta & \sin\theta & 0 \\ -\sin\theta & \cos\theta & 0 \\ 0 & 0 & -1 \end{bmatrix} \begin{pmatrix} x \\ y \\ z \end{pmatrix}$$

and the resulting trace is $-1 + 2\cos\theta$. In this category, we have symmetry elements like σ (rotation by 0°, followed by reflection), S_n (rotation by $360°/n$ followed by reflection) and i (inversion, rotation by 180°, followed by reflection). However, one need not remember these formulae, since we know that inversion reverses the signs of all three coordinates and hence $\chi(i) = -3$. Similarly, reflection in a plane keeps two coordinates in the plane intact and reverses the sign of the third, so that $\chi(\sigma) = 1$. This is true for every reflection, whether it is in a vertical, horizontal or dihedral plane.

We now apply these concepts to the vibrational modes of water. Against every symmetry operation we need to write the number of unshifted atoms and the contribution of the symmetry element per atom and then multiply the two. This gives a $3N$-dimensional representation, including three translations, three rotations and the remaining three are vibrations. At this stage, you must check that the character for E is $3N$, i.e., 9 in this case. Translational motion is represented by the movement of the centre of mass in the three directions, and hence its representation is the same as that of any arbitrary vector, i.e., $\chi(R)$. For rotational motion, the characters are identical to those for translation in the case of proper rotations, but change sign for the improper rotations. Thus, we may obtain the reducible representation for vibrational modes by taking away the reducible representations for translational and rotational motion, i.e., twice $\chi(R)$ for proper rotations and zero for improper rotations, from the total representation of order $3N$. Alternatively, you may add the irreducible representations for x, y and z given in the character table, which correspond to translational motion in the three directions, and sum R_x, R_y and R_z for rotational motion, and then subtract these from the total representation. In either case, the same result should be obtained. The example for water is worked out in Table 6.2.

Table 6.2. Normal modes of water

	$1 + 2\cos\theta$		$-1 + 2\cos\theta$	
C_{2v}	E	$C_2(z)$	$\sigma_v(xz)$	$\sigma_v'(yz)$
$\chi(R)$	3	−1	1	1
n_R	3	1	1	3
Γ_{3N}	9	−1	1	3
$\Gamma_{trans} + \Gamma_{rot}$	6	−2	0	0
Γ_{vib}	3	1	1	3
$\Gamma_{stretch}$	2	0	0	2
Γ_{bend}	1	1	1	1

Using the reduction formula from the Great Orthogonality Theorem, we get $2A_1 \oplus B_2$ on resolving Γ_{vib}. For example, $n(A_1)=\frac{1}{4}(3\times1+1\times1+1\times1+3\times1)=2$ and $n(B_2)=\frac{1}{4}(3\times1-1\times 1-1\times1+3\times1)=1$. Similarly, it can be shown that $n(A_2) = n(B_1) = 0$. Again check that the sum of the normal modes equals $3N - 6$.

We may further separate out the stretching vibrations and the bending mode. A non-cyclic molecule containing N atoms has $N - 1$ bonds and the same number of stretching vibrations. Since three atoms are required to define a bond angle, there are $N - 2$ bond angles, and therefore $N - 2$ bending modes. Four atoms define a dihedral angle, and so there are $N - 3$ torsions. The total number of vibrational modes is thus $(N - 1)$ stretching modes + $(N - 2)$ bending modes + $(N - 3)$ torsion modes, making the total $3N - 6$, as required. For a linear molecule, one of the rotational degrees of freedom gets converted to a degenerate vibration and there are $3N - 5$ modes of vibration. The various kinds of vibrational motions are depicted below:

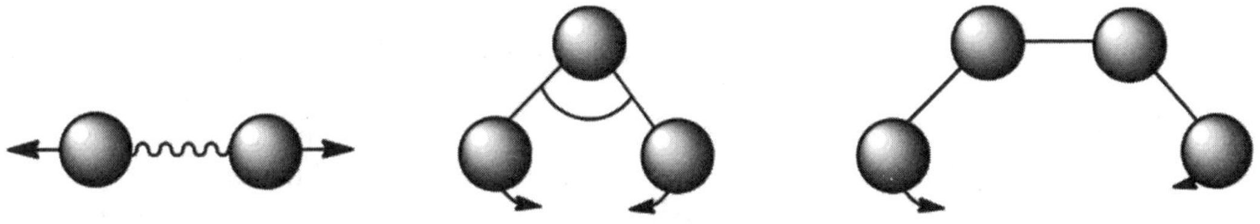

Bond stretching Valence angle bending Torsion

A triatomic molecule like H_2O therefore has two stretching modes and one bending mode of vibration. Similarly, the linear molecule carbon dioxide also has two stretching modes of vibration and one doubly degenerate bending mode.

In order to understand the motions, it is convenient to change to *internal coordinates*. These constitute the bond lengths and bond angles. Since stretching modes involve the bonds only, and if we label the two O–H bonds as r_1 and r_2, they transform as follows:

Internal coordinates of water

$$E\begin{pmatrix} r_1 \\ r_2 \end{pmatrix}=\begin{pmatrix} 1 & 0 \\ 0 & 1 \end{pmatrix}\begin{pmatrix} r_1 \\ r_2 \end{pmatrix}$$

$$C_2\begin{pmatrix} r_1 \\ r_2 \end{pmatrix}=\begin{pmatrix} 0 & 1 \\ 1 & 0 \end{pmatrix}\begin{pmatrix} r_1 \\ r_2 \end{pmatrix}$$

$$\sigma_v(xz)\begin{pmatrix} r_1 \\ r_2 \end{pmatrix}=\begin{pmatrix} 0 & 1 \\ 1 & 0 \end{pmatrix}\begin{pmatrix} r_1 \\ r_2 \end{pmatrix}$$

$$\sigma_{v'}(yz)\begin{pmatrix} r_1 \\ r_2 \end{pmatrix}=\begin{pmatrix} 1 & 0 \\ 0 & 1 \end{pmatrix}\begin{pmatrix} r_1 \\ r_2 \end{pmatrix}$$

since the two bonds exchange positions under C_2 and $\sigma_v(xz)$ operations. The reducible representation for the stretching motions is therefore given as Γ_{stretch} in Table 6.2. The stretching modes reduce to $A_1 \oplus B_2$ using the reduction formula. Subtraction from Γ_{vib} gives Γ_{bend}, which is the only remaining mode (A_1). It can be seen that θ transforms as A_1.

Only for the case of molecules with one central atom (which include molecules like H_2O, H_2S, CO_2, NH_3, XeF_4, XeF_6, etc., but not molecules like HCCH, H_2O_2, etc.), Γ_{stretch} can be obtained by subtracting the totally symmetric irreducible representation (A_1 in this case) from the number of unshifted atoms. The number of unshifted bonds is one less than the number of unshifted atoms, since the central atom is not included and it transforms as the totally symmetric representation.

6.2.5 Selection rules

In the previous chapter, we saw that transitions in the infrared region are governed by the selection rule $\Delta \upsilon = \pm 1$ for a harmonic oscillator (normal modes are harmonic) and, additionally, the dipole moment of the molecule must change during the vibration. Inspection of the normal modes for water shows that the dipole moment oscillates in each of these normal modes and hence they are all infrared active. However, this kind of analysis is only possible for simple molecules, and for the majority of polyatomic molecules, it is often difficult to determine by inspection alone whether the dipole moment oscillates during a particular vibrational mode. Here, too, symmetry can come to our rescue. The selection rules are governed by the magnitude of the transition dipole moment $\left|\vec{M}_{lm}\right|$. As stated in Chapter 2, this vector quantity can be expanded into its components in the three Cartesian directions, i.e.,

$$\left|\vec{M}_{lm}\right| = \hat{i}\mu_{xlm} + \hat{j}\mu_{ylm} + \hat{k}\mu_{zlm}$$

where $\hat{i}$, $\hat{j}$ and $\hat{k}$ are unit vectors along the three Cartesian directions (not to be confused with the operator symbol). The three components are therefore scalar quantities. If it can be shown that all the components vanish, the transition is said to be forbidden. It remains to demonstrate that at least one of the components is non-vanishing to be able to say that a vibrational mode of a molecule is infrared active. Consider, for example, the μ_{xlm} component. It is a triple integral involving the excited and ground state vibrational wave functions, and the x-component of the dipole moment operator. For the $\upsilon'' \rightarrow \upsilon'$ transition, the x-component of the transition dipole moment is given by

$$\int_{-\infty}^{\infty} \psi_{\upsilon'} \hat{\mu}_x \psi_{\upsilon''}\, dQ$$

where $\psi_{\upsilon'}$ and $\psi_{\upsilon''}$ are, respectively, the excited and ground state vibrational wave functions, $\hat{\mu}_x$ is the operator for the x-component of the dipole moment and Q is the normal coordinate. Only if the product in the integral is completely symmetrical will the integral survive (see Chapter 2). This means that it must transform as (or contain) the totally symmetric irreducible representation of the point group to which the molecule belongs (e.g., A_1 in C_{2v}, Σ_g^+ for $D_{\infty h}$, Σ^+ for $C_{\infty v}$, A_g for C_{2h}, etc.)

The symmetry of the integral depends on the symmetry properties of the three quantities in the product. Consider first the wave functions. We have seen that the selection rule for harmonic vibrations is $\Delta \upsilon = \pm 1$, and that, at room temperature, most of the molecules are in the ground vibrational level. Therefore, the most important transition is the fundamental, corresponding to $\upsilon = 0 \rightarrow 1$. The ground state wave function is given by

$$\psi_0 = N_0 H_0(y) e^{-y^2/2} = N_0 e^{-\alpha Q^2/2}$$

where $\alpha = \sqrt{\mu_{red} k} / \hbar$.

The sign of ψ_0 is therefore independent of the sign of Q, which appears as its square in the expression for the wave function. *It follows that ψ_0 transforms as the totally symmetric irreducible representation of the relevant point group.*

The excited state wave function is

$$\psi_1 = N_1 H_1(y) e^{-y^2/2} = N_1 (2\sqrt{\alpha} Q) e^{-\alpha Q^2/2}$$

since $H_1(y) = 2y$. Because of the presence of the Q term (the only vector quantity) in the expression, ψ_1 has the same symmetry as the normal coordinate Q, i.e., *ψ_1 transforms as the same symmetry as the normal mode.*

Thus, for water, we may surmise that the symmetries of the ground and excited state wave functions are as under

Mode	Type	Normal mode	ψ_0	ψ_1
1	Symmetric stretch	A_1	A_1	A_1
2	Bending	A_1	A_1	A_1
3	Asymmetric stretch	B_2	A_1	B_2

Our analysis for water applies equally to other symmetrical bent triatomic molecules of the type AB_2, such as H_2S, SO_2, NO_2, etc.

We now look at the symmetry properties of the dipole moment. Since $\vec{\mu}_x = \sum q_i \vec{x}_i$, where q_i is the electronic charge, the x-component behaves like x, or rather the translational vector in the x-direction, T_x. Usually, these are given in the character table. For example, the x written in the B_1 row of C_{2v} tells us that the translational vector for the x direction is a basis for the B_1 irreducible representation. An examination of the translational vectors confirms this.

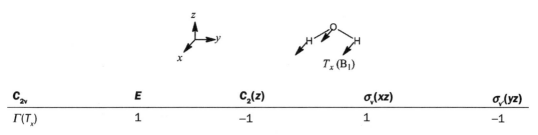

T_x (B_1)

C_{2v}	E	$C_2(z)$	$\sigma_v(xz)$	$\sigma_v(yz)$
$\Gamma(T_x)$	1	−1	1	−1

T_x is symmetrical with respect to E and $\sigma_v(xz)$, but changes sign under C_2 and $\sigma_v(yz)$ operations, and so it transforms as B_1. Similarly, it can be shown that T_y and T_z transform as B_2 and A_1, respectively.

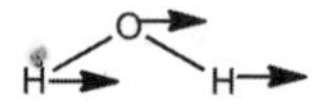

$T_y\,(B_2)$

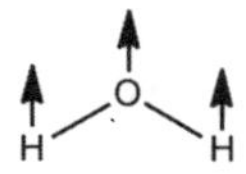

$T_z\,(A_1)$

Therefore, the components of $\hat{\mu}$ transform in C_{2v} as $\Gamma(\mu_x) = B_1$, $\Gamma(\mu_y) = B_2$ and $\Gamma(\mu_z) = A_1$.

Having determined all the irreducible representations, we are now equipped to determine the symmetries of the quantities under the integral sign. These are given by the direct products $\Gamma(\upsilon')\otimes\Gamma(\mu_x)\otimes\Gamma(\upsilon'')$, $\Gamma(\upsilon')\otimes\Gamma(\mu_y)\otimes\Gamma(\upsilon'')$ and $\Gamma(\upsilon')\otimes\Gamma(\mu_z)\otimes\Gamma(\upsilon'')$, respectively, where $\Gamma(\upsilon')$ and $\Gamma(\upsilon'')$ refer to the irreducible representations for the excited and ground state vibrational wave functions. If any of the direct products is the totally symmetric representation, A_1, then the transition is allowed.

The direct products for non-degenerate point groups are obtained by multiplying together the characters $\chi(R)$ for the two representations. The following general rules hold:

(a) the direct product of any species with the totally symmetric species leaves it unchanged, and

(b) the direct product of any species with itself gives the totally symmetric species.

Let us first consider the asymmetric stretch vibration (ν_3), whose irreducible representation is B_2. For this vibration, the direct product $\Gamma(\upsilon')\otimes\Gamma(\mu_x)\otimes\Gamma(\upsilon'')$ is $B_2\otimes B_1\otimes A_1 = B_2\otimes B_1 = A_2$. Thus, the x-component of the electric field cannot excite this transition.

C_{2v}	E	$C_2(z)$	$\sigma_v(xz)$	$\sigma_{v'}(yz)$
B_2	1	−1	−1	1
B_1	1	−1	1	−1
A_1	1	1	1	1
$B_2\otimes B_1\otimes A_1 = A_2$	1	1	−1	−1

Similarly, it can be shown that the direct product $\Gamma(\upsilon')\otimes\Gamma(\mu_y)\otimes\Gamma(\upsilon'')$ is $B_2\otimes B_2\otimes A_1$ which transforms as A_1 (since B_2 multiplied by itself gives the total symmetric representation), and $\Gamma(\upsilon')\otimes\Gamma(\mu_z)\otimes\Gamma(\upsilon'')$ is $B_2\otimes A_1\otimes A_1 = B_2$. The appearance of A_1 as one of the direct products confirms that this transition is infrared active. Further, since the y-component of the transition dipole moment is A_1, we may state that this transition is *dipole allowed, with transition dipole along* y or that this transition is y *polarized.*

It can be similarly shown that the symmetric stretch and bending modes (both A_1) are infrared active, with transition dipole along z. We also make the following observation: since $\Gamma(\upsilon'')$ is always totally symmetric, the triple product depends on the other two quantities. In order that their product is the totally symmetric representation, they must belong to the same irreducible representation, for then the product is a square, which is always totally symmetric. We may state the selection rule differently as: the fundamental transition is allowed if the symmetry of the normal mode (and hence $\upsilon = 1$) is the same as that of x, y or z. This leads us to another important rule. For molecules with a centre of symmetry, the normal modes are either of u or g representation. The translational vectors, and hence the dipole moment components, are always of u representation because the vectors change sign on inversion. Since only those modes that have the same symmetry as the translational vectors can be

IR active, it follows that *g* modes are IR inactive. Note that this is the extent to which we can commit ourselves – we cannot say that all *u* modes are IR active. In large symmetrical molecules with a centre of symmetry, there may be several *u* modes, but there can only be three translational modes, so some of the vibrational modes will be IR inactive.

A case in point is the ethylene molecule (D_{2h}), which has three mutually perpendicular C_2 axes. The Cartesian coordinates are selected to coincide with these three axes. As stated for water, the most important one, i.e., the one containing the carbon atoms is labelled *z*, the one perpendicular to it in the molecular plane is labelled *y*, and the remaining axis perpendicular to the molecular plane becomes *x*.

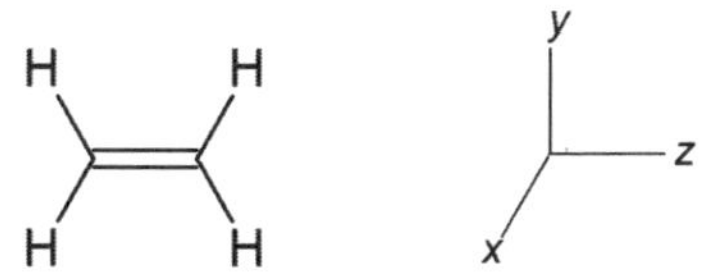

The symmetry analysis has also been performed in Table 6.3. Since the molecule has six atoms, there are 18 total degrees of freedom and 12 vibrational modes. The reduction formula gives

$$\Gamma_{\text{vib}} = 3\text{A}_g \oplus \text{B}_{2g} \oplus 2\text{B}_{3g} \oplus \text{A}_{1u} \oplus 2\text{B}_{1u} \oplus 2\text{B}_{2u} \oplus \text{B}_{3u}$$

According to the previous discussion, only the modes with '*u*' designation can be infrared active, and from the character table, we can see that the B_{1u}, B_{2u} and B_{3u} modes can be excited, respectively, by the *z*-, *y*- and *x*-components of the electric vector, leaving A_u, which does not have a matching component in the electric dipole. For this reason, this is called a 'silent' vibration.

Table 6.3. Character table for point group D_{2h}

D_{2h}	E	$C_2(z)$	$C_2(y)$	$C_2(x)$	i	$\sigma(xy)$	$\sigma(xz)$	$\sigma(yz)$	Basic components
A_g	1	1	1	1	1	1	1	1	x^2, y^2, z^2
B_{1g}	1	1	−1	−1	1	1	−1	−1	R_z *xy*
B_{2g}	1	−1	1	−1	1	−1	1	−1	R_y *xz*
B_{3g}	1	−1	−1	1	1	−1	−1	1	R_x *yz*
A_u	1	1	1	1	−1	−1	−1	−1	
B_{1u}	1	1	−1	−1	−1	−1	1	1	*Z*
B_{2u}	1	−1	1	−1	−1	1	−1	1	*Y*
B_{3u}	1	−1	−1	1	−1	1	1	−1	*X*
$\chi(R)$	6	2	0	0	0	0	2	6	
n_R	3	−1	−1	−1	−3	1	1	1	
Γ_{3N}	18	−2	0	0	0	0	2	6	
$\Gamma_{\text{trans}} + \Gamma_{\text{rot}}$	6	−2	−2	−2	0	0	0	0	
Γ_{vib}	12	0	2	2	0	0	2	6	

Transitions are also classified as parallel or perpendicular depending on whether the direction of the transition moment is parallel or perpendicular to the symmetry axis of the molecule. In the case of water, since $|\vec{M}_{zv'v''}|$ is non-zero for the symmetric stretch (ν_1) and bending (ν_2) modes, these two are parallel vibrations, while the asymmetric stretching mode (ν_3) is a perpendicular vibration because it is excited by the y (perpendicular to the symmetry axis z) component of the electric field. Notice the simplification provided by symmetry. It is now easy to assign a vibration as parallel or perpendicular in order to interpret the vibration–rotation spectra, where you may recall that the selection rules for the rotational quantum number J differ for the two kinds of modes for linear and symmetric top molecules.

The other piece of information that we obtain is regarding the plane of vibration. In this case, the plane of the molecule is yz. All vibrational modes that have +1 character for this plane are in-plane vibrations; otherwise they are out-of-plane vibrations. Both A_1 and B_2 have +1 character for this $\sigma(yz)$ plane, and hence, all the normal modes of vibration of water are in-plane.

Ammonia (NH_3)

Ammonia belongs to the C_{3v} point group. The symmetry elements are shown below:

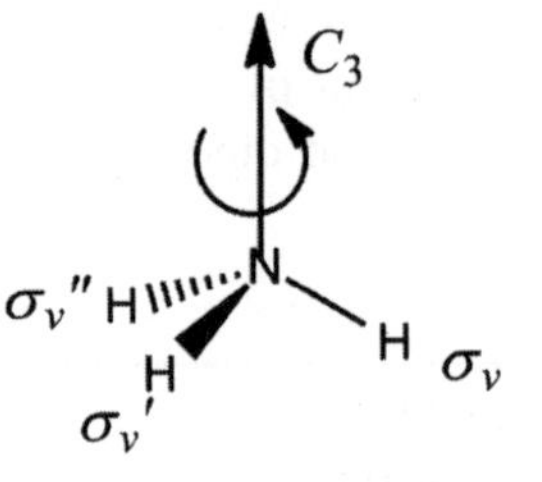

The character table, along with the vibrational analysis, is given below:

C_{3v}	E	$2C_3$	$3\sigma_v$		
A_1	1	1	1	z	$x^2 + y^2, z^2$
A_2	1	1	−1	R_z	
E	2	−1	0	$(x, y), (R_x, R_y)$	$(x^2 - y^2, xy)(xz, yz)$
$\chi(R)$	3	0	1		
n_R	4	1	2		
Γ_{3N}	12	0	2		
$\Gamma_{trans} + \Gamma_{rot}$	6	0	0		
Γ_{vib}	6	0	2		
$\Gamma_{stretch}$	3	0	1		

Since the C_3 axis passes through the nitrogen atom only, only nitrogen is unshifted during this operation. The σ_v axes pass through nitrogen and one hydrogen, and so the number of unshifted atoms is 2 for this operation. The number of vibrational degrees of freedom of ammonia is 3 × 4 – 6 = 6, and hence, Γ_{vib} in the table has the correct dimension. Reduction of Γ_{vib} gives:

$$n(A_1) = \frac{1}{6}(6\times1+2\times0\times1+3\times2\times1)) = 2$$

$$n(A_2) = \frac{1}{6}(6\times1+2\times0\times1+3\times2\times(-1)) = 0$$

$$n(E) = \frac{1}{6}(6\times2+2\times0\times(-1)+3\times2\times0) = 2$$

Therefore,

$$\Gamma_{vib} = 2A_1 \oplus 2E$$

of which $\Gamma_{stretch} = A_1 \oplus E$. Note that the E representations are doubly degenerate, and hence, the total dimension is six, as required. Since z transforms as A_1 and (x, y) as E, all vibrations are infrared active. The infrared spectrum of ammonia is given in Table 6.4.

Table 6.4. The infrared spectrum of ammonia

ω/ cm^{-1}	Symmetry	Band assignment
3337	A_1	ν_1, N-H symmetric stretch
950	A_1	ν_2, N-H symmetric deformation
3444	E	ν_3, N-H asymmetric stretch
1627	E	ν_4, H-N-H bend

The most symmetric A_1 bands are numbered ν_1 and ν_2, the higher wavenumber one numbered first. Similarly, the two E bands are numbered ν_3 and ν_4. As stated previously, vibrations with wavenumbers > 2500 cm^{-1} involve stretching motions of hydrogens and are hence ascribed to N-H stretching bands, the symmetric stretch to A_1 and the asymmetric stretch to E. The other two lower wavenumber bands are bending modes (Figure 6.6).

Regarding the band assignments, we make the observation that none of these are pure vibrations. Modes of the same symmetry have some amount of mixing. There is some amount of mixing of both the A_1 modes, for example, and the bond stretching mode is not pure stretching – it has some degree of symmetrical deformation too. However, the mixing is not expected to be large because of the large difference in the wavenumbers, and ν_1 is primarily bond stretching and similarly ν_2 is primarily a symmetric bending mode though they are of the same symmetry. Likewise, the same holds true for ν_3 and ν_4, too.

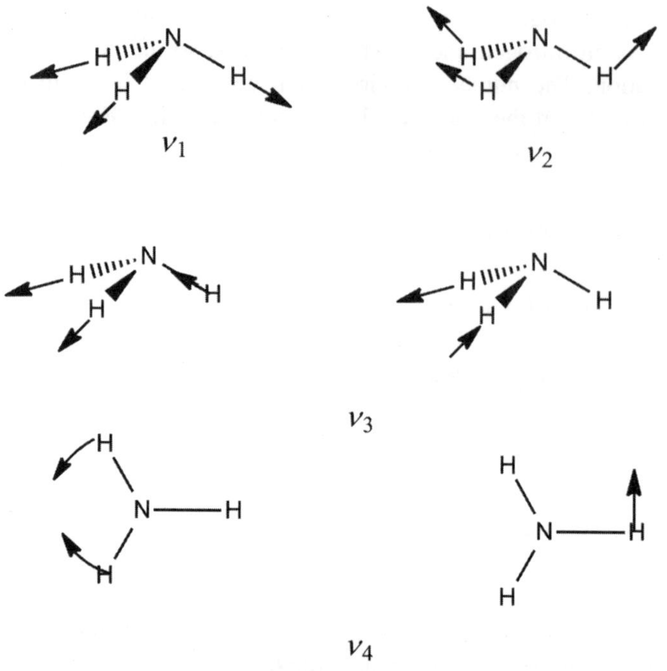

Figure 6.6. Modes of vibration of ammonia

6.2.6 Higher vibrational wave functions

We have only considered the fundamental vibrations up to now. As discussed in the previous chapter, in case the temperature is high or the vibrational frequency low, higher vibrational levels are also occupied. Moreover, anharmonicity permits transitions to these higher levels. We now have a look at the symmetries of these levels. We had found that the $\upsilon = 0$ level is totally symmetric and the $\upsilon = 1$ level has the same symmetry as the normal mode. The wave function of the next higher vibrational level is given by ψ_2, which may be written as

$$\psi_2 = N_2 H_2(\sqrt{\alpha}Q)e^{-\alpha Q^2}$$

Since $H_2(y) = 4y^2 - 2$, this wave function transforms as the totally symmetric representation. The next wave function,

$$\psi_3 = N_3 H_3(\sqrt{\alpha}Q)e^{-\alpha Q^2}$$

again transforms as the symmetry representation of the normal coordinate, since $H_3(y)$ is an odd function of y, i.e., $(8y^3 - 12y)$. We had already observed in the previous chapter that the Hermite polynomials of odd order are odd functions and those of even order are even functions. Odd functions transform as the normal coordinate and even functions as the totally symmetric representation. Thus, alternate vibrational wave functions are totally symmetric. This situation is represented in Figure 6.7 for water.

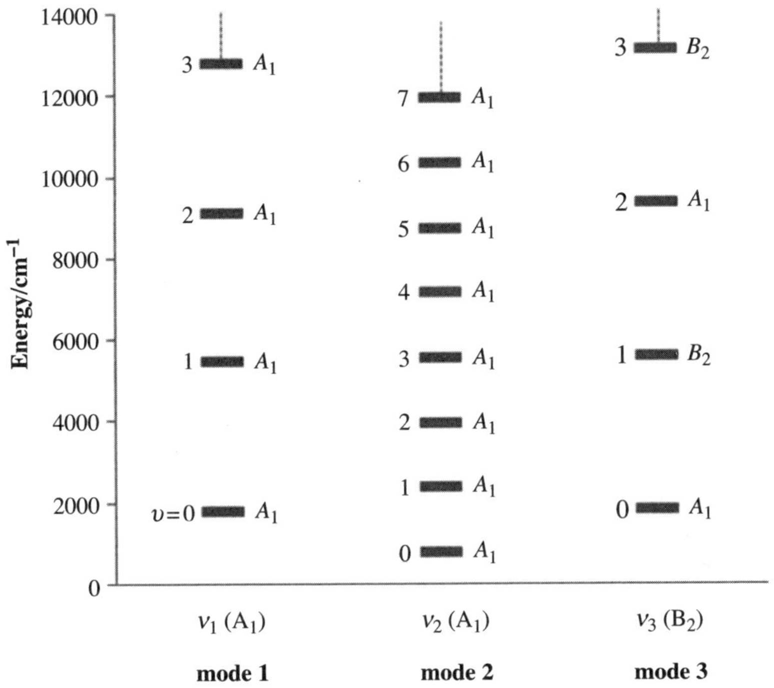

Figure 6.7. Vibrational energy levels for water

6.2.7 Combination bands

Wc had seen in the previous chapter that the $\Delta \upsilon = \pm 1$ selection rule of the harmonic oscillator breaks down for real molecules and anharmonicity allows larger quantum jumps of $\Delta \upsilon = \pm 2$, ± 3, …, etc., called overtones. In the case of polyatomic molecules, it also allows two normal modes to be simultaneously excited, resulting in *combination bands*. Before we discuss the selection rules for combination bands, let us describe a uniform notation for excited states. As stated previously, vibrations are numbered ν_1, ν_2, … according to their symmetry, starting with the most symmetrical ones (the ones that have irreducible representations that appear first in the character table). Among vibrations with the same symmetry, the higher frequency vibration is numbered first. Thus, for water, the symmetric stretch is denoted ν_1, the bending mode ν_2 and the asymmetric stretch ν_3. The vibrational quantum numbers are also numbered in the same order. For example, (0,0,0) refers to the ground vibrational state with all three quantum numbers equal to zero.

Since the normal modes are independent, the total wave function is the product of the normal modes, i.e.,

$$\psi_{\text{vib}} = \prod_{i=1}^{3N-5/6} \psi_{\upsilon_i}(Q_i)$$

Hence, the ground state is totally symmetrical, since all the wave functions transform as A_1 for $\upsilon = 0$.

The energy is the sum of the energies in the various normal modes, i.e.,

$$E_{\text{vib}} = \sum_{i=1}^{3N-5/6} h\nu_i\left(\upsilon_i + \frac{1}{2}\right)$$

Hence, the zero-point energy is the sum of the zero-point energies in each vibrational mode, i.e.,

$$E_{zp} = \frac{1}{2}\sum_{i=1}^{3N-5/6} h\nu_i$$

For water, this is equal to $E_{zp} = \frac{1}{2}(h\nu_1 + h\nu_2 + h\nu_3)$.

The three fundamentals are referred to as (1,0,0), (0,1,0) and (0,0,1). Their symmetry representations are the same as the normal mode that is excited, i.e., A_1, A_1 and B_2, respectively. The first overtones, (2,0,0), (0,2,0) and (0,0,2), likewise all transform as A_1 because the excited vibrational levels have even quantum numbers. The frequencies of the transitions are $\sim 2\nu_1$, $\sim 2\nu_2$ and $\sim 2\nu_3$, respectively (slightly lower, because of anharmonicity).

Transitions of the type (0,1,0) → (0,2,0) are hot bands. The energies of the initial and final states are $\frac{1}{2}h\nu_1 + \frac{3}{2}h\nu_2 + \frac{1}{2}h\nu_3$ and $\frac{1}{2}h\nu_1 + \frac{5}{2}h\nu_2 + \frac{1}{2}h\nu_3$, respectively, and the frequency of transition is $\sim\nu_2$. Both the initial and final states are of A_1 symmetry, and the transition is allowed and z-polarized, despite breaking the harmonic oscillator rule.

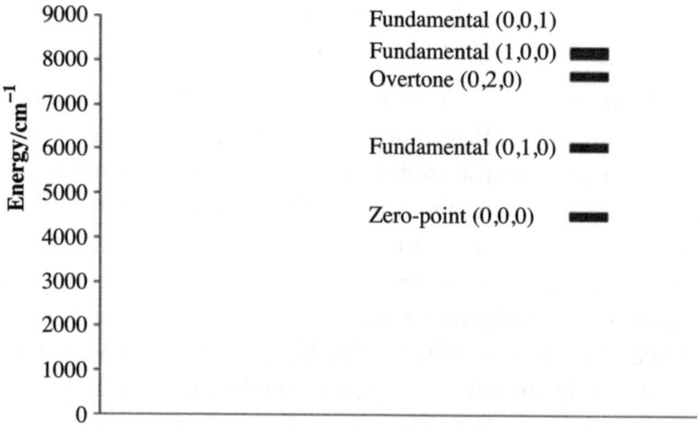

Figure 6.8. Combination bands of water

Combination bands of the type (1,1,0), (1,0,1) and (0,1,1) are also possible with energies $h(\nu_1 + \nu_2)$, $h(\nu_1 + \nu_3)$ and $h(\nu_2 + \nu_3)$ respectively. Consider the (0,0,0) → (1,0,1) transition of water, which involves the simultaneous excitation of the symmetric and asymmetric stretches. The ground state is A_1. The excited (1,0,1) state transforms as $A_1 \otimes B_2 = B_2$ as the two excited states belong to these symmetries. We have already seen that μ_y transforms as B_2 and hence the transition is allowed and y polarized.

Another combination band is (0,0,0) $\rightarrow$ (2,1,0) with energy $h(2\nu_1 + \nu_2)$, symmetry A_1 and z-polarized. Combination bands are also of the type (0,1,0) $\rightarrow$ (1,0,0) with energy $h(\nu_1 - \nu_2)$.

Before we end the discussion on the infrared spectrum of water, we give the observed wavenumbers of the three fundamental modes, 3657, 1595 and 3756 cm^{-1}, respectively. As expected, the bending mode has the least energy and the asymmetric stretch mode the highest, although, compared to carbon dioxide, the difference in the wavenumbers of the symmetric and asymmetric stretches is much smaller. Group Theory tells us that states of the same symmetry mix, the amount of mixing depending inversely on the difference in their energies. In this case, the symmetric stretch and bending modes have the same symmetry designations and can interact. However, the energy difference is quite large, and the 1595 cm^{-1} mode is mostly H-O-H bend with very little mixing from the symmetric stretch, and the 3657 cm^{-1} mode is primarily the symmetric stretch. For D_2O, the corresponding wavenumbers are 2669, 1178 and 2788 cm^{-1}, because of the higher isotopic mass of deuterium. The asymmetric stretch mode (B_2) is a pure mode.

Reproduced below (Figure 6.9) is the background IR spectrum of air, which mostly consists of N_2 and O_2, which are IR inactive. The main IR active molecules present in air are CO_2 and H_2O, and these, together with other molecules like ozone and sulphur dioxide, constitute the greenhouse gases. Both the stretching modes of water are infrared active, and they appear as close bands at 3657 and 3756 cm^{-1} in Figure 6.9.

We shall take up the case of other molecules, including linear ones, after we have discussed Raman spectroscopy in the next chapter.

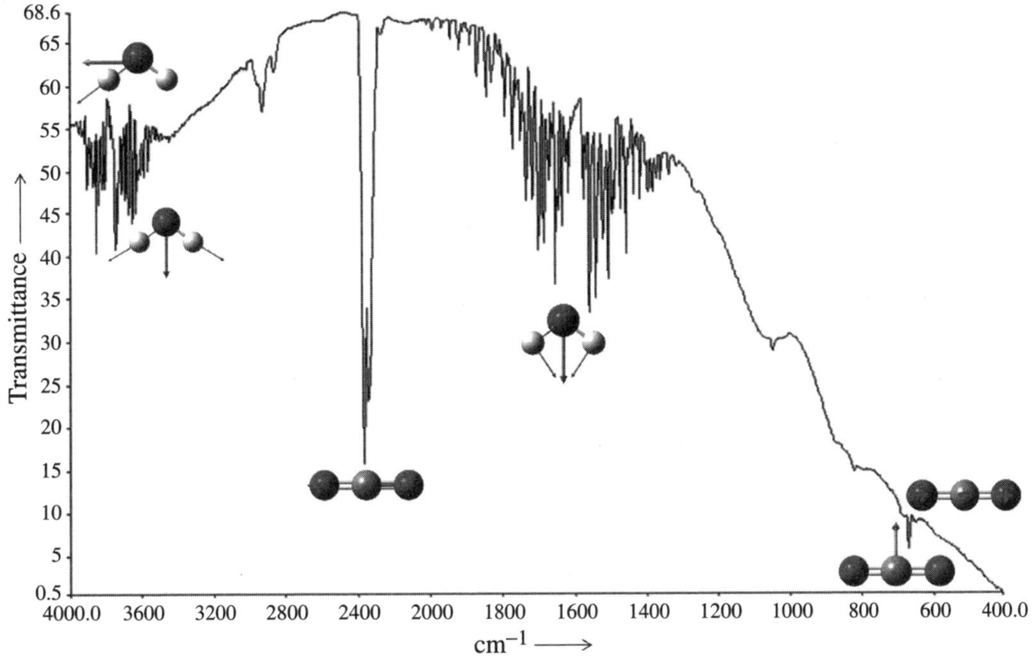

Figure 6.9. Background infrared spectrum of air

6.3 SUMMARY

- Normal mode analysis can help identify the vibration modes for simple molecules.
- For larger molecules, Group Theory can aid prediction of the number and kind of vibrational modes and their infrared activity.
- Each normal mode can be considered independent and treated as an individual diatomic molecule.
- Vibrations are classified as parallel or perpendicular depending on whether the dipole moment change is parallel or perpendicular to the principal axis.
- The selection rules for vibration–rotation spectra differ for parallel and perpendicular modes.

6.4 EXERCISES

1. (a) Describe what is meant by a fundamental (normal) mode.
   (b) If m is the mass of a B atom and M that of an A atom, show that the eigenvalues of the force constant matrix of a linear molecule BAB are

   $$\lambda_1 = 0, \lambda_2 = k/m, \lambda_3 = k/m + 2k/M$$

   where $\lambda = 4\pi^2\tilde{v}^2c^2$.
   How many translational and how many genuine vibrations are found?
   (c) Assign the observed wavenumbers for CO_2 (667, 1334 and 2350 cm^{-1}) to the normal modes in (b). Calculate values for the bond stretching force constant, k, from the above data. Why are the values for k obtained from the two equations different?
   (d) Classify the normal modes according to the symmetry species of the point group to which the molecule belongs.
   (e) Write down the expressions for the kinetic and potential energies in terms of the normal coordinates.
2. (a) Consider that the atoms of HCN are constrained to move only along the bond axis. Find the normal modes of vibration of HCN.
   (b) Repeat for NNO and OCS.
3. (a) Explain the difference between a combination band and an overtone.
   (b) Why are hot bands temperature dependent?
4. (a) How many normal modes of vibration does CO_2 have? Sketch their atomic displacements.
   (b) CO_2 concentrations in the atmosphere are of importance because CO_2 is a greenhouse gas – it absorbs strongly in the infrared region of the electromagnetic spectrum and is responsible for the 'trapping' of outgoing radiation from the earth. Which of the vibrations you have drawn are infrared active?
   (c) The fundamental wavenumbers of the vibrations of CO_2 are well known: the infrared active ones are $\omega_2 = 667\ cm^{-1}$ and $\omega_3 = 2349\ cm^{-1}$. What wavelengths of radiation will be absorbed by these bands? Sketch the rotational structure of the absorption spectrum you would expect to see from each, and explain the differences.

5. Analysis of the rotational fine structure from the IR spectra of H-^{12}C≡^{12}C-H and D-^{12}C≡^{12}C-D gives the rotational constants as 1.181 cm^{-1} and 0.851 cm^{-1}, respectively. Determine the C≡C and C-H bond lengths, stating any assumptions you make.
6. (a) Show that the normal modes of the planar molecule BF_3 (D_{3h}) transform as $A_1' + A_2'' + 2E'$.
   (b) For each normal mode, determine whether the $\upsilon = 0 \rightarrow 1$ transition is allowed and classify the allowed transitions as parallel or perpendicular.
   What kind of rotational fine structure would you expect to see for each of these allowed transitions?
   (c) In the IR spectrum of BF_3, a band is observed which appears to show PQR branches; the spacing of the lines in the P and R branches is approximately 0.699 cm^{-1}. Use these data to estimate the B-F bond length in BF_3 (see Chapter 4).
   (d) What would you expect the rotational fine structure of a perpendicular band from BF_3 to look like? Be as quantitative as you can.
7. To a fair approximation, ethene can be considered to be a prolate symmetric top with the axes as shown below:

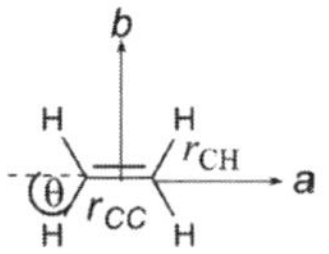

   (a) Write down expressions for the three moments of inertia in terms of the distances shown and the masses m_H and m_C.
   Hence, show that $I_c = I_a + I_b$ (the test for a planar molecule).
   (b) The spacing of the lines in the P- and R-branches of a parallel band are found to be 1.68 cm^{-1}, and the spacing between the Q branches in a perpendicular band are found to be 8.06 cm^{-1}. Assuming that the spacing in the P and R branches gives an approximate value for $\tilde{B}$, and using the known value of r_{CH} of 0.1071 nm, find the angle θ.
8. (a) HF_2^- and DF_2^- are both linear molecules. Both have a fundamental frequency at 675 cm^{-1}. Sketch the mode in each case and give reasons for your answer.
   (b) Consider a non-degenerate vibration. Assign the quantum number "υ" to the upper level for the overtone which
   (i) is totally symmetric,
   (ii) has the same symmetry as the normal coordinate 'Q' of the vibration.
9. Which of the following molecules may show absorption in the infrared: H_2, HCl, CO_2, H_2O, CH_3CH_3, CH_4, CH_3Cl, N_2, N_3^-?
10. The infrared spectrum of N_2O shows fundamental bands for three normal modes of vibration. Assuming that the structure is linear, how does the spectrum distinguish between N-N-O and N-O-N? Sketch the normal modes.
11. Make a set of sketches representing all the possible modes of vibration of the carbon dioxide molecule (which is linear) and indicate which of these modes would be active in the infrared.
12. The frequency of the O-H stretching in CH_3OH is at $\omega = 3300$ cm^{-1}. Predict ω for the O-D stretch in CH_3OD.

13. How many normal modes of vibration are there for the following molecules? (a) N_2 (b) H_2O (c) C_2H_4 (d) C_2H_2
14. The infrared spectrum of the carbon monoxide–haemoglobin complex gives a peak at about 1950 cm^{-1}, which is due to the carbonyl stretching frequency.
    (a) Compare this value with the fundamental frequency of free CO, which is 2143.3 cm^{-1}. Comment on the difference.
    (b) Convert this frequency to kJ mol^{-1}.
    (c) What conclusion can you draw from the fact that there is only one band present?

7

The Raman Effect

Comparing to Compton Effect,
"Ah, but my effect will play a great role for chemistry and molecular structure!"
—Sir C. V. Raman

7.1 INTRODUCTION

We now move on to the next topic, i.e., Raman spectroscopy, discovered in 1928 by the great Indian physicist, Sir C. V. Raman, who was awarded the Nobel Prize in 1930 'for his work on the scattering of light and for the discovery of the effect named after him'. All the spectroscopies studied so far were based on the absorption of light. In contrast, Raman spectroscopy is based on *scattering* of light. Hence, the Bohr condition is not required for observing the Raman effect.

When a molecule is illuminated by visible or ultraviolet radiation, it gets excited to a *virtual* state from which it may drop to the same, lower or higher energy level. If it drops to the same energy level, the scattered radiation has the same frequency as the incident radiation (elastic scattering), and this is called *Rayleigh scattering*, responsible for the blue colour of the sky and the red colour of sunset. Lord Rayleigh showed that the intensity of the scattered radiation is inversely proportional to the fourth power of the wavelength, i.e., $I_{\text{Rayleigh}} \propto 1/\lambda^4$, so blue light (shorter wavelength) is scattered more than red light (longer wavelength). When we view sunset through several layers of the atmosphere, what we observe is red light, because the blue part of the visible (white) light is scattered away sideways.

Alternatively, the molecule may drop to a higher energy with absorption of energy from the incident radiation, leaving the scattered light at lower energy (inelastic scattering) and it is then observed at lower wavenumber. Such scattering is called *Stokes scattering*. In case the molecule is already in an excited internal energy level, it may return to the ground state, and in that case the scattered radiation will be at higher wavenumber than the incident radiation. This kind of scattering is known as *anti-Stokes scattering*, and collectively Stokes and anti-Stokes scattering account for *Raman scattering* (Figure 7.1). Anti-Stokes lines are much less intense than Stokes lines, and very sophisticated detectors and intense monochromatic radiation in the form of lasers is required to detect them. If, for example, the intensity of the incident light is 1 in some arbitrary unit, then the intensity of the Rayleigh scattered light is of the order of 10^{-4}, that of the Raman scattered Stokes light is of the order of 10^{-7}, and the anti-Stokes lines are even weaker in intensity.

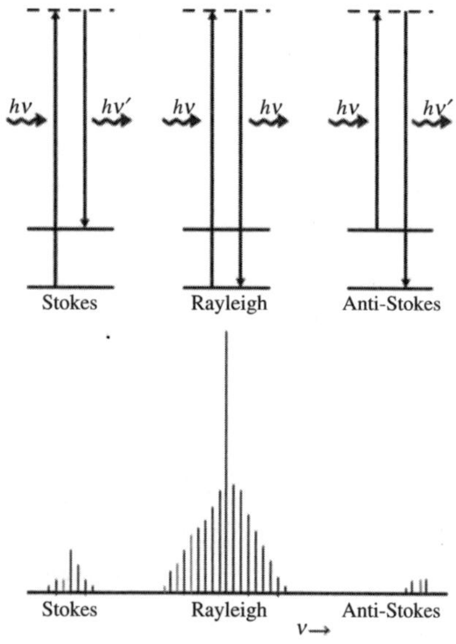

Figure 7.1. Transitions involved in Stokes, Rayleigh and anti-Stokes lines in Raman spectroscopy

7.2 CLASSICAL TREATMENT OF THE RAMAN EFFECT

As in the previous chapters, we first discuss the classical treatment of the Raman effect and then the modifications brought about by quantum mechanics. Classically, the Raman scattering is interpreted in terms of the polarizability. Hence, we first discuss polarizability.

7.2.1 Polarizability

When an electric field is applied to a molecule, the electrons move towards the positive electrode, causing a temporary dipole moment to be induced in the molecule. Evidently, the induced dipole moment is directly proportional to the strength of the electric field, i.e.,

$$\vec{\mu}_{\text{ind}} = \alpha \vec{E} \tag{7.1}$$

where α is the proportionality constant, called the polarizability of the molecule. Larger the value of α, larger is the induced dipole moment for the same strength of the electric field. The polarizability is thus the ease with which the electron cloud of a molecule can be distorted by an external electric field. Molecules with large atoms, covalent and multiple bonds are more likely to get their charge clouds distorted, and hence have higher values of the polarizability.

Since the electric field oscillates with the frequency ν_0 of the incident radiation, we may write

$$\vec{E} = \vec{E}_0 \cos(2\pi\nu_0 t)$$

$$\Rightarrow \vec{\mu}_{\text{ind}} = \alpha \vec{E}_0 \cos(2\pi\nu_0 t) \tag{7.2}$$

and thus the induced dipole moment oscillates with the same frequency as the radiation.

7.2.2 Molecular rotations

Figure 7.2 shows the variation of the component of the oscillating dipole moment for a rotating diatomic molecule.

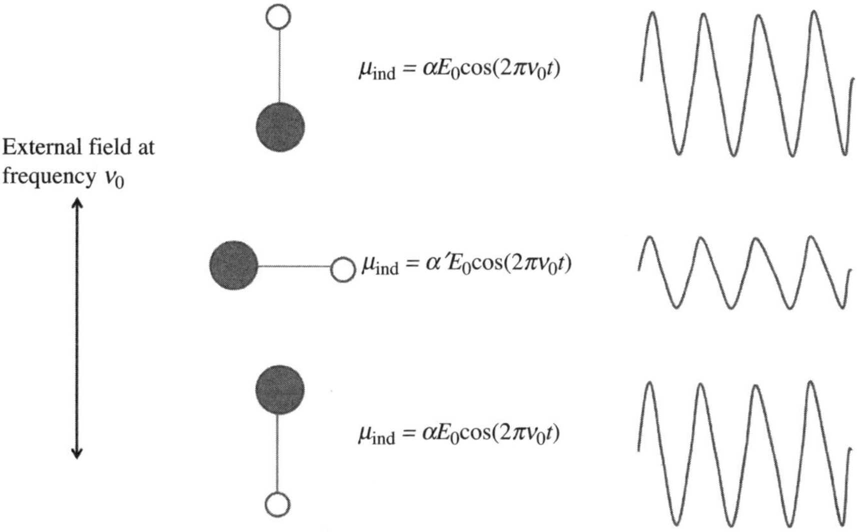

Figure 7.2. Polarizability changes during rotation of a diatomic molecule

It is seen that the polarizability returns to its original value after a rotation of the molecule by 180°, so that it oscillates at twice the frequency of rotation. For a rotating molecule, the polarizability is thus a time-varying term that depends on the frequency of rotation of the molecule, ν_{rot}.

$$\alpha = \alpha_0 + \alpha_{rot}\cos(2\pi(2\nu_{rot})t) \tag{7.3}$$

The factor of two is introduced because the polarizability returns to its initial value *twice* during each rotation by 360°. Insertion in equation (7.2) and using the trigonometric identity

$$\cos(A)\cos(B) = \frac{1}{2}(\cos(A-B) + \cos(A+B))$$

gives for the induced dipole moment

$$\begin{aligned}\vec{\mu}_{ind} &= (\alpha_0 + \alpha_{rot}\cos(2\pi 2\nu_{rot} t))\vec{E}_0\cos(2\pi\nu_0 t)\\ &= \alpha_0\vec{E}_0\cos(2\pi\nu_0 t) + \frac{\alpha_{rot}\vec{E}_0}{2}(\cos(2\pi(\nu_0 - 2\nu_{rot})) + \cos(2\pi(\nu_0 + 2\nu_{rot})))\end{aligned} \tag{7.4}$$

Incident photons with frequency ν_0 will give rise to scattered photons at the same frequency (ν_0) (Rayleigh scattering), lower frequency ($\nu_0 - 2\nu_{rot}$) (Stokes scattering) and ($\nu_0 + 2\nu_{rot}$) (anti-Stokes scattering) as a result of rotational scattering. Therefore, the occurrence of Rayleigh and Raman scattering is correctly accounted for by the classical theory.

7.2.3 Molecular vibrations

In a similar manner, in the case of molecular vibrations, the polarizability is a time-varying term that depends on the vibrational frequency of the molecule, ν_{vib} (Figure 7.3). The polarizability may be expanded as a Maclaurin series in the displacement, x.

$$\alpha = \alpha_{x=0} + \left(\frac{d\alpha}{dx}\right)_{x=0} x + \frac{1}{2!}\left(\frac{d^2\alpha}{dx^2}\right)_{x=0} x^2 + \ldots \tag{7.5}$$

Only the first two terms are significant, since x is very small (see Chapter 5). The first term $\alpha_{x=0}$ is simply the equilibrium polarizability, α_0, which is a constant.

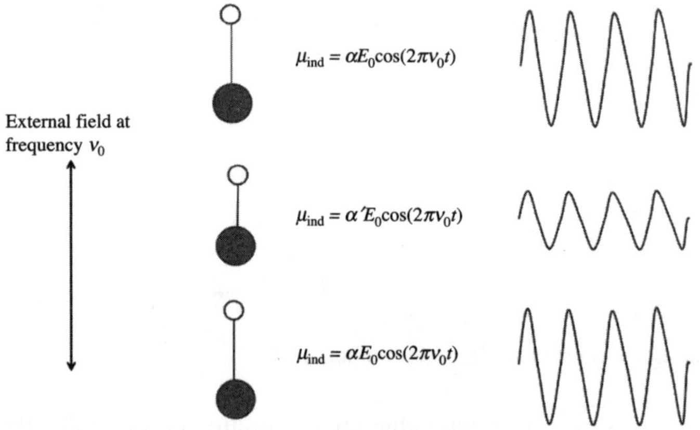

Figure 7.3. Polarizability changes during vibration of a diatomic molecule

Since the displacement coordinate oscillates with the vibration frequency, $x = x_0 \cos 2\pi\nu_{vib} t$, and $\vec{E} = \vec{E}_0 \cos 2\pi\nu_0 t$, the induced dipole moment is given by

$$\vec{\mu}_{ind} = \left(\alpha_0 + \left(\frac{d\alpha}{dx}\right)_{x=0} x_0 \cos(2\pi\nu_{vib} t)\right)\vec{E}_0 \cos(2\pi\nu_0 t)$$

$$= \alpha_0 \vec{E}_0 \cos(2\pi\nu_0 t) + \frac{1}{2}\left(\frac{d\alpha}{dx}\right)_{x=0} x_0 \vec{E}_0 \left(\cos(2\pi(\nu_0 - \nu_{vib})) + \cos(2\pi(\nu_0 + \nu_{vib}))\right) \tag{7.6}$$

The first term corresponds to Rayleigh scattering, as the frequency is unchanged, the second to Stokes scattering and the third to anti-Stokes scattering. The observation of three types of scattering at the three frequencies

ν_0 Rayleigh scattering
$\nu_0 - \nu_{vib}$ Stokes scattering
$\nu_0 + \nu_{vib}$ anti-Stokes scattering

is thus easily explained by classical theory. Also, the first term corresponding to Rayleigh scattering depends on the equilibrium polarizability (α_0) which is always positive and never zero for any molecule, and Rayleigh scattering is therefore always observed. The derivative of the polarizability with respect to the displacement can, however, take negative, positive, even zero values, depending on whether the polarizability decreases, increases or remains unchanged with increasing displacement. The condition, therefore, for a vibration to be Raman active is that $(d\alpha / dx)_{x=0}$ be non-zero, or, in other words, the polarizability must change during a vibration for it to be Raman active. Moreover, the intensities of the Stokes and anti-Stokes Raman bands depend on this quantity, which is smaller than the polarizability itself. This explains the lower intensities of the Raman bands vis-à-vis the Rayleigh bands.

7.3 QUANTUM THEORY

Though the classical theory satisfactorily explains the occurrence of Rayleigh and Raman scattering, and the latter is a thousand times less intense than the former, it is unable to explain the unequal intensities of Stokes and anti-Stokes lines. Rather it predicts a Stokes/anti-Stokes intensity ratio of $[(\nu_0 - \nu_{\text{vib}}) / (\nu_0 + \nu_{\text{vib}})]^4$, according to which the Stokes lines should be less intense. The observation is just the opposite (Figure 7.1). According to quantum theory, Stokes lines occur due to excitation of the molecule from its ground vibrational level to a virtual level by the exciting radiation, after which the molecule returns to a higher vibrational level with net absorption of energy. The opposite is the case for anti-Stokes lines, which occur due to the drop of an initially vibrationally excited molecule to the ground vibrational level with emission of energy to the exciting radiation. Since the relative probabilities of the two processes depend on the populations of the two levels, which depend on the Boltzmann distribution, the Boltzmann factor has to be introduced, leading to the equation

$$\frac{I_{\text{S}}}{I_{\text{AS}}} = \left(\frac{\nu_0 - \nu_{\text{vib}}}{\nu_0 + \nu_{\text{vib}}} \right)^4 e^{h\nu_{\text{vib}}/k_B T} \tag{7.7}$$

In principle, the observed ratio of intensities of Stokes (I_{S}) and anti-Stokes (I_{AS}) lines can help determine the temperature of the sample, since temperature is the only unknown in equation (7.7). However, the wavelength dependence of the intensities and the heating of the sample by the high energy laser radiation used for observation of Raman lines prevent quantitative determination of temperatures.

If we invoke quantum theory, we must consider the selection rules for the transitions. The foregoing discussion allows us to reach the conclusion that there must be a change in the polarizability for a rotation or vibration to be Raman active. Before we apply this selection rule, we introduce the polarizability ellipsoid.

7.3.1 Polarizability ellipsoid

In equation (7.1), we had expressed the polarizability of a diatomic molecule as a proportionality constant between the induced dipole moment and the electric field. However, it is no ordinary proportionality constant. Both the induced dipole moment and electric field are vector quantities, and hence $\vec{\vec{\alpha}}$ is a tensor (represented by a double arrow). A tensor is defined as an operator that changes one vector to another. We rewrite equation (7.1) correctly as

$$\vec{\mu}_{\text{ind}} = \vec{\vec{\alpha}} \vec{E} \tag{7.8}$$

Tensor quantities have nine components, and equation (7.8) can be rewritten as

$$\begin{pmatrix} \mu_x \\ \mu_y \\ \mu_z \end{pmatrix} = \begin{pmatrix} \alpha_{xx} & \alpha_{xy} & \alpha_{xz} \\ \alpha_{yx} & \alpha_{yy} & \alpha_{yz} \\ \alpha_{zx} & \alpha_{zy} & \alpha_{zz} \end{pmatrix} \begin{pmatrix} E_x \\ E_y \\ E_z \end{pmatrix} \tag{7.9}$$

The polarizability matrix, like the moment of inertia matrix encountered in microwave spectroscopy and the force constant matrix in vibrational spectroscopy, is symmetric, i.e., for example, $\alpha_{xy} = \alpha_{yx}$. There are thus six independent components. By analogy with the moment of inertia tensor, we define a *polarizability ellipsoid*, which is a three-dimensional surface, whose distance from the electrical centre of the molecule is $1/\sqrt{\alpha_i}$, where α_i is the magnitude of the polarizability along the line joining a point on the surface of the ellipsoid with the electrical centre of the molecule. As for the moment of inertia, diagonalization of the polarizability matrix gives three eigenvalues, α_a, α_b and α_c and the corresponding axes *a*, *b* and *c* of the ellipsoid (molecule based axes), so that we can define the polarizability ellipsoid as a three-dimensional 'quadratic' surface, the points on which satisfy the equation

$$\alpha_a a^2 + \alpha_b b^2 + \alpha_c c^2 = 1 \tag{7.10}$$

since, on diagonalization, we obtain

$$\begin{pmatrix} \mu_a \\ \mu_b \\ \mu_c \end{pmatrix} = \begin{pmatrix} \alpha_a & 0 & 0 \\ 0 & \alpha_b & 0 \\ 0 & 0 & \alpha_c \end{pmatrix} \begin{pmatrix} E_a \\ E_b \\ E_c \end{pmatrix}$$

In general, the polarizability ellipsoid is *anisotropic*. It depends on the orientation of the molecule with respect to the electric field.

We examine the shapes of the polarizability ellipsoids of some simple molecules:

Hydrogen, H_2

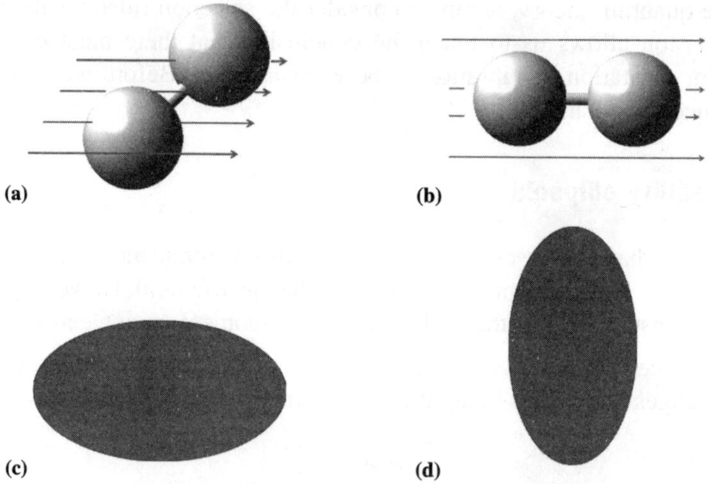

Hydrogen is a linear molecule, and is more polarizable along the bond axis (b) than perpendicular to it (a). The ellipsoid therefore has its minor axis (is flattened) along the bond axis (remember we are plotting the inverse of the square root of the polarizability, so the ellipsoid will have its minor axis along the direction where the polarizability is largest) (d). The other two directions are equivalent (c).

Carbon tetrachloride

This is a spherical top molecule. All directions are equivalent and $\alpha_a = \alpha_b = \alpha_c$. Its polarizability ellipsoid is therefore a sphere.

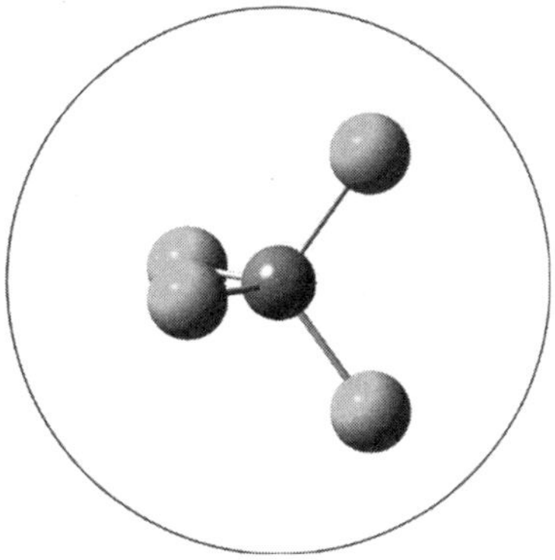

Chloroform

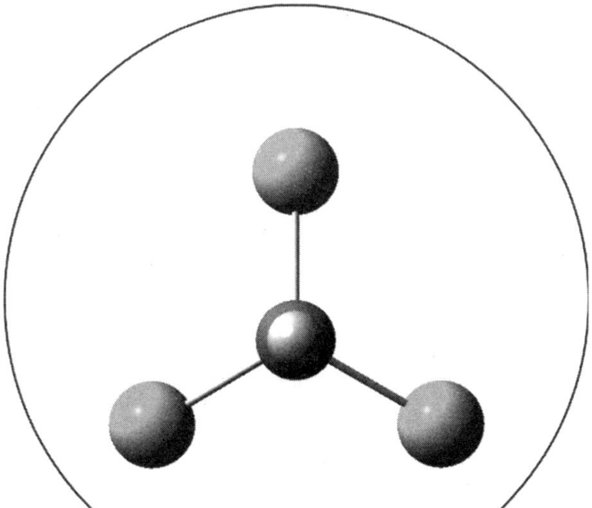

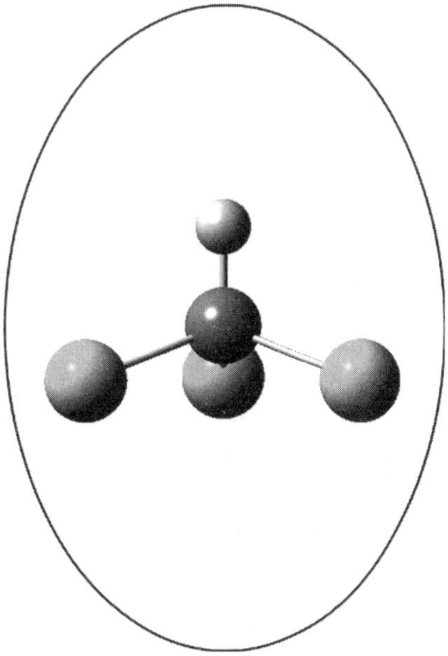

This is a symmetric top molecule. The molecule is less polarizable along the C-H bond (greater across the symmetry axis because of the large chlorines), so the unique axis is *a*, the polarizability ellipsoid has its major axis along this direction. The other two directions are equivalent (right).

Water

Finally, we take up the case of an asymmetric top molecule, water, for which $\alpha_a \neq \alpha_b \neq \alpha_c$, as shown in Figure 7.4, where we have shown three views of the molecule.

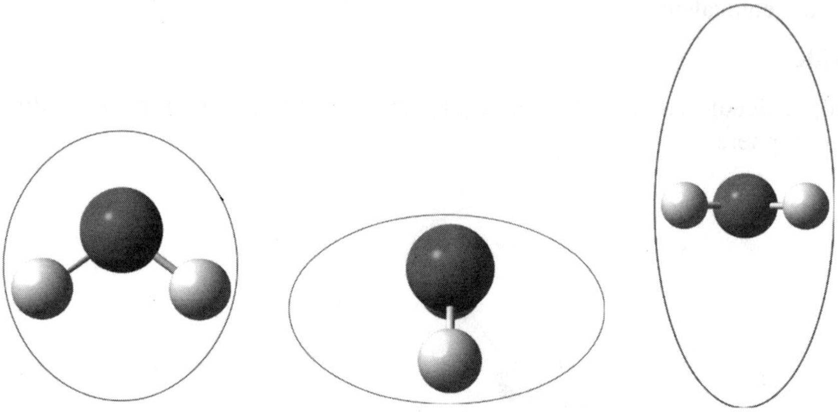

Figure 7.4. Three views of the water molecule and its polarizability ellipsoid

7.3.2 Selection rules

We may now state that the gross selection rule for Raman activity is that the polarizability ellipsoid should either change in shape or in size during the rotation or vibration. Notice that the polarizability has the dimensions of volume.

7.3.2.1 Rotational Raman

Linear molecules

The polarizability ellipsoid shown for hydrogen is obviously anisotropic. Therefore, when the molecule undergoes end-over-end rotation, the polarizability ellipsoid presents a different view as the angle of rotation is varied. Hence hydrogen (and also chlorine, carbon dioxide and heteronuclear diatomics like CO and HCl) is rotational Raman active. For the first time in our study of spectroscopy, we have encountered a method to study the rotations and determine the bond lengths of homonuclear diatomic molecules, which are neither microwave nor infrared active. We have the first evidence of the usefulness of the Raman technique.

The J selection rule for a linear molecule is $\Delta J = 0, \pm 2$. The $\Delta J = 0$ selection rule refers to Rayleigh scattering. This selection rule is to be contrasted with the $\Delta J = \pm 1$ selection rule for microwave spectroscopy. Examination of the polarizability ellipsoid for hydrogen shows that it presents the same appearance twice during a complete rotation. On this basis, the classical treatment leads to three frequencies of the emitted photon, i.e., ν_0, $\nu_0 \pm 2\nu_{rot}$. However, a more appropriate explanation is based on the conservation of the two-photon angular momenta of the incident and emitted photons. The $\Delta J = \pm 1$ selection rule appeared in microwave spectroscopy because the photon has an intrinsic angular momentum of 1, and can thus increase or decrease the rotational angular momentum by one unit depending on whether absorption or emission takes place. In the case of Raman spectroscopy, two photons are involved – the incident and the emitted photon. In case the two have parallel angular momenta, there is no net change in the angular momentum and the J quantum number, and we have Rayleigh scattering. If, on the other hand, the two photons have antiparallel spins, there is a net change of two units of angular momenta, and we have the $\Delta J = \pm 2$ selection rule, and consequent appearance of Stokes and anti-Stokes lines.

As stated in the previous chapter, the $\Delta J = \pm 2$ selection rule leads to O and S branches. The wavenumbers of the S branch are

$$\tilde{\nu}_S = F(J+2) - F(J) = \tilde{B}(J+2)(J+3) - \tilde{B}J(J+1) = \tilde{B}(4J+6) \tag{7.11}$$

where J is the rotational quantum number of the lower state. The Raman frequencies of the spectral lines are therefore

$$\tilde{\nu}_{\text{Raman}} = \tilde{\nu}_0 \pm \tilde{B}(4J+6) \tag{7.12}$$

In deriving equation (7.12), we have ignored centrifugal distortion. As we had seen in Chapter 4, except at very high values of the quantum number, this term is much smaller than the main rotational term. The precision of the Raman instrument is not such as to warrant inclusion of this term.

Equation (7.12) shows that the first line ($J = 0$) is offset from the Rayleigh line ($\tilde{\nu}_0$) by $6\tilde{B}$. Subsequent lines are separated from each other by $4\tilde{B}$. We can thus determine the bond lengths of diatomic molecules from the value of $\tilde{B}$ obtained from their Raman spectra. Polyatomic linear molecules like carbon dioxide and acetylene show similar behaviour, except that their rotational constants are smaller than those for the diatomics because of their larger mass, and hence their rotational lines are more closely spaced. Moreover, since there is only one rotational constant and the number of bond lengths to be determined is larger, isotopic substitution is required (see Chapter 4).

The contours of the O and S branches depend on the relative populations of the rotational energy levels and are similar to those observed in the PR contours in infrared spectroscopy (Figure 7.5).

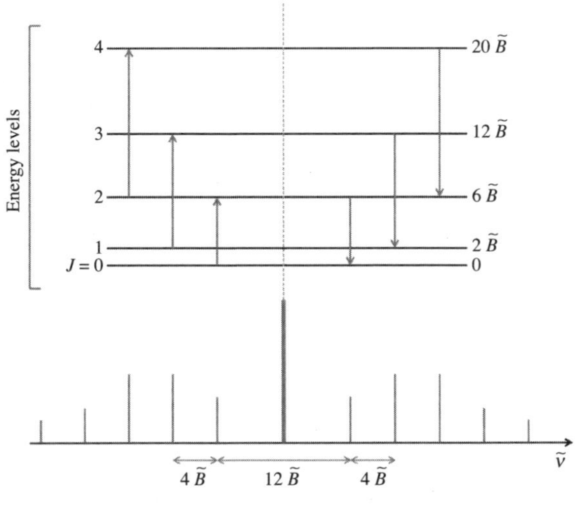

Figure 7.5. Rotational Raman spectrum of a linear molecule

Example 7.1 The first three Stokes lines in the rotational Raman spectrum of $^{12}C^{16}O_2$ are found at 2.34 cm^{-1}, 3.90 cm^{-1} and 5.46 cm^{-1}. Calculate the C=O bond length in CO_2.

Solution

According to the above discussion, the first line is displaced from the Rayleigh line by $6\tilde{B}$ and subsequent lines are spaced by $4\tilde{B}$. Hence

$$6\tilde{B} = 2.34 \text{ cm}^{-1}, \text{ or } \tilde{B} = 0.39 \text{ cm}^{-1}$$
$$4\tilde{B} = (3.90 - 2.34) = 1.56 \text{ cm}^{-1}, \tilde{B} = 0.39 \text{ cm}^{-1}$$
$$4\tilde{B} = (5.46 - 3.90) = 1.56 \text{ cm}^{-1}, \tilde{B} = 0.39 \text{ cm}^{-1}$$

Since $I_b = \frac{2.799274 \times 10^{-46}}{\tilde{B}}$ kg m^2 (Chapter 4), we obtain I_b =7.18 × 10^{-46} kg m^2.

For CO_2, the centre of mass is at the carbon atom, and using $I_b = \sum_i m_i r_i^2$, we obtain $I_b = m_O r_{CO}^2 + m_C 0^2 + m_O r_{CO}^2 = 2m_O r_{CO}^2$

or

$$r_{CO} = \sqrt{\frac{I_b}{2m_O}} = \sqrt{\frac{7.18 \times 10^{-46}}{2 \times 15.9949 \times 1.6605 \times 10^{-27}}} = 1.16 \times 10^{-10} \text{ m} = 116 \text{ pm}$$

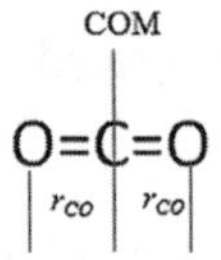

Example 7.2 Calculate the initial state (i.e., J'') corresponding to the most intense line in the rotational Raman spectrum of $^{12}C^{16}O_2$ at 25 °C.

Solution

From equation (4.31),

$$J_{\max} = \sqrt{\frac{k_B T}{2\tilde{B}hc}} - \frac{1}{2} = \sqrt{\frac{1.381 \times 10^{-23} \times 298}{2 \times 0.39 \times 6.626 \times 10^{-34} \times 2.998 \times 10^{10}}} - \frac{1}{2}$$
$$= 15.8 \approx 16, \text{ the nearest whole number.}$$

Symmetric top molecules

Because of the axial symmetry, such molecules show similarity to linear molecules. As seen from the polarizability ellipsoid of chloroform, rotations along the bond axis do not produce change in the polarizability ellipsoid, but end-over-end rotations produce a change in polarizability. As shown in Chapter 4, the rotational terms for a prolate symmetric top molecule are given by

$$F(J,K) = \tilde{B}J(J+1) + (\tilde{A} - \tilde{B})K^2$$
$$(J = 0, 1, 2, \ldots; K = \pm J, \pm(J-1), \ldots) \tag{7.13}$$

The rotational selection rule for a symmetric top is $\Delta K = 0$. Since K is the quantum number for axial rotation, this condition is consistent with the fact that the polarizability ellipsoid does not change its appearance with this rotation, and such a rotation is Raman inactive. The selection rule for J is $\Delta J = 0, \pm1, \pm2$, except for $K = 0$ states, when $\Delta J = \pm2$ only. This means that the $\Delta J = 0, \pm1$ selection rule will not be observed for $J = 0$, for in that case $K = 0$. Thus, there are two cases for absorption:

(i) $\Delta J = +1, \Delta K = 0$ (R branch)

$$\tilde{\nu}_R = F(J+1,K) - F(J,K) = \tilde{B}(J+1)(J+2) + (\tilde{A} - \tilde{B})K^2$$
$$-\left[\tilde{B}J(J+1) + (\tilde{A} - \tilde{B})K^2\right] = 2\tilde{B}(J+1)$$
$$(J = 1, 2, \ldots \text{ but } J \neq 0) \tag{7.14}$$

Lines will be observed at $\nu_0 \pm 2\tilde{B}(J''+1)$, where J'' refers to the *lower* value (Figure 7.6, top).

(ii) $\Delta J = +2, \Delta K = 0$ (S branch)

$$\tilde{\nu}_S = F(J+2,K) - F(J,K) = \tilde{B}(J+2)(J+3) + (\tilde{A} - \tilde{B})K^2$$
$$-\left[\tilde{B}J(J+1) + (\tilde{A} - \tilde{B})K^2\right] = \tilde{B}(4J+6)$$
$$(J = 0, 1, 2, \ldots) \tag{7.15}$$

Lines are observed at $\tilde{\nu}_0 \pm \tilde{B}(4J''+6)$, where J'' refers to the *lower* value (Figure 7.6, middle).

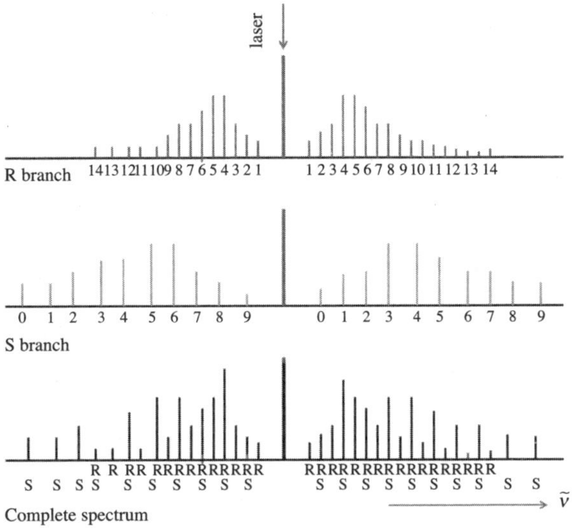

Figure 7.6. Raman rotational lines of a symmetric top molecule

As seen in Figure 7.6, the intensity structure is complex, but the line spacing is $2\tilde{B}$ because of the R-lines.

Spherical top molecules
This class of molecules has completely isotropic (spherical) polarizability ellipsoids, which appear the same under any rotation. Hence, their rotations are Raman inactive.

Asymmetric top molecules
Such molecules display complicated Raman rotation spectra and these are interpreted, as in microwave spectroscopy, in terms of prolate or oblate symmetric tops, depending on which class a particular molecule resembles.

7.3.2.2 Vibrational Raman

The selection rules for vibrational Raman spectra are the same as those observed for a harmonic oscillator, i.e., $\Delta\upsilon = \pm 1$. The overtones and combination bands are generally too weak to be observed in the Raman. As for rotational Raman spectroscopy, the additional requirement is that the polarizability ellipsoid must change either in shape or size during a vibration. Let us consider the molecules we have studied previously.

We first examine the appearance and size of the polarizability ellipsoid of hydrogen as it vibrates. Stretching the bond results in greater polarizability of the bond, and hence a smaller polarizability ellipsoid. Similarly, compression results in a larger polarizability ellipsoid. In contrast to rotation, there is change in the *size* and not the shape of the polarizability ellipsoid. This vibration is therefore Raman active. Stokes lines arise from the $\upsilon = 1 \leftarrow 0$ transitions, and anti-Stokes lines from $\upsilon = 1 \rightarrow 0$, and because of the Boltzmann factor, the latter are much weaker.

The rotational selection rule for vibration-rotation is the same as that for pure rotational Raman spectra, i.e., $\Delta J = 0, \pm 2$, and the spectrum shows an OQS pattern, in contrast to the PR contour observed in the infrared spectroscopy of heteronuclear diatomic molecules.

Linear polyatomics, like carbon dioxide and acetylene, have a larger number of vibrational degrees of freedom. For example, carbon dioxide has four modes of vibration, two of them degenerate (Chapter 6). Consider first the symmetric stretching mode. This is similar to the vibration of hydrogen and the polarizability ellipsoid shows variation in size with the vibration. Obviously, $(d\alpha / dQ)_{Q=0}$ is large and positive (Figure 7.7) and this vibration is Raman active.

We next consider the antisymmetric stretch mode (ν_3). The polarizability ellipsoid changes in appearance, the elongated bond becoming more polarizable, and the compressed one less polarizable. Because of the changing shape of the polarizability ellipsoid, we expect this vibration also to be Raman active. Experimentally, however, it is found to be Raman inactive. We seek an explanation for this observation. In each phase of vibration, the polarizability ellipsoid shows the same appearance. If we plot the magnitude of the polarizability as a function of the normal coordinate, a graph similar to the one shown in Figure 7.7 is obtained. The polarizability increases in each phase of vibration and goes through a minimum at $Q = 0$. Thus $(d\alpha / dQ)_{Q=0}$ is zero and the vibration is Raman inactive.

We next consider the doubly degenerate bending mode, ν_2. Bending the molecule causes the polarizability ellipsoid to change its appearance to that of a bent triatomic molecule like water. Here, too, the same arguments as those for the ν_3 mode reveal that the mode is Raman inactive,

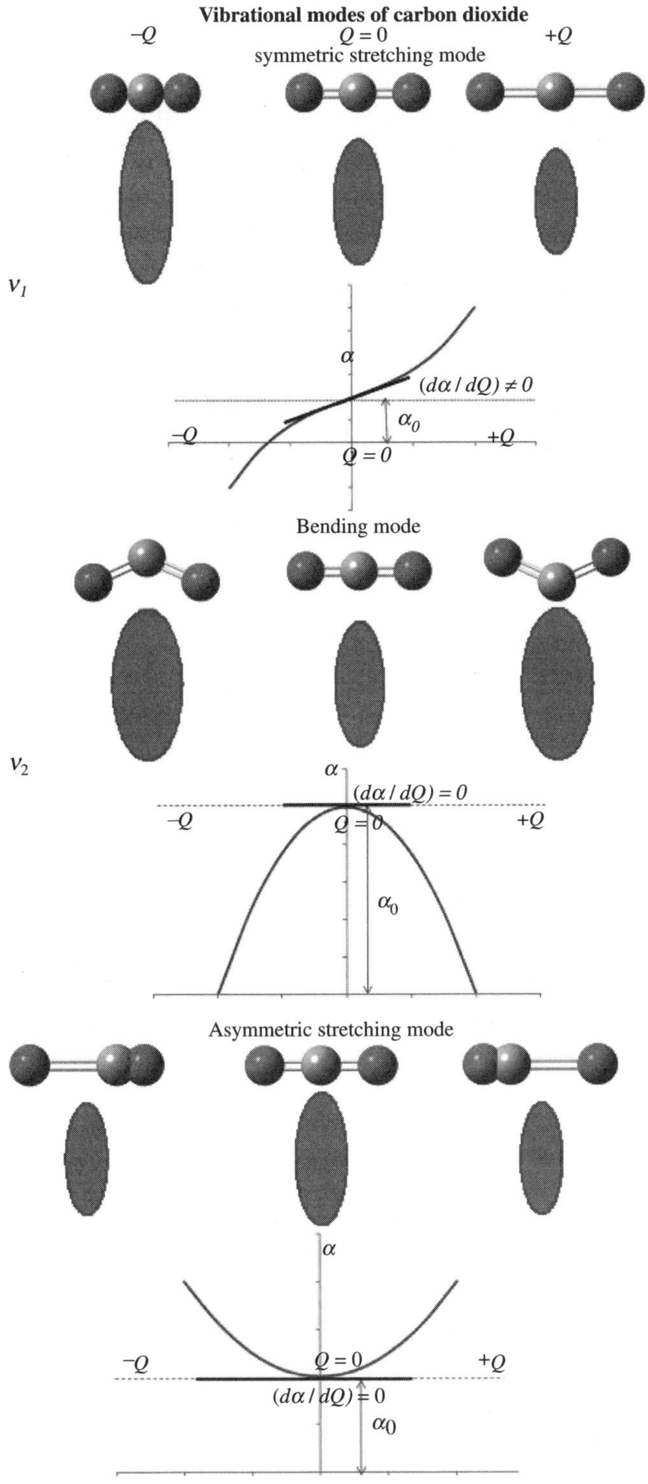

Figure 7.7. Changes in the polarizability ellipsoid of the carbon dioxide molecule during its vibrations

since the polarizability goes through a maximum at the equilibrium geometry, and is similar at the two extremes of motion.

Only the symmetric stretch mode turns out to be Raman active. From the above analysis, we deduce that vibrations that have a *symmetric* polarizability change do not appear in Raman.

For the asymmetric top molecule, H_2O, all modes are found to be Raman active (Figure 7.8). We had also found these modes to be infrared active. This is generally true for all asymmetric top molecules – all their vibration modes are both infrared and Raman active.

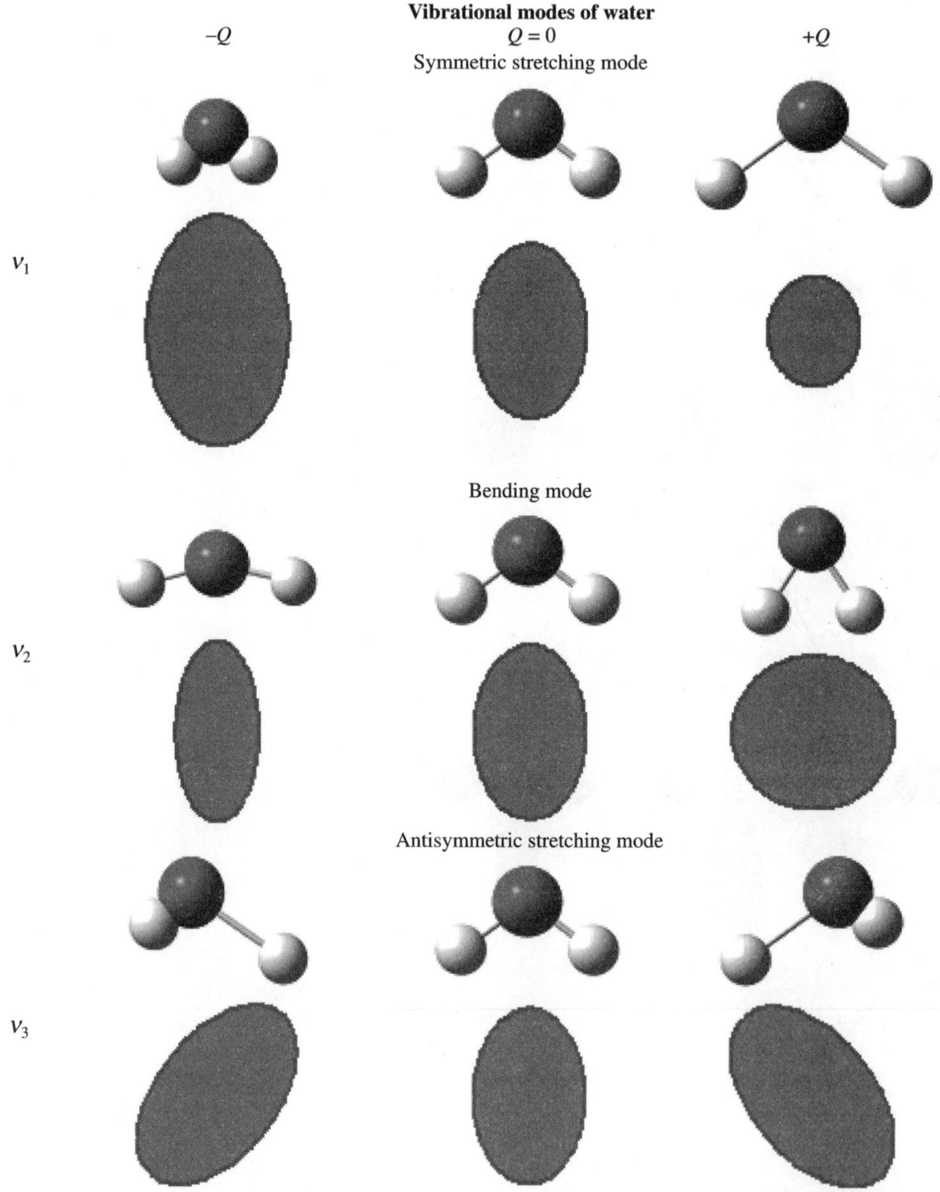

Figure 7.8. Changes in the polarizability ellipsoid during the normal vibrations of the water molecule

7.4 RULE OF MUTUAL EXCLUSION

The infrared and Raman spectra of water and carbon dioxide are summarized in Table 7.1.

Table 7.1. Infrared and Raman spectra of water and carbon dioxide

Band	Type	Water			Carbon dioxide		
		$\tilde{\nu}$ / cm^{-1}	Infrared*	Raman*	$\tilde{\nu}$ / cm^{-1}	Infrared	Raman
ν_1	Symmetric stretch	3657	s	s	1337	Inactive	Active
ν_2	Bending	1595	vs	w	667	Active	Inactive
ν_3	Asymmetric stretch	3756	vs	w	2349	Active	Inactive

*s = strong; vs = very strong; w = weak

For water, we find that the symmetric stretching mode is strong in Raman and the antisymmetric stretching mode is very strong in the infrared, but weak in Raman. The two techniques are therefore complementary in nature – combined together, they provide all information pertaining to the vibration-rotations of a molecule.

For carbon dioxide, the symmetric stretching mode is Raman active, while the other two modes are Raman inactive, but infrared active. This behaviour is not particular to carbon dioxide, but applies to all molecules having a centre of symmetry. This is the basis of the *Rule of Mutual Exclusion* that states that no normal mode can be both infrared and Raman active in a molecule that possesses a centre of symmetry.

We have a ready explanation for this rule from Group Theory. We have observed that a Raman band is inactive if the induced dipole moment $\vec{\mu}_{\text{ind}} = 0$. Evidently, if all the elements of the derivative polarizability tensor are zero, the vibration is Raman inactive. For the simple molecules discussed above, inspection of the changing shape of the polarizability ellipsoid during a vibration can help in determining if the mode is Raman active or not. A more general method involves the use of Group Theory.

In Chapter 6, we had arrived at a simple method for determining the symmetries of the vibrational modes of some molecules and deduction of their infrared activities. Just as a vibration is infrared active if it belongs to the same symmetry species as a component of the dipole moment, i.e., to the same species as either x, y or z, for Raman activity, the analogous selection rule is

A vibration is Raman active if it belongs to the same symmetry species as a component of polarizability, i.e., to one of the binary products x^2, y^2, z^2, xy, xz, yz or to a combination of products such as $x^2 + y^2$, $x^2 - y^2$.

The polarizability components α_{xx}, α_{xy} (and their derivatives) transform the same way as the six functions x^2, xy,… transform under the symmetry operations of the molecular point group. The water vibrations are $2A_1 \oplus B_2$. All three are infrared active because z transforms as A_1 and y as B_2. All are also Raman active because x^2, y^2 and z^2 transform as A_1 and yz as B_2.

When we consider the centrosymmetric molecule, carbon dioxide, we first need to find out how the normal modes transform. Since it is a linear molecule and belongs to an infinite point group ($D_{\infty h}$), we apply the method we applied to water, but with a difference. For C_n and S_n, $\chi(R) = \pm 1 + 2\cos\phi$, as the case may be, because rotation could be by any arbitrary angle ϕ.

Secondly, the molecule has only two rotational degrees of freedom (along the x, y axes) as it is linear. From the character table, we see that together they transform as Π_u. We simply remove the characters of Π_u from Γ_{3N} to account for rotation, and $\chi(R)$ to account for the three translation modes, leaving us with the four-dimensional representation for Γ_{vib}. Since this molecule is symmetric with a single central atom, we can find Γ_{stretch} by subtracting the totally symmetric representation from the characters for the number of unshifted atoms. This is found by inspection to be $\Sigma_g^+ \oplus \Sigma_u^+$ (the former has all 1's as characters for all symmetry elements, while the latter has 1's for the proper rotations and –1's for the improper rotations. Adding the two gives 2 for all proper rotations and 0 for improper ones, as required). The remaining bending mode corresponds to the two-dimensional irreducible representation Π_u, as seen from the $D_{\infty h}$ character table.

$D_{\infty h}$	E	$2C_\infty^\varphi$	...	$\infty\sigma_v$	I	$2S_\infty^\varphi$	...	∞C_2
$\chi(R)$	3	$1 + 2\cos\phi$		1	–3	$-1 + 2\cos\phi$		–1
n_R	3	3		3	1	1		1
Γ_{3N}	9	$3 + 6\cos\phi$		3	–3	$-1 + 2\cos\phi$		–1
Γ_{trans}	3	$1 + 2\cos\phi$		1	–3	$-1 + 2\cos\phi$		–1
Γ_{rot}	2	$2\cos\phi$		0	–2	$2\cos\phi$		0
Γ_{vib}	4	$2 + 2\cos\phi$		2	2	$-2\cos\phi$		0
Γ_{stretch}	2	2		2	0	0		0
Γ_{bend}	2	$2\cos\phi$		0	2	$-2\cos\phi$		0

The Σ_u^+ and Π_u modes are infrared active because z belongs to the former and (x,y) to Π_u. The Σ_g^+ mode (symmetric stretch) is Raman active since $(x^2 + y^2)$ and z^2 are bases for this irreducible representation.

The third molecule we investigated in the previous chapter is ethylene, for which

$$\Gamma_{\text{vib}} = 3A_g \oplus B_{2g} \oplus 2B_{3g} \oplus A_{1u} \oplus 2B_{1u} \oplus 2B_{2u} \oplus B_{3u}$$

The A_g, B_{2g} and B_{3g} modes are all Raman active because (x^2, y^2, z^2), xz and yz, respectively, transform as them. The B_{1u}, B_{2u} and B_{3u} modes are infrared active and z, y and x polarized, respectively. The remaining mode (A_{1u}) is called a 'silent' vibration because it cannot be observed either in the infrared or in Raman.

The above observations lead us to the following generalizations:

- For all the molecules, the squares (x^2, y^2, z^2 or combinations thereof) of the Cartesian bases belong to the totally symmetric representation (A_1, Σ_g^+ or A_g, as the case may be).
- 'u' modes cannot be Raman active, and, similarly, 'g' modes cannot be infrared active.

The latter observation is the basis of the Rule of Mutual Exclusion. Since the translational components x, y and z are always *ungerade* functions (they change sign on inversion), only 'u' vibrations can have infrared activity. Their squares, however, are always *gerade*, since $u \otimes u = g$, and, hence, only the 'g' vibrations can have Raman activity.

The squares of the translational vectors always transform as the totally symmetric representation of the point group of the molecule (A_1, A_g, Σ_g^+, as the case may be) since all −1's become +1 on squaring. Further, all molecules have a symmetric stretching mode, which belongs to the totally symmetric representation. Since this mode is always Raman active, because x^2, y^2 and z^2 or their combinations belong to this representation, we can say that all molecules have at least one Raman active mode. This is not generally true for infrared activity, where we saw that homonuclear diatomic molecules do not have any infrared active mode.

In fact, all symmetrical modes of centrosymmetric molecules are Raman active and have strong intensities, whereas the asymmetric bands are infrared active. We can therefore recognize symmetric bands in Raman by their intensity. The symmetry of Raman bands can also be determined by studying their polarization.

7.5 POLARIZATION OF RAMAN LINES

A great deal of information about the symmetry of Raman bands can be obtained by studying their polarization. Consider a beam of light incident in the x-direction. Its electrical and magnetic field vectors, therefore, point in the y- and z-directions (Figure 7.9). We shall consider linearly polarized radiation, i.e., one with all its electric field vectors in one direction only, say the y-direction, since all Raman experiments today use linearly polarized radiation.

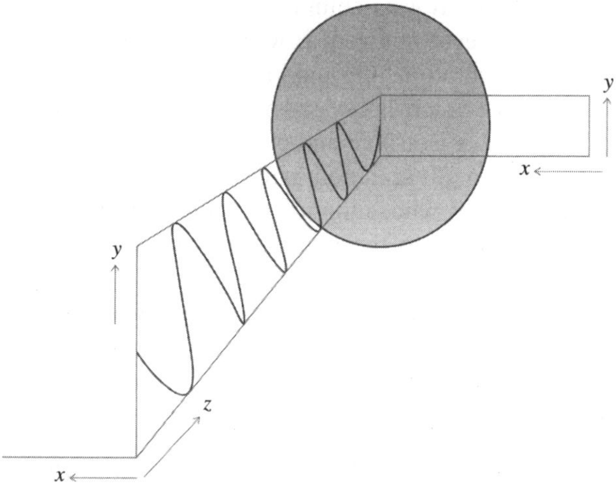

Figure 7.9. Polarization of Raman lines

If such radiation falls on a molecule which has an isotropic polarizability ellipsoid (this molecule could be any of the spherical top molecules of T_d or O_h symmetry), the induced dipole moment and scattered radiation would also have the same polarization as the incident beam, since the polarizability ellipsoid has no preferred direction and hence does not change the orientation of the electric field. The scattered radiation is thus *polarized*.

Now consider a molecule with an anisotropic polarizability ellipsoid, e.g., CH_3Cl. The molecule is more polarizable in the C-Cl bond direction, and hence the induced dipole moment and polarization of the scattered radiation is greater along this direction. The anisotropic polarizability rotates the polarization of the scattered beam. However, in a sample of gaseous molecules, the C-Cl bonds are oriented in all possible directions, and the resulting scattered radiation is completely *depolarized.*

The polarization of a scattered beam is determined by computing its depolarization ratio, defined as $\rho_p = I_\perp / I_{//}$. The *p* subscript signifies that we are talking about initially polarized radiation. The intensity of the scattered radiation is measured perpendicular and parallel to the incident radiation. In the present example, $I_{//}$ is the *y*-direction since the incident beam is *y* polarized. If the direction of observation is *z*, $I_\perp$ is measured along the *x*-direction, i.e., $\rho_p = I_x / I_y$.

The measurement of the depolarization ratio can help identify the symmetry of a vibration. This is best illustrated with an example. Consider the symmetric stretch vibration of a spherical top molecule like carbon tetrachloride. In this mode, all C-Cl bonds compress and elongate at the same time. The polarizability ellipsoid does not change its shape (which remains spherical) during the vibration, but its size varies, as the molecule becomes more polarizable when it expands and less polarizable on compression. Not surprisingly, this mode of vibration is called a breathing mode. Since the polarizability ellipsoid is isotropic, the scattered radiation remains polarized, and $I_\perp = 0$, so that ρ_p is also zero.

Now consider another mode of vibration of the same molecule, the asymmetric stretch mode, in which one bond elongates while the others compress, and vice versa. The molecule takes on the shape of a symmetric top molecule in each phase of vibration. Because of the anisotropic polarizability ellipsoid, the scattered radiation is completely depolarized, $I_\perp = I_{//}$ and $\rho_p = 1$. We shall see shortly that this value is never realized. However, it is important to note that, in the former case of the symmetric vibration, the polarization depends on the direction of polarization of the incident radiation (which is dependent on the laboratory or space coordinates) since the polarizability ellipsoid is isotropic and cannot alter the direction of polarization. In the latter case, the polarization of scattered radiation depends on the molecular coordinates (the direction of the C-Cl bond whose motion is opposite to the other C-Cl bonds), and the direction is completely random, so the polarization averages to zero because of molecular motion.

The depolarization ratio therefore depends on the symmetry of the polarizability ellipsoid or its tensor. In general, the elements of the polarizability tensor depend upon the choice of molecular coordinates. Diagonalization can rotate the coordinates to the axes of the polarizability ellipsoid. The sum of eigenvalues of a matrix is equal to its trace, so that we have

$$\alpha_a + \alpha_b + \alpha_c = \alpha_{xx} + \alpha_{yy} + \alpha_{zz} \tag{7.16}$$

Therefore, without bothering to diagonalize the matrix, we can obtain the sum of eigenvalues. We define

$$\bar{\alpha} = \frac{1}{3}\left(\alpha_{xx} + \alpha_{yy} + \alpha_{zz}\right) \tag{7.17}$$

as a measure of the average polarizability in the three directions. This is the symmetric component of the polarizability ellipsoid. The asymmetric component depends on the magnitudes of the off-diagonal elements of the polarizability ellipsoid, and is given by

$$\gamma^2 = [(\alpha_{xx} - \alpha_{yy})^2 + (\alpha_{yy} - \alpha_{zz})^2 + (\alpha_{xx} - \alpha_{zz})^2 + 6(\alpha_{xy}^2 + \alpha_{yz}^2 + \alpha_{xz}^2)] \tag{7.18}$$

It can be shown that, for polarized incident radiation, ρ_p is given by

$$\rho_p = \frac{3[(\gamma)^2]}{45[(\bar{\alpha})^2] + 4[(\gamma)^2]} \tag{7.19}$$

According to equation (7.19), the minimum value zero of ρ_p is obtained when $\gamma^2 = 0$ (no asymmetry) and in that case, ρ_p is also zero and the radiation is said to be polarized. This situation is realized for the completely isotropic polarizability ellipsoids of spherical top molecules (those belonging to T_d and O_h point groups). In practice, an absolutely zero value is seldom realized due to intermolecular interactions, which somewhat distort the polarizability ellipsoid.

Mathematically, when $\bar{\alpha} = 0$, ρ_p attains its maximum value of ¾. The diagonal elements of the polarizability matrix are always positive, and hence their average $\bar{\alpha}$ is also positive and finite. Hence, the maximum value of ¾ can never be realized since $\bar{\alpha}$ cannot be zero.

In the case of Raman spectroscopy, the transition dipole moment $\left|\vec{M}_{\upsilon'\upsilon''}\right|$ for a $\upsilon' \leftarrow \upsilon''$ transition of a diatomic molecule is given by

$$\left|\vec{M}_{\upsilon'\upsilon''}\right| = \int_{-\infty}^{\infty} \psi_{\upsilon'} \hat{\alpha} \psi_{\upsilon''} \, dx \tag{7.20}$$

where x is the displacement coordinate. As for the dipole moment operator in Chapter 5, the polarizability operator may be expanded in a Maclaurin series in the normal coordinate

$$\hat{\alpha} = \hat{\alpha}_{x=0} + \left(\frac{d\hat{\alpha}}{dx}\right)_{x=0} x + \frac{1}{2!}\left(\frac{d^2\hat{\alpha}}{dx^2}\right)_{x=0} x^2 + \ldots \tag{7.21}$$

The value $\alpha_{x=0}$ is simply the equilibrium polarizability, α_0, which is a constant. Because of the orthogonality of the vibrational wave functions, this term vanishes unless $\upsilon' = \upsilon''$, i.e., it is applicable only to Rayleigh scattering. The condition, therefore, for a vibration to be Raman active is that $\left(d\hat{\alpha}/dx\right)_{x=0}$ (hereafter called α') must be different from zero, or, in other words, the polarizability must change during a vibration for it to be Raman active.

For Raman scattering, expression (7.19) remains the same, except that the isotropic and anisotropic components are replaced by their derivatives $\bar{\alpha}'$ and $(\gamma')^2$ with respect to the normal coordinate, i.e.,

$$\rho_p = \frac{3[(\gamma')^2]}{45[(\bar{\alpha}')^2] + 4[(\gamma')^2]} \tag{7.22}$$

For Rayleigh scattering, equation (7.19) implies that the scattered radiation is polarized ($0 \le \rho_p \le$ ¾) because the diagonal elements of the polarizability matrix are always positive and their average $\bar{\alpha}$ is also positive. The same cannot be said about its derivative, which can be positive, negative or even zero.

Our study of carbon dioxide vibrations had revealed that only in the case of the symmetrical stretching vibration is $\bar{\alpha}'$ different from zero (Figure 7.7). For the other two modes, this derivative was found to be zero. Hence, for the symmetric stretch vibration of carbon dioxide, $0 \le \rho_p \le$ ¾, and the band is polarized. For the other two modes, since $\bar{\alpha}'$ is zero, it follows that

$\rho_p = ¾$ and the bands are depolarized. We now look for conditions under which $\bar{\alpha}'$ is non-zero and the Raman band is polarized.

Using equation (7.17), we can write

$$\bar{\alpha}' = \frac{1}{3}\int_{-\infty}^{\infty} \psi_{v'} \left(\alpha'_{xx} + \alpha'_{yy} + \alpha'_{zz}\right) \psi_{v''}\, dx \tag{7.23}$$

We have already shown that the ground state vibrational wave function $\psi_{v''}$ belongs to the totally symmetric representation of the point group to which the molecule belongs. The diagonal polarizability components transform as x^2, y^2 and z^2. Being squares, they also transform as the totally symmetric representation (all −1's become +1 on squaring), i.e., A_1, A_g or Σ_g^+, as the case may be. We have already shown two of the quantities in the integral of equation (7.23) to belong to the totally symmetric representation. Unless $\psi_{v'}$ also transforms as the totally symmetrical representation, this integral, and hence $\bar{\alpha}'$ will be zero. We thus make the statement '*Raman lines are polarized only when a vibration transforms as the totally symmetric representation of the point group of the molecule*'. In all other cases, $\rho_p = ¾$ and the Raman lines are depolarized. Since ρ_p depends on the magnitude of $\bar{\alpha}'$ for polarized lines, the Raman line is said to be polarized if $0 \le \rho_p \le ¾$. In case the incident radiation is not linearly polarized, the analogous expression is $0 \le \rho_n \le 6/7$, where the subscript *n* indicates natural (unpolarized) incident radiation.

Hence, the polarization of Raman bands can help in the assignment of Raman spectra. Polarized Raman lines indicate that the corresponding mode transforms as the totally symmetric representation of the point group of the molecule.

For water, both the symmetric stretch and bending modes transform as the totally symmetric representation A_1 and the corresponding Raman lines are polarized. For carbon dioxide, only the symmetric stretch is predicted to be Raman active and polarized. The experimental Raman spectrum of carbon dioxide, however, displays two main peaks at 1285 cm^{-1} and 1388 cm^{-1} instead of the single line expected. The two peaks are almost equally intense. Since only one Raman active band is predicted for carbon dioxide, the other band must be an overtone or a combination band, but such bands usually have low intensities. The high intensity of the band is due to a phenomenon known as *Fermi resonance*.

7.6 FERMI RESONANCE

According to perturbation theory, states of the same symmetry can mix, the amount of mixing being inversely proportional to the difference in energies of the two states. Thus, we had earlier stated that the symmetric stretch and bending modes of water can interact because they have the same symmetry (A_1), but the amount of interaction is small because of the large difference in their energies. Such interaction results in the lower of the two energy levels getting further stabilized at the expense of the higher energy level, which gets destabilized. In the case of carbon dioxide, there is only one fundamental of Σ_g^+ symmetry. Where does the other state arise from? The answer lies in the overtone of the bending mode (ν_2). This mode is infrared active, but its first overtone is Raman active. In fact, the first overtones of all vibrations are Raman active. To get the symmetry species for the overtone, we take the direct product $\Pi_u \otimes \Pi_u$, which gives a four-dimensional reducible representation. The direct products of irreducible representations are also available in Group multiplication tables, where the entry against $\Pi \otimes \Pi$

is $\Sigma^+ + [\Sigma^-] + \Delta$. The representation in the square parenthesis is the antisymmetric component and need not concern us here since first overtones are always symmetric. We also have $u \otimes u = g$, so that we have for the overtone of a Π_u vibration, two symmetric states, $\Sigma_g^+ + \Delta_g$. Since this contains the Σ_g^+ state, it is Raman active and can interact with the fundamental symmetric stretch mode (Figure 7.10). Before we proceed further, we may state that the direct product of any irreducible representation with itself always gives or contains the totally symmetric representation. For this reason, *all first overtones are Raman active and polarized.*

Returning to carbon dioxide, we find that $2\tilde{\nu}_2 \approx 2 \times 667 = 1334\ \text{cm}^{-1}$ and $\tilde{\nu}_1 \approx 1337\,\text{cm}^{-1}$. We have a case of 'accidental' near degeneracy, and the interaction is so strong that the overtone is shifted to 1285 cm^{-1} and the fundamental to 1387 cm^{-1} (Figure 7.10). The overtone 'borrows' intensity from the fundamental to become almost as intense. The large energy gap of 102 cm^{-1} between the two indicates strong interaction (Figure 7.10). In most cases of Fermi resonance, the interaction is not so strong, and the overtone remains much less intense than the fundamental.

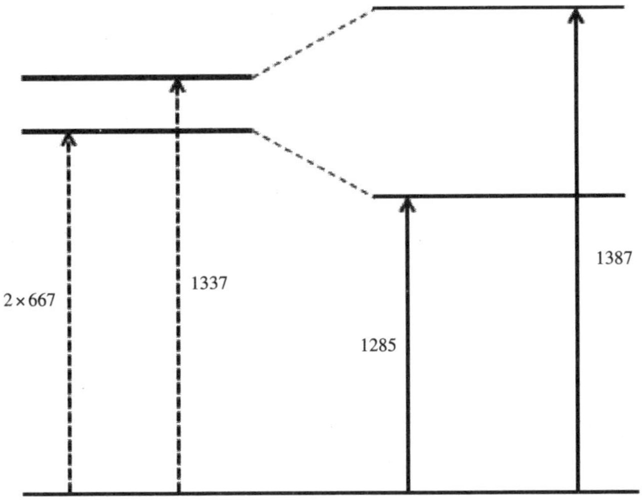

Figure 7.10. Mixing of vibrational energy levels

The experimental spectrum also shows two hot bands and a small peak at 1369 cm^{-1} corresponding to the $^{13}CO_2$ isomer in its natural abundance (~1%). Isotopic substitution at carbon should not influence $\tilde{\nu}_1$ because this wavenumber is independent of the mass of carbon $\omega_1 = \dfrac{1}{2\pi c}\sqrt{\dfrac{k}{m_O}}$, since the carbon is stationary during this mode. However, the bending mode is shifted down to 647 cm^{-1} on isotopic substitution. The gap between $\tilde{\nu}_1$ and $2\tilde{\nu}_2$ therefore increases, leading to a smaller shift on interaction and the fundamental $\tilde{\nu}_1$ moves to 1369 cm^{-1}.

It would appear that Fermi resonance could complicate the analysis of infrared and Raman spectra of many molecules, but, since only vibrations of the same symmetry can interact, there are limited examples of this phenomenon. One typical example of Fermi resonance is found in the vibrational spectra of aldehydes, where the C-H stretching mode of the CHO group interacts with the first overtone of the deformation vibration of the CHO group, 2δ(CHO), at ~2 × 1400 cm^{-1}, resulting in a Fermi doublet with branches at approximately 2830 and 2730 cm^{-1}. It is important in Fermi resonance for the two interacting levels to be localized in the same part of the molecule.

Another example is that of the linear molecule cyanogen chloride Cl-CN, which has a fundamental (Π) bending mode at 378 cm^{-1}, whose first overtone is slightly lower than ~2 × 378 cm^{-1} = 756 cm^{-1} because of anharmonicity. Fermi resonance with the C-Cl stretching frequency (Σ^+) at 744 cm^{-1} results in a splitting of the frequency into two strong bands at 783 and 714 cm^{-1} in the infrared spectrum.

Another aspect of the Raman spectrum of carbon dioxide is the absence of alternate rotational lines, as a result of which the rotational spacing is double that predicted. This is due to the influence of nuclear spin that also affects the infrared and Raman spectra of several other centrosymmetrical molecules.

7.7 INFLUENCE OF NUCLEAR SPIN

Besides electrons, nuclei too have spins, designated by the nuclear spin quantum number, *I*. Unlike electron spin, however, the nuclear spin quantum number can take on integral or half-integral values, depending on the number of nucleons present in a particular nucleus. The spin quantum numbers of some nuclei are given in Table 7.2.

Table 7.2. Some nuclei, their spin quantum numbers and statistics

Spin *I*	Nuclei	Statistics
0	^{12}C, ^{14}C, ^{16}O, ^{18}O, ^{32}S	Bose–Einstein
½	^{1}H, ^{3}H, ^{13}C, ^{13}N, ^{15}N, ^{15}O, ^{31}P, ^{19}F	Fermi–Dirac
1	^{2}H (D), ^{14}N	Bose–Einstein
3/2	^{11}B, ^{33}S, ^{35}Cl, ^{37}Cl	Fermi–Dirac
5/2	^{17}O	Fermi–Dirac
3	^{10}B	Bose–Einstein

It can be seen that nuclei with odd number of nucleons (odd atomic mass numbers) have half-integral spins and those with even number of nucleons have integral values of the nuclear spin quantum number. This has an important bearing on their wave functions. You may recall that for electronic wave functions we had stated that the total wave function should be antisymmetric to electron exchange because electrons are fermions and obey Fermi-Dirac statistics. This is because they have half-integral spins. On the other hand, particles with integral (and zero) spins are called bosons and obey Bose-Einstein statistics, which requires their total wave function to be symmetric. From Table 7.2, we see that many nuclei are bosons.

We first examine how the nuclear spin affects the vibration–rotation spectrum of carbon dioxide. Carbon dioxide has a centre of symmetry and, under inversion, the two oxygens exchange position. From Table 7.2, we see that the normal isotope of oxygen, ^{16}O (and also ^{18}O) has a nuclear spin quantum number of zero, so carbon dioxide is a boson, and its total wave function must be symmetric to *nuclear* exchange.

The total wave function

$$\psi = \psi_{el}\psi_{n} = \psi_{el}\psi_{vib}\psi_{rot}\psi_{ns} \tag{7.24}$$

where we have ignored the translational wave function, which is always symmetric to nuclear exchange, and added the term ψ_{ns} to the expression given in Chapter 4 (where we had assumed that this is equal to one) in order to account for nuclear spin. We now look at the symmetry of each term in the expression:

Electronic: Carbon dioxide has a singlet $^1\Sigma_g^+$ ground state (all electrons paired). We shall discuss molecular term symbols in the next chapter. At this stage, it is sufficient to know that molecules with all electrons paired have singlet ground states. Our study of the helium atom electronic states in atomic spectroscopy (Chapter 3) had led us to the conclusion that singlet states have symmetric spatial wave functions in order to comply with the Pauli Exclusion Principle, since their spin wave functions are antisymmetric. Hence, the ground electronic wave function is symmetric (*s*) to nuclear exchange.

Vibrational: Consider first the ground vibrational state (0,0,0) of carbon dioxide, which has the largest population at room temperature. As we have already seen in Chapter 5, this state is always symmetrical to nuclear exchange.

Rotational: In contrast to vibrational states, many rotational levels are occupied at room temperature. The symmetry of associated Legendre polynomials dictates that rotational levels with odd values of the J quantum number are odd (*a*) and those with even values of J are even (Chapter 4).

Nuclear Spin: Since $I = 0$ for oxygen, $\psi_{ns} = 1$.

Therefore, in order for the total wave function to be symmetric to nuclear exchange, as required for bosons, only even J values can exist for the ground state of carbon dioxide: all odd J states are 'missing'. This is reflected in the rotational spectrum of carbon dioxide, where alternate rotational lines are found to be missing. Figure 7.11 shows the Stokes lines in the rotational Raman spectrum of carbon dioxide.

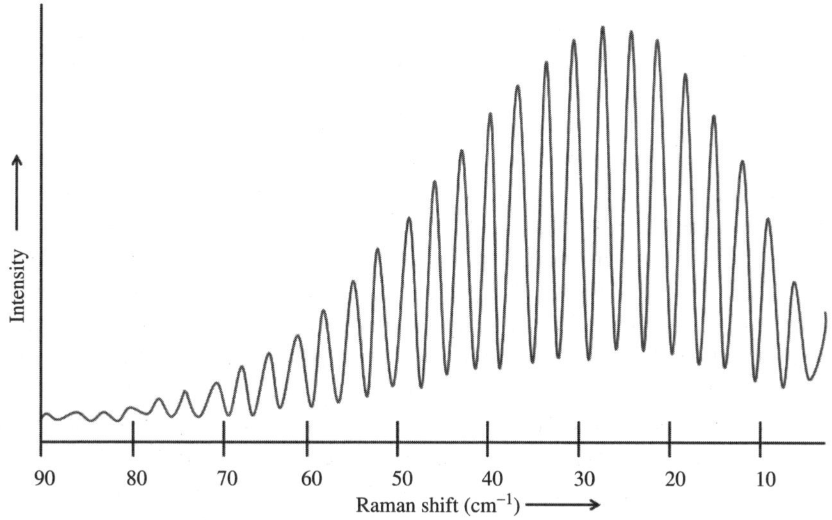

Figure 7.11. Rotational Raman spectrum of carbon dioxide

The observed transitions are $J = 0 \rightarrow 2$, $2 \rightarrow 4$, $4 \rightarrow 6$, etc. The missing states cause every other line in the spectrum to be absent. In the spectrum shown in Figure 7.11, the first line is too close to the Rayleigh line to be observed. It is displaced from the incident radiation by $6\tilde{B}$, whereas the other lines are spaced by $8\tilde{B}$ instead of $4\tilde{B}$. This effect would not be observed if the two oxygen atoms were not subject to the Pauli principle, i.e., if we remove the centre of symmetry. Indeed, the isotopologue $^{18}OC^{16}O$ has been observed to retain all its rotational lines.

Example 7.3 The rotational structure in the Raman spectrum of carbon dioxide, CO_2, is offset from the wavenumber of the incident radiation by 2.3622 cm^{-1}, 5.5118 cm^{-1}, 8.6614 cm^{-1}, ... Calculate the rotational constant of carbon dioxide.

Solution

The first line is displaced from the incident radiation by $6\tilde{B} = 2.3622$. Hence $\tilde{B} = 0.3937$ cm^{-1}.

For the (1,0,0) excited vibrational state of carbon dioxide (symmetric stretch excited), the situation remains the same since the excited vibrational level has the same symmetry as the excited vibrational mode, which is Σ_g^+ in this case. Hence, only even J values exist for this state too. This mode is active in Raman, where the combined selection rule $\Delta\upsilon = \pm 1$, $\Delta J = \pm 2$ (OS branches) makes vibration-rotation transitions possible between the even J levels of both states.

Let us now consider the (0,0,1) excited state, in which the antisymmetric stretch of carbon dioxide is excited. In this case, the excited level belongs to Σ_u^+ symmetry, because this is the symmetry of this vibration mode. The excited vibrational level now has odd (u) symmetry, so that the *odd* J levels only exist for this state. This vibration is infrared active, where the selection rules $\Delta\upsilon = \pm 1$, $\Delta J = \pm 1$ (PR branches) are consistent with transitions from even ground state rotational levels to odd excited state rotational levels. The observed rotational transitions are $0 \rightarrow 1$, $2 \rightarrow 3$, $4 \rightarrow 5$, etc., and the rotational line spacing is $4\tilde{B}$.

A similar situation is expected for the diatomic molecule, oxygen. However, dioxygen has a triplet $^3\Sigma_g^-$ ground state, and hence an antisymmetrical spatial wave function. The arguments for carbon dioxide now lead us to the conclusion that in the ground vibrational level of oxygen, only *odd* values of J exist. Note that the same holds for $^{18}O_2$, but not, for example, for $^{16}O^{17}O$ or $^{16}O^{18}O$, in which the centre of symmetry is lost. Thus, in a spectrum of oxygen with natural isotopic ratios, lines with even J will appear very weakly due to the asymmetrical isotopologues $^{16}O^{18}O$, etc. The pure rotational Raman spectrum will show lines corresponding to $1 \rightarrow 3$, $3 \rightarrow 5$, $5 \rightarrow 7$, etc. Rotational transitions from even J values are missing, leading to a rotational line spacing of $8\tilde{B}$ instead of the expected $4\tilde{B}$ for all the transitions, except the first $1 \rightarrow 3$ transition, for which the spacing from the Rayleigh line is $10\tilde{B}$. A note of warning here: failure to take this into account results in calculated bond lengths which are too small by a factor of $\sqrt{2}$.

The nuclear spin effect also manifests itself in the rotational spectrum of hydrogen, where an alternation in intensities is observed. Hydrogen is a fermion ($I = ½$) and its total wave function must therefore be antisymmetric to nuclear exchange. It has a singlet ($^1\Sigma_g^+$) ground state, and hence its electronic wave function is symmetric. Its ground state vibrational wave function is also symmetric. This implies that the product of the rotational and nuclear spin wave functions must be antisymmetric. Thus, even J values are possible with antisymmetric nuclear wave functions and odd J values for symmetric nuclear wave functions. Exactly the same arguments apply as those for electron spin (Chapter 3) in determining the odd and even nuclear spin wave functions. If we label the two nuclei A and B, we obtain the following nuclear spin wave functions

$$(s)\quad \psi_{\mathrm{ns}} = \begin{cases} \alpha(A)\alpha(B) \\ \frac{1}{\sqrt{2}}\left[\alpha(A)\beta(B)+\beta(A)\alpha(B)\right] \\ \beta(A)\beta(B) \end{cases} \tag{7.25a}$$

$$(a)\quad \psi_{\mathrm{ns}} = \frac{1}{\sqrt{2}}\left[\alpha(A)\beta(B)-\beta(A)\alpha(B)\right] \tag{7.25b}$$

There are thus *three* symmetric and *one* antisymmetric nuclear spin wave functions. Only odd values of the rotational quantum number (J) can be associated with the former and even J values with the antisymmetric nuclear wave function. We therefore expect a 1:3 intensity alteration in the rotational spectrum of hydrogen.

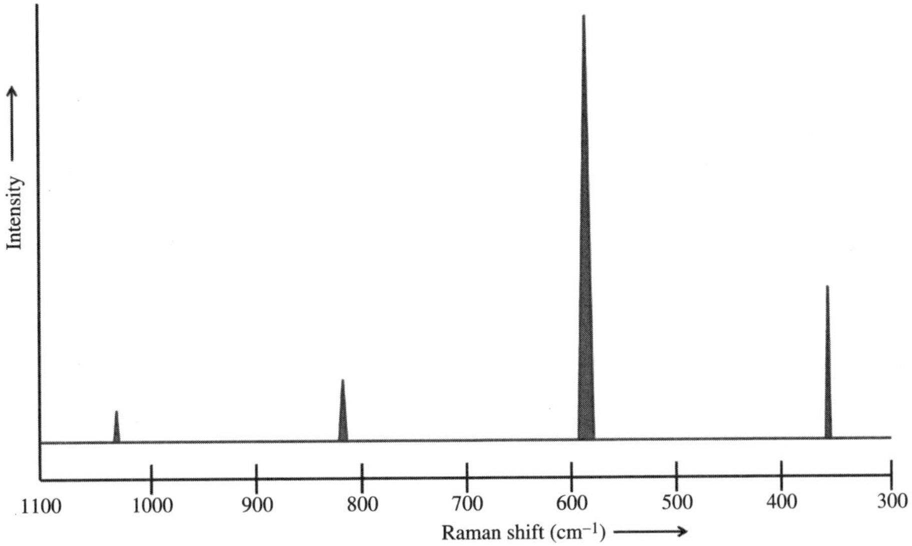

Figure 7.12. Rotational Raman spectrum of hydrogen

Figure 7.12 gives a schematic rotational Raman spectrum of hydrogen, where one can see that the 1:3 ratio is not exactly observed. The reason is the superimposition of the rotational level degeneracies ($g_J = 2J + 1$) and the Boltzmann distribution effect.

Table 7.3 gives the rotational wave function parities, and symmetries and degeneracies of the corresponding nuclear spin states (g_N). The total statistical weight is the product of the two degeneracies, which does not follow a 1:3 ratio. The final column gives the rotational energy E in Kelvin (E_{rot} / k_B).

Table 7.3. Nuclear spin statistics of $^1H_2$

J	g_J	Parity	Symmetry	Nuclear spin states	Total statistical weight	E/K
0	1	+	s	1	1	0.0
1	3	–	a	3	9	170.5
2	5	+	s	1	5	509.9
3	7	–	a	3	21	1015.1
4	9	+	s	1	9	1081.7

Therefore, hydrogen exists in two states, *ortho* hydrogen and *para* hydrogen. The state with the higher spin multiplicity is called *ortho* and the other state is called *para*. Thus, *ortho* hydrogen can exist only in odd J levels, and has a zero-point rotational energy. *Para* hydrogen exists in even J levels only. The two forms cannot be interconverted except by molecular collisions involving nuclear exchange, and are hence distinct from each other.

The last column lists E_{rot} / k_B, which is an estimate of the temperature at which the rotational energy equals the thermal energy (Chapter 4). For hydrogen, the values are high because of its low molecular mass and consequently low moment of inertia and high rotational constant. At very low temperatures (below 100 K), hydrogen exists only in the *para* form. However, at high temperatures (above 1000 K), both forms co-exist, with the *ortho* form being roughly three times as much as the *para* form.

The other $I = ½$ nuclei listed in Table 7.2 behave similarly. Acetylene, in which the two hydrogens exchange, also shows similar behaviour because the carbons have zero nuclear spin. However, because of its higher mass, acetylene has more closely spaced energy levels, so that it is only at very low temperatures that it will exist predominantly in the *para* form. The 1:3 intensity ratio is easily discernible in the ro-vibrational spectra of acetylene and the close-up of a few lines labelled with their lower J value is shown in Figure 7.13.

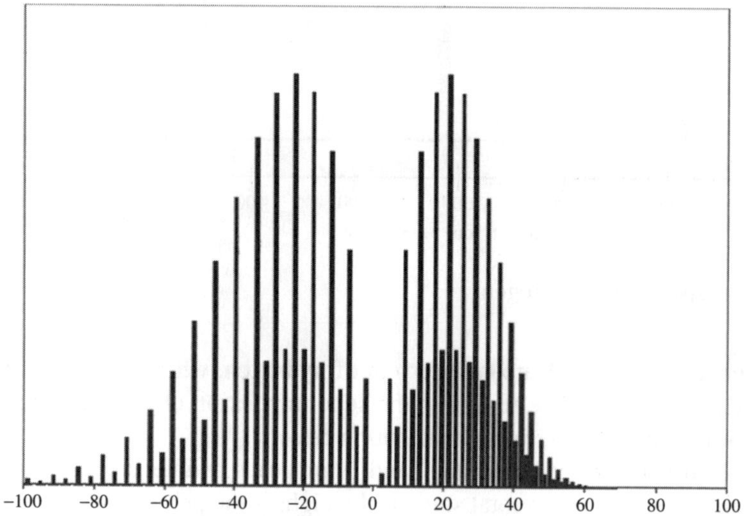

Figure 7.13. Vibration–rotation spectrum of acetylene

Finally, we discuss some molecules with $I = 1$ ($^{14}N_2$, $^{2}H_2$). These are bosons and their total wave functions should be symmetric. In general, for a homonuclear diatomic molecule, there are $(2I + 1)(I + 1)$ symmetric and $(2I + 1)I$ antisymmetric nuclear spin wave functions, so that there are six symmetric (*ortho*) and three antisymmetric (*para*) wave functions for $I = 1$ nuclei. The *ortho* forms of the molecules are associated with even J quantum numbers and the *para* ones with odd J values. At high temperatures, the *ortho*:*para* ratio is roughly 2:1, and at low temperatures, it is even higher because the $J = 0$ level is associated with this form. Thus, there is a 2:1 alternation in intensities in favour of the even J values.

For the diatomic molecules discussed above, the situation does not change for the excited vibrational levels, because these molecules have only one vibrational degree of freedom, which is the symmetric stretch and belongs to Σ_g^+ symmetry. It is Raman active with a $\Delta J = 0, \pm 2$ selection rule, which allows only odd – odd and even – even transitions.

We now apply these concepts to structure elucidation of some simple molecules. There are many points to consider before doing a complete assignment of the vibrational spectrum.

7.8 STRUCTURE DETERMINATION USING COMBINED INFRARED AND RAMAN DATA

We are now in a position to perform structure elucidation of some simple molecules using combined infrared and Raman spectroscopy and Group Theory. The following considerations can help assign the spectra:

Number of fundamental vibrational modes

First of all, the number of fundamental vibrational modes should be ascertained. These are $(3N - 5)$ for linear molecules and $(3N - 6)$ vibrational modes for bent molecules. Some of these may be degenerate, so the actual number of vibrational frequencies may be less. Usually the strongest bands in the infrared and Raman are fundamentals. The remaining have to be classified as hot bands (by their temperature dependence), overtones or combination bands. Fermi resonance may also cause complications.

Q branches

The presence of PR contours (no Q branch) in the stretching modes indicates that the molecule is linear. Perpendicular modes of linear molecules also show PQR contours, but the absence of Q branches in any band indicates linearity.

Rule of mutual exclusion

If no common bands are found in infrared and Raman spectra, the molecule may have a centre of symmetry. However, one must be careful about this diagnosis, since it is possible that a transition may be Raman or infrared active but may be too weak to be observed. For example, bending modes are rarely observed in Raman, though they may be Raman active.

Assignment of vibrational frequencies

Once the structure of the molecule is established, the observed vibrational frequencies can be assigned. A nonlinear non-cyclic molecule of N atoms has N–1 stretching modes and N–2 bending modes. The remaining N–3 are torsions. Bending modes occur at the lowest frequencies, while asymmetric stretches are highest. Modes above 2500 cm^{-1} always involve stretching of H atoms. Use group vibrational frequencies to exploit isotopic frequency shifts $\left(\text{use } \omega = \frac{1}{2\pi c}\sqrt{\frac{k}{\mu_{red}}}\right)$. Strong modes in Raman are symmetric modes and are polarized. These belong to the totally symmetric representation of the molecule's point group.

Vibrational dipole moments and polarizability

The strengths of the infrared or Raman bands can be qualitatively correlated with changes in dipole moments or polarizabilities during the vibration. For example, the more symmetric the vibration, the more Raman active is the mode.

7.8.1 Examples of spectral assignments

Nitrous oxide (N_2O)

$\tilde{\nu}$ / cm^{-1}	Infrared	Raman	Band assignment
580	PQR m (temp. dep.)	m	$2\nu_2 \leftarrow \nu_2$ (hot band)
589	PQR s	vw	ν_2 bend (Π)
1167	PR m	vw	$2\nu_2$ overtone
1285	PR vs	vs	ν_1 symmetric stretch (Σ^+)
2223	PR vs	s	ν_3 anti-symmetric stretch (Σ^+)

(i) There are three atoms, so the number of vibrational modes is 3 (bent) or 4 (linear). In a linear molecule, the bending modes are degenerate, so there should be three distinct frequencies. The two extra bands must be hot bands, overtones or combination bands.

(ii) The strongest bands are 589 (s), 1285 (vs) and 2223 cm^{-1} (vs). They must correspond to fundamentals. These must be, respectively, the bending, symmetric stretching and antisymmetric stretching modes, based on their relative values. The bending mode is also perpendicular. Its PQR structure confirms this assignment.

(iii) Look for PR contours in the *fundamentals*. The ν_1 and ν_3 fundamentals have PR structure. The missing Q branch indicates that the molecule is linear.

(iv) Since there are common bands in IR and Raman, the molecule does not have a centre of symmetry. The only possible structure for a linear molecule with no centre of symmetry is N–N–O.

(v) The remaining two bands at 580 and 1167 cm^{-1} must be respectively the hot band and overtone of the bending mode (ν_2). The fact that the first band is temperature dependent indicates that it is a hot band. This is confirmed by its position (slightly less than ν_2). The second band is ~$2\nu_2$ and must be an overtone.

Acetylene (C_2H_2)

$\tilde{\nu}$ / cm^{-1}	Infrared	Raman	Band assignment	Mode
612	–	vw	ν_4, bend (Π_g)	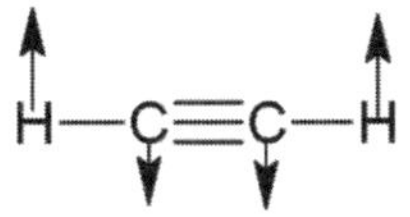 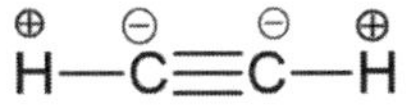
729	PQR	–	ν_5, bend (Π_u)	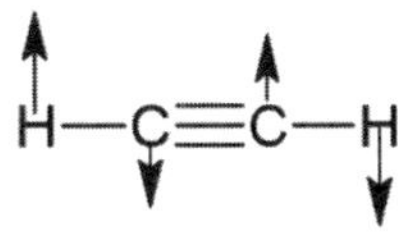 
1974	–	vs	ν_2, symmetric stretch (Σ_g^+)	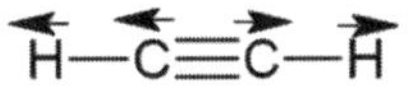
3287	PR	–	ν_3, asymmetric stretch (Σ_u^+)	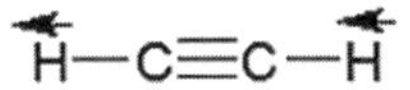
3374	–	s	ν_1, symmetric stretch (Σ_g^+)	

(i) There are four atoms, so there should be 7 or 6 fundamentals, depending on whether the molecule is linear or non-linear. The fact that there are only 5 indicates that the molecule is probably linear and the bending modes are degenerate. A molecule with four atoms will have three stretching modes and two bending modes. Both bending modes must be degenerate.

(ii) The bands at 612 and 729 cm^{-1} are bending modes because their wavenumbers < 1000 cm^{-1}, and the PQR contour of the band at 729 cm^{-1} confirms this.

(iii) The Q branch is missing in the 3287 cm^{-1} band. This confirms that the molecule is linear.

(iv) The bands that are IR active are Raman inactive and vice versa. Hence the molecule has a centre of symmetry. The only possible structure is H-C≡C-H.

(v) The band at 1974 cm^{-1} is very strong in Raman; hence, it must be the symmetric C≡C stretch since its wavenumber is less than 2500 cm^{-1}. The high wavenumber bands at 3287 cm^{-1} and 3374 cm^{-1} are the C-H stretches. The former is IR active, and so it must be the asymmetric stretch. The other band is Raman active, so it must be the symmetric C-H stretch.

We now take up a few more examples to illustrate the power of Group Theory in deducing the selection rules of infrared and Raman transitions.

Chloroform ($CHCl_3$)

Chloroform, like ammonia (Chapter 6), belongs to the C_{3v} point group.

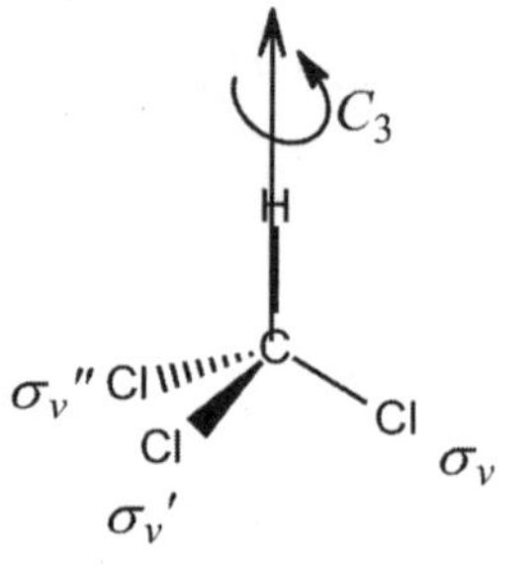

A similar analysis shows that
$\Gamma_{vib} = 3A_1 \oplus 3E$
because there are three additional degrees of freedom due to the extra atom. As for ammonia, all modes are infrared and Raman active. The computed Raman and infrared spectra of chloroform are shown in Figure 7.14, wherein it is clear that symmetric modes are strong in Raman and asymmetric modes are strong in infrared, and hence the complementary nature of the two kinds of spectroscopies is evident from the data. This is also obvious from Table 7.4, which lists the infrared and Raman bands.

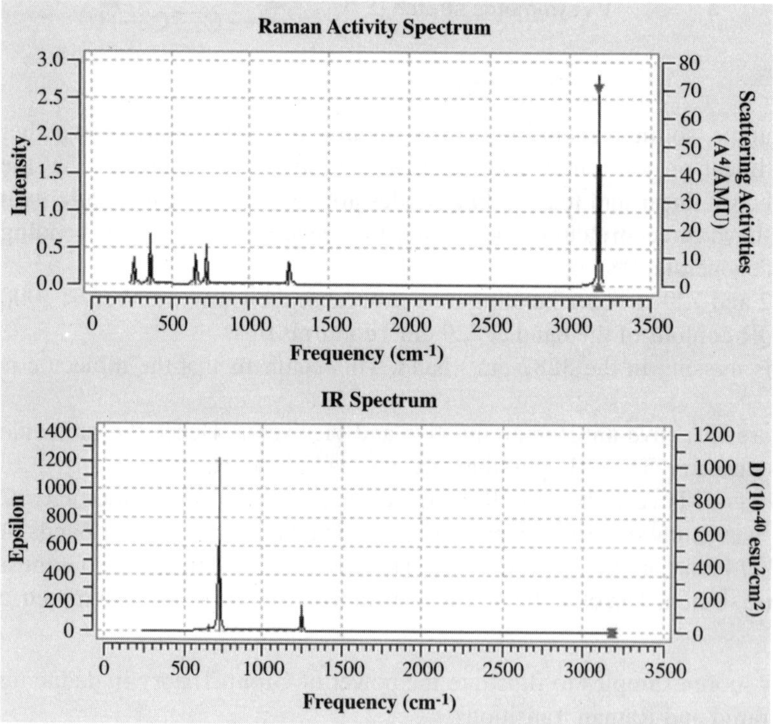

Figure 7.14. Computed Raman and IR spectrum of chloroform (B3LYP/6–311+G(d,p) calculation using Gaussian 09W)

Table 7.4. Vibrational spectrum of chloroform

ω/cm^{-1}	Symmetry	Infrared	Raman	Band assignment*
3034	A_1	m	w	ν_1, C–H str.
680	A_1	s	s	ν_2, sym. str. CCl_3
363	A_1	liq	m	ν_3, sym. deform. CCl_3
1220	E	vs	w	ν_4, bend
774	E	vs	w	ν_5, d–str.
261	E	liq	w	ν_6, d–deform.

*str. = stretch; sym. = symmetric; deform. = deformation

Carbon tetrachloride (CCl_4)

CCl_4 was commonly used as an organic solvent until its severe carcinogenic properties were discovered. This molecule belongs to the tetrahedral point group (T_d), whose symmetry elements are shown below:

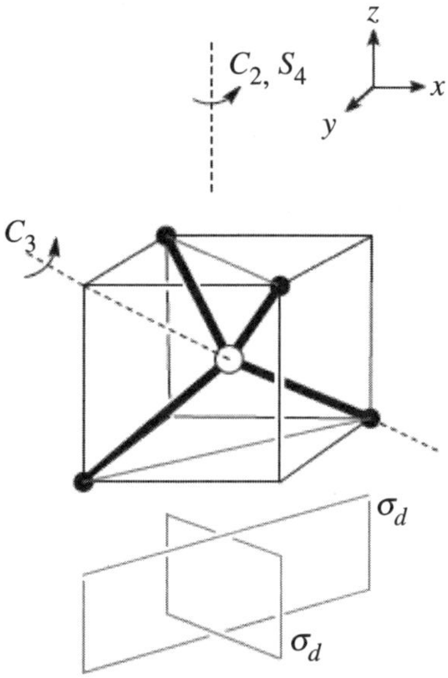

T_d	E	$8C_3$	$3C_2$	$6S_4$	$6\sigma_d$
$\chi(R)$	3	0	−1	−1	1
n_R	5	2	1	1	3
Γ_{3N}	15	0	−1	−1	3
$\Gamma_{trans} + \Gamma_{rot}$	6	0	−2	0	0
Γ_{vib}	9	0	1	−1	3
$\Gamma_{stretch}$	4	1	0	0	2
Γ_{bend}	5	−1	1	−1	1

$\Gamma_{stretch}$ gives $A_1 \oplus T_2$. Thus, there is one singly degenerate and one triply degenerate stretching mode, giving a total of four modes. The A_1 vibration is only Raman active (and polarized), while T_2 is both IR and Raman active. The remaining five are composed of $E \oplus T_2$. E is only Raman active. Hence, all modes are Raman active and four bands should be observed in Raman: A_1, E and two T_2, and only two in infrared corresponding to T_2.

The observed wavenumbers are given in Table 7.5.

Table 7.5. The observed wavenumbers in the vibrational spectrum of CCl_4

Symmetry species	No.	ω/cm^{-1}	Infrared[a]	Raman[b]
A_1	ν_1	459	Inactive	pol
E	ν_2	217	Inactive	depol
T_2	ν_3	776	789 (vs)	depol
			768 (vs)	depol
T_2	ν_4	314	w	depol

[a]vs = very strong; w = weak
[b]pol = polarized; depol = depolarized

The observed spectra are in conformity with expectations based on Group Theory. However, instead of four bands, five bands are observed in the infrared (Figure 7.15). This is again due to our old friend Fermi Resonance. A combination band ($\nu_1 + \nu_4$) is expected at the wavenumber 773 cm^{-1}. This has the symmetry $A_1 \otimes T_2$, which is also T_2 and so it can interact with the fundamental at 776 cm^{-1}, resulting in a Fermi doublet with very strong peaks at 789 and 768 cm^{-1} in the infrared. This splitting is also observed in the Raman spectrum of liquid carbon tetrachloride (Figure 7.15).

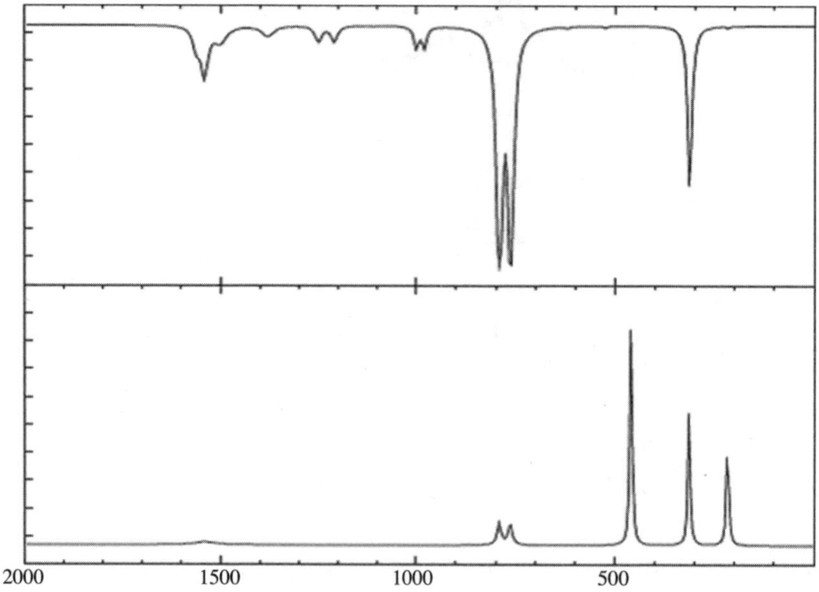

Figure 7.15. Infrared (top) and Raman spectra of CCl_4

Figure 7.16 shows the Raman spectrum with the intensity measured perpendicular and parallel to the incident radiation. The band at 459 cm^{-1} corresponds to the totally symmetric vibration (A_1) and is polarized ($\rho_p = 0.02$). The other two bands at 218 and 314 cm^{-1} correspond to the non-totally symmetric vibrations and are depolarized ($\rho_p = 0.75$).

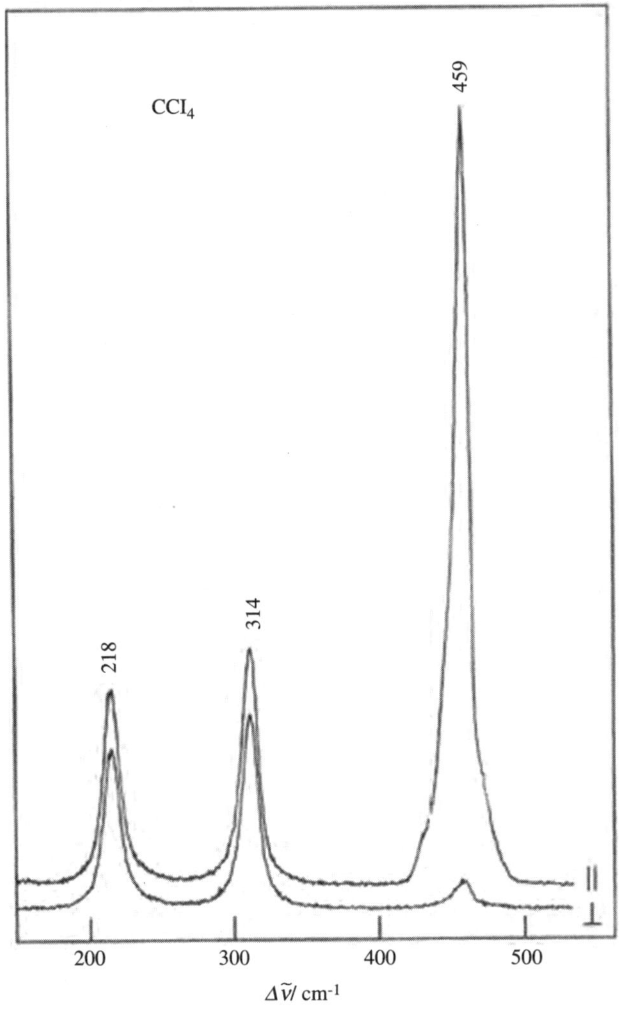

Figure 7.16. Raman spectrum of CCl_4 with the intensity measured parallel and perpendicular to the incident radiation

Though this kind of analysis is possible for simple molecules, the going becomes more tough as the number of atoms, and hence the number of vibrational modes, increases. One exception is the highly symmetrical molecule, fullerene (C_{60}), which has 20 hexagonal and 12 pentagonal rings. This molecule is a rare example of a molecule belonging to the icosahedral point group (I_h), which has an order $h = 120$. Yet, this molecule displays a relatively simple Raman spectrum (Figure 7.17) with only ten lines, though the molecule has 174 vibrational degrees of freedom. Eight of these belong to the five fold degenerate H_g representations and two to A_g.

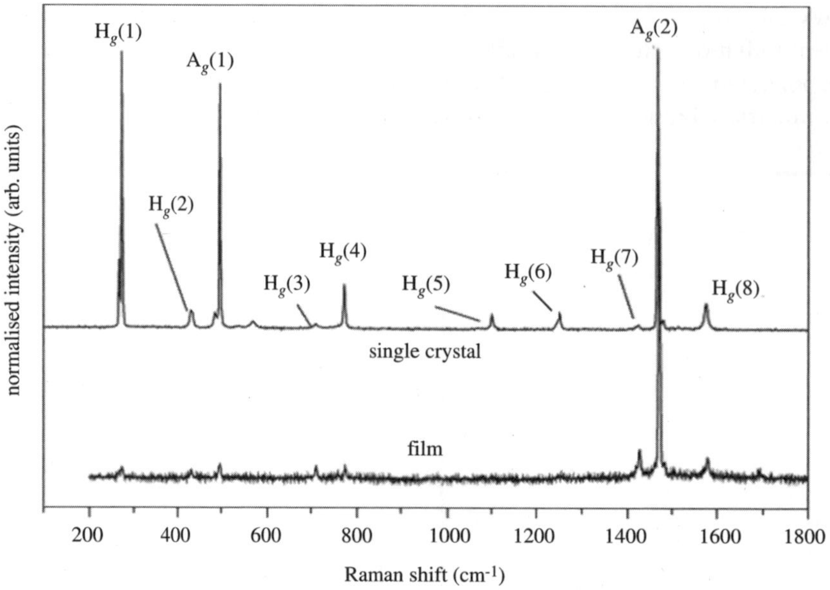

Figure 7.17. Raman spectrum of a C_{60} single crystal (1064 nm) and a polycrystalline C_{60} film (488.0 nm) recorded at 90 K for two different lasers. The spectra were normalized to a total height of unity [spectrum reprinted from Kuzmany et al. (2004) with permission]

The infrared spectrum is even simpler – there are just four allowed bands, each of them a triply degenerate F_{1u} representation (Figure 7.18). The high symmetry of the molecule makes its spectra particularly simple to interpret, and Group Theory helped characterize the molecule using its infrared and Raman spectra.

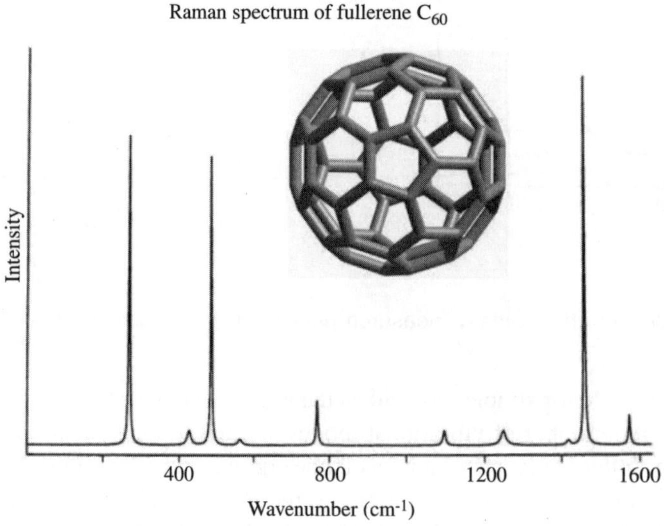

Figure 7.18. Absorbance FTIR spectrum of film of C_{60} on KBr substrate [Bethune et al. (1991), reprinted with permission]

However, such a simplification is not possible for most molecules. Fortunately, for organic compounds, the vibrational frequencies of individual functional groups are relatively independent of the environment of the group, and the idea of group frequencies is of immense help in the assignment of spectra. In the next section, group frequencies of certain commonly encountered groups in organic chemistry are discussed.

7.9 GROUP FREQUENCIES

In Chapter 5, we saw that, for diatomics, two factors determine the band position: the force constant of the bond (k) and the reduced mass (μ_{red}). Once these two bond properties are known, the band position can be calculated. Alternatively, if the band position is known, the force constant of the bond can be calculated, provided the reduced mass is known. In order to interpret the Raman and infrared spectra of organic molecules, the idea of group frequencies is useful. It is assumed that the k and μ_{red} values for a specific type of bond or functional group are largely unaffected by the other vibrations in a molecule. These characteristic frequencies can then be located in the spectrum to identify the functional groups present in the molecule.

Some group frequencies of organic molecules are listed in Table 7.6. In this table, stretching vibrations are abbreviated with the letter *nu* (ν), bending vibrations with the symbol *delta* (δ) and torsions by the symbol *rho* (ρ).

Table 7.6. Some group frequencies for functional groups in organic chemistry

Vibration		Wavenumber (cm^{-1})	Infrared intensity	Raman intensity[1]
O–H stretch	ν(O–H)	3650–3000	s	w
N–H stretch	ν(N–H)	3500–3300	m	m
C–H stretch of alkynes	ν(≡C–H)	3350–3300	s	w
C–H stretch of alkenes	ν(=C–H)	3100–3000	m	s
C–H stretch of alkanes	ν(-C–H)	3000–2750	s	s
C≡C stretch of alkynes	ν(C≡C)	2250–2100	variable, absent in symmetrical alkynes	vs
C=C stretch of alkenes	ν(C=C)	1750–1450	variable	vs-m
C–C stretch of aliphatic chains and cycloalkanes	ν(C–C)	1150–950	m-w	s-m
CC stretch of aromates (substituted benzenes)	ν(CC)	1600, 1580 1500, 1450 1000	m-s m-s variable, w	s-m m-w s
C=O stretch	ν(C=O)	1870–1650	vs	s-w

Vibration		**Wavenumber (cm^{-1})**	**Infrared intensity**	**Raman intensity**[1]
Anti-symmetric C–O–C stretch	ν_{as}(COC)	1150–1060	s	w
Symmetric C–O–C stretch	ν_s(COC)	970–800	variable, w	s-m
CH_2 bending vibrations, anti-symmetric CH_3 bend	$\delta(CH_2)$, $\delta_{as}(CH_3)$	1470–1400	m	m
Symmetric CH_3 bend	$\delta_s(CH_3)$	1380	s-m	m-w
Symmetric bending of aliphatic chains (C_n with $n = 3,\ldots,12$) = chain expansion	δ_s(CC)	425–150	variable, w	s-m
Amide I band (CONH stretch)	Amide I	1670–1630	s	m-w
Amide III band (C–N stretch, coupled with opening of the CNH angle)	Amide III	1350–1250	w-s	m-w

[1]s, strong; m, medium; w, weak; v, very; s, sym; as, asymmetric.
Note: Intensities relative to most intense band: vs 100–90%, s 90–75%, m-s 75–65%, m 65–35%, m-w 35–25%, w 25–10%, vw 10–0%

For a functional group, the force constant, and hence the band position, also depends on the following factors:

- *Hydrogen bonding:* It can weaken the bond and hence shift the band position. For example, the free O-H stretch is observed above 3500 cm^{-1}, but shifts to lower wavenumber on hydrogen bonding.
- *Substituents:* Electron donating and electron withdrawing substituents can respectively increase or decrease the electron density in a particular bond, causing a change of force constant and hence the band position.
- *Hybridization:* Comparison of the ν(C-H) for alkynes, alkenes and alkanes shows that it decreases with decreasing s character of the hybridized carbon atom, which changes from 50% (sp) to 33% (sp^2) to 25% (sp^3), showing that higher the s character of the hybridized carbon atom, higher is the force constant of the C-H bond.

Besides these observations, we have the following generalizations:

- Some vibrations are inherently weak in IR and strong in Raman. Examples are C≡C, C=C, P=S, S–S, C–S stretching vibrations, which are stronger in Raman. This is because these bonds have more covalent character.

- On the other hand, O–H and N–H stretches are stronger in the infrared since these bonds have more ionic character.
- Stretching bonds of multiple bonds are more intense in Raman than those of single bonds. The intensity decreases in the order C≡C > C=C > C–C.

Similar group frequencies cannot be listed for inorganic molecules, in which the spectra are more influenced by symmetry and the band shifts are larger. Moreover, there is greater overlap of functional groups, as seen by the overlap of the symmetric stretch regions of the sulphate and carbonate groups.

Symmetry also governs the selection rules, as seen in the previous chapter. For instance, the appearance of the Raman spectrum is strongly influenced not only by the nature of substituents on the benzene ring, but also by their number and relative positions.

7.10 COMPARISON OF IR AND RAMAN SPECTROSCOPIES

The main difference between IR and Raman spectroscopy is that between absorption and scattering. Symmetrical molecules like N_2 can be studied with Raman spectroscopy. The two techniques are complementary, i.e., what information cannot be obtained from IR spectroscopy can be easily deciphered from Raman. The Raman spectrum of air is dominated by a strong band at ~2350 cm^{-1} corresponding to nitrogen (80% of air) and a weaker band at ~1560 cm^{-1} due to oxygen. Contrast this with the IR spectrum which mainly shows the asymmetric bands of water vapour and carbon dioxide (Figure 6.9).

Raman offers certain advantages over IR spectroscopy:

- Normal glass or quartz containers can be used in Raman spectroscopy. For IR, salts like KBr/ NaCl/ CsI/ CaF_2 have to be used because glass is not transparent to IR radiation.
- Water is a poor Raman scatterer and hence aqueous solutions of inorganic compounds, hygroscopic, air-sensitive and biomolecules can be investigated by Raman spectroscopy.
- Low frequencies can be observed in Raman. Thus metal–ligand stretching frequencies of coordination compounds, which are in the 100–700 cm^{-1} wavenumber range, can be studied using Raman, but not with IR.
- Raman uses laser beams of very small diameter 1–2 mm. Hence smaller samples can be measured.
- Raman spectra are less cluttered though almost all fundamentals are Raman active. Hence, peak overlap is less likely.
- A disadvantage is that Raman instruments are generally more expensive and hence Raman spectroscopy has not been exploited as much as IR.
- Since Raman uses a strong laser source, there is danger of the sample getting burnt or decomposed.

7.11 ADVANCED RAMAN TECHNIQUES

We shall briefly mention two advanced Raman techniques – Resonance Raman and Coherent anti-Stokes Raman Spectroscopy (CARS). Raman intensities are usually weak and both these techniques are used to enhance intensities. In conventional Raman, the molecule is excited to a virtual level. In resonance Raman scattering, the excited wavelength is chosen to nearly coincide with an excited electronic level. We shall see in the next chapter that only a few vibrational modes contribute to the most intense scattering, and hence the spectrum is greatly simplified. This technique is used for examining metal ions in biological macromolecules, such as the iron in haemoglobin and cytochromes and the cobalt in vitamin B.

In CARS, laser beams of different frequencies are made to mix together to give rise to beams of the Stokes frequency, so that the coherent emission is at the anti-Stokes frequency, enhancing its intensity. This technique can be used to study Raman spectra in the presence of incoherent competing radiation, such as that in flames. The Raman spectrum is then interpreted in terms of the different temperatures in different regions of the flame.

For a long time, Raman experiments were being performed with lasers operating at visible frequencies, because of the inverse fourth power dependence of the intensity on the wavelength. Lately, however, near infrared (NIR) spectrometers using the Nd-YAG laser operating at 9398 cm^{-1} have been applied. The loss in intensity is made up by using the Fourier transform technique. The advantage is that the same FT instrument can be used for data acquisition of both the infrared and Raman spectra. Another advantage is that the problem of heating and fluorescence from the sample is avoided.

Raman spectroscopy finds uses in several fields, including forensic science, airport security and restoration of paintings.

7.12 SUMMARY

- The IR and Raman techniques are complementary to each other.
- In the Raman effect, the important property of the molecule that interacts with the electromagnetic radiation is its polarizability, and a dipole moment is induced in the molecule.
- Covalent bonds produce strong Raman lines, whereas ionic bonds show strong IR absorptions.
- The Rule of Mutual Exclusion states that for molecules with a centre of symmetry, bands that are IR active are Raman inactive and vice versa.
- Group Theory is a great aid in the assignment of spectra.
- The Raman technique offers many advantages over infrared spectroscopy. Aqueous solutions can be studied because of the weak bands of water. Glass and quartz cells can be used in Raman spectroscopy.
- Raman spectroscopy is being utilized for a variety of purposes now.

7.13 EXERCISES

1. What are the selection rules for rotational and vibrational Raman spectroscopy? The anti-Stokes lines of the pure *rotational* Raman spectrum of a molecule are of roughly the same intensity as the Stokes lines, but the anti-Stokes lines of the *vibrational* Raman spectrum are generally much weaker than the Stokes lines. Account for this observation.

2. The wavelength of the incident radiation in a Raman spectrometer is 484.895 cm^{-1}. What is the wavenumber and wavelength of the scattered Stokes radiation for the $J = 4 \leftarrow 2$ transition of $^{16}O_2$? $[\tilde{B}(^{16}O_2) = 1.4457 cm^{-1}]$.
3. Suppose that three conformations are proposed for the nonlinear molecule H_2O_2 (**1**, **2** and **3** below). The infrared spectrum of gaseous H_2O_2 has bands at 870, 1370, 2869 and 3417 cm^{-1}. The Raman spectrum of the same sample has bands at 877, 1408, 1435 and 3407 cm^{-1}. All bands correspond to fundamental vibrational wavenumbers and you may assume that the 870 and 877 cm^{-1} bands arise from the same normal mode as do the 3417 and 3407 cm^{-1} bands.
   (a) If H_2O_2 were linear, how many bands would H_2O_2 have?
   (b) Give the symmetry point group for each of the proposed conformations of nonlinear H_2O_2.
   (c) Determine which of the proposed conformations is inconsistent with the spectroscopic data. Explain your reasoning.

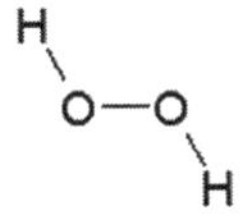

1

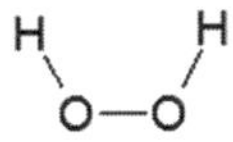

2

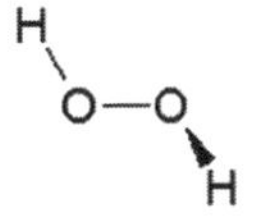

3

4. The planar molecule BF_3 is an oblate symmetric top.
   (a) Given that the *b*-axis lies along a B-F bond, draw a diagram to show where the *a*- and c-axes lie, and show that $I_a = 2I_b = 2I_c$.
   (b) Lines in the rotational Raman spectrum of BF_3 are spaced at intervals of 0.699 cm^{-1}. What is the BF bond length?
5. Explain the difference between Stokes and anti-Stokes lines in Raman spectroscopy.
6. The measured frequency shifts in the pure rotational Raman spectrum of oxygen are given below.
   (a) Determine the numbering of the rotational lines.
   (b) Evaluate the rotational and centrifugal distortion constants for the ground state of oxygen.

   Observed wavenumber (cm^{-1}) shifts of the rotational lines in the Raman spectrum of oxygen

   14.381 25.876 37.369 48.855 60.337 71.809 83.267 94.712 106.143 117.555 128.949

   (c) Determine also the vibrational frequency from your calculated values of $\tilde{B}$ and $\tilde{D}$.
7. Show that the wavenumbers of the Stokes lines in the rotational Raman spectrum of a linear molecule are given by

$$\tilde{\nu}(J) = \tilde{\nu}_i - 2\tilde{B}(2J + 3)$$

   where $\tilde{B}$ is the rotational constant, $\tilde{\nu}_i$ is the wavenumber of the incident radiation, and $J = 0, 1, 2, 3\ldots$.
8. The rotational Raman spectrum of $^{14}N_2$ is observed using light of wavenumber 29697.2 cm^{-1}. Stokes lines were observed at the following wavenumbers (in cm^{-1}):

   29685.2 29677.3 29669.4 29661.4

   Calculate the value of $\tilde{B}$ for $^{14}N_2$ and $^{15}N_2$.

9. (a) Calculate the moments of inertia of $^{12}C^{32}S_2$ and $^{16}O^{12}C^{32}S$ from the following data.
   Bond lengths:
   $r_{C\text{–}S}$ in CS_2: 154.7 pm
   $r_{C\text{–}S}$ in OCS: 157.2 pm
   $r_{C\text{–}O}$ in OCS: 112.5 pm
   The centre of mass of OCS is 53.8 pm away from the C atom.
   (b) Comment on the observation that the line spacings in the rotational Raman spectrum of CS_2 and OCS are roughly similar. The nuclear spin quantum number of ^{32}S is zero.
   (c) Outline how the bond lengths of OCS may be determined using rotational Raman spectroscopy.
10. A homonuclear diatomic molecule X_2 has first and second rotational Raman Stokes lines that are shifted from the wavenumber of the incident light by 14.457 cm^{-1} and 26.023 cm^{-1}. Deduce:
   (a) the nuclear spin of X,
   (b) the symmetry of the electronic state of X_2,
   (c) the rotational constant of X_2.
11. The infrared spectrum of an AB_2 molecule shows a band at 667 cm^{-1} (PQR) and another at 2350 cm^{-1}. The Raman spectrum shows a band at 1340 cm^{-1}. Deduce the structure of the molecule and assign the three frequencies.
12. (a) Find the symmetry species for the normal modes of *trans* (planar) N_2F_2.
   (b) Which are IR and which are Raman active?
   (c) Which are in-plane?
   (d) What would be the IR and Raman activity of (i) each first overtone, (ii) each combination of fundamentals.
   (e) What do you understand by the 'Rule of Mutual Exclusion' in relation to the infrared and Raman spectra of molecules?
13. (a) Describe the forms of the normal modes of vibration of a linear symmetrical XY_2 molecule and deduce, in terms of dipole moment and polarizability changes, which of the modes are infrared and/or Raman active. How are the selection rules modified if the molecule is linear but unsymmetric of the type XYY or XYZ? How do infrared vibration-rotation band contours and the degree of polarization of Raman lines aid in distinguishing bond-stretching from angle-bending frequencies in the vibrational spectra of linear molecules?
   (b) Interpret the observed infrared and Raman frequencies for the linear N_2O molecule given below in terms of the symmetry of the molecule (i.e., YXY or XYY) and the normal vibration frequencies?

Frequency(cm^{-1})	Infrared	Raman
2,223.5	vs, PR	s, depol.
1,285	vs, PR	vs, pol.
1,167	m, PR	vw
588	s, PQR	–

   Assign the bands to fundamentals, overtones or combinations and state the number of quanta of each fundamental involved in the overtones and combination bands. Calculate the zero-point energy of the molecule.

14. Explain the term Fermi resonance taking the example of the Raman spectrum of carbon dioxide. For this molecule, why is this phenomenon not observed in the infrared spectrum? Given that the bending mode of carbon dioxide has Π_u symmetry while the symmetric stretch has Σ_g^+ symmetry, explain the interaction of the overtone of the former with the fundamental of the latter.
15. Answer the following:
    (a) Pauli's exclusion principle is irrelevant for the nuclear spin factor in the total wave function of a heteronuclear diatomic molecule. Why?
    (b) $\Delta J = 0, \pm 2$ for rotational Raman transitions in a linear molecule. Why?
    (c) What is meant by the polarizability ellipsoid of a molecule?
    (d) From the following list, identify those which may show pure rotational Raman spectra: CH_2Cl_2, CH_3Cl, CH_4, C_2H_6, SF_6, H_2, HCl. Give reasons for your answer. Which of these molecules are also microwave active?
    (e) To intensify anti-Stokes lines in Raman spectra, will you heat the sample or cool it?
    (f) How can you distinguish between overtones and hot bands?
    (g) What are the two main effects of Fermi resonance?
16. The pure rotational Raman spectra of benzene and deuterobenzene have been analyzed to give the rotational constants 0.18960 cm^{-1} for C_6H_6 and 0.15681 cm^{-1} for C_6D_6. Determine the C–C and C–H bond lengths in benzene.
17. What is meant by the term *polarizability*? State the selection rules for Raman scattering, and show that they give rise to the rotational Stokes spectrum consisting of lines at wavenumbers

$$\tilde{\nu} = \tilde{\nu}_i - \tilde{B}(4J+6)$$

where $\tilde{\nu}_i$ is the wavenumber of the incident radiation. What is the corresponding expression for anti-Stokes rotational transitions?
18. Determine the number of vibrational modes that will be observed in (a) an infrared experiment and (b) a Raman experiment for hydrazine (H_2N-NH_2) in each of the following conformations:

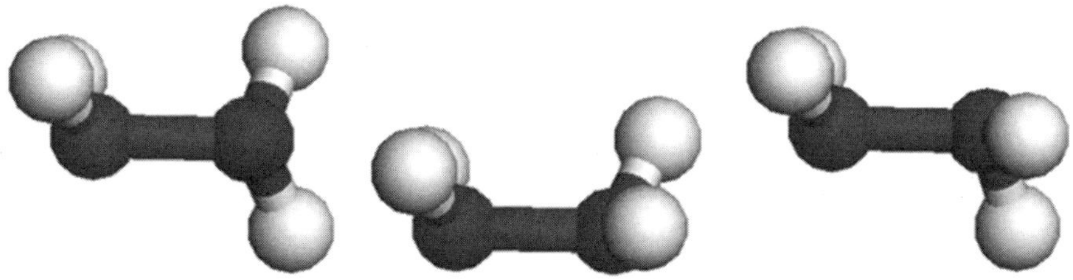

19. Some general trends that can be observed from Table 7.6 are
    - Vibrations involving hydrogen atoms usually occur at higher wavenumbers
    - ν(C–C) < ν(C=C) < ν(C≡C)
    - stretching vibrations have higher group frequencies than the corresponding bending vibrations.

    Explain why.
20. The planar molecule 1,3-butadiene might have either the s-*cis* or s-*trans* configuration, in which the CH_2 groups lie either on the same side or on opposite sides of the carbon-carbon single bond, respectively.

Investigation of the rotational Raman spectrum of butadiene shows a simple spectrum with lines of nearly equal spacing. Although both possible forms are asymmetric tops, the rotational spectrum of one form would resemble the simple spectrum of a symmetric top. Which form does the rotational Raman spectrum favour? Given that the A, B and C values are 0.377, 0.150 and 0.136 cm^{-1}, respectively, for *trans* butadiene, and 0.680, 0.202 and 0.156 cm^{-1}, respectively, for *cis* butadiene.

21. Elements A and B can combine to form a triatomic molecule AB_2 of molecular mass 76. The ground state of A is $^3P_0$ and its atomic number is less than 10. The ground state of B is $^3P_2$.
The molecule gives the following spectra:

$\tilde{\nu}$ / cm^{-1}	
397	IR (gas), strong, PQR contour
656	Raman (liq), strong, polarized
796	Raman (liq), weak, polarized
878	IR (gas), very weak, no definite contour
1523	IR (gas), strong, PR contour
2184	IR (gas), weak, PR contour

(a) Identify the molecule, justifying your choice. Explain the atomic term symbols.
(b) The molecule could have one of the following structures (i) symmetric linear, (ii) asymmetric linear, (iii) symmetric bent, and (iv) asymmetric bent. Which structure is consistent with this data? Give reasons for ruling out the other structures.
(c) Suggest assignments for the frequencies, listing them as fundamentals, overtones and combination bands. Explain the contour of each band.
(d) Classify the bands in terms of the point group of the molecule.
(e) Discuss the possibility of *Fermi resonance* in this case and list, giving reasons, the possible frequencies which can interact because of Fermi resonance.

22. Acetylene shows two C-H stretching vibrations: a symmetrical one at 3374 cm^{-1} (H—C≡C—H) and an unsymmetrical one at 3287 cm^{-1} (H—C≡C—H).
(a) Which of these vibrations will be Raman active and why?
(b) Calculate the Raman wavelength for a 453.8 nm exciting line.
(c) Calculate the IR active frequency.

23. When CCl_4 is irradiated with the 435.8 nm mercury line, Raman lines are obtained at 439.9, 441.8, 444.6 and 450.7 nm. Calculate the Raman frequencies of CCl_4 (expressed in wavenumber).

24. What relative intensities of the Raman lines on the short and long wavelength side of 435.8 nm would be expected for the Raman lines at 27 °C of $SOCl_2$ that correspond to the vibrational transition of 283 cm^{-1}?

8

Electronic Spectroscopy of Diatomic Molecules

Evans boldly put 50 atm of ethylene in a cell with 25 atm of oxygen.
The apparatus subsequently blew up,
but luckily not before he obtained the spectra shown in figure 8.
— A. J. Merer and R. S. Mulliken, *Chem. Rev.* 69, 645 (1969)

8.1 INTRODUCTION

The next higher energy spectroscopic regions are the visible and ultraviolet regions, which together constitute the region where most electronic transitions are observed. Just as we observed that lower energy rotational transitions accompany vibrational transitions in the infrared, in the UV / Vis region, both vibrational and rotational transitions occur along with electronic transitions. Electronic spectra are therefore broad, with one electronic transition spread over several vibrational transitions, which provide a coarse structure and rotational transitions a fine structure.

8.2 THE BORN–OPPENHEIMER APPROXIMATION

Since the electronic, vibrational and rotational transition energies span several orders of magnitude

$$\Delta E_{\text{el}} \approx 10^3 \Delta E_{\text{vib}} \approx 10^6 \Delta E_{\text{rot}} \tag{8.1}$$

we can consider them essentially independent of each other and apply the Born–Oppenheimer approximation, which states that "Since the energies of the various motions are very different, motions of a diatomic molecule may be considered as independent." We have seen the approximation breaking down, as in centrifugal distortion, but we shall initially ignore these complications because, compared with electronic energies, these deviations are negligible. However, the anharmonicity can certainly not be ignored.

Hence the total energy is given by:

$$E \approx E_{\text{el}} + E_{\text{vib}} + E_{\text{rot}} \tag{8.2}$$

Since transition energies are usually reported in wavenumber (reciprocal cm) units, we may write equivalently:

$$T \approx T_e + G(\upsilon) + F(J) \tag{8.3}$$

where T is the total energy term ($= E/hc$), T_e is the electronic energy term ($= E_{el}/hc$), and the other terms are the vibrational and rotational terms. The transition energy is the difference in the total energies of the excited and ground states.

$$\tilde{\nu}_{spect} = \Delta T_e + \Delta G(\upsilon) + \Delta F(J) \tag{8.4}$$

Since the rotational term is much smaller than the electronic and vibrational terms, initially we choose to ignore this and express the total energy as the sum of only the electronic and vibrational terms. As before, we denote the excited state with a single prime and the ground state with a double prime, so that

$$\begin{aligned}\tilde{\nu}_{spect} \approx \Delta T_e + \Delta G(\upsilon) = (T_e' - T_e'') + \omega_e'\left(\upsilon' + \tfrac{1}{2}\right) - x_e'\omega_e'\left(\upsilon' + \tfrac{1}{2}\right)^2 \\ -\left\{\omega_e''\left(\upsilon'' + \tfrac{1}{2}\right) - x_e''\omega_e''\left(\upsilon'' + \tfrac{1}{2}\right)^2\right\}\end{aligned} \tag{8.5}$$

Figure 8.1 illustrates the various quantities in the expression. Conventionally, the ground electronic state is represented by the symbol X and the excited states of the same multiplicity by the symbols A, B, etc., in order of increasing energy. Excited states of different multiplicity are labelled with lower case letters a, b, c, etc. In polyatomic molecules (but not diatomic molecules) it is customary to add a tilde (e.g., $\tilde{A}$) to these empirical labels to prevent possible confusion with the symmetry species label.

The ground and excited electronic states are represented by Morse potential energy curves. These are plotted for the more general case, where an electronic transition causes a weakening of the bond, and hence an increase in the internuclear separation, i.e., $r_e' > r_e''$.

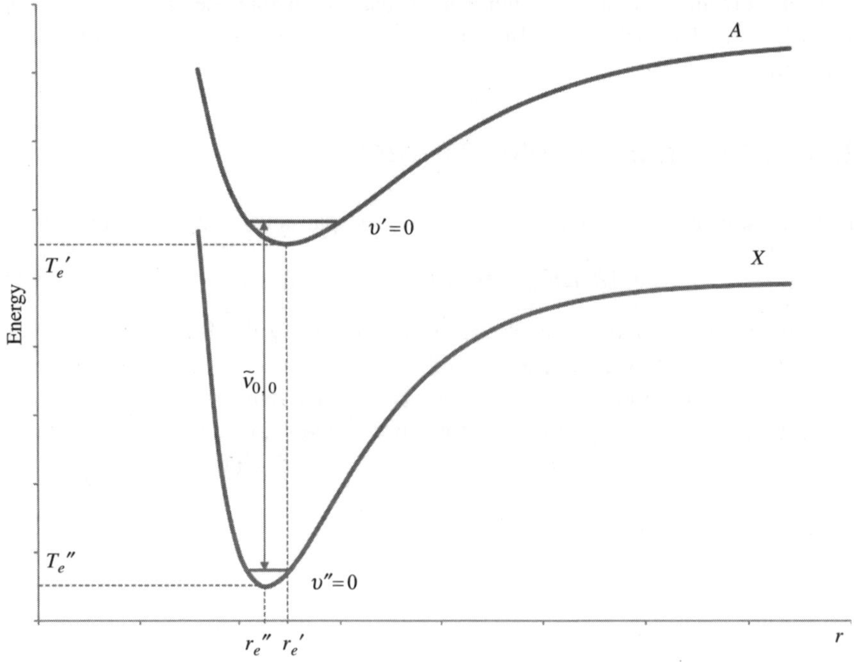

Figure 8.1. Electronic transitions

Since there are five unknowns in equation (8.5), i.e., x_e', ω_e', x_e'', ω_e'' and ΔT (= $T_e' - T_e''$), if at least five lines in the spectrum can be resolved, all five unknowns can be determined. Thus, electronic spectroscopy can provide information not only about the ground state but also about the excited electronic states, which is particularly valuable, since excited states are extremely unstable, and therefore short-lived.

8.3 VIBRATIONAL COARSE STRUCTURE

Thus, electronic transitions depend on the selection rules for electronic and vibrational transitions. Unlike microwave and infrared spectroscopies, all molecules have electronic spectra. There is essentially no selection rule for vibrational transitions from one electronic state to another – all are allowed (Figure 8.2). Each transition is labelled as (υ', υ''), i.e., the *upper state* is written first in the brackets. The first five lines from the left in Figure 8.2 are thus labelled (0,0), (1,0), (2,0), (3,0) and (4,0), respectively, and are called a υ' *progression* since υ' increases (progresses) by 1 for each line in the set. Another υ' progression, this time starting from $\upsilon'' = 1$ [(1,1), (2,1), (3,1), (4,1), (5,1)] and a υ'' progression [(1,0), (1,1), (1,2), (1,3) and (1,4)] are also shown in the figure.

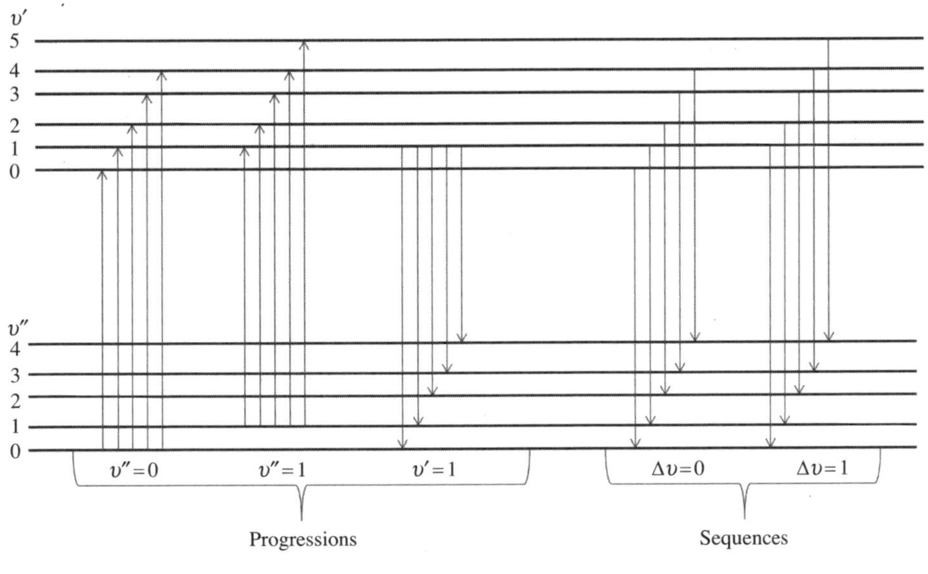

Figure 8.2. Vibrational progressions and sequences

Sequences are transitions with fixed values of $\Delta\upsilon$. Two such emission sequences are also shown in the figure. Fortunately, absorption spectra are much simpler, since at room temperature only the $\upsilon'' = 0$ level is populated (Chapter 5), and hence only the first υ' progression originating from $\upsilon'' = 0$ is important in absorption spectroscopy (Figure 8.3).

Even this simplification can result in a large number of (υ',0) transitions. This has the potential to complicate electronic spectra, but the Franck–Condon principle simplifies matters.

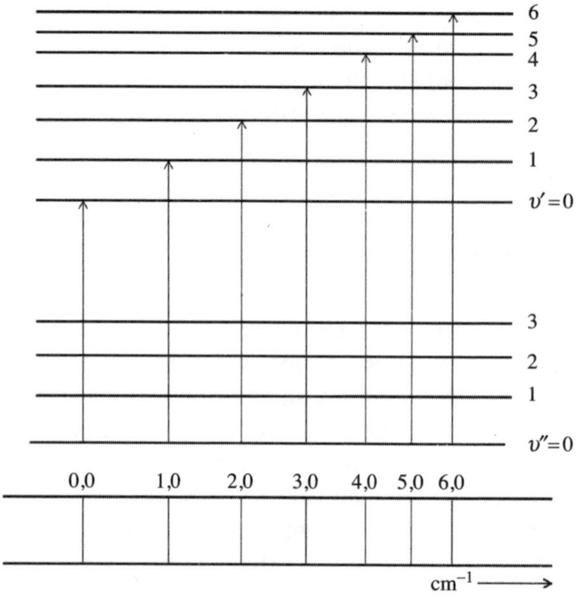

Figure 8.3. The (υ',0) progression

8.4 FRANCK–CONDON PRINCIPLE

This principle states that "Since electronic transitions occur very rapidly ($<10^{-15}$ s), vibration and rotation of the molecule do not change the internuclear distance appreciably during the transition." According to this principle, electronic transitions occur so fast that the nucleus gets no opportunity to relax during the transition. The logic for this rule is that the nucleus is more than 2000 times heavier than the electron. It cannot respond quick enough during electronic transitions, which are thus vertical transitions. Suppose a transition takes place from the $\upsilon'' = 0$ level of the ground electronic state. Since the $\upsilon' = 0$ level is the lowest vibrational energy level of the excited electronic state, one would expect the transition $\upsilon' = 0 \leftarrow \upsilon'' = 0$ to be the most intense. However, except in certain situations, this is not always the case. The excited state usually has weaker bonding than the ground state because electronic transitions are usually from bonding to antibonding orbitals, resulting in decreased bond orders and increased bond length (Figure 8.1) in the excited state. In order for the $\upsilon' = 0 \leftarrow \upsilon'' = 0$ transition to take place, the bond length would have to continuously increase during the transition. The Franck–Condon principle states that this does not happen because the nuclei are more massive and hence much slower than electrons. Thus, electronic transitions are vertical (no change in internuclear distance during the transition) and end at a vibrational level of the excited state where the electron has a significant probability of being found. From our study of vibrational spectroscopy (Chapter 5), we saw that the probability distribution in the Morse potential energy curve is as shown in Figure 8.4.

We had discussed the probability distribution in detail in that chapter. We had found that for $\upsilon = 0$, the probability is maximum at $x = 0$, i.e., $r = r_e$. As the vibrational quantum number increases, finding the molecule at the equilibrium internuclear distance becomes less and less

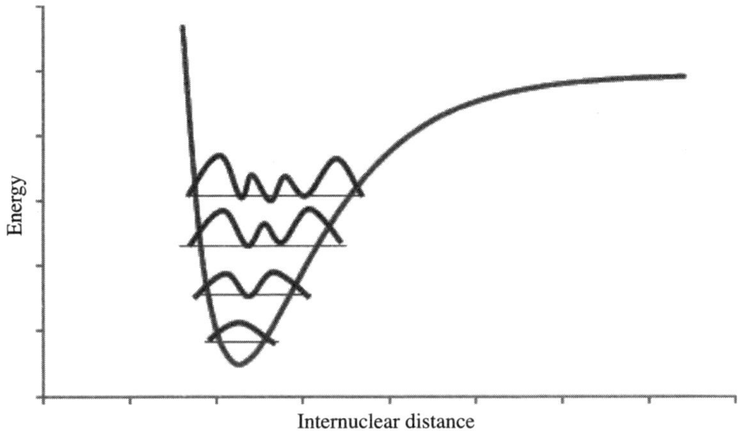

Figure 8.4. Probability distribution for a diatomic molecule

probable and it is more likely to be found at the turning points. A vertical transition may terminate at $\upsilon' = 0$ if the vertical arrow leads to the middle of this state, else it may terminate at a higher υ' value at the turning points where there are again peaks in the probability distribution curves.

Three situations may arise. In the first, there is no appreciable change in the bonding in the excited state, as, for example, when an electron is excited from a nonbonding or weakly bonding level. The equilibrium internuclear distance for the excited state is then almost the same as that of the ground state, i.e., $r_e' \approx r_e''$. A vertical transition leads to $\upsilon' = 0$ [Figure 8.5 (a)], with the resulting spectrum as shown at the bottom of the figure. The (0,0) transition corresponds to a pure electronic transition with energy ΔT_e since there is no accompanying vibrational energy change. This is also called an *adiabatic* transition, as it is the lowest energy electronic transition between two electronic states. Besides the (0,0) transition, a few (υ', 0) transitions for small υ' may also be observed, with progressively decreasing intensities.

The second situation arises when the electron is excited from a bonding to an antibonding orbital, in which case $r_e' > r_e''$ because of the weaker bonding in the excited state. In that case, Figure 8.5(b) shows that a vertical transition terminates at one of the higher energy vibrational levels of the excited electronic state, and this is the most intense transition. A few transitions to neighbouring levels are also observed, and the spectrum bears the usual Gaussian shape.

Finally, a situation may arise where $r_e' >> r_e''$, i.e., the electron leaves a strongly bonding orbital. In that case, a few transitions to high vibrational quantum numbers of the excited state may be observed, followed by a continuum. Figure 8.5(c) shows that the spacing between adjacent vibrational lines rapidly decreases until it finally becomes zero and a continuum results. The vertical transition terminates at a high quantum number, which is not bounded on the right-hand side. When a vibration is executed, there is no barrier at the right to stop the vibration, and the internuclear distance can increase indefinitely, leading to bond dissociation. The onset of the continuum marks the point at which the two atoms have just separated. As seen in Chapter 5, the minimum energy required to break a bond is called its dissociation energy. More energy than the dissociation energy can be continuously absorbed, since the excess energy goes as kinetic energy of the two separated atoms, and kinetic energy is not quantized. Hence, a continuum is observed.

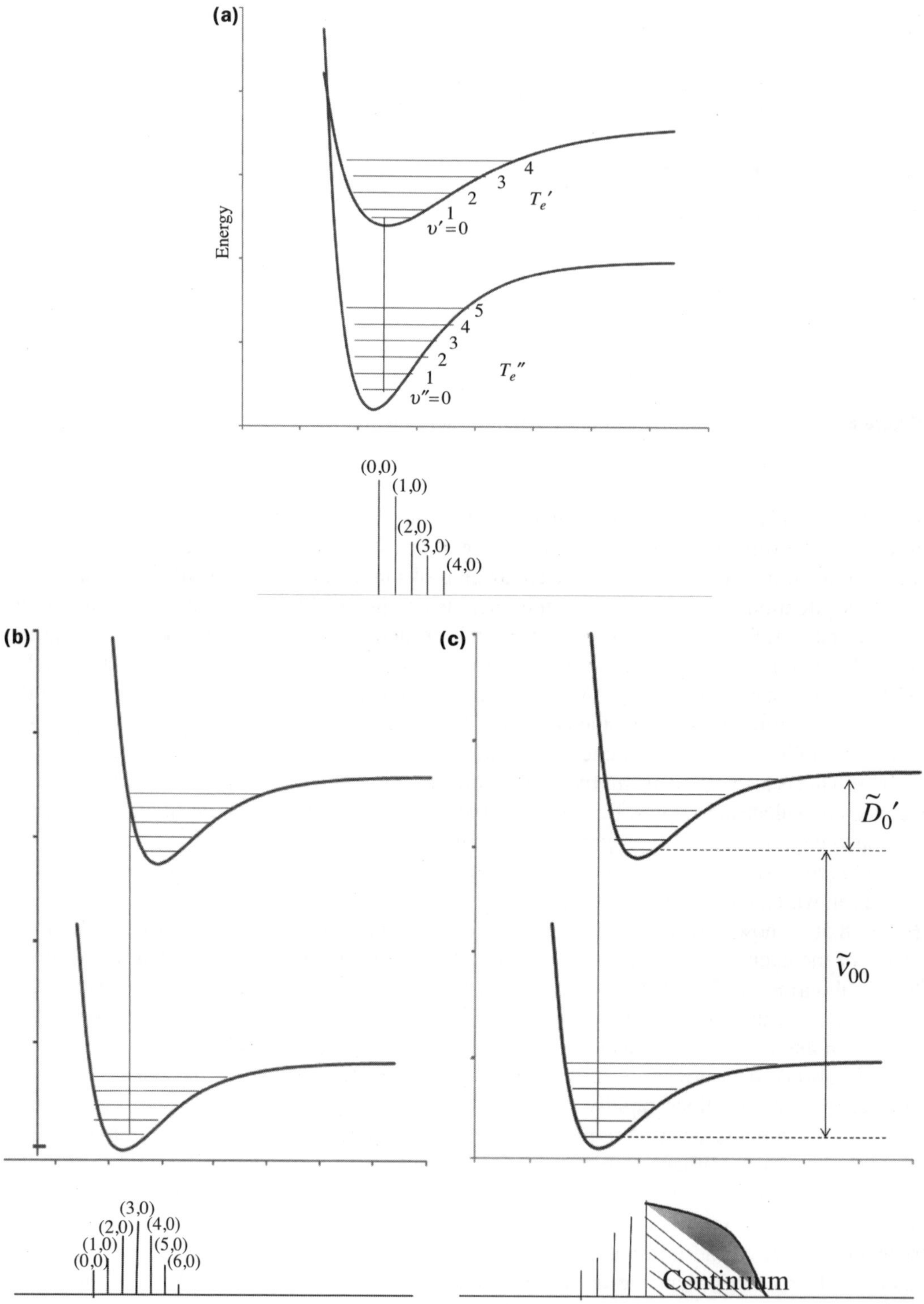

Figure 8.5. Franck–Condon transitions

8.5 DISSOCIATION ENERGY

The dissociation energy is defined as the least energy required to break a bond. In Chapter 5, we had estimated this quantity for the ground state of HCl using infrared data and found that the estimated value is much larger than the experimental value. This was ascribed to the fact that only the first few lines are observable in the infrared, and hence we cannot account for higher anharmonicity terms that become important at higher vibrational quantum numbers.

No such limit exists for electronic transitions. We shall see in this section that electronic spectroscopy allows us to obtain not only the dissociation energy of the ground state, but also that of the excited state. The latter is important as most excited states are unstable and hence short-lived (~10^{-9} s).

From Figure 8.5, we can see that the point at which dissociation sets in is equal to the dissociation energy of the excited state plus the energy of the (0,0) transition. Hence

$$\tilde{\nu}_{\text{continuum}} = \tilde{D}_0' + \tilde{\nu}_{0,0} \tag{8.6}$$

The latter is nothing but the pure electronic transition energy.

Even if the continuum limit is not reached in a spectrum [Figure 8.5(b)], the dissociation energy can still be estimated. We had seen in Chapter 5 that anharmonicity decreases the separation between adjacent vibrational lines with increasing vibrational quantum numbers. Using equation (8.5) and referring to Figure 8.5(b), the separation between adjacent lines originating from the same ground state level υ'' is given by

$$\Delta\tilde{\nu} = \Delta G(\upsilon') = G(\upsilon'+1) - G(\upsilon') = \omega_e' - 2x_e'\omega_e'(\upsilon'+1) \tag{8.7}$$

If we plot the separation $\Delta\tilde{\nu}$ versus $\upsilon' + 1$, a graph of the type given in Figure 8.6 is obtained. The slope of the graph is $-2x_e'\omega_e'$ and the intercept is ω_e', from which the vibrational frequency and anharmonicity constant for the excited state can be easily extracted. Such a plot is called a Birge–Sponer plot [Figure 8.6(a)]. Extrapolation to $\Delta\tilde{\nu} = 0$ gives $\upsilon_{\text{max}}+1$, the quantity estimated in Chapter 5 for the ground electronic state of HCl using infrared data. In that chapter, we had assumed a linear relationship based on a few lines, since we had very few lines to base our calculations on. Electronic spectroscopy gives far more lines and we see that $\Delta\tilde{\nu}$ does not depend linearly on υ if the second and higher order anharmonicity constants are significant [Figure 8.6(a)]. Rather a curve is obtained, and $\tilde{\nu}$ converges faster than that estimated by a linear relationship. Hence, we had obtained a higher value of the dissociation energy than the true value using our limited data from infrared spectroscopy.

Most importantly, the sum of all these separations is the dissociation energy of the excited electronic state. Since there are very many levels involved, the sum may be replaced by an integral and we may write

$$\tilde{D}_0' = \sum_{\upsilon'=0}^{\upsilon'_{\text{max}}} \Delta\tilde{\nu} = \int_0^{\upsilon'_{\text{max}}} \Delta\tilde{\nu}\, d\upsilon \tag{8.8}$$

Hence, the area under the graph provides the dissociation energy. This area is the total area of the large triangle in Figure 8.6(a). Even if the Birge–Sponer extrapolation is nonlinear, as is the case when the higher order anharmonicity constants are significant, a simple computer

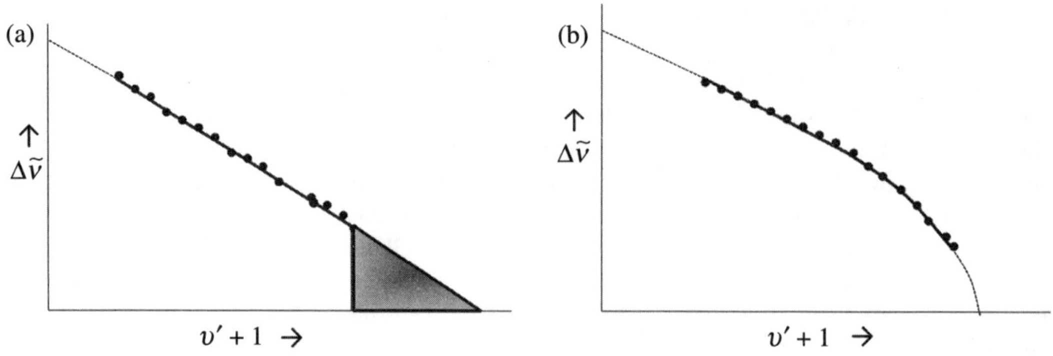

Figure 8.6. Birge–Sponer extrapolation (a) negligible second order anharmonicity constants; (b) second-order anharmonicity constants not negligible

routine (using, for example, Simpson's rule for integration) can be used to extrapolate and estimate the area under the curve.

Alternatively, a plot of $\Delta\tilde{\nu}$ versus $\tilde{\nu}$ directly gives by extrapolation $\tilde{\nu}_{\text{continuum}}$ as the point where $\Delta\tilde{\nu}$ becomes zero.

Once the excited state dissociation energy has been estimated, obtaining the ground state dissociation energy is a simple matter. The sketch on the left of Figure 8.7 shows another quantity E_{ex}. When the ground state dissociates, the products are ground state atoms. However, the excited state dissociates to one or both excited state atoms. The difference in energy is E_{ex}, which is the excitation energy of the atoms. We may thus write from Figure 8.7

$$\tilde{\nu}_{\text{continuum}} = \tilde{D}_0'' + E_{ex} \tag{8.9}$$

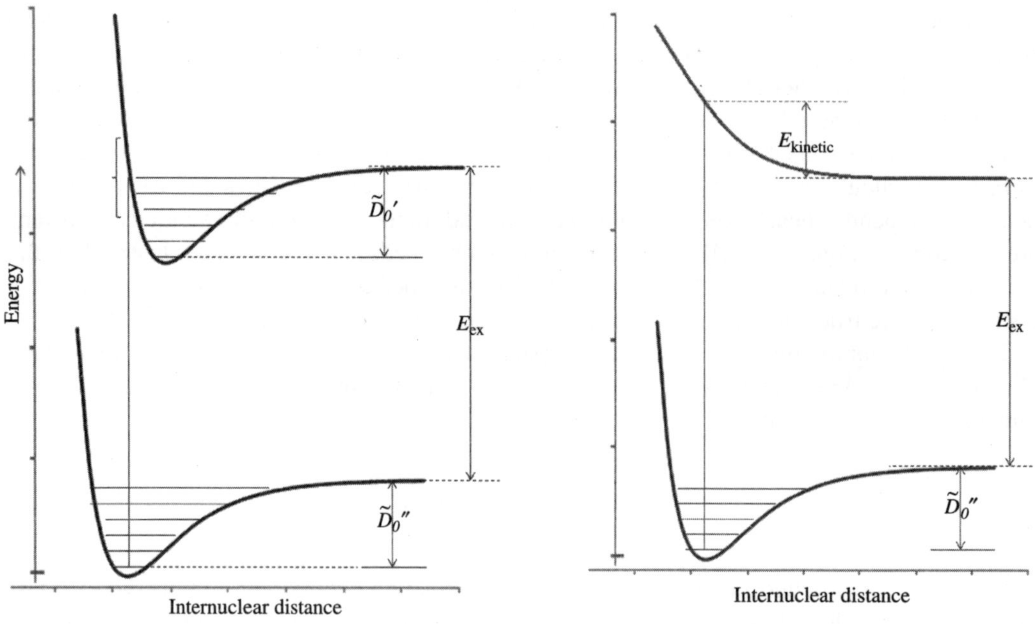

Figure 8.7. Dissociation upon excitation to stable upper state and continuous upper state

Usually, a rough value of $\tilde{D}_0''$ is known from thermochemical data. This allows us to estimate E_{ex}. This value is then compared with the atomic excitation energies obtained from atomic spectra. One of these may match with the estimated value, in which case the exact value from atomic spectra is inserted in equation (8.9) to get the exact ground state dissociation energy.

Example 8.1 The absorption frequencies for the Schumann–Runge bands of O_2 include (near the dissociation limit):

υ'	$\tilde{\nu}/cm^{-1}$
16	56,720
17	56,852
18	56,955
19	57,030
20	57,083
21	57,115

(a) Plot the differences $\Delta\tilde{\nu} = \tilde{\nu}_{\upsilon'+1} - \tilde{\nu}_{\upsilon'}$ versus $\tilde{\nu}$ and determine the wavenumber at which the continuum begins by extrapolating to $\Delta\tilde{\nu} = 0$.

(b) The lower electronic state of the transition is the ground state and dissociates to ground state oxygen atoms (^{3}P); the excited state involved dissociates to a ^{3}P-oxygen atom and a ^{1}D-oxygen atom. From this and the atomic oxygen energies, O(^{1}D)-O(^{3}P) = 15,868 cm^{-1}, calculate the dissociation energy, $\tilde{D}_0''$, for ground state oxygen.

Solution

(a) As shown below, a plot of $\Delta\tilde{\nu} = \tilde{\nu}_{\upsilon'+1} - \tilde{\nu}_{\upsilon'}$ versus $\tilde{\nu}$ yields a straight line whose slope is –0.271 cm^{-1} and intercept is 15557 cm^{-1}. Extrapolation to $\Delta\tilde{\nu} = 0$ gives $\tilde{\nu}_{\text{continuum}} = 57406\ \text{cm}^{-1}$.
Incidentally, since the intercept is 15557 cm^{-1}, this corresponds to ω_e', and $x_e = 0.271/(2 \times 15557) = 0.0087$.

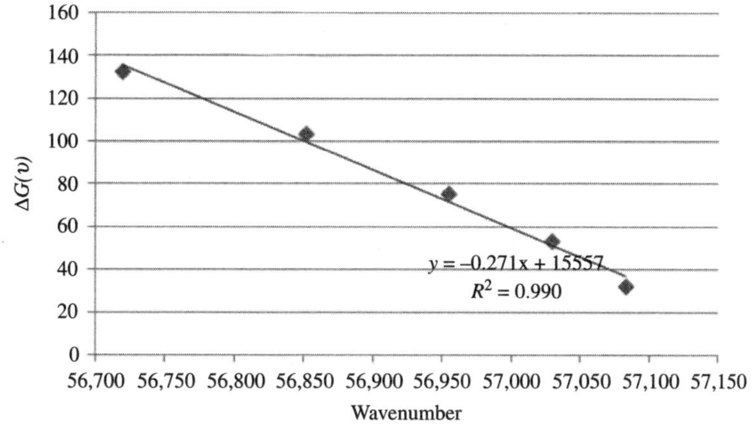

(b) From equation (8.9), $\tilde{D}_0'' = \tilde{\nu}_{\text{continuum}} - E_{ex} = 57406 - 15868 = 41538\ \text{cm}^{-1}$.

Example 8.2 The electronic spectrum of the I_2 molecule shows the $\tilde{\nu}_{00}$ line at 15642 cm^{-1} and the convergence limit at 20020 cm^{-1}.

The dissociation products of the upper electronic state are one iodine atom in its ground state $\left({}^2P_{3/2}\right)$ and one iodine atom in the excited state $\left({}^2P_{1/2}\right)$. From the atomic spectrum of iodine, the excitation $\left({}^2P_{1/2}\right) \leftarrow \left({}^2P_{3/2}\right)$ has a value of 7600 cm^{-1}.

Calculate the dissociation energy of the iodine molecule in its (a) ground state, and (b) excited state.

Solution

From equation (8.6), $\tilde{\nu}_{\text{continuum}} = \tilde{D}_0' + \tilde{\nu}_{0,0}$. It is given that $\tilde{\nu}_{\text{continuum}} = 20020\ \text{cm}^{-1}$ and $\tilde{\nu}_{0,0} = 15642\ \text{cm}^{-1}$. Therefore, $\tilde{D}_0' = 4378\ \text{cm}^{-1}$.

From equation (8.9), $\tilde{\nu}_{\text{continuum}} = \tilde{D}_0'' + E_{\text{ex}}$ and $E_{\text{ex}} = 7600\ \text{cm}^{-1}$. Hence, $\tilde{D}_0'' = 12420\ \text{cm}^{-1}$.

In some cases, more than one spectroscopic dissociation limit is found corresponding to dissociation into two or more different states of products. The separation between the excitation energies often corresponds with the separations between only one set of excited states of the atoms observed spectroscopically. The nature and the energies of the excited products are thus also obtained from electronic spectra.

The sketch on the right of Figure 8.7 shows another way in which electronic excitation can lead to dissociation. In this case, the excited state is unstable (unbound) as it has no energy minimum, and every transition leads to a continuum because the molecule spontaneously dissociates in the excited state.

It may also happen that a few normal transitions are observed, followed by a continuum, and then the regular pattern is restored. This phenomenon is known as *predissociation* and occurs when the Morse curve of a stable excited state intersects with the potential energy curve of another unstable excited state, as shown in Figure 8.8.

A transition to any of the upper vibrational levels labelled *a*–*c* would lead to a normal vibrational-electronic spectrum, along with the rotational fine structure. Two such bands are shown to the left of Figure 8.9. However, transitions to the levels *d*–*f* may cause the molecule to cross over to the continuous potential energy in the space of one vibration, causing dissociation and a continuum to appear. Such transitions are termed 'radiationless transfer' and occur faster than the time taken for a complete rotation (~10^{-10} s), but are slower than the time taken for a vibration. Since the molecule is unable to complete a rotation during the crossover, all rotational fine structure will be destroyed, but the vibrational fine structure will be retained. In case the crossover is also faster than the vibrational transition, a complete continuum will be observed (Figure 8.9). The remaining transitions (*g*–*j*) shown in Figure 8.8 proceed normally since the crossover point is crossed, leading to transitions such as the last two in Figure 8.9.

8.5.1 The Franck–Condon factor: Quantum mechanical treatment

So far we have not discussed the selection rules, apart from the fact that all vibrational transitions are allowed, subject to intensity distribution according to the Franck–Condon principle.

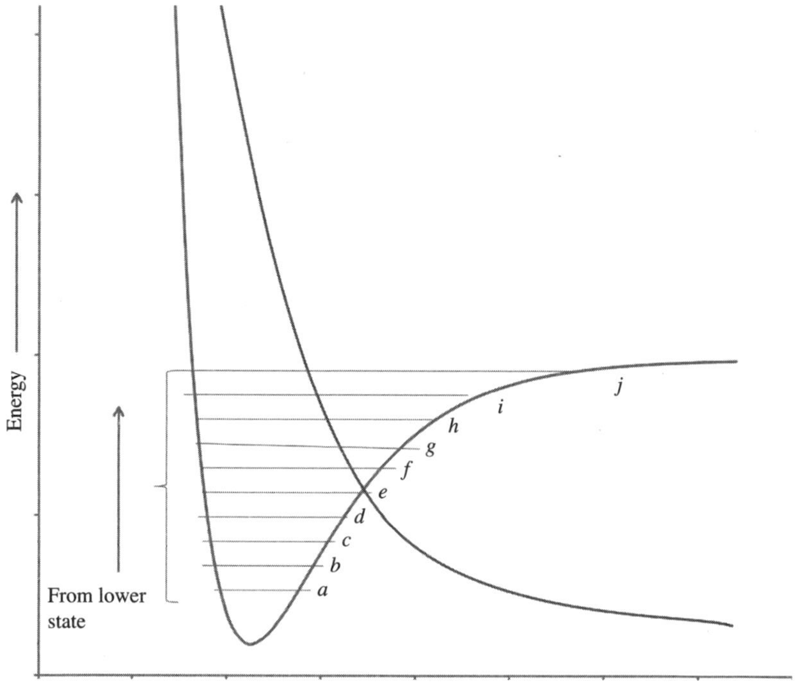

Figure 8.8. Energy levels in predissociation

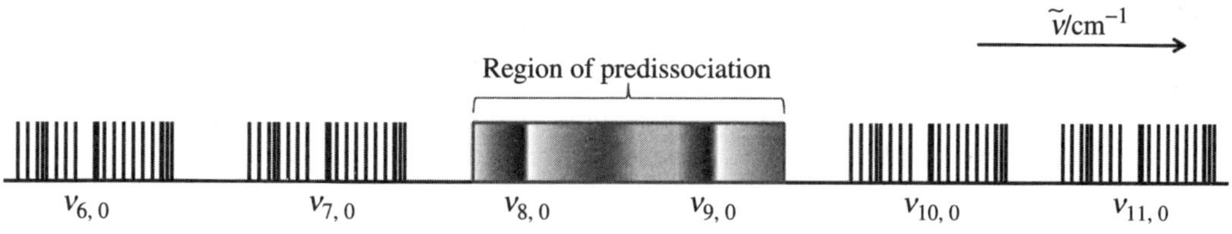

Figure 8.9. Predissociation

We now derive this condition quantitatively. The intensity of any transition is proportional to the square of the transition dipole moment:

$$\left|\vec{M}_{lm}\right| = \int_{-\infty}^{+\infty} \psi_m^* \hat{\mu} \psi_l \, d\tau$$

where $\hat{\mu}$ is the dipole moment operator and both ψ_m and ψ_l are functions of the electronic and vibrational states of the upper and lower levels, $\psi_{ei'\upsilon'}$ and $\psi_{ei''\upsilon''}$ respectively. At this stage, we ignore the rotational component of the total wave function.

Since now we are dealing with wave functions that are products of the electronic and nuclear wave functions, the dipole moment operator can also be separated into its electronic and nuclear components:

$$\hat{\mu} = \hat{\mu}_{el} + \hat{\mu}_n$$

Applying the Born–Oppenheimer approximation, we can write $\psi_{el'\upsilon'}$ as a product $\psi_{el'}\psi_{\upsilon'}$, and the transition dipole moment becomes:

$$\left|\vec{M}_{lm}\right| = \int \psi_m^* \hat{\mu} \psi_l \, d\tau = \int\int \psi_{\mathrm{el}'}^* \psi_{\upsilon'}^* (\hat{\mu}_{\mathrm{el}} + \hat{\mu}_{\mathrm{n}}) \psi_{\mathrm{el}''} \psi_{\upsilon''} d\tau_{el} d\tau_{\mathrm{n}} = \int \psi_{\mathrm{el}'} \hat{\mu}_{\mathrm{el}} \psi_{\mathrm{el}''} \, d\tau_{\mathrm{el}} \int \psi_{\upsilon'} \psi_{\upsilon''} \, d\tau_{\mathrm{n}}$$

$$+ \int \psi_{\mathrm{el}'} \psi_{\mathrm{el}''} \, d\tau_{\mathrm{el}} \int \psi_{\upsilon'} \hat{\mu}_{\mathrm{n}} \psi_{\upsilon''} \, d\tau_{\mathrm{n}} \tag{8.10}$$

The second term vanishes because eigenfunctions of the same (electronic) operator with different eigenvalues are orthogonal. The same cannot be said about the vibrational functions in the first term because they are eigenfunctions of different Hamiltonians–one for the ground electronic state and the other for the excited electronic state. Both these states are like different molecules because their internuclear distances and force constants are different. Hence, we rewrite equation (8.10) with only the first term as

$$\left|\vec{M}_{lm}\right| = \vec{R}_{\mathrm{el}} \langle \upsilon' | \upsilon'' \rangle \tag{8.11}$$

where the electronic dipole moment $\vec{R}_{\mathrm{el}} = \int \psi_{\mathrm{el}'} \hat{\mu}_{\mathrm{el}} \psi_{\mathrm{el}''} \, d\tau_{el}$ and $\langle \upsilon' | \upsilon'' \rangle = \int \psi_{\upsilon'}^* \psi_{\upsilon''} d\tau_n$ is known as the Franck–Condon factor.

The intensity of the transition is proportional to $\left|\vec{M}_{lm}\right|^2$, and thus to $\langle \upsilon' | \upsilon'' \rangle^2$, which is the Franck–Condon factor, consistent with the classical approach. It can be shown that the Franck–Condon factor integrated over all states is equal to unity, showing that the electronic transition is spread over different vibrational states, and the total intensity of a transition is given by the integrated absorption coefficient.

We next discuss the $\vec{R}_{\mathrm{el}}$ term and the consequent electronic selection rules, but before that we have to take a brief look at the electronic structure of diatomic molecules.

8.6 ELECTRONIC STRUCTURE OF DIATOMIC MOLECULES

The only molecule which lends itself to a complete quantum mechanical description is the hydrogen molecule ion, H_2^+, which is a one-electron system, but here, too, the Born–Oppenheimer approximation, which treats electronic motion separately from nuclear motion, has to be invoked, and the molecule has to be cast in the cylindrical coordinate system in order to simplify calculations.

For all other molecules, some approximations are required in order to solve their electronic structures. There are basically two methods – the valence bond method and the molecular orbital method to deal with electronic structure. Here we shall describe the molecular orbital method, which is based on the variation theorem and the linear combination of atomic orbitals (LCAO).

The molecular orbitals of H_2^+ are formed by linear combinations of the 1s orbitals of both hydrogens. If we label the two hydrogen atoms H_A and H_B, the two normalized linear combinations can be written as

$$\psi_{\pm} = \frac{1}{\sqrt{2}} \left(\psi_{1sA} \pm \psi_{1sB} \right) \tag{8.12}$$

The positive combination is called a bonding orbital because it increases the electron density in the region between the two nuclei. Since the probability density is given by the square of the wave function, we square this wave function to get

$$\psi_+^2 = \frac{1}{2}(\psi_{1sA} + \psi_{1sB})^2 = \frac{1}{2}\psi_{1sA}^2 + \psi_{1sA}\psi_{1sB} + \frac{1}{2}\psi_{1sB}^2$$

which shows that the electron density in the region between the two nuclei (the middle term) is increased with this wave function. Likewise, the negative combination decreases the electron density in the region between the two nuclei, leading to an antibonding orbital. All homonuclear diatomic molecules have $D_{\infty h}$ symmetry because they have a centre of symmetry. The positive combination is a g function because it does not change sign on inversion, while the negative combination is a u function (Figure 8.10). The former is symmetrical ellipsoid in shape and is commonly called $1s\sigma_g$ since it is formed from the 1s orbitals. Similarly, the antibonding orbital is called $1s\sigma_u$*. The asterisk (*) indicates that the orbital is antibonding.

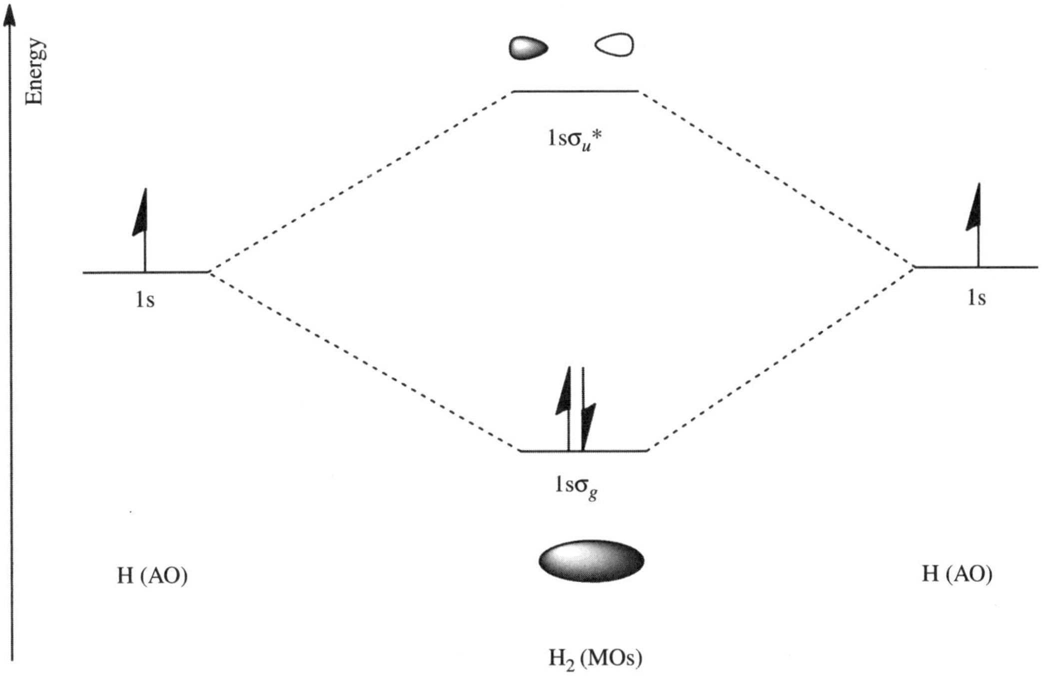

Figure 8.10. Formation of $1s\sigma_g$ and $1s\sigma_u$* orbitals from two atomic 1s orbitals

We have seen that when two orbitals of the same symmetry interact, they give rise to two energy levels, one lower in energy and the other higher in energy than the participating orbitals. The lowering in energy of the bonding orbital is because now the electron feels the attraction of two nuclei instead of one in the atomic orbital. The extent to which the orbitals are lowered or raised in energy depends on the energy difference of the two interacting orbitals: smaller the energy difference, higher is the interaction. In this case, the two orbitals are of the same energy and so the interaction is very strong. Note that two *atomic* orbitals combine to give the same number of *molecular* orbitals and that the antibonding orbitals have a node perpendicular to the internuclear axis. The variation of the orbital energies with internuclear distance is depicted in Figure 8.11.

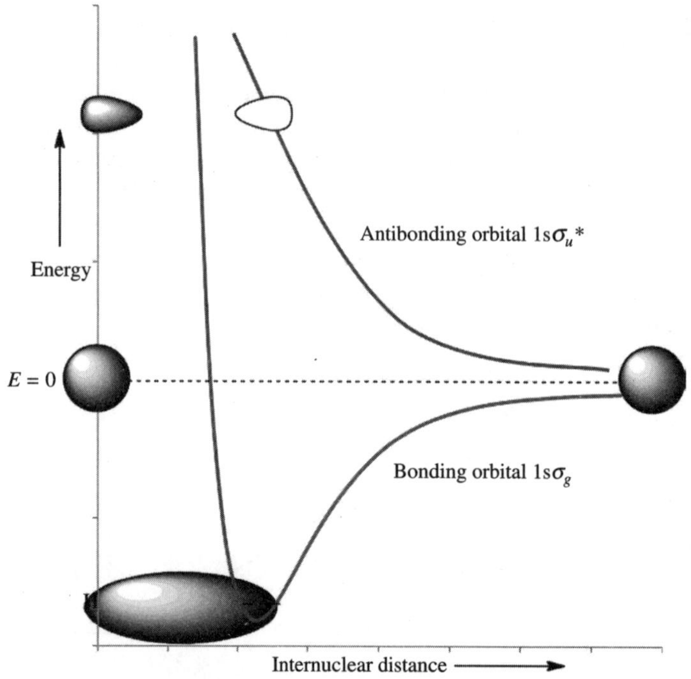

Figure 8.11. Variation of energy with internuclear distance

The interaction of the 2s orbitals similarly produces the $2s\sigma_g$ and $2s\sigma_u$* orbitals since the 2s and 1s orbitals have the same symmetry.

When we come to 2p orbitals, we find they are of two types – the ones that lie along the internuclear axis, which is the *z*-axis, and the others in the perpendicular *x* and *y* directions.

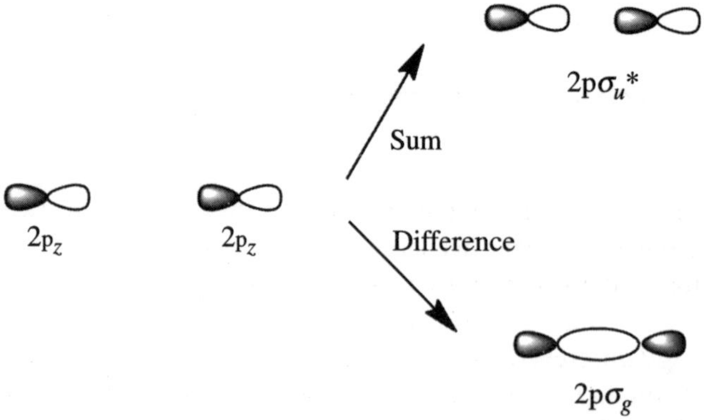

Figure 8.12. Formation of $2p\sigma_g$ and $2p\sigma_u^*$ from two atomic $2p_z$ orbitals

Combination of the two atomic $2p_z$ orbitals gives $2p\sigma_g$ and $2p\sigma_u{}^*$ orbitals in a similar manner (Figure 8.12). Note that here the difference gives the bonding orbital.

The other two 2p orbitals cannot form direct overlaps and only lateral (sideways) overlap is possible. The positive overlap yields the bonding ($2p\pi_u$) and negative overlap the antibonding ($2p\pi_g{}^*$) orbitals. When two $2p_y$ orbitals combine, the result is as shown in Figure 8.13. Note that in this case the bonding orbitals are of *u* symmetry. A common confusion in students' minds is that they assume that all *g* orbitals are bonding and *u* antibonding. The bonding or

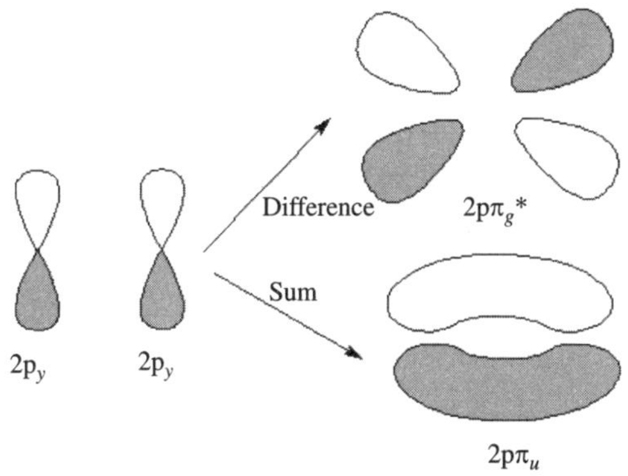

Figure 8.13. Formation of $2p\pi_u$ and $2p\pi_g{}^*$ orbitals from two atomic $2p_y$ orbitals

antibonding characteristic depends on whether the overlap is positive or negative, and not on the *gerade* or *ungerade* properties of the wave function.

On the right of Figure 8.14, we have shown the molecular orbital diagram of the hydrogen molecule. In our study of atomic spectroscopy, we had found that for the hydrogen atom, the energy of an orbital depends only on its *n* quantum number, and hence the 2s and 2p orbitals are degenerate. Comparison of Figures 8.10 and 8.12 reveals that the overlap of the dumbbell-shaped $2p_z$ orbitals is more effective than that of the spherical 2s orbitals, and so the $2p\sigma_g$ orbital is stabilized to a greater extent than the $2s\sigma_g$ orbital, and its corresponding antibonding orbital is therefore also destabilized more. Lateral overlap is least effective, and hence the $2p\pi_u$ orbital is higher in energy than $2p\sigma_g$. Note that the π orbitals are doubly degenerate (together they comprise the orbitals formed from overlap of $2p_y$ and $2p_x$ orbitals). A filled set of π orbitals will contain four electrons. The node in a π_u orbital is in the plane which contains the internuclear axis and is not perpendicular to this axis and between the two atoms as is the node in a σ_u orbital. The π_u orbital is therefore bonding. The π_g orbital, on the other hand, is antibonding because it has, in addition to the node in the plane of the bond axis, another at the bond mid-point, perpendicular to the bond axis. The bonding and antibonding characters of the π orbitals have just the opposite relationship to their *g* and *u* dependence as have the σ orbitals.

The diagram on the left-hand side of Figure 8.14 is for fluorine and oxygen. For hydrogen atom, the 2s and 2p orbitals are of the same energy. As more electrons are added, the 2s and 2p orbitals no longer remain degenerate and the gap between them increases with increasing atomic number (see also Chapter 3). Figure 8.15 shows the radial distribution curves for 2s and 2p orbitals. The maximum probability of finding a 2p electron is closer to the nucleus than that for a 2s electron, but the 2s electron has a small peak closer to the nucleus, as a result of which it is able to *penetrate* more. This decreases its energy with respect to the 2p electrons due to its stronger coulombic attraction to the positively charged nucleus. This effect increases with increasing atomic number, since the nucleus' positive charge also increases. Thus, among the second-row elements, oxygen, fluorine and neon, having the highest atomic numbers, have

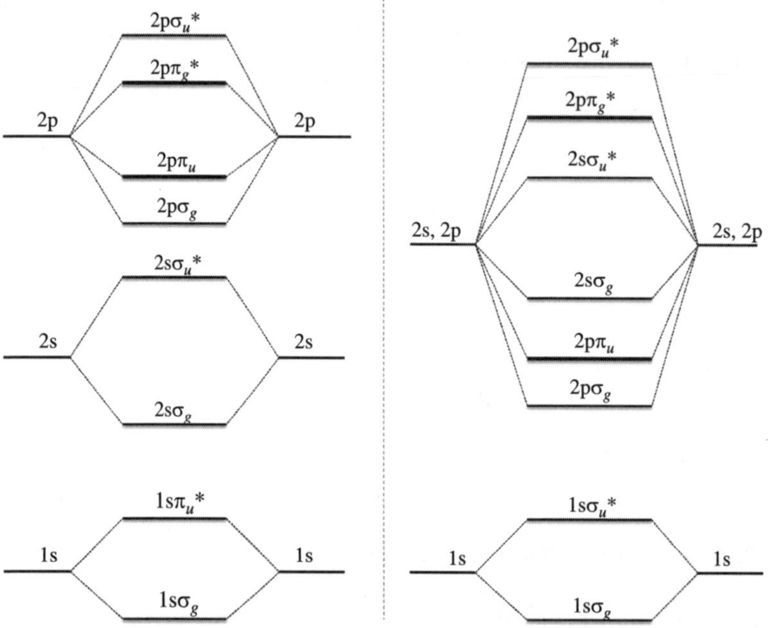

Figure 8.14. Molecular orbital diagram for O_2 and F_2 (left) and H_2 (right)

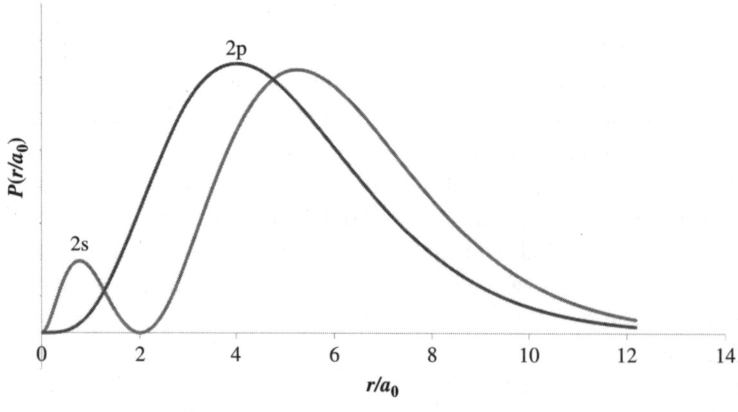

Figure 8.15. Radial distribution curves for 2s and 2p orbitals

large 2s-2p gaps, and no mixing of these orbitals occurs, resulting in the energy level diagram shown on the left-hand side of Figure 8.14. This energy level order is valid for all diatomic molecules having a total of more than 14 electrons.

For Li_2 through N_2, a slightly different energy level diagram operates. Because of the closeness of the 2s and 2p orbital energies, the 2sσ_g and 2pσ_g molecular orbitals are also close in energy. Being of the same symmetry (σ_g), the two orbitals can interact. We have seen that

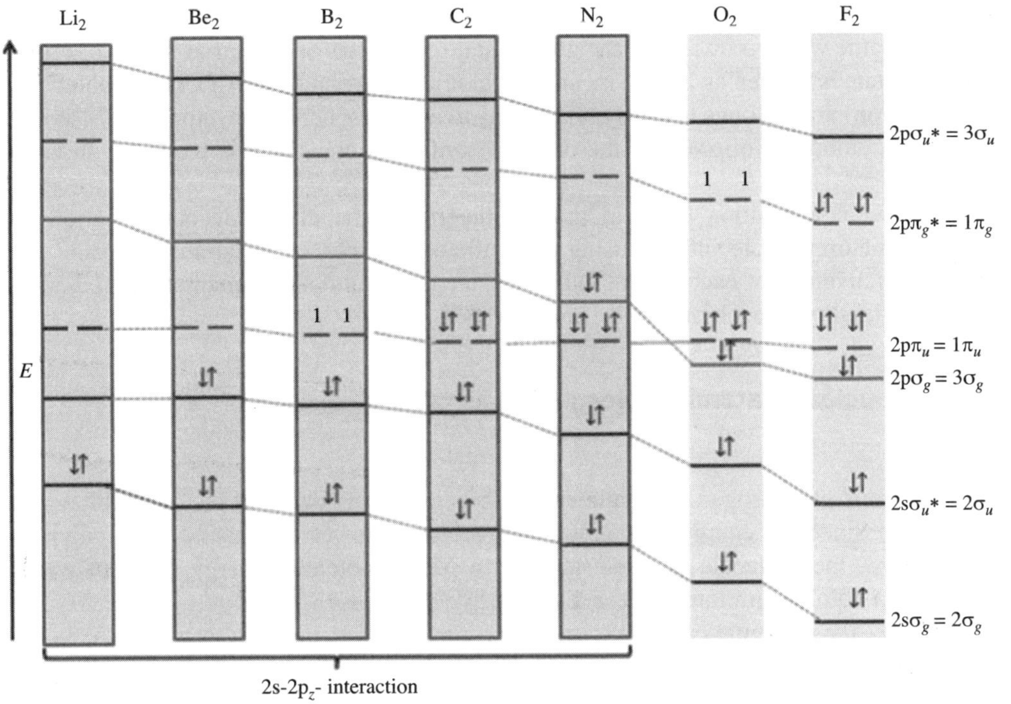

Figure 8.16. Molecular orbital energy level diagrams for the diatomic molecules of the second row elements

two orbitals of the same symmetry interact, the lower energy orbital getting further lowered in energy and the upper orbital increasing in energy. The $2s\sigma_g$ orbital thus gets stabilized and the $2p\sigma_g$ destabilized, so much so that it gets higher in energy than $2p\pi_u$ (see Figure 8.16).

The strength of the interaction is larger, closer the interacting orbitals are to each other in energy. Figure 8.16 gives the molecular energy level diagrams for diatomic molecules of the period 2 elements. We see a crossover between N_2 and O_2 when the $2p\pi_u$ and $2p\sigma_g$ molecular orbitals interchange their relative orders.

From the above discussion, we may conclude that it is not right to associate a particular atomic orbital with a molecular orbital, since, as a result of s-p mixing, $2s\sigma_g$ has contribution from $2p\sigma_g$ and vice versa, and none of the two orbitals is pure. In fact, all orbitals of the same symmetry will mix to some extent depending on their relative energies. Instead of associating the atomic orbital with a molecular orbital, the molecular orbitals of the same symmetry are numbered consecutively, starting with the lowest energy one. Also, an asterisk is not required to signify an antibonding orbital, since antibonding orbitals may have some bonding character, and vice versa. The number is added as a prefix to the symbol. In this numbering scheme $1s\sigma_g$, $2s\sigma_g$ and $2p\sigma_g$ are numbered $1\sigma_g$, $2\sigma_g$ and $3\sigma_g$, respectively, just as the 1s, 2s, 3s, etc., atomic orbitals are numbered in order of increasing energy. Thus, the numerical prefix is similar to the principal quantum number n in atoms. As n increases through a given symmetry set, so does the orbital energy, the orbital increases in size and consequently concentrates charge density further from the nuclei, and finally the number of nodes increases as n increases, as in atomic orbitals.

The similarity with atomic orbitals does not end here. Electron configurations and states are written in the same way. For example, the ground state of H_2^+ has one electron in its $1\sigma_g$ orbital and hence its state is written as X $^2\Sigma_g^+$, meaning that it is the ground state (*X*), is a doublet (one unpaired electron) and belongs to Σ_g^+ symmetry (as in atomic spectroscopy, the state is written with a capital symbol, as opposed to the electron configuration, which is written with lower case symbols).

We are now in a position to build up and determine the electronic configurations of homonuclear diatomic molecules by using the Aufbau principle, as for atoms. We shall also discuss the effectiveness of each orbital in binding the nuclei and make qualitative predictions regarding the stability of each molecular configuration.

8.6.1 Homonuclear diatomic molecules

Hydrogen, H_2

The two electrons of hydrogen molecule can both be accommodated in the $1\sigma_g$ orbital with opposite spins. Since the energy of this bonding orbital is lower than that of the individual hydrogen atoms, the molecule is stable and has a Morse potential energy diagram with a minimum at the equilibrium internuclear distance.

As for atoms, the molecule can be excited by promotion of an electron to one of the higher energy orbitals shown in Figure 8.16. As in atomic spectroscopy, the excited state may also emit a photon corresponding to the difference in energies of the excited and ground states and return to the ground state. The similarity with atomic spectroscopy ends here, since, unlike atoms, where line spectra are observed, in molecular spectroscopy, bands are observed because of changes in the vibrational energy of the molecule which accompany the change in electronic energy.

Helium, He_2

The electronic configuration of He_2 is $(1\sigma_g)^2(1\sigma_u)^2$. Thus, there are two electrons in the bonding orbital and two in an antibonding orbital. The net bond order is zero. However, when two orbitals combine, the antibonding orbital is destabilized to a greater extent than the bonding orbital is stabilized. This can be seen from a variational treatment of the wave function in equation (8.12), whereby the coefficients in the linear combination $\psi = c_A\psi_A + c_B\psi_B$ are varied such that the energy is minimized. The variation principle gives the condition

$$\begin{pmatrix} H_{AA} - E & H_{AB} - ES_{AB} \\ H_{BA} - ES_{AB} & H_{BB} - E \end{pmatrix} \begin{pmatrix} c_A \\ c_B \end{pmatrix} = \begin{pmatrix} 0 \\ 0 \end{pmatrix} \tag{8.13}$$

where $H_{AA} = \int \psi_A{}^* \hat{H} \psi_A \, d\tau$, $H_{BB} = \int \psi_B{}^* \hat{H} \psi_B \, d\tau$, $H_{AB} = H_{BA} = \int \psi_A{}^* \hat{H} \psi_B \, d\tau$, $S_{AB} = \int \psi_A{}^* \psi_B \, d\tau$. H_{AA} and H_{BB} are simply the energies of the participating atomic orbitals. S_{AB} is the overlap integral. The condition (8.13) is satisfied if either the determinant or the coefficients are zero. The latter is a trivial solution, since if the coefficients are zero, there is no wave function. Thus

$$\begin{vmatrix} H_{AA} - E & H_{AB} - ES_{AB} \\ H_{BA} - ES_{AB} & H_{BB} - E \end{vmatrix} = 0 \tag{8.14}$$

For the homonuclear case, $H_{AA} = H_{BB}$. Solution of equation (8.14) gives the energies

$$E_{\pm} = \frac{H_{AA} \pm H_{AB}}{1 \pm S_{AB}} \tag{8.15}$$

Since S_{AB} is positive, the denominator $(1 - S_{AB})$ is smaller for the antibonding orbital, and hence the antibonding effect is larger than the bonding effect, explaining why the He_2 molecule does not exist.

Lithium, Li_2

The Li_2 molecule with the electron configuration $(1\sigma_g)^2(1\sigma_u)^2(2\sigma_g)^2$ marks the beginning of what can be called the second quantum shell in analogy with the atomic case. Since the $1\sigma_u$ antibonding orbital approximately cancels the binding obtained from the $1\sigma_g$ bonding orbital, the bonding in Li_2 can be described as arising from the single pair of electrons in the $2\sigma_g$ orbital. This is a general case. Only the valence electrons interact significantly during bond formation. The inner or core electrons can be ignored, as they remain largely unperturbed. Thus, all alkali metals with one electron in an s orbital in their valence shell show similar behaviour to hydrogen (Figure 8.17). The bond order is 1 since there are two electrons in a bonding orbital.

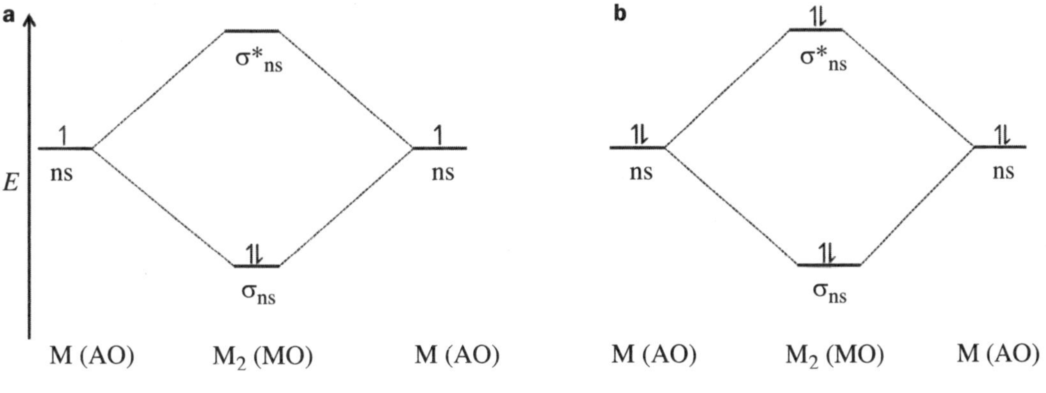

Figure 8.17. Molecular orbital energy-level diagrams for (a) alkali metal and (b) alkaline earth metal diatomic (M_2) molecules

Beryllium, Be_2

The electron configuration of Be_2 is $(1\sigma_g)^2(1\sigma_u)^2(2\sigma_g)^2(2\sigma_u)^2$ and the molecule is not predicted to exist as its net bond order is zero. This is true for all alkaline earth metals, which bear a superficial resemblance to helium because of their $(ns)^2$ configuration (Figure 8.17).

Boron, B_2

Adding two electrons to the Be_2 configuration gives the electron configuration $(1\sigma_g)^2(1\sigma_u)^2(2\sigma_g)^2(2\sigma_u)^2(1\pi_u)^2$. The two $1\pi_u$ electrons could both occupy one of the π_u orbitals of the degenerate set with paired spins or they could be assigned one to each of the π_u orbitals and have parallel spins. Hund's principle applies to molecules as well as to atoms, and the configuration with single occupation of both π_u orbitals with parallel spins is thus predicted to be the most stable. The formal bond order is 1 and the molecule is predicted to be paramagnetic, having a triplet ground state.

Carbon, C_2

Dicarbon should be diamagnetic as two more electrons complete the four electrons in the $1\pi_u$ orbital. The bond order of 2 signifies a strong bond. The ground state is therefore $X^1\Sigma_g^+$. However, many of the transitions originate from an excited $a^3\Pi_u$ state with the electron configuration $(1\sigma_g)^2(1\sigma_u)^2(2\sigma_g)^2(2\sigma_u)^2(1\pi_u)^3(3\sigma_g)^1$, which is only 716 cm^{-1} above the ground state, and is therefore significantly populated. In fact, the so-called Swan bands $d^3\Pi_g \leftrightarrow a^3\Pi_u$ are observed in the emission spectra of comets and absorption spectra of stellar atmospheres, including that of the sun.

Nitrogen, N_2

Given a total of 14 electrons, the electron configuration is $(1\sigma_g)^2(1\sigma_u)^2(2\sigma_g)^2(2\sigma_u)^2(1\pi_u)^4(3\sigma_g)^2$. The $(1\sigma_g)^2$ and $(1\sigma_u)^2$, as well as $(2\sigma_g)^2$ and $(2\sigma_u)^2$, contributions cancel each other, and we are left with six electrons in bonding $1\pi_u$ and $3\sigma_g$ orbitals, resulting in a triple bond.

Oxygen, O_2

From here on, the order of orbitals changes. Filling the orbitals in order of increasing energy, the 16 electrons of O_2 are described by the configuration $(1\sigma_g)^2(1\sigma_u)^2(2\sigma_g)^2(2\sigma_u)^2(3\sigma_g)^2(1\pi_u)^4(1\pi_g)^2$. Two electrons in the antibonding $1\pi_g$ orbital cancel the contribution from two electrons in $1\pi_u$, and we are left with four electrons, i.e., a double bond.

One interesting feature of the electronic configuration of O_2 is that, like B_2, its outer orbital is not fully occupied. This prediction of molecular orbital theory regarding the electronic structure of O_2 has an interesting consequence. The oxygen molecule should be paramagnetic because of the resultant spin angular momentum possessed by electrons. This is demonstrated experimentally by the observation that liquid oxygen is attracted to the poles of a strong magnet. This is considered a triumph of Molecular Orbital (MO) theory, since Valence Bond (VB) theory predicted a diamagnetic ground state for oxygen.

Fluorine, F_2

The $1\pi_g$ orbital gets completely filled in the electron configuration $(1\sigma_g)^2(1\sigma_u)^2(2\sigma_g)^2(2\sigma_u)^2(3\sigma_g)^2(1\pi_u)^4(1\pi_g)^4$. The net bond order is 1.

8.6.2 Heteronuclear diatomic molecules

In a similar fashion, the molecular orbital energy diagrams for heteronuclear diatomic molecules can be constructed. As for homonuclear diatomics, the wave function is a linear combination of atomic orbitals. For a homonuclear diatomic, the squares of the coefficients for the two atoms must be the same by symmetry. However, if we consider a heteronuclear molecule like HF, then clearly there will be a net displacement of electrons towards the more electronegative atom. For HF, the two participating orbitals are the valence orbitals of each, i.e., 1s on hydrogen and 2p on fluorine. We know that the energy of the former is –13.6 eV. That of the latter is –18.6 eV, so in the formation of the linear combination

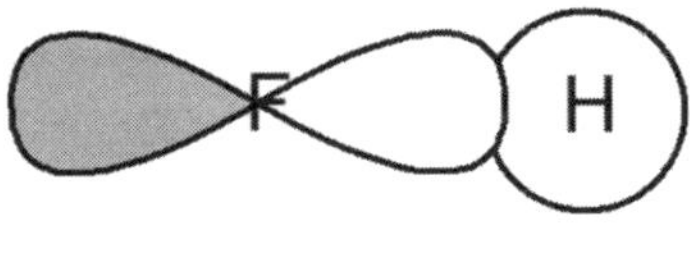

$$\psi = c_1 s_H + c_2 p_{z_F}$$
$$= c_1 \psi_H + c_2 \psi_F$$

the two will not contribute equally to the bonding interaction ($c_1 \neq c_2$).

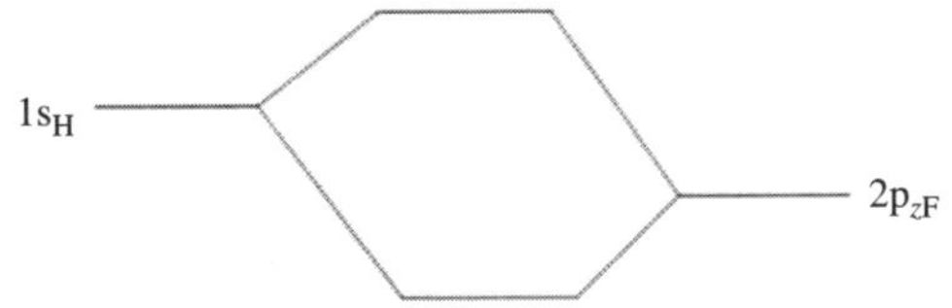

The secular equations from the variation principle, equation (8.13), also give the optimized coefficients. This equation is relatively easy to solve for the homonuclear case when $H_{AA} = H_{BB}$ and leads to two sets of coefficients, $\psi_+ = \frac{1}{\sqrt{2}}(\psi_A + \psi_B)$ for E_+ and $\psi_- = \frac{1}{\sqrt{2}}(\psi_A - \psi_B)$ for E_- (equation 8.12). The former is bonding because of the build-up of electron density in the region between the two nuclei (Figure 8.10) and the other is antibonding.

For the heteronuclear case, we make the simplifying assumption $S_{AB} = 0$, which leads to a quadratic equation whose solution is

$$E = \frac{(H_{AA} + H_{BB}) \pm \sqrt{(H_{AA} - H_{BB})^2 + 4H_{AB}^2}}{2}$$
$$= \frac{H_{AA} + H_{BB}}{2} \pm \frac{H_{AA} - H_{BB}}{2}\sqrt{1 + \frac{4H_{AB}^2}{(H_{AA} - H_{BB})^2}} \tag{8.16}$$

For $H_{AB} << |H_{AA} - H_{BB}|$, this expression may be expanded in a Taylor series

$$(1+x)^{1/2} = 1 + x/2 + x^2/4 + \ldots$$

Taking only the first term in the expansion, we obtain

$$E \approx \frac{H_{AA} + H_{BB}}{2} \pm \frac{H_{AA} - H_{BB}}{2}\left(1 + \frac{2H_{AB}^2}{(H_{AA} - H_{BB})^2}\right)$$

which gives

$$E_- = H_{BB} - \frac{H_{AB}^2}{H_{AA} - H_{BB}},\ E_+ = H_{AA} + \frac{H_{AB}^2}{H_{AA} - H_{BB}}$$

If we assume that $H_{AA} < H_{BB}$, we find that E_+ is similar to the lower energy orbital, H_{AA}, and E_- to the higher of the two atomic orbitals. Further, the lower energy orbital gets further lowered and the higher energy orbital raised in energy due to the interaction. The interaction is greatest when $H_{AA} \approx H_{BB}$.

For HF, the values of E_+ and E_- are −19.6 and −12.9 eV, respectively. The two wave functions are

$$\psi_+ = 0.33\psi_H + 0.94\psi_F$$
$$\psi_- = 0.94\psi_H - 0.33\psi_F$$

Therefore, the partial charge on fluorine in the ground state is $0.94^2 = 0.88$ or 88%, since both electrons are in the E_+ level.

For heteronuclear diatomic molecules, therefore, the following points must be kept in mind. As for homonuclear molecules, only the valence electrons need to be considered. Since now the two atoms constituting the bond are nonequivalent, the valence orbitals of one of these will be lower in energy. *The more electronegative the atom, the lower are its energy levels*. We have also seen that when two orbitals combine, the lower one gets further lowered in energy and the upper one rises in energy. Hence, the bonding molecular orbital will have a greater contribution from the more electronegative atom and the antibonding molecular orbital will have more of the less electronegative atom. Thus, the electron density will be more localized at the more electronegative atom, which explains the polarity of such bonds.

Carbon monoxide, CO

This 'skewed' molecular orbital energy level diagram for CO is shown in Figure 8.18. Heteronuclear diatomic molecules belong to the $C_{\infty v}$ point group. Thus, there is no g or u notation. For example, for CO, the electron configuration is $(1\sigma^+)^2(2\sigma^+)^2(3\sigma^+)^2(4\sigma^+)^2(1\pi)^4(5\sigma^+)^2$, which is the same as that for the isoelectronic nitrogen molecule without the *g* or *u* notation. In general, the energy level diagram for a heteronuclear diatomic molecule resembles that of its isoelectronic homonuclear counterpart. The electron configuration results in a triple bond as for nitrogen.

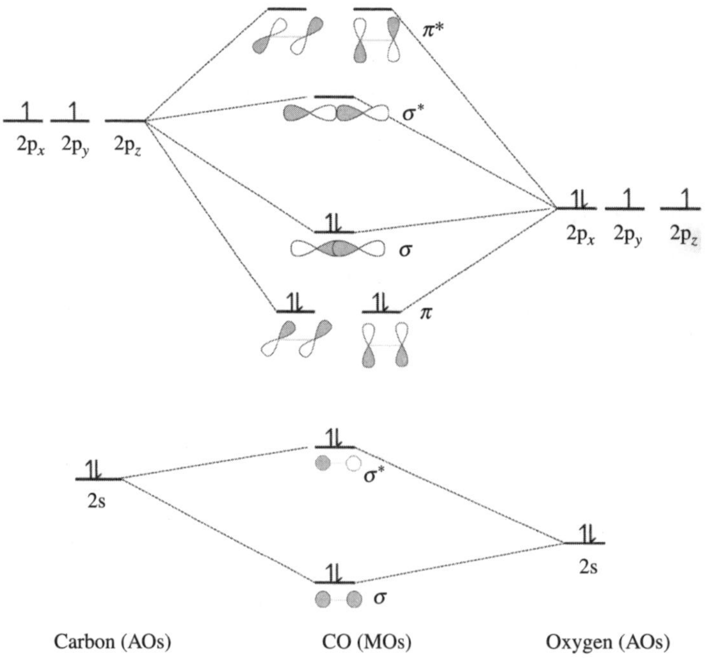

Figure 8.18. Molecular orbital energy-level diagram for carbon monoxide

Nitric oxide, NO

The molecular orbital energy-level diagram for NO is similar to that for the O_2 molecule (see Figure 8.14) since the total number of electrons is more than 14. Because there is one electron less, there is only one electron in the degenerate π^* orbitals. This antibonding orbital has greater contribution from the less electronegative nitrogen. We can say that the NO radical is nitrogen centred.

Hydrogen chloride, HCl

The valence electrons to be considered are $(1s)^1$ on hydrogen and $(3s)^2(3p)^5$ on chlorine, similar to HF, where the 2p orbitals of F were involved. Chlorine, being much more electronegative than hydrogen, only has its 3p electrons close in energy to the hydrogen 1s orbital. The 3s electrons are too low in energy to participate in bonding. Of the three 3p orbitals, only the $3p_z$ orbital, which lies along the internuclear axis, has the right orientation to overlap with hydrogen and form a bond. The other two 3p orbitals do not participate in bonding and, like 3s, they are nonbonding orbitals. The bonding orbital has a greater contribution from the more electronegative chlorine atom, while the corresponding antibonding orbital has a greater contribution from the hydrogen. Thus, we have the following electron configurations:

$$\text{H} \quad (1s)^1$$
$$\text{Cl} \quad [\text{core}](3s)^2(3p_x)^2(3p_y)^2(3p_z)^1$$

Overlap of the H 1s orbital and the Cl $3p_z$ orbital results in a bonding σ and an antibonding σ^* orbital. The two electrons from the H 1s and the Cl $2p_z$ orbitals are accommodated in the bonding orbital. Hence, the electron density resides more near the Cl atom and this explains the polarization of the H–Cl bond to give $H^{\delta+}–Cl^{\delta-}$ (Figure 8.19).

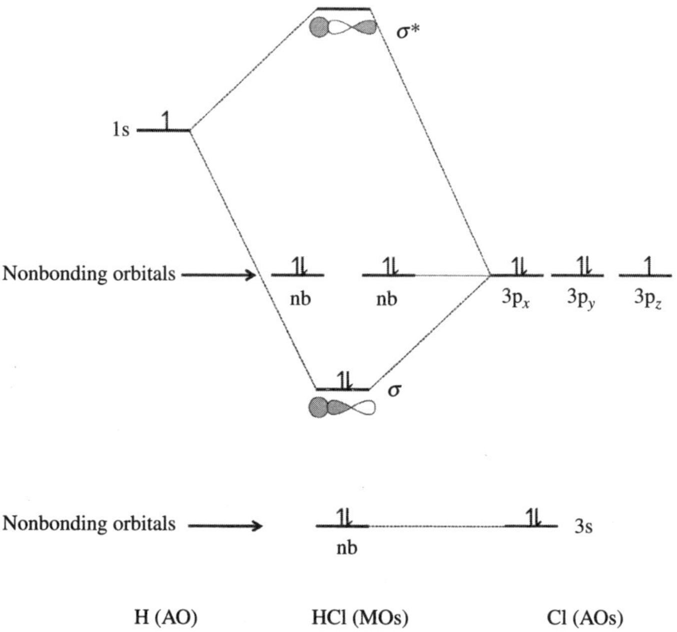

Figure 8.19. Molecular orbital energy-level diagram for HCl

8.7 ELECTRONIC ANGULAR MOMENTUM IN DIATOMIC MOLECULES

Just as an atomic state is characterized by the term symbol $^{2S+1}L_J$, molecular states are characterized by $^{2\Sigma+1}\Lambda_\Omega$. For homonuclear molecules, we shall not consider the component of the total angular momentum on the internuclear axis (Ω) and instead use the parity (*g* or *u*) as a subscript. In describing molecular states, the symbols used are the Greek symbols corresponding to the analogous atomic symbols. Thus, the single electron orbital angular quantum number l is replaced by λ and $\Lambda = \Sigma\lambda_i$ is analogous to $L = \Sigma l_i$. However, unlike atomic states, where the degeneracy of each l value is $2l + 1$, in molecular states, the degeneracy of each λ is restricted to 2 for all values of λ except $\lambda = 0$, for which the degeneracy is 1 only. We thus have

Symbol	σ	π	δ	ϕ
λ	0	±1	±2	±3

In order to avoid confusion with the state symbol (Σ), we shall continue using S for the spin quantum number. The quantum number λ represents the component of the angular momentum along the internuclear axis. From Figure 8.12, we can see that this is zero for σ orbitals, but π orbitals are perpendicular to the internuclear axis, and hence have non-zero values of λ.

Two additional subscripts and superscripts are used in term symbols. The first concerns the parity of the wave function (*u* or *g*), which is specified for homonuclear diatomic molecules. Heteronuclear molecules do not have centres of inversion and thus *g* and *u* are not applicable. The second parameter indicates the symmetry of the electronic wave function with respect to reflection in a plane containing the nuclei, and is applied only to Σ wave functions. If an electronic orbital wave function is symmetric with respect to reflection (Σ^+) it must, by the Pauli principle, be associated with an antisymmetric electron spin wave function, and vice-versa.

The molecular state corresponding to a configuration can be obtained in a manner analogous to atomic states. The electron configurations may be divided into the following categories.

8.7.1 Completely filled shells

For a homonuclear diatomic molecule with completely filled shells (closed shells) and no open shells, such as $(\sigma)^2$ or $(\pi)^4$, the electron spins all cancel and therefore the total spin angular momentum is zero, $S = 0$ (and therefore the multiplicity equals 1, a singlet). The total molecular orbital angular momentum is also equal to zero, $\Lambda = 0$ (a Σ state). The parity for a molecular state with all filled shells is always gerade, or *g*. The +/– superscript applies only to Σ states, and labels the symmetry of the electronic wave function with respect to reflection in a plane containing the nuclei. This is also + for a closed-shell configuration. Therefore, the complete molecular term symbol for the case is $^1\Sigma_g^+$. Examples are ground states of H_2, Li_2, C_2, N_2 and F_2.

8.7.2 Single unpaired electron

For a molecule with one electron in an open shell, such as $(\sigma^+)^1$, the total spin angular momentum is the same as the spin of the single electron, $S = s_1 = ½$. The multiplicity is thus $2S + 1 = 2$, i.e., a doublet state. The total molecular orbital angular momentum is the absolute value of the molecular orbital angular momentum of the single electron, $\Lambda = |\lambda_1|$. The parity for the molecular state is the same as the parity of the molecular orbital occupied by the contributing electron.

Three unpaired electrons in a π orbital are equivalent to one unpaired electron because of the electron–hole formalism that we learnt about in atomic spectroscopy. In the former configuration, there are three electrons and one 'hole' (vacancy), while in the latter it is one electron and three holes. Both yield the same states. This greatly simplifies matters when considering three unpaired electrons.

The B_2^+ ion

As an example, consider the B_2^+ ion. The electron configuration is $(1\sigma_g)^2(1\sigma_u)^2(2\sigma_g)^2(2\sigma_u)^2(1\pi_u)^1$. Only the last part of the configuration, $(1\pi_u)^1$, contributes. The total spin angular momentum is $S = s_1 = ½$. Thus, the multiplicity is $2S + 1 = 2$, a doublet. Since the electron is in a π orbital, the molecular orbital angular momentum is $\lambda_1 = \pm 1$. The total molecular orbital angular momentum is therefore $\Lambda = |\lambda_1| = 1$. This corresponds to a Π state and so a term symbol of $^2\Pi$. When parity is included, since the single electron that contributes to the term symbol is in a $1\pi_u$ molecular orbital, the parity is ungerade (u). Therefore, the complete molecular term symbol for B_2^+ is $^2\Pi_u$. There is no + or – in Π states.

The O_2^- ion

For this ion, the electron configuration is $(1\sigma_g)^2(1\sigma_u)^2(2\sigma_g)^2(2\sigma_u)^2(3\sigma_g)^2(1\pi_u)^4(1\pi_g)^3$. We use the electron–hole formalism, as in atomic states, to get the $^2\Pi_g$ state.

The NO molecule

NO, which in its ground state has an electronic configuration $\ldots(5\sigma^+)^2(1\pi)^4(2\pi)^1$, has a single unpaired electron. Its spin multiplicity is thus 2 (a doublet state), and it has total orbital angular momentum of 1 (an unpaired π electron). Its term symbol is thus $^2\Pi$, or more completely $^2\Pi_{1/2}$ or $^2\Pi_{3/2}$, depending on whether the spin and orbital angular momentum components are parallel or anti-parallel. There is no centre of symmetry, so we have written the Ω value instead of the parity. This is analogous to the J quantum number in atomic states and hence takes the values $|\Delta + S|, \ldots |\Lambda - S| = 3/2, 1/2$.

8.7.3 Two unpaired electrons

For a molecule with two electrons contributing in one or more open shells, the total spin angular momentum is determined in the same way as that for an atom. That is, the total spin angular momentum takes the values $S = |s_1 + s_2|, \ldots. |s_1 - s_2|$. The total molecular orbital angular momentum Λ takes the following values for a molecule with two unpaired electrons,

$$\Lambda = |\lambda_1 + \lambda_2|, |\lambda_1 - \lambda_2|$$

Note that Λ can only take these two values and none in-between. Only the absolute values of λ_1 and λ_2 are considered.

The parity is determined by the direct products of the parities of the two unpaired electrons.

Example:

Electron configuration $(\sigma)^1(\pi)^1$

As an example, consider an excited state of a molecule with electron configuration $(\sigma)^1(\pi)^1$. The total spin angular momentum is $S = |s_1 + s_2|, \ldots |s_1 - s_2| = 1$ and 0 since both $s = ½$, and we have possible triplet and singlet states.

The molecular orbital angular momentum for the electron in the σ orbital is $\lambda_1 = 0$, and the molecular orbital angular momentum for the electron in the π orbital is $\lambda_2 = \pm 1$. The total molecular orbital angular momentum is therefore

$$\Lambda = |\lambda_1 + \lambda_2|, |\lambda_1 - \lambda_2| = |0+1|, |0-1| = 1, 1$$

Since $\Lambda = 1$, this is a Π state. The term symbols possible for this electron configuration are therefore $^1\Pi$ and $^3\Pi$.

To get the parity of this state (if it is a homonuclear diatomic molecule), we must take the product of the parities of the electrons contributing to the term symbol. For example, suppose that the electron configuration is $(\sigma_g)^1(\pi_u)^1$. The overall parity is then $g \otimes u = u$, and the complete molecular term symbols are $^1\Pi_u$ and $^3\Pi_u$.

The oxygen molecule

The ground state electronic configuration of O_2 is $\ldots(5\sigma_g)^2(1\pi_u)^4(2\pi_g)^2$. We now have two unpaired electrons in degenerate orbitals. It is a situation similar to that encountered in atomic spectroscopy when we considered the carbon $(2p)^2$ configuration, where the Pauli exclusion principle forbade some of the possible states. We may solve the problem analogously by writing out the microstates, but Group Theory greatly simplifies matters. Since we are looking at a $(\pi)^2$ configuration, we take the direct product $\Pi \otimes \Pi$ of the irreducible representations. The tables of direct products list this as $\Sigma^+ \oplus [\Sigma^-] \oplus \Delta$. Here, the quantity within square brackets [ ] indicates the antisymmetrized product. The total electronic wave function must be antisymmetric with respect to electron exchange, so the symmetric spatial products must be paired with antisymmetric (singlet) spin functions, and the antisymmetric spatial function with the symmetric (triplet) spin function. The parity is *g*. The term symbols arising from the $(\pi)^2$ configuration are therefore $^1\Sigma_g^+$, $^3\Sigma_g^-$ and $^1\Delta_g$. According to Hund's rules, the triplet state is lowest in energy. Of the two remaining states, the Δ state has higher orbital multiplicity and hence has lower energy. Hence, the order of energies is

$$^3\Sigma_g^- < {}^1\Delta_g < {}^1\Sigma_g^+$$

and the ground state of oxygen (and B_2) is a triplet.

The states of molecular hydrogen

In this case, there are two electrons. We thus expect to find singlet and triplet states, depending on whether the electron spins are paired or parallel. In the ground state, both electrons are in the same $1\sigma_g$ orbital and must form a singlet state. We have $\lambda_1 = \lambda_2 = 0$ and $\Lambda = 0$: the state is thus $^1\Sigma$. We also specify the orbital geometry; since both electrons are in the $1\sigma_g$ orbital, we write $^1\Sigma_g$. Finally, since the wave function of the electron is unchanged upon reflection in the plane of symmetry, the full term symbol for the ground state of H_2 is $X(1\sigma_g)^2\,{}^1\Sigma_g^+$.

We consider the first three possible singly excited states: $\{(1\sigma_g)^1(1\sigma_u)^1\}$, $\{(1\sigma_g)^1(2\sigma_g)^1\}$ and $\{(1\sigma_g)^1(1\pi_u)^1\}$ (see Figure 8.14). Since there are two unpaired electrons, the spin quantum number S can take on values 0 and 1, so we get singlet and triplet states.

$(1\sigma_g 1\sigma_u)$. Since both electrons are σ electrons, $\Lambda = \lambda_1 + \lambda_2 = 0$. One constituting orbital is *even* and the other *odd* so we have $A(1\sigma_g 1\sigma_u)\ {}^1\Sigma_u^+$ and $a(1\sigma_g 1\sigma_u)\ {}^3\Sigma_u^+$.

$(1\sigma_g 2\sigma_g)$. We have again the ${}^1\Sigma$ and ${}^3\Sigma$ states, since both electrons are σ, but the overall state is now *even* since both constituting orbitals are *even* and *symmetrical*. Thus, the states are $B(1\sigma_g 2\sigma_g)\ {}^1\Sigma_g^+$ and $b(1\sigma_g 2\sigma_g)\ {}^3\Sigma_g^+$.

$(1\sigma_g 1\pi_u)$. Now $\Lambda = \lambda_1 + \lambda_2 = 1$, since one electron is in a π state and the other in a σ state; the overall states are $C(1\sigma_g 1\pi_u)\ {}^1\Pi_u$ and $c(1\sigma_g 1\pi_u)\ {}^3\Pi_u$.

The energies of the three excited singlet states are $A(1\sigma_g 1\sigma_u){}^1\Sigma_u^+ < B(1\sigma_g 2\sigma_g){}^1\Sigma_g^+ < C(1\sigma_g 1\pi_u){}^1\Pi_u$ (Figure 8.20).

Similar states are obtained by excitation to 3s and 3p states, 4s and 4p states, etc. In addition, for n = 3, 4,... electrons can also be excited to the nd orbital.

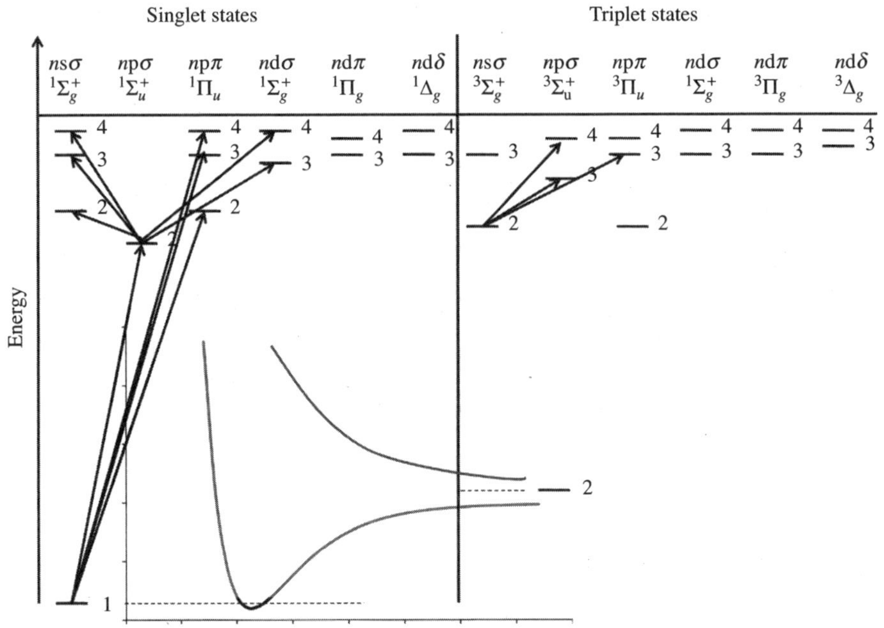

Figure 8.20. Singlet and triplet states of the hydrogen molecule

8.8 SELECTION RULES

For diatomic molecules, the following selection rules operate:

(i) $\mathbf{\Delta\Lambda = 0, \pm 1}$

This is rationalized in terms of conservation of angular momentum of the molecule plus photon, which has a spin angular momentum of $\hbar$. For $\Delta\Lambda = 0$, the rotational angular momentum changes by 1 ($\Delta J = \pm 1$). Thus, transitions $\Sigma \leftrightarrow \Sigma$, $\Sigma \leftrightarrow \Pi$ and $\Pi \leftrightarrow \Pi$, etc., are allowed, but $\Sigma \leftrightarrow \Delta$, for example, is not allowed.

(ii) **$\Delta S = 0$**
As in atomic spectroscopy, the spin quantum number cannot change during an electronic transition. Transitions that change the multiplicity are weak for molecules formed of light atoms, where spin–orbit coupling is small. For heavier atoms, spin–orbit coupling effects are larger, and transitions for which $\Delta S \neq 0$ can be observed.

(iii) **$\Delta\Omega = 0, \pm 1$**
The total angular momentum is similar to the J quantum number in atomic spectroscopy and a similar selection rule operates.

(iv) **$+ \leftrightarrow +$ and $- \leftrightarrow -$**
Σ^+ states can go only to other Σ^+ states (or to Π states), while Σ^- can go only to Σ^- (or Π). Thus, $\Sigma^+ \leftrightarrow \Sigma^+$ and $\Sigma^- \leftrightarrow \Sigma^-$.

(v) **$g \leftrightarrow u$ (where applicable).**

The H_2 molecule
Figure 8.21 displays the potential energy diagrams for various excited electronic states of hydrogen. Notice that the $^3\Sigma_u^+$ state is unstable without an energy minimum because of the equal number of bonding and antibonding electrons.

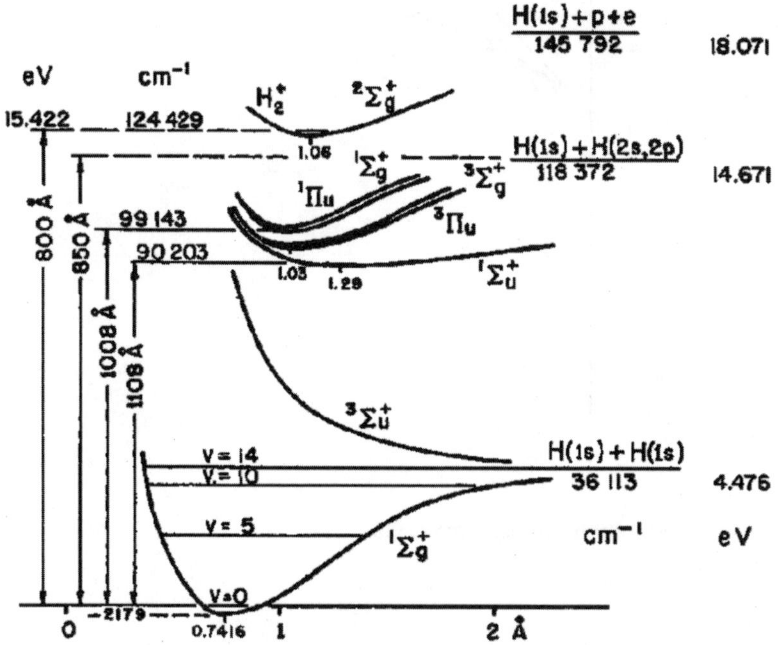

Figure 8.21. Potential energy curves of the ground state and the lowest excited states of molecular hydrogen [Field et al. (1966); reprinted with permission]. The total energy is plotted as a function of the separation of the two hydrogen nuclei. Excitation, dissociation and ionization energies are given relative to the $v = 0$ level of the ground state

Because of the $\Delta S = 0$ selection rule, transitions between singlet and triplet states are forbidden.

The first allowed electronic dipole transitions from the ground state $X^1\Sigma^+_g$ are to the $B^1\Sigma^+_u$ and $C^1\Pi_u$ states. They occur at energies between 11 and 14 eV (i.e., at UV wavelengths, λ ~0.1 μm) and are known as the H_2 Lyman and Werner bands.

The O_2 molecule

The potential energy curves for molecular oxygen are given in Figure 8.22.

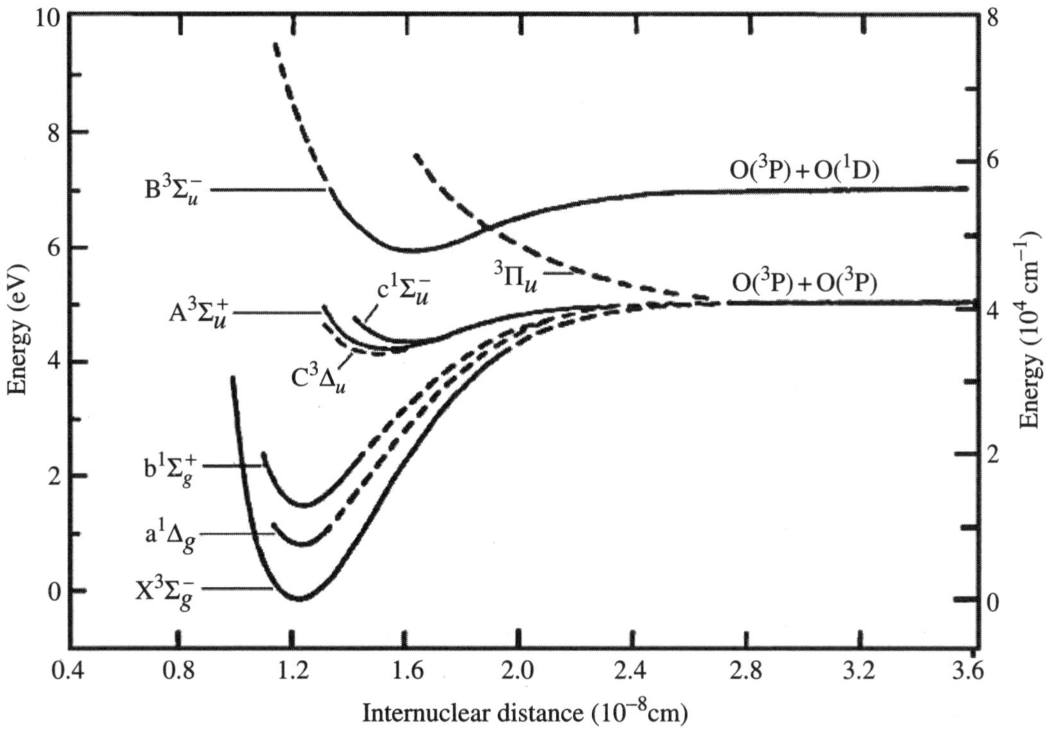

Figure 8.22. Potential energy curves for molecular oxygen [Krupenie (1972); reprinted with permission]

As O_2 is homonuclear (no dipole moment), no vibration–rotation spectrum would be expected within the *X* state. Thus, O_2 would not be expected to absorb at wavelengths longer than ~200 nm and only electronic transitions with shorter wavelengths are possible. The only allowed electronic electric dipole transition from the ground state is $B\ ^3\Sigma_u^- \leftarrow X\ ^3\Sigma_g^-$ as transition to $^3\Sigma_u^+$ is forbidden due to change in reflection symmetry ($- \not\leftrightarrow +$). Given the shapes of the *B* and *X* potential energy curves, this absorption would be expected to be very temperature dependent (stronger at higher temperatures), since $r_e' >> r_e''$ and higher temperatures would push the molecule to a higher v'' level, increasing the overlap between the ground and excited states.

The CO molecule

For CO, the allowed electronic electric dipole transitions are given in Figure 8.23. As the molecule lacks a centre of symmetry, there is no *g* or *u* notation.

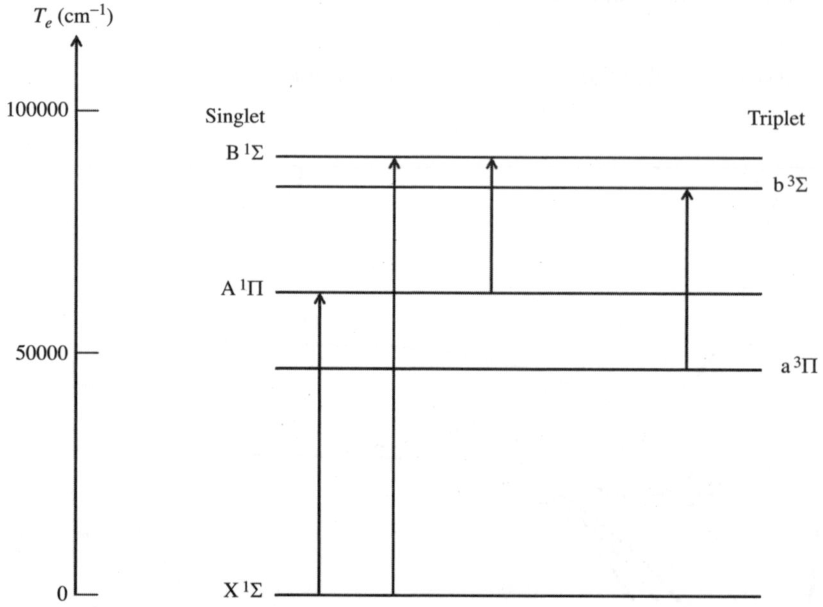

Figure 8.23. Electronic transitions in CO

8.9 ROTATIONAL FINE STRUCTURE OF ELECTRONIC–VIBRATIONAL TRANSITIONS

So far we have ignored the rotational fine structure. If we ignore centrifugal distortion and use the Born–Oppenheimer approximation, the total energy of a diatomic molecule is

$$T = T_e + G(\upsilon) + \tilde{B}J(J+1)$$

while changes in the total energy are given by

$$\Delta T = \Delta T_e + \Delta G(\upsilon) + \Delta\tilde{B}J(J+1)$$

and the wavenumber of a spectroscopic line corresponding to such a change is:

$$\tilde{\nu}_{\text{spect}} = \tilde{\nu}_{\upsilon',\upsilon''} + \Delta\tilde{B}J(J+1)$$

where $\tilde{\nu}_{\upsilon',\upsilon''}$ represents the wavenumber of electronic–vibrational transition.

The selection rule for J depends on the type of electronic transition that the molecule undergoes. If both the lower and upper electronic states are $^1\Sigma$ states (with no electronic angular momentum about the internuclear axis, and $S = 0$), the selection rule is

$$\Delta J = \pm 1 \text{ (only for } ^1\Sigma \leftrightarrow {}^1\Sigma \text{ transitions)}$$

whereas for all other transitions (provided that at least one of the states has $S \neq 0$) the rule is

$$\Delta J = 0, \pm 1$$

In the latter case, there is an added restriction that a state with $J = 0$ cannot undergo a transition to another $J = 0$ state: $J = 0 \leftrightarrow J = 0$ is not allowed. Once more conservation of angular momentum can be used to explain these selection rules.

Thus, for transitions between $^1\Sigma$ states, only P and R branches will appear, while for other transitions Q branches will appear in addition. For the difference of the two electronic states, we have

$$\tilde{\nu}_{\text{spect}} = \tilde{\nu}_{\upsilon',\upsilon''} + \tilde{B}'J'(J'+1) - \tilde{B}''J''(J''+1)$$

giving the P, R and Q branches as follows:

1. P branch: $\Delta J = -1$, $J'' = J' + 1$

$$\tilde{\nu}_{\text{P}} = \tilde{\nu}_{\upsilon',\upsilon''} - (\tilde{B}' + \tilde{B}'')(J'+1) + (\tilde{B}' - \tilde{B}'')(J'+1)^2$$

   where $J' = 0, 1, 2, \ldots$
2. R branch: $\Delta J = +1$, $J' = J'' + 1$

$$\tilde{\nu}_{\text{R}} = \tilde{\nu}_{\upsilon',\upsilon''} + (\tilde{B}' + \tilde{B}'')(J''+1) + (\tilde{B}' - \tilde{B}'')(J''+1)^2$$

   where $J'' = 0, 1, 2, \ldots$
   These two equations can be combined into

$$\tilde{\nu}_{\text{P,R}} = \tilde{\nu}_{\upsilon',\upsilon''} + (\tilde{B}' + \tilde{B}'')m + (\tilde{B}' - \tilde{B}'')m^2$$

where $m = \pm 1, \pm 2, \ldots$ with positive m values for the R branch ($\Delta J = +1$) and negative values for the P branch ($\Delta J = -1$). Note that m cannot be zero (as this would correspond to $J' = -1$ for the P branch), so that no line from the P and R branches appears at the band origin, $\tilde{\nu}_{\upsilon',\upsilon''}$.

These equations are similar to those derived for ro-vibrational bands in infrared spectroscopy. However, there the $\tilde{B}$ values referred to ground and excited *vibrational* levels and the difference was ascribed to the change in the average bond length on vibration. Here, the $\tilde{B}$ values are for different *electronic* levels, and the difference in the average bond lengths is much larger. For electronic transitions, bond orders are expected to change and thus bond lengths are likely to be markedly different between different electronic states (e.g., CO). The spreading out or bunching up of transitions, which occurs in vibrational–rotational spectra, may thus be more marked for electronic transitions, and 'band heads' now become evident at low J. Frequently these are more visible in spectra than the band origin, resembling Q branches in infrared spectra.

Figure 8.24 shows the resulting spectra for $\tilde{B}' < \tilde{B}''$ and a 10% difference in the magnitudes of $\tilde{B}'$ and $\tilde{B}''$. P branch lines appear at the *low wavenumber* side of the band origin, and the spacing between the lines increases with m, as in infrared. The R branch appears on the *high wavenumber* side, and the line spacing decreases rapidly with m. The point at which the R branch separation decreases to zero is known as the *band head.*

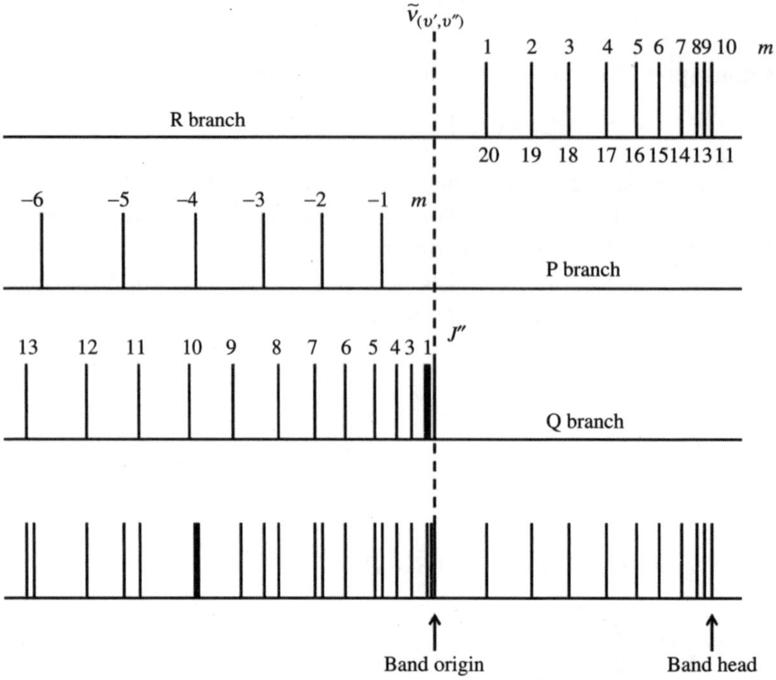

Figure 8.24. Rotational fine structure of a vibration–electronic transition

(3) Q branch: $\Delta J = 0$, $J' = J'' \neq 0$ (since $J = 0 \leftrightarrow J = 0$ is not allowed)

$$\tilde{\nu}_{\mathrm{Q}} = \tilde{\nu}_{v',v''} + (\tilde{B}' - \tilde{B}'')J'' + (\tilde{B}' - \tilde{B}'')J''^2$$

The Q branch lines lie on the low wavenumber side of the origin.

8.9.1 The Fortrat diagram

If we rewrite the expressions for the P, R and Q lines with continuous variables p and q, we obtain

$$\tilde{\nu}_{\mathrm{P,R}} = \tilde{\nu}_{v',v''} + (\tilde{B}' + \tilde{B}'')p + (\tilde{B}' - \tilde{B}'')p^2$$

$$\tilde{\nu}_{\mathrm{Q}} = \tilde{\nu}_{v',v''} + (\tilde{B}' - \tilde{B}'')q + (\tilde{B}' - \tilde{B}'')q^2$$

Each equation represents a *Fortrat parabola*, where p takes both positive and negative values, while q is always positive (Figure 8.25). The band head is clearly at the vertex of the P, R parabola. We calculate the position of the vertex by differentiation:

$$d\tilde{\nu}_{\mathrm{P,R}} / dp = (\tilde{B}' + \tilde{B}'') + 2(\tilde{B}' - \tilde{B}'')p = 0$$

or

$$p = -(\tilde{B}' + \tilde{B}'')/[2(\tilde{B}' - \tilde{B}'')] \text{ for band head}$$

If $\tilde{B}' < \tilde{B}''$, the band head occurs at *positive* p values (i.e., in the R branch). Conversely, if $\tilde{B}' > \tilde{B}''$, the band head is found at *negative* p values (i.e., in the P branch). A 10% difference between $\tilde{B}'$ and $\tilde{B}''$ gives $p = 11$.

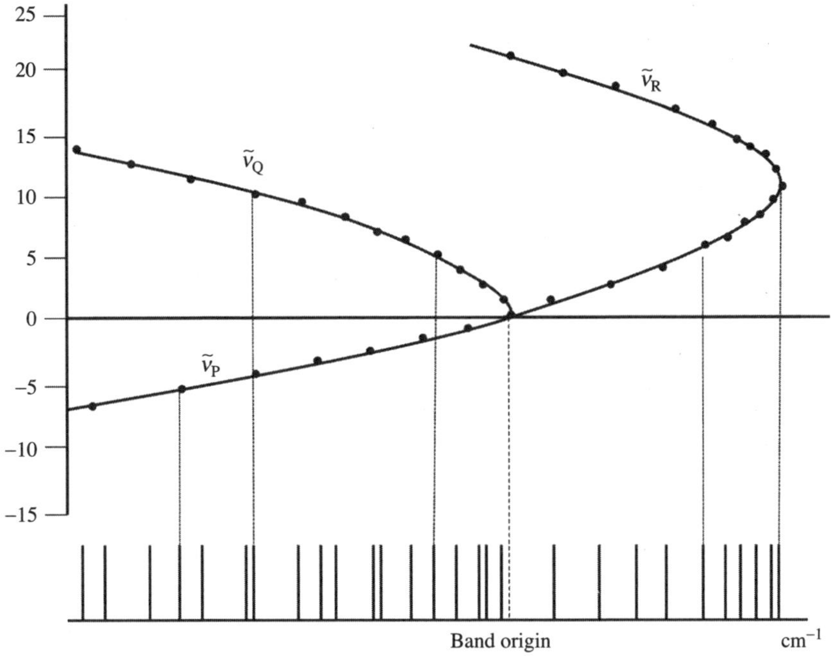

Figure 8.25. The Fortrat diagram

8.10 PHOTOELECTRON SPECTROSCOPY (PES)

While studying atomic spectroscopy, we had discussed photoelectron spectroscopy. We had found that

(1) A photoelectron spectrum is a plot of the number of electrons ejected versus the kinetic energy when a high-intensity radiation, usually from a He I lamp operating at 58.4 nm, which has an energy of 21.22 eV, impinges on an atom or molecule. The resulting spectrum allows one to calculate the ionization energy using the relation

$$IE = h\nu_0 - E_k$$

where IE is the ionization energy, $h\nu_0$ is the energy of the incident radiation, and E_k is the kinetic energy of the ejected electrons.

(2) The ionization energies are the negative of the orbital energies. These orbital energies can be directly correlated with those computed using quantum mechanics. This is Koopmans' theorem.
(3) The area under a peak is proportional to the degeneracy of the level from which the electron is ejected.

Thus, photoelectron spectroscopy provides a ready correlation between theoretical and experimental quantities. The energy level diagrams reported in this chapter have either been found from photoelectron spectroscopy or electron spectroscopy, and some have been computationally determined.

In Chapter 3, we had deciphered the PES spectrum of argon. In this chapter, we explore what further information about molecules can be gained from their photoelectron spectra.

Since energy is conserved, the energy of the incident beam is split into the ionization energy of the molecule, the internal energy of the molecule (vibration/rotation) and the kinetic energies of the electron and molecule ion. Since the latter is much heavier than the electron, the bulk of the kinetic energy is taken by the electron, and we may safely ignore the last term.

$$h\nu_0 = IE + E_{M^+} + E_k(e^-) + E_k(M^+) \tag{8.17}$$

Since molecules have other ways besides changes in electronic energy to change their energy, an extra term E_{M^+} has to be included in equation (8.17). We therefore expect photoelectron spectroscopy to provide information about vibrational energy changes of the molecule ion (Figure 8.26). The *adiabatic* ionization energy is the least energy required to ionize the molecule. This is the (0,0) transition from the ground state of the molecule to the

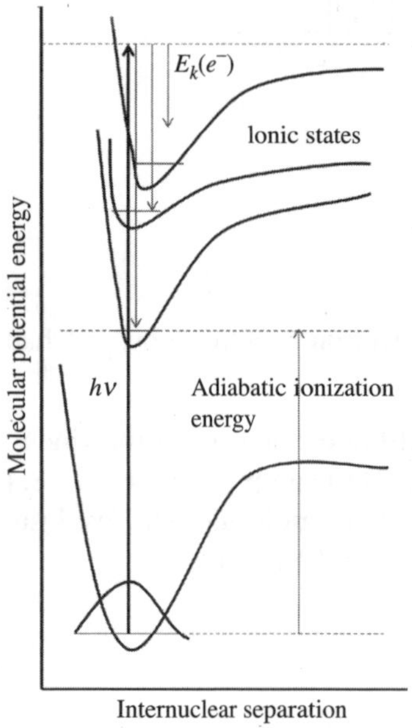

Figure 8.26. Photoelectron spectroscopy

ion. More energy than this goes in vibrationally exciting the ion. Thus, the adiabatic ionization energy corresponds to the first (least ionization energy) line in the photoelectron spectrum.

We illustrate this with an example. Figure 8.27 shows a schematic photoelectron spectrum of the hydrogen molecule. The only occupied orbital in hydrogen is $1\sigma_g$ and the only ionization possible is from this orbital, leading to the H_2^+ molecule ion with a $^2\Sigma_g^+$ state. Such a band is observed (Figure 8.27) with the first line at 15.41 eV. This is then the adiabatic ionization energy of hydrogen. A series of lines is also observed with almost constant spacing between individual lines. This must correspond to the vibrational spacing in the ionized molecule, as can be seen from the diagram on the right of Figure 8.27. The σ_g orbital is strongly bonding, and removal of an electron from this orbital would lead to weakening of the H-H bond (its bond order reduces to 0.5 from 1.0), and consequent increase in the H-H bond length, as shown in the diagram on the right. A vertical (Franck–Condon) transition leads to one of the higher vibrational levels as the most intense transition. The corresponding ionization is therefore called *vertical* ionization energy.

The photoelectron transition ($^1\Sigma_g^+ \rightarrow {}^2\Sigma_g^+$) would appear to be forbidden because of the change in spin multiplicity, but it must be remembered that the product is the ionized state of the molecule. The spin conservation rule must apply to the cation plus the ionized electron. Thus, if the electron that escapes is of α spin, the molecule would be left with an electron of β spin, effectively corresponding to a singlet state.

The line structure is more clearly shown in Figure 8.27. As a result of the weakening of the bond, its bond length increases and its vibrational frequency decreases. The ionized level has more closely spaced vibrational lines and a shallower potential energy well because its

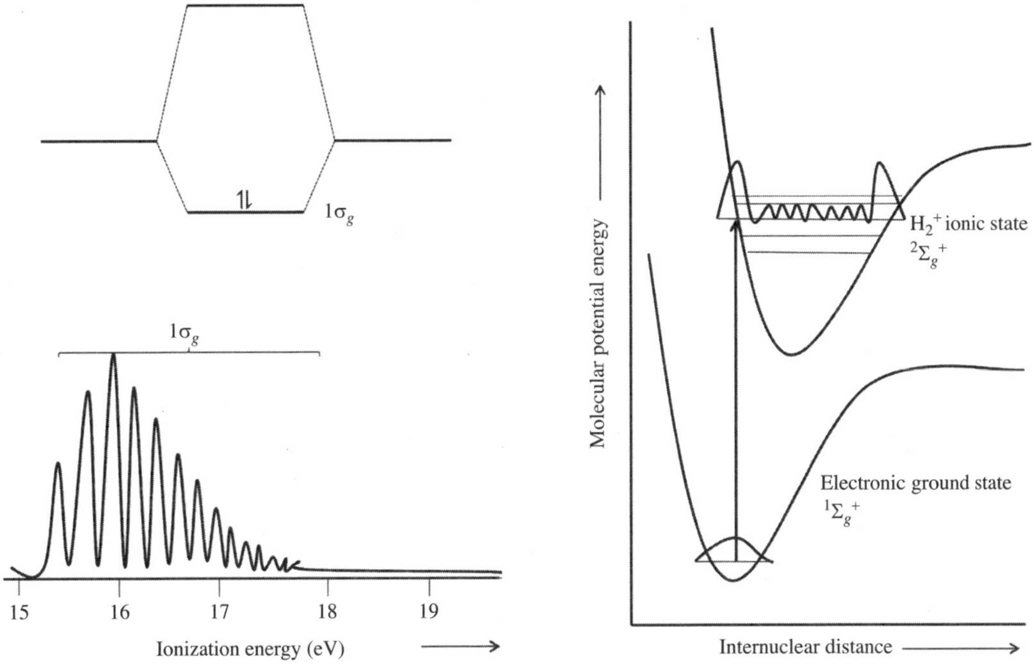

Figure 8.27. Photoelectron spectrum of H_2

dissociation energy is smaller. In the present example, the (2,0) transition corresponds to the vertical ionization energy, which is always higher than the adiabatic ionization energy.

Figure 8.26 shows that if the molecule and its ionization product have the same internuclear distance, the vibrational structure would be sparse, and if a large change in the internuclear distance occurs, considerable vibrational structure would be observed. Thus, if the electron that ionizes is from a bonding or an antibonding orbital, the vibrational structure gives information about the vibrational frequency, and hence the strength of bonding, in the ionized molecule. It must be appreciated that photoelectron spectroscopy provides information not only about the neutral molecule but also about its ionization product, which is difficult to observe otherwise because of its extremely short lifetime.

We shall take up another example, that of nitrogen. Referring to Figure 8.16, we find that the electron configuration of molecular nitrogen is $(1\sigma_g)^2(1\sigma_u)^2(2\sigma_g)^2(2\sigma_u)^2(1\pi_u)^4(3\sigma_g)^2$. The He I radiation has sufficient energy to ionize the valence electrons only. The easiest to ionize electrons are the $3\sigma_g$ electrons. The next higher ionization potential is for the $1\pi_u$ electrons, followed by $2\sigma_u$. Ionization leaves the molecule in its cationic state (N_2^+).

> **Ionization from $3\sigma_g$.** The resulting N_2^+ has the electron configuration $(1\sigma_g)^2(1\sigma_u)^2(2\sigma_g)^2(2\sigma_u)^2(1\pi_u)^4(3\sigma_g)^1$. All the levels up to π_u are completely filled, so we consider only the unpaired electron in the $3\sigma_g$ orbital. This gives $^2\Sigma_g^+$ for the molecular term symbol of the ion. Its adiabatic ionization energy is 15.58 eV.
>
> **Ionization from $1\pi_u$.** In this case, the resulting electron configuration is $(1\sigma_g)^2(1\sigma_u)^2(2\sigma_g)^2(2\sigma_u)^2(1\pi_u)^3(3\sigma_g)^2$. As for O_2^-, this gives rise to the $^2\Pi_u$ state and the ionization energy is 16.69 eV.
>
> **Ionization from $2\sigma_u$.** This must result from ionization from the $2\sigma_u$ orbital, leaving $(1\sigma_g)^2(1\sigma_u)^2(2\sigma_g)^2(2\sigma_u)^1(1\pi_u)^4(3\sigma_g)^2$. The unpaired electron is in the σ_u orbital, and so the state of the nitrogen ion is $^2\Sigma_u^+$. The adiabatic ionization energy is 18.76 eV.

Two striking points emerge from Figure 8.28:

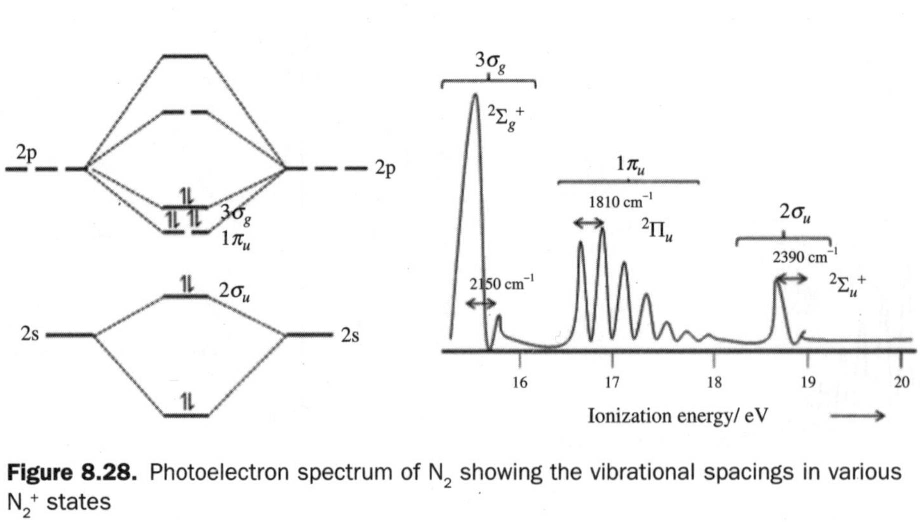

Figure 8.28. Photoelectron spectrum of N_2 showing the vibrational spacings in various N_2^+ states

1. There is hardly any vibrational structure in the first and last peaks, which signifies that the bonding does not change very much by ionization of electrons from these orbitals. These orbitals are weakly bonding or antibonding.
2. There is considerable vibrational structure in the central peak, which implies a strongly bonding or antibonding orbital.

Moreover, the area under the central peak is double that of the other two, confirming that this must be a doubly degenerate π orbital. This observation is a direct validation of s-p mixing, which alters the energy level order of nitrogen compared to oxygen and fluorine.

A further piece of information obtained from the band sub-structure in photoelectron spectroscopy concerns the vibrational spacing of each *ionic* state. As seen from Figure 8.28, the spacing in the $^2\Sigma_g^+$ state is 2150 cm^{-1}. The fundamental vibrational wavenumber for the N_2 molecule is 2379 cm^{-1}. Ionization decreases the vibrational frequency, implying that the bond has weakened slightly on ionization, which must therefore have been from a *weakly bonding* orbital ($3\sigma_g^+$).

The next band shows vibrational spacing of 1810 cm^{-1}. This implies that the electron must have been ionized from a *strongly bonding* orbital. This, together with the considerable vibrational structure, confirms that the strength of bonding in nitrogen is due to the π electrons.

In the last band, the vibrational frequency remains almost unchanged. Since there is a slight increase in the wavenumber, this orbital may be classified as *very weakly antibonding*.

This assignment validates all our previous discussions regarding the bonding in diatomic molecules. According to elementary bonding theories, the $3\sigma_g$ orbital is considered a bonding orbital and contributes to the bond order of 3 for nitrogen. However, it interacts with the $2\sigma_g$ orbital and gets antibonding character. A similar interaction between $2\sigma_u$ and $3\sigma_u$ reduces the

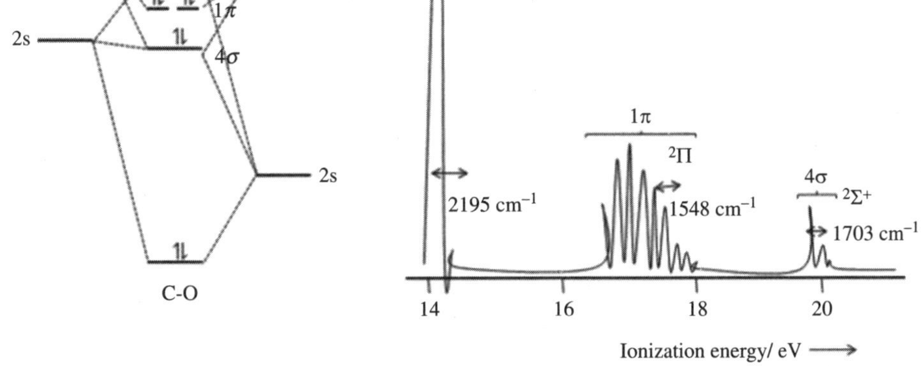

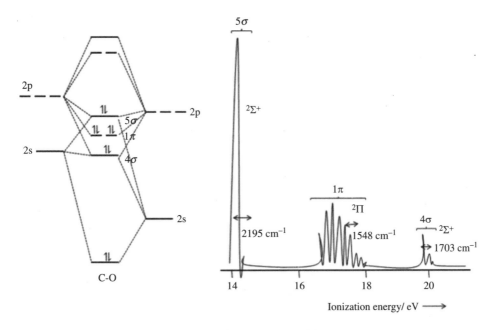

Figure 8.29. Photoelectron spectrum of carbon monoxide

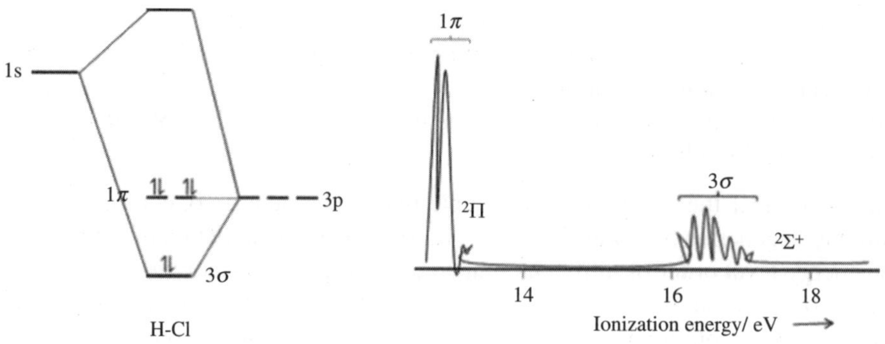

Figure 8.30. Photoelectron spectrum of HCl

antibonding character of $2\sigma_u$ and makes it almost a nonbonding orbital. This interaction is because of the small gap between the interacting orbitals and is known as s-p mixing. Photoelectron spectroscopy is the strongest validation for this mixing.

The lesson to be learnt is that an orbital is antibonding or bonding not by its symmetry designation but by the interaction it has with other orbitals.

The photoelectron spectrum of carbon monoxide (Figure 8.29) is reminiscent of that of nitrogen and confirms the adage that molecules with the same number of electrons and atoms will have similar structures and chemical properties (isoelectronic principle).

Similarly, the photoelectron spectrum of HCl (Figure 8.30) has two bands corresponding to nonbonding 1π MOs (negligible vibrational structure) and the 3σ bonding MO (vibrational structure).

8.11 SUMMARY

- The total energy of a molecule can be approximated by the sum of its electronic, vibrational and rotational energies.
- Changes in vibrational energy appear as a coarse structure and those in rotational energies as a fine structure in the electronic spectra.
- The Franck–Condon principle governs the intensity distribution of vibrational lines in electronic spectra.
- The dissociation energies of the ground and excited states may be determined from electronic spectra. In addition, predissociation may also occur.
- Photoelectron spectroscopy provides information about the orbital energies in molecules.

8.12 EXERCISES

1. The first five vibrational energy levels of HI are at 1144.83, 3374.90, 5525.51, 7596.66 and 9588.35 cm^{-1}. Calculate the dissociation energy, $\tilde{D}_0$, in reciprocal centimetre and electron volt from a Birge–Sponer plot.

2. (a) What is the Franck–Condon principle? Explain assuming the Morse equation for the potential energy curves.
   (b) Although there are no quantum mechanical restrictions on the change in the vibrational quantum number during an electronic transition, the probabilities and intensities of such vibrational changes during electronic transitions are not the same. Explain.
   (c) Show that if the ground and excited vibrational states have identical potential energy curves, the (0,0) transition is the most intense.
   (d) Consider the transition from one excited electronic state, with bond length r_e'', to another with bond length r_e'. Calculate the overlap between the vibrational ground states of each of the two electronic states, and show that the intensity of the (0,0) transition is greatest when the two equilibrium bond lengths are the same. Let the force constants in the two electronic states be the same.
   (e) In the case of Br_2, $r_e'' = 228$ pm and there is an upper state with $r_e' = 266$ pm. Taking the vibrational wavenumber as 250 cm^{-1}, calculate the Franck–Condon factor, and hence show that the intensity of the (0,0) transition is only 5×10^{-10} of what it would have been if the potential energy curves were directly above one other.
3. (a) Explain how the dissociation energies of a diatomic molecule in its ground and excited state can be calculated from a study of its electronic spectrum.
   (b) Explain the term 'predissociation'.
4. Mecke (1923) observed absorption maxima at the following wavenumbers in the visible spectrum of iodine vapour. These peaks are due to excitation from the ground vibrational level of the ground electronic state ($\upsilon'' = 0$) to upper vibrational levels of quantum number υ', of an upper electronic state.

υ'	$\tilde{\nu}$	υ'	$\tilde{\nu}$	υ'	$\tilde{\nu}$	υ'	$\tilde{\nu}$
56	19,821.2	55	19,797.2	51	19,691.3	50	19,657.8
46	19,515.9	45	19,473.3	41	19,292.7	40	19,245.2
36	19,024.1	35	18,961.6	31	18,697.8	30	18,626.6
26	18,320.9	25	18,239.1	21	17,892.8	20	17,801.6

   (a) Derive an expression for $\Delta\tilde{\nu}$, the difference between the wavenumbers for the transitions to vibrational levels υ' and $(\upsilon' + 1)$ for the electronically excited state. With the aid of a graphical plot, evaluate ω_e and x_e', the equilibrium vibration frequency and anharmonicity factor for the iodine molecule in the electronically excited state.
   (b) The wavenumber for the transition $(\upsilon'' = 1) \rightarrow (\upsilon' = 26)$ was found to be 18107.5 cm^{-1}. Hence evaluate ω_e'', the equilibrium vibration frequency for the iodine molecule in its ground electronic state.
   (c) Derive an expression for the wavenumber of a peak in the $\upsilon'' = 0$ series in terms of υ', ω_e', x_e' and E_{el}, where E_{el} is the electronic excitation energy. Evaluate E_{el} by substituting in known values for $\upsilon' = 26$. Assume that $x_e\omega_e''/4$ is negligibly small and that ω_e'' is equal to ω_0''.
5. (a) Write down the complete expression for the transition moment integral of an electronic transition and explain the significance of each integral in it.

(b) Show that for diatomic molecules the following selection rules operate:
   (i) In an electronic transition, the spin function is unchanged.
   (ii) Electronic transitions are allowed only between states of different symmetry.
   (iii) Vibrational–rotational transitions that are possible are those for which $\Delta J = \pm 1$, $\Delta \upsilon = \pm 1$.

Are these selection rules strictly valid or do they hold only under certain assumptions? What are these assumptions? Give the exceptions, if any, to these rules.

6. The orbital energies of the 1s, 2s and 2p orbitals of the nitrogen atom are −425.28 eV, −25.72 eV and −15.44 eV, respectively.
   (a) Sketch an MO diagram for N_2 that includes the 1s, 2s and 2p orbitals of nitrogen.
   (b) Indicate the electronic configuration and the term symbol for the ground state of N_2.
   (c) Indicate the electronic configuration and the term symbols for the states of N_2^+ that correspond to ionizing one electron out of the four highest occupied MOs of N_2. The photoelectron spectrum of N_2 in the 15–20 eV region shows three major peaks, which correspond to ionization out of the three highest occupied MOs of N_2. The ionization potentials corresponding to the three vertical ionizations are 15.57, 16.69 and 18.75 eV, and the vibrational spacings are 2191 cm^{-1}, 1850 cm^{-1} and 2397 cm^{-1}, respectively.
   (d) Perform an assignment of the photoelectron spectrum of N_2 based on the MO diagram.
   (e) Comparing the observed vibrational spacings to the N_2 ground state value of $\tilde{\nu}$ = 2331 cm^{-1}, what can you say about the bonding or antibonding nature of the three highest filled orbitals of N_2?
7. (a) The diagram below shows the high-resolution ultraviolet photoelectron spectrum of carbon monoxide, CO, in the gas phase (the upper traces were obtained at higher signal amplification). The spectrum was recorded using He(I) radiation at 21.2 eV.

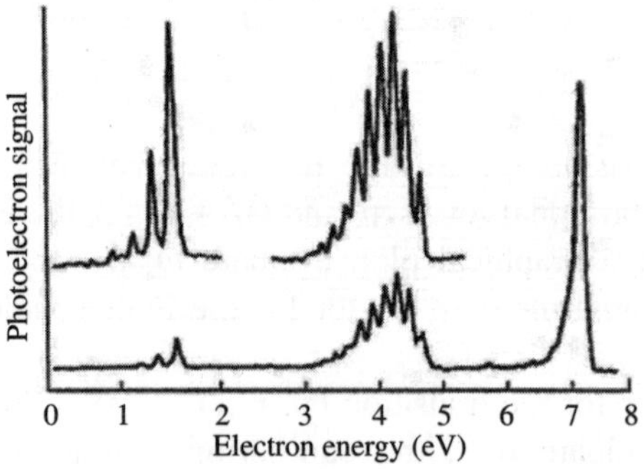

   (b) Explain the principles of the technique of ultraviolet photoelectron spectroscopy.
   (c) Draw a molecular orbital diagram for CO, and use it to explain the appearance of the photoelectron spectrum shown; include a discussion of why the band at around 4 eV shows more structure than the other two bands do. The spacing of the peaks within the band at around 4 eV is ~ 0.18 eV, whereas for the band at around 1.5 eV, the spacing is ~ 0.21 eV. Explain the origin of this difference.
   (d) Use the spectrum to estimate the adiabatic ionization energy of CO.

(e) Explain how Koopmans' Theorem can be used to deduce information about the electronic structure of CO.

8. The photoelectron spectrum of CO, using 21.22 eV He radiation, shows:
   (a) Band I with ionization energy 14.00 eV, intense, no fine structure and vibrational splitting of 2195 cm^{-1},
   (b) Band II with ionization energy 16.5 eV, moderately intense, considerable fine structure and vibrational splitting of 1548 cm^{-1},
   (c) Band III with ionization energy of 19.7 eV, weak, some fine structure and vibrational splitting of 1703 cm^{-1}.

   Explain the nature of the appearance of these bands. Assign these bands to the appropriate molecular orbitals of CO from which the electrons have been ejected. Give some reasoning for this assignment.

   Classify the orbitals into bonding, antibonding and nonbonding.

   Note: The fundamental frequency of the CO molecule is 2154 cm^{-1}.
9. What is the Franck–Condon principle in electronic spectroscopy? How does it predict the intensities of vibrational progressions for the case when (i) $r_e' < r_e''$ (ii) $r_e' \approx r_e''$ (iii) $r_e' > r_e''$? Illustrate your answer with diagrams.
10. Why are *band heads* more commonly observed in the rotational structure of vibronic ('vibrational–electronic') bands than in ro-vibrational bands within a single electronic state?
11. The term symbol for the ground state of N_2^+ is $^2\Sigma_g^+$. What is the total spin and total orbital angular momentum of the molecule? Show (using a molecular orbital diagram) that the term symbol agrees with the electronic configuration that would be predicted using the Aufbau principle.
12. The dissociation energies of I_2 and Cl_2 are 12452 and 19972 cm^{-1}, respectively. Use these to deduce the likely dissociation products of ICl ($^3\Pi$).
13. The molar internal energy of formation of BrCl(g) at 0 K is –0.64 kJ mol^{-1} and the bond dissociation energies of bromine and chlorine are 239 kJ mol^{-1} and 190 kJ mol^{-1}, respectively. Electronic transitions from the ground vibrational and electronic states of BrCl to vibrational levels υ' of an excited state are observed at the following wavenumbers:

State υ'	**3**	**4**	**5**	**6**	**7**	**8**	**9**	**10**
$\tilde{\nu}/cm^{-1}$	17343	17533	17713	17881	18036	18180	18310	18428

   (a) Calculate the dissociation energy of BrCl in its electronic ground state.
   (b) Deduce the electronic states of the halogen atoms produced in the dissociation of BrCl from the observed electronic state. BrCl dissociates in its ground state to produce ground state atoms and the $^2P_{1/2} \leftarrow {^2P_{3/2}}$ excitation energies of Cl and Br are 881 cm^{-1} and 3685 cm^{-1}.
14. A transition between the O_2 states $^3\Sigma_g$ and $^1\Delta_g$ violates three selection rules. Name them.
15. The electronic configuration of the B_2 molecule is

$$(1\sigma_g)^2(1\sigma_u)^2(2\sigma_g)^2(2\sigma_u)^2(1\pi_u)^2$$

   Find all the electronic states derivable from this configuration. Give their term symbols in terms of the symmetry species of the point group to which the molecule belongs. Select the ground state term symbol and justify your selection.

9

Electronic Spectroscopy of Polyatomic Molecules

The more accurate the calculations become,
the more the concepts tend to vanish into thin air.
—R. S. Mulliken, *J.C.P.* 43, S2 (1965)

9.1 INTRODUCTION

The complexity increases as the number of atoms in a molecule increases. However, some simplifying assumptions aid the interpretation of molecular spectra of polyatomic molecules. In the previous chapter, we had enumerated the selection rules for diatomic molecules. These can be broadly listed as follows:

1. *Spin selection rule*: $\Delta S = 0$.
2. *Orbital selection rules*: $\Delta\Lambda = 0, \pm 1$; Parity: $g \leftrightarrow u$, but $g \leftrightarrow g$ and $u \leftrightarrow u$ forbidden; Reflection: $+ \leftrightarrow +, - \leftrightarrow -$, but $+ \not\leftrightarrow -$.

Though the spin and parity selection rules still apply to polyatomic molecules, the other selection rules are particular to linear molecules. However, we may still divide our selection rules for polyatomic molecules into the spin and orbital selection rules.

The reason for treating polyatomic molecules separately is that their electronic spectra are usually investigated in solution, in which collisions with solvent blur the spectra. The rotational fine structure is completely lost, and only a few features of the vibrational structure are seen, depending on the magnitude of the interaction between the solute and the solvent. It is found that the more polar the solvent, the less clear is the vibrational structure. For example, the vibrational structure in the ultraviolet (UV) spectrum of acetone is clearly seen in hexane solution but not in water solution.

9.2 INTENSITIES OF ELECTRONIC TRANSITIONS

As we saw in Chapter 1, the basis of experimental spectroscopy is the Lambert–Beer law

$$A = \log\frac{I_0}{I} = \varepsilon C l \tag{9.1}$$

The observed value of the molar absorption coefficient ε is an indicator of the 'allowedness' of a transition. If a transition is spin forbidden, ε is generally in the range $10^{-5} - 10^{0}$ dm^3 mol^{-1} cm^{-1}, whether the transition is orbitally allowed or not. However, as we saw in Chapter 3, for heavy atoms, the spin quantum number S does not remain a 'good' quantum number. Instead, the total angular momentum quantum number J becomes 'good' and the spin-only selection rule breaks down. In such cases, higher values of ε may be observed for 'spin-forbidden' transitions.

If a transition is orbitally forbidden (but spin allowed), ε is in the range $10^0 - 10^3$ dm^3 mol^{-1} cm^{-1}. A 'fully allowed' transition is one which is allowed by both spin and orbital selection rules, and the ε values for such transitions range from 10^3 to 10^5 dm^3 mol^{-1} cm^{-1}. Note that the concentration, C, is expressed in mol dm^{-3} and l in cm in expression (9.1).

We have also seen that electronic transitions are usually broad and spread over several vibrational transitions due to the Franck–Condon principle. The value of ε referred to above is for the maximum of the band, also written as ε_{max}. In terms of ε_{max}, the intensity of electronic transitions is classified as strong, medium or low depending on whether ε_{max} is of the order 10^4–10^5, 10^3 or $< 10^3$, respectively.

The value of ε_{max} may be a good indicator of the 'allowedness' of transitions of the same type in different molecules, e.g., the n$\rightarrow\pi^*$ transitions of the carbonyl group in a series of ketones, but it is not such a good parameter to gauge the strengths of bands of different shapes. In such cases, the integrated absorption coefficient is a better parameter to compare intensities of transitions. This quantity is given by

$$\bar{A} = \int_{\text{band}} \varepsilon(\tilde{\nu})d\tilde{\nu} = \frac{1}{Cl}\int_{\text{band}} \log\frac{I_0}{I}d\tilde{\nu} \tag{9.2}$$

The integral represents the area under the absorption curve plotted as a function of the wavenumber. It is proportional to the square of the transition moment integral, as described in Chapter 2.

The intensities of electronic transitions are often represented by a dimensionless quantity, the *oscillator strength* (f), defined as the ratio of the observed integrated absorption coefficient of a transition to the theoretical value for a fully allowed transition, i.e.,

$$f = \frac{\bar{A}_{\text{obs}}}{\bar{A}_{\text{theor}}} \tag{9.3}$$

9.2.1 Calculation of oscillator strength

The quantity in the denominator of equation (9.3) is the integrated absorption coefficient for a *fully allowed electronic transition*. The question is: how do we define a fully allowed electronic transition? In the days before the advent of quantum mechanics, the electron was thought to be bound to the nucleus by a Hooke's law kind of force. Hence, the behaviour of the electron responsible for the absorption of radiation was described by a harmonic potential similar to the harmonic oscillator. Our study of vibrational transitions in Chapter 5 had revealed that, for a one-dimensional harmonic oscillator, the transition dipole moment for a $\upsilon = 1 \leftarrow 0$ transition is given by the expression

$$\left|\vec{M}_{01x}\right| = \frac{1}{\sqrt{2\alpha}}\left(\frac{d\mu}{dx}\right)_{x=0} \tag{9.4}$$

where $\alpha = \sqrt{\mu_{red} k / \hbar}$ and the harmonic oscillator is oscillating in the x direction. For an electron bound to a nucleus, the dipole moment is given by $\mu = ex$, so that $(d\mu / dx)_{x=0} = e$, where e is the electronic charge. Since the mass of the electron is much smaller than that of the proton ($m_e << m_p$), in the expression for the reduced mass

$$\frac{1}{\mu_{red}} = \frac{1}{m_e} + \frac{1}{m_p}$$

the second term is much smaller than the first and can be ignored. We thus have $\mu_{red} \approx m_e$. Also, $k = 4\pi^2 c^2 \omega_e^2 \mu_{red} \approx 4\pi^2 c^2 \omega_e^2 m_e$. Therefore, $\alpha = \sqrt{\mu_{red} k / \hbar} \approx 2\pi c \omega_e m_e / \hbar$ and

$$\left|\vec{M}_{01x}\right| = e\sqrt{\frac{\hbar}{4\pi c \omega_e m_e}} \tag{9.5}$$

As the transition dipole moment is a vector quantity, for a three-dimensional harmonic oscillator, we have

$$\left|\vec{M}_{01}\right|^2 = \left|\vec{M}_{01x}\right|^2 + \left|\vec{M}_{01y}\right|^2 + \left|\vec{M}_{01z}\right|^2 \tag{9.6}$$

There is nothing special about the x direction: all directions will make similar contributions, and hence, $\left|\vec{M}_{01}\right|^2$ is simply thrice the contribution of the x-component, i.e.,

$$\left|\vec{M}_{01}\right|^2 = \frac{3e^2\hbar}{4\pi c \omega_e m_e} \tag{9.7}$$

The integrated absorption coefficient is given by equation (2.35)

$$\bar{A} = \frac{8\pi^3 N_A}{3hc(4\pi\varepsilon_0)(\ln 10)} \omega_e \left|\vec{M}_{01}\right|^2 = \frac{\pi N_A e^2}{c^2 m_e (4\pi\varepsilon_0) \ln 10} \tag{9.8}$$

Substitution of the various quantities with $1/(4\pi\varepsilon_0) = 8.987 \times 10^9$ N m^2 C^{-2} gives

$$\bar{A} = 2.315 \times 10^9 \text{ m mol}^{-1} = 2.315 \times 10^8 \text{ dm}^3 \text{ mol}^{-1} \text{ cm}^{-2} \tag{9.9}$$

Hence, the oscillator strength is given by

$$f = \frac{\left(\int_{\text{band}} \varepsilon(\tilde{\nu}) d\tilde{\nu}\right)_{\text{obs}}}{2.315 \times 10^8} = 4.320 \times 10^{-9} \int_{\text{band}} \varepsilon(\tilde{\nu}) d\tilde{\nu} \tag{9.10}$$

where the constant has the units dm^{-3} mol cm, and hence, this formula applies if $\varepsilon(\tilde{\nu})$ is expressed in the usual dm^3 mol^{-1} cm^{-1} units. In case the SI units for $\varepsilon(\tilde{\nu})$ (m^2 mol^{-1}) are used, the factor in equation (9.9) becomes 4.320×10^{-10} mol m^{-1}. Many electronic absorption bands have f values near unity, indicating that the transitions are fully allowed. On the other hand, some electronic bands with very much smaller values of f are also found, and these are assigned to 'forbidden' bands.

Example 9.1 Many molecules in solution have absorption bands whose width at half height (FWHM, see Chapter 1) is ~5000 cm^{-1}. The band area may then be calculated approximately as that of a rectangle of length ε_{max} and breadth 5000 cm^{-1}. On this basis, calculate the oscillator strength:

(a) of a transition giving rise to a band with $\varepsilon_{max} = 10^2$ m^2 mol^{-1},

(b) of the benzene ${}^1A_{1g} \to {}^1B_{2u}$ transition at 256 nm, for which $\varepsilon_{max} = 160$ dm^3 mol^{-1} cm^{-1} and the width at half-height is 4000 cm^{-1}.

Solution

(a) $\bar{A} = (100\ m^2\ mol^{-1})(5000\ cm^{-1}) = 5\times10^5\ m^2\ mol^{-1}cm^{-1} = 5\times10^7 m\ mol^{-1}$

Therefore, $f = (4.32 \times 10^{-10}\ mol\ m^{-1})(5 \times 10^7\ m\ mol^{-1}) = 0.0216$

(b) $\bar{A} = (160\ dm^3\ mol^{-1}\ cm^{-1})(4000\ cm^{-1}) = 64\times10^4\ dm^2\ mol^{-1}\ cm^{-2}$ and

$f = (4.32 \times 10^{-9}\ dm^{-3}\ mol\ cm)(64 \times 10^4\ dm^3\ mol^{-1}\ cm^{-2}) = 2.76 \times 10^{-3}$

Since the numerical values of f in both cases are small, both the transitions are forbidden.

We now have a look at the factors that make a transition forbidden or allowed.

9.3 TYPES OF ELECTRONIC TRANSITIONS

In the previous chapter, we had seen that the orbitals in a diatomic molecule can be classified as σ, σ^*, π, π^*, etc. In addition, atoms such as N, O, S and Cl are found to have nonbonding electrons. This nomenclature strictly applies only to linear molecules ($D_{\infty h}$ or $C_{\infty v}$). No other molecular point group has notations such as σ or π for its irreducible representations. However, the nomenclature continues to be used loosely for nonlinear molecules, where σ or π refers to the local symmetry of a bond, where σ is an orbital which lies along that bond axis and π has a node there. In denoting an orbital, the correct symmetry label should be used according to the point group of the molecule, as we shall show later.

In this notation, the orbitals of a polyatomic molecule such as formaldehyde can be depicted as in Figure 9.1. The four main transitions are $\sigma\to\sigma^*$, n$\to\sigma^*$, n$\to\pi^*$ and $\pi\to\pi^*$ (Figure 9.1). Unlike gas phase spectra, for solution phase spectra, the ground state is written first. We consider these four transitions in turn. Their approximate regions in the electromagnetic spectrum are shown in Figure 9.2.

Figure 9.2 shows that absorption occurs below the working range of standard spectrometers (200 – 900 nm) by molecules in which the only transitions possible are $\sigma\to\sigma^*$, such as alkanes. Absorption shifts to the working range if certain groups called *chromophores* are present. Chromophores contain empty π^* orbitals to which n, σ and π electrons can be promoted. Typical chromophores include C=O, C=C, N=N and NO_2. Other substituents, which are not themselves chromophores, sometimes modify the absorption of molecules containing

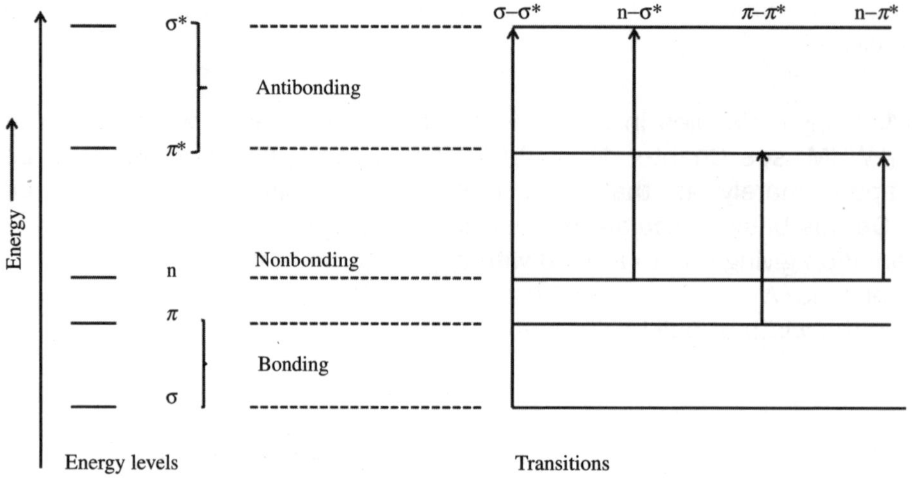

Figure 9.1. Molecular electronic energy levels

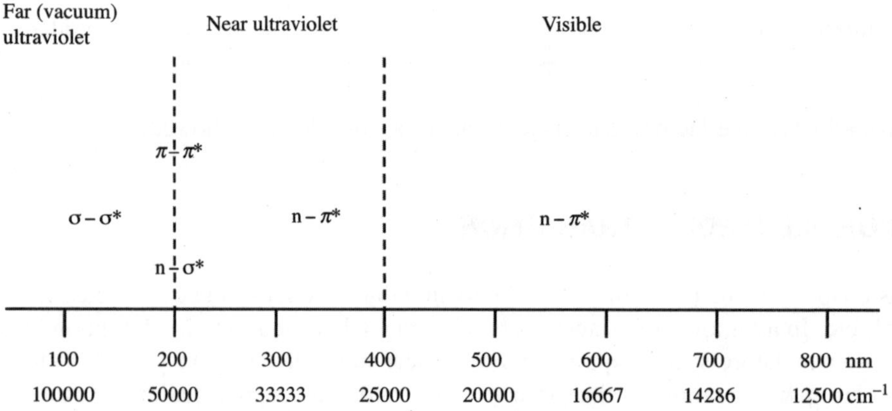

Figure 9.2. Regions of the electronic spectrum and the electronic transitions observed

chromophores. Such molecules are called *auxochromes*. Some examples of auxochromes are OH, Cl, NH_2 and CH_3. They act by pushing electrons in or out of the chromophore, thereby lowering or raising the energy of the π and π^* orbitals and affecting spectra. The following terms (Figure 9.3) are sometimes associated with the action of auxochromes:

(a) *Hyperchromic*: Increase in intensity
(b) *Hypochromic*: Decrease in intensity
(c) *Bathochromic (or red)*: Shift to longer wavelength
(d) *Hypsochromic (or blue)*: Shift to shorter wavelength

Often, hyperchromic effects are associated with bathochromic shifts, resulting from an electron donating auxochrome like $-CH_3$, and hypochromic effects with hypsochromic shifts caused by electron withdrawing substituents such as Cl.

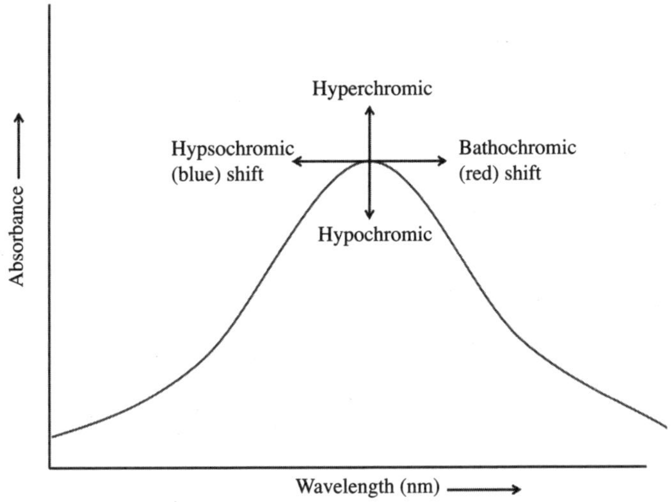

Figure 9.3. Terms used in reference to electronic spectra

$\sigma \rightarrow \sigma^*$ transitions

Figures 9.1 and 9.2 show that these transitions involve the largest energies, which are two to three times larger than those for the other transitions. The corresponding wavelengths are small (λ < 150 nm). This region is not accessible to conventional spectroscopy since both glass and quartz absorb radiation in this region. The region below 200 nm is often called the vacuum UV region because of absorption by molecular oxygen in this region, due to which evacuated chambers are required. Molecules with no unsaturated bonds or heteroatoms, such as alkanes, cannot display any other absorption and their spectra are investigated in this region.

n $\rightarrow \sigma^*$ transitions

These transitions involve promotion of a nonbonding electron to a σ^* orbital. The transition wavelengths are between 150 – 250 nm. The molar absorption coefficients are low. Polar solvents stabilize the nonbonding electrons by solvating them through hydrogen bonding, thus increasing the absorption energy. In molecular spectroscopy terms, the transitions undergo a *blue* shift (with reference to visible radiation, blue is towards the higher energy or shorter wavelength) or *hypsochromic* shift. The number of molecules that display n$\rightarrow\sigma^*$ absorptions is rather limited (Table 9.1), because the wavelengths are below the practical working limit of standard spectrometers.

n $\rightarrow \pi^*$ transitions

These transitions are observed as weak bands in unsaturated molecules that contain heteroatoms such as oxygen, nitrogen and sulphur. The ε_{max} values range from 10 to 100 dm^3 mol^{-1} cm^{-1}. In case of aldehydes and ketones, these transitions are due to excitation of a nonbonding electron on oxygen to the π^* orbital of the carbonyl moiety, and generally occur in the range between 270 and 300 nm.

$\pi \rightarrow \pi^*$ transitions

In this case, the selection rules, such as those based on symmetry concepts, determine whether a transition to a particular π^* orbital is allowed or forbidden. The ε_{max} values usually range from

Table 9.1. Some examples of absorption due to n→σ* transitions

Compound	λ_{max} (nm)	ε_{max}
H_2O	167	1480
CH_3OH	184	150
CH_3Cl	173	200
CH_3I	258	365
$(CH_3)_2S$	229	140
$(CH_3)_2O$	184	2520
CH_3NH_2	215	600
$(CH_3)_3N$	227	900

1000 to 10,000 $dm^3\ mol^{-1}\ cm^{-1}$. In a molecule containing several π orbitals, several possibilities of π→π* transitions exist, and the extent to which each transition is allowed or forbidden varies, leading to spectra in which several bands of varying intensities are observed, as for benzene. The effect of substituents is quite marked and varies both with the position and nature of substitution in benzene.

In unsaturated molecules containing heteroatoms, the n→π* and π→π* transitions are the most important spectroscopic transitions. We therefore consider them in more detail and compare the effects of conjugation and solvent on their wavelengths and intensities.

9.3.1 Effect of conjugation

Increase in conjugation causes the π→π* band to shift to longer wavelengths (Figure 9.4).

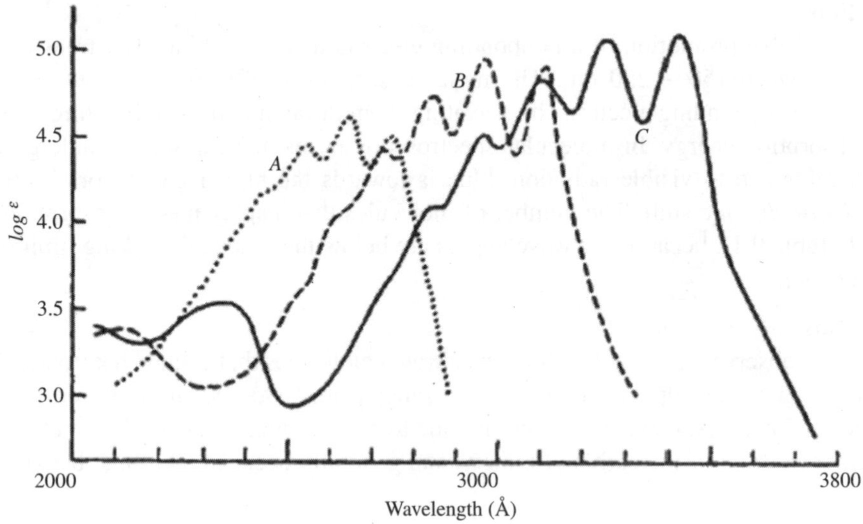

Figure 9.4. UV/Vis spectra of the polyenes $CH_3(CH{=}CH)_nCH_3$, where $n = 3$, 4 and 5 [Reproduced from Nayler & Whiting (1955) with permission from the Royal Society of Chemistry]

The isolated C=C chromophore (ethylene) causes absorption at around 190 nm due to a $\pi \rightarrow \pi^*$ transition. The absorption band is intense ($\varepsilon \approx 10{,}000$ dm^3 mol^{-1} cm). When two isolated double bonds are brought into conjugation, both levels are shifted (Figure 9.5). A hypochromic and bathochromic shift is observed.

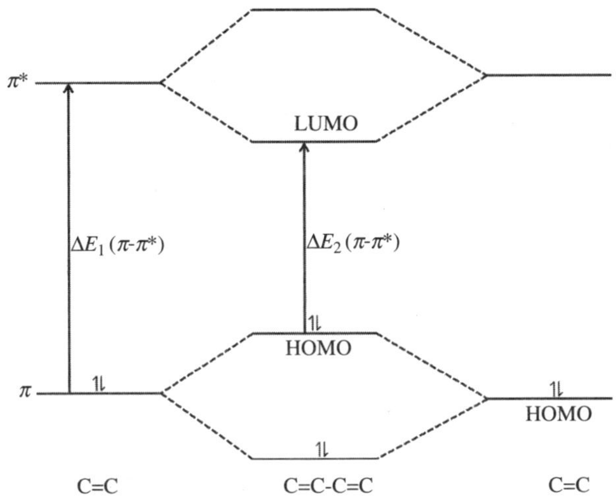

Figure 9.5. MO level correlation diagram to illustrate the effect of conjugation on the $\pi \rightarrow \pi^*$ transition of unsaturated hydrocarbons

As for $\pi \rightarrow \pi^*$ transitions, increased conjugation shifts $n \rightarrow \pi^*$ transitions to longer wavelengths. The lowest unoccupied molecular orbital (LUMO) π^* orbital is shifted downwards (Figure 9.6), and the $n \rightarrow \pi^*$ absorption shifts to lower energy, resulting in a bathochromic shift of the $n \rightarrow \pi^*$ transition to 300 – 350 nm, and the ε_{max} value also increases up to 10-fold. Extended conjugation increases this effect. For example, acetone (propanone) shows a weak band due to the $n \rightarrow \pi^*$ transition at 279 nm ($\varepsilon_{max} = 15$) in hexane. This band shifts to 327 nm, and its ε_{max} value also increases to 98 due to extended conjugation in mesityl oxide (4-methyl-3-penten-2-one).

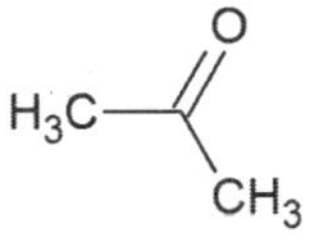

Acetone

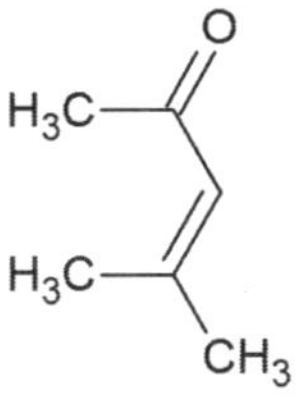

Mesityl oxide

When sufficient conjugation is present, the energy of the LUMO π^* orbital is lowered to the point where v_2 (Figure 9.6) corresponds to the visible region of the spectrum (~500 nm rather than 290 nm), a feature which is exploited in tunable dye lasers.

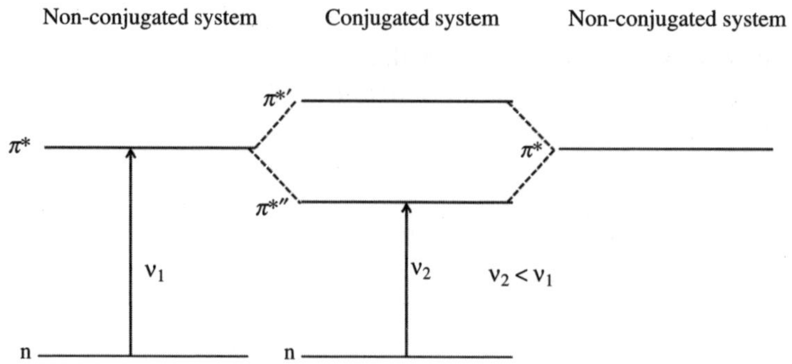

Figure 9.6. Effect of conjugation on n→π* transitions

9.3.2 Effect of solvent

Change to a more polar solvent causes a red or bathochromic shift in the π→π* bands. This is due to the fact that a hydrolytic solvent stabilizes a π–π* excited state more than the ground state (Figure 9.7).

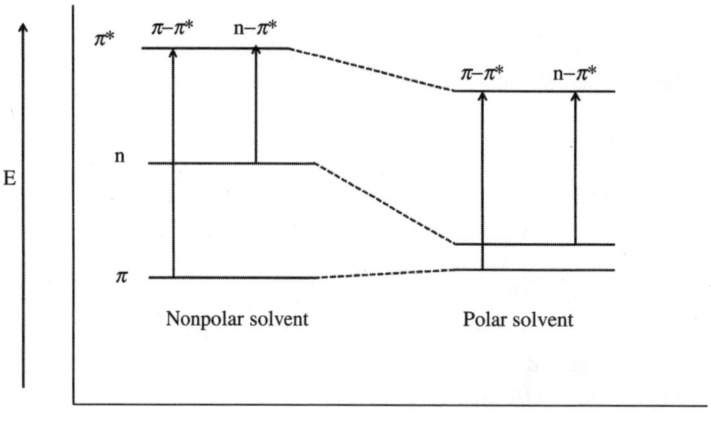

Figure 9.7. Effect of polar solvents on n→π* and π→π* transitions

The effect of solvent on the position of bands due to n→π* transitions is usually opposite to that for π→π* transitions (Figure 9.7). Hydrolytic solvents stabilize the *n* level, causing a hypsochromic shift. Hydrogen bonding solvents, such as water, methanol or ethanol, cause stabilization of the lone pair through hydrogen bonding. The effect is more pronounced if the solvent has a larger dielectric constant. Wavelength shifts up to 30 nm are observed. For example, the λ_{max} for acetone shifts from 279 nm to 257 nm when the solvent is changed from hexane to water. For mesityl oxide, a shift from 327 nm to 305 nm is similarly observed. The amount of shift also depends on the dipole moment of the solute. As the dipole moment of the solute increases, so does the shift. For example, for flavoxanthin, the λ_{max} changes from 448.5 to 446.5 nm only when the solvent is changed from hexane to the more polar methanol, whereas the same change in solvent brings about a shift of 12 nm in the wavelength from 279 to 267 nm for acetone, which has a higher dipole moment than flavoxanthin.

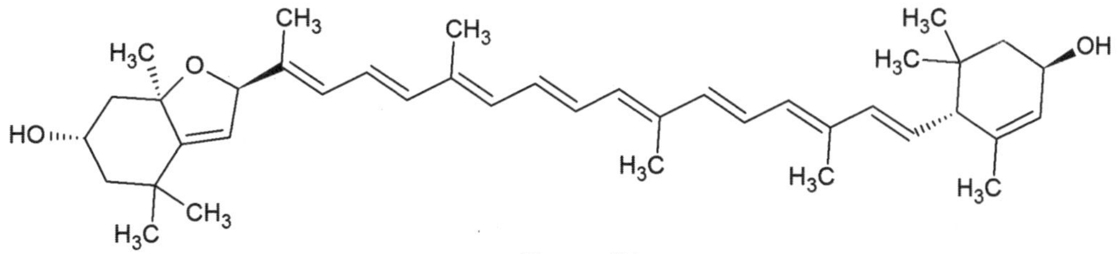

Flavoxanthin

The pH of the medium also has a profound effect on n→π^* transitions, causing them in some cases to disappear completely. For example, the band at 300 nm due to an n→π^* transition in pyridine disappears on decreasing the pH below its pK_a, due to protonation of the nitrogen lone pair.

Auxochromes such as OH and Cl donate electrons to the π system and increase the energies of both the π and π^* orbitals, but the n orbitals remain unaffected. This increases the energy of the $n\rightarrow\pi^*$ transition, leading to a hypsochromic shift.

Several theories to explain the observed phenomena for $\pi\rightarrow\pi^*$ transitions were propounded in the previous century. Some simple systems to which these concepts can be applied are linear and cyclic conjugated π systems.

9.4 THEORIES OF $\pi\rightarrow\pi^*$ TRANSITIONS

Almost all the empirical theories of π-electron systems are based on the concept of $\sigma - \pi$ separability, i.e., the π-electron system can be considered independent of the σ-electron framework.

$$\begin{aligned}\hat{H}_{el} &= \hat{H}_\sigma + \hat{H}_\pi \\ \psi_{el} &= \psi_\sigma + \psi_\pi \\ E_{el} &= E_\sigma + E_\pi\end{aligned} \tag{9.11}$$

This is so because the σ electrons lie in the plane of the molecule, where the π orbitals have a node, and hence, their mutual interaction is not possible. As Figure 9.1 shows, the longest wavelength (smallest energy) transition of unsaturated systems containing no heteroatoms is the $\pi - \pi^*$ transition. We therefore need to focus only on the π-electron system to interpret their UV and visible spectra.

Among the several methods based on the concept of $\sigma - \pi$ separability are the Hückel molecular orbital (HMO) method and the free electron model (FEM) method. The HMO method and its later modification Pariser-Parr-Pople (PPP) method are among the first methods to be widely applied to conjugated systems. Considering the drastic approximations involved, these methods yield good, intuitive results, which help in interpreting the results of more sophisticated methods that are currently available. In fact, these methods are only of historical importance now. Advancements in computational techniques and computer technology have now made it possible to perform accurate computations on even large systems such as proteins.

Their accuracy now transcends even experiment. The earlier methods have not lost their importance, however, because they laid the foundation for the later sophisticated methods of computational chemistry.

9.4.1 Hückel molecular orbital (HMO) theory

We begin with the simplest π electron system, ethylene (ethene).

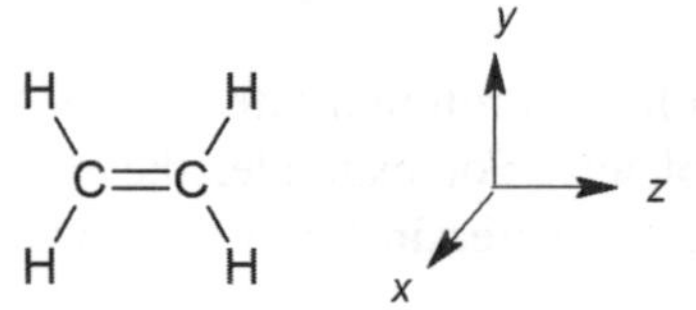

The molecular plane is *yz*. Since we are considering only the π electrons, the hydrogens are ignored as they are capable of σ bonding only. We also ignore the 1s electrons of carbon, which are part of the core, having very negative energies. In the language of valence bond theory, the valence 2s, $2p_y$ and $2p_z$ orbitals participate in sp^2 hybridization. Each carbon atom is involved in σ bonding with the other carbon atom and two hydrogens in the molecular plane. This leaves one electron in the $2p_x$ orbital perpendicular to the plane of the molecule. The overlap of the $2p_x$ orbitals of the two carbon atoms results in a π bond between the two. If we label the two $2p_x$ orbitals χ_1 and χ_2, the LCAO of the *j*th MO can be described by

$$\phi_j = \sum_{r=1}^{2} c_{jr}\chi_r \tag{9.12}$$

where the χ_r are called the basis orbitals.

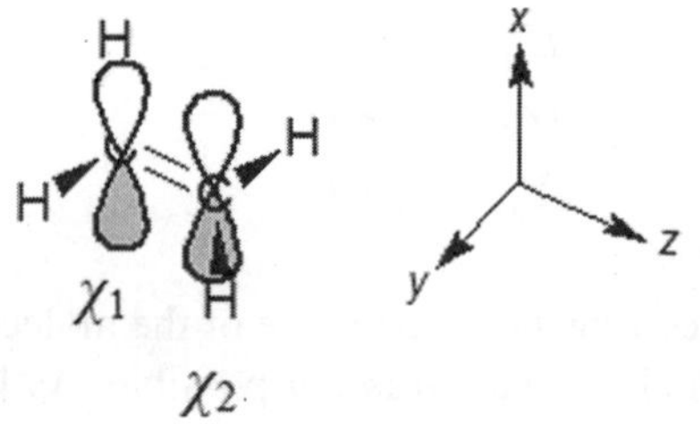

The secular equations are given, as in equation (8.13), by

$$\begin{vmatrix} H_{11} - ES_{11} & H_{12} - ES_{12} \\ H_{21} - ES_{21} & H_{22} - ES_{22} \end{vmatrix} = 0$$

where $H_{rr} = \int \chi_r^* \hat{H}_\pi \chi_r \, d\tau, H_{rs} = \int \chi_r^* \hat{H}_\pi \chi_s \, d\tau, S_{rr} = \int \chi_r^* \chi_r \, d\tau$ and $S_{rs} = \int \chi_r^* \chi_s \, d\tau$.

Integrals of the type H_{rr} correspond to the π electron energy of an electron in the atomic orbital χ_r, i.e., the 2p orbital of carbon. Due to the various approximations involved, we cannot write an explicit expression for $\hat{H}_\pi$ and so it is impossible to solve for quantities such as H_{rr}. The HMO method simplifies matters by substituting α_r for H_{rr}, where α, called the *coulomb* integral, is an empirical parameter to be determined so as to give best agreement with experiment. Since both carbon atoms are equivalent, $H_{11} = H_{22} = \alpha$. In HMO, even if the atoms are non-equivalent, the coulomb integrals of all carbon atoms are designated α.

Integrals of the type H_{rs} over different atoms are put equal to zero if the two atoms are not directly bonded. If the two atoms are bonded to each other, this quantity is put equal to another empirical parameter, $\beta_{C\text{-}C}$, or simply β, where β is the *resonance* integral. A frequently used value for β is –2.4 eV.

The terms S_{rr} and S_{rs} are the *overlap* integrals. If the basis orbitals are normalized, $S_{rr} = 1$. S_{rs} is the overlap between orbitals on *different* atoms and is hence not necessary equal to zero on account of orthogonality. In fact, it has a value of 0.3 for overlap of p orbitals comprising π orbitals. However, HMO theory sets a value of zero to all overlap integrals S_{rs} $(r \neq s)$. This rather drastic approximation does not have a large effect on the calculated energies, but greatly simplifies the calculations.

Inserting these values, the secular determinant (9.13) can now be written as

$$\begin{vmatrix} \alpha - \varepsilon & \beta \\ \beta & \alpha - \varepsilon \end{vmatrix} = 0$$

where we have written ε for the orbital energies. We divide throughout by β

$$\begin{vmatrix} \dfrac{\alpha - \varepsilon}{\beta} & 1 \\ 1 & \dfrac{\alpha - \varepsilon}{\beta} \end{vmatrix} = 0$$

and put $x = (\alpha - \varepsilon)/\beta$, whence

$$\begin{vmatrix} x & 1 \\ 1 & x \end{vmatrix} = 0 \qquad (9.13)$$

or $x^2 - 1 = 0$. Therefore, the two roots are $x = \pm 1 = (\alpha - \varepsilon)/\beta$ or $\varepsilon = \alpha \pm \beta$. Equation (9.13) is usually referred to as the *HMO determinantal equation*.

In this expression, α is the energy of an electron in a 2p orbital on carbon atom, which is a bound state, and hence, its value is negative. Likewise, β also refers to an electron in a bound region and is also negative. This implies that $\alpha + \beta$ corresponds to the lower energy orbital (Figure 9.8).

In general, since $\varepsilon = \alpha - x\beta$ and β is negative, the more negative values of x correspond to lower energy orbitals. We now start filling up the electrons in the molecular orbitals according to the Aufbau principle. In the ground state of ethylene, both electrons are in the lower energy orbital and the total energy is thus $E_\pi = 2\alpha + 2\beta$. Thus, an energy lowering of 2β occurs on bond formation, since the energies of the two isolated carbon atoms is 2α. Solving for the coefficients of the MOs, we substitute the first value of x (= –1) in the expression

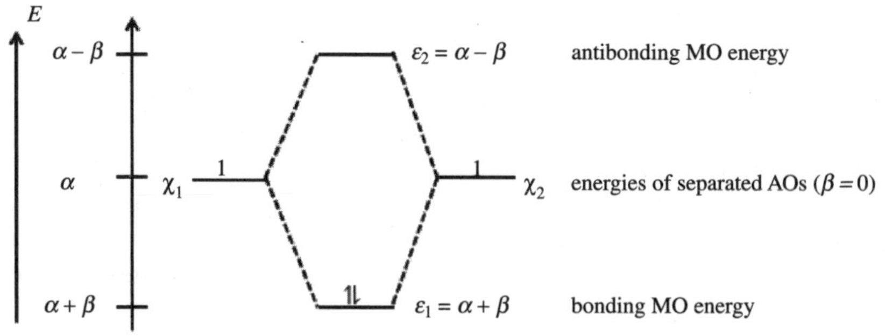

Figure 9.8. HMO energy scheme for ethylene

$$\begin{pmatrix} x & 1 \\ 1 & x \end{pmatrix}\begin{pmatrix} c_{11} \\ c_{12} \end{pmatrix} = \begin{pmatrix} 0 \\ 0 \end{pmatrix}$$

to find that $c_{11} = c_{12}$ or the normalized wave function for the lower energy (bonding) orbital is

$$\phi_1 = \frac{1}{\sqrt{2}}(\chi_1 + \chi_2) \tag{9.14a}$$

In a similar fashion, we obtain

$$\phi_2 = \frac{1}{\sqrt{2}}(\chi_1 - \chi_2) \tag{9.14b}$$

for the second (antibonding) MO. This wave function has a node (it changes sign between χ_1 and χ_2), and hence, there is depletion of charge density in the region between the two carbon atoms. The contours of the two orbitals obtained from an *ab initio* calculation are depicted below

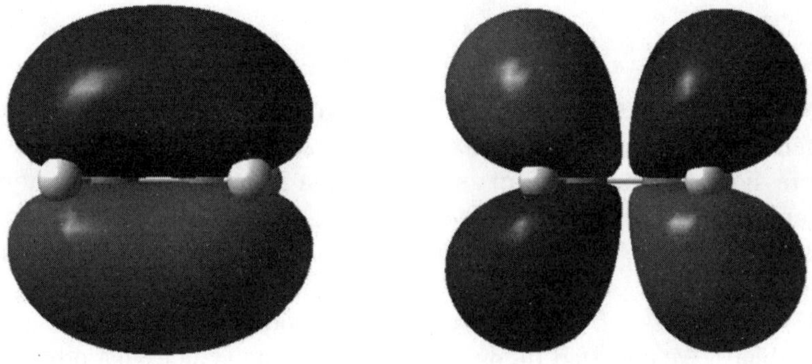

ϕ_1 ϕ_2

In the excited state, one electron is promoted from the lower energy orbital to the higher energy one:

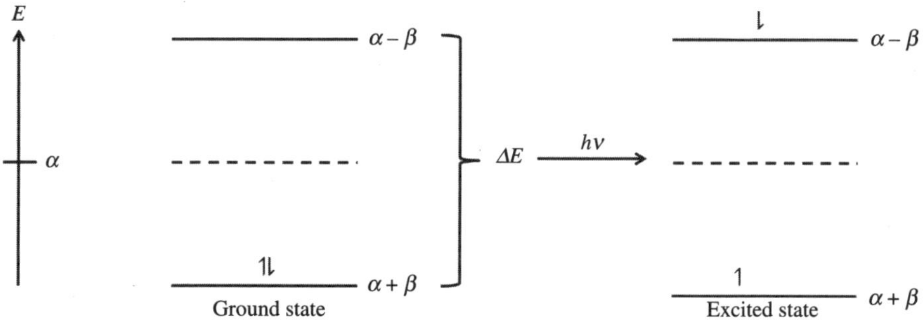

The total energy is back to 2α. The transition energy is $(\alpha - \beta) - (\alpha + \beta) = -2\beta$. Using a value of −2.4 eV for β, $\Delta E = 4.8$ eV. This gives

$$\lambda = hc/\Delta E = (6.626\times10^{-34}\ \mathrm{J\,s}\times2.998\times10^{8}\ \mathrm{m\,s^{-1}})/(4.8\ \mathrm{eV}\times1.602\times10^{-19}\ \mathrm{J\,eV^{-1}})$$
$$= 1.240\times10^{-6}/\Delta E = 2.583\times10^{-7}\,\mathrm{m} = 258\ \mathrm{nm}$$

The observed wavelength is 171 nm, which is much shorter than the calculated value. However, considering the number of approximations involved and the uncertain value of β (values ranging from −0.78 to −2.7 eV have been used), there is at least qualitative agreement.

As stated earlier, wavelengths <200 nm fall in the vacuum UV region, which is not easily accessible. Conjugation shifts the wavelength to longer values, as we shall see for 1,3-butadiene.

9.4.1.1 Effect of conjugation

The next even polyene to be considered is the four carbon atom system, 1,3-butadiene, which may exist as a *cis* or *trans* isomer.

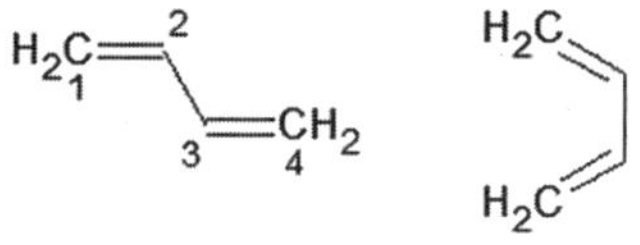

s-trans ***s-cis***

Simple HMO does not distinguish between the two isomers: it is only concerned with the connectivity between atoms. For this reason, the HMO determinant is also called a *topological* matrix. Using our discussion for ethylene, we may now write the HMO topological matrix for 1,3-butadiene

$$\begin{pmatrix} \alpha & \beta & 0 & 0 \\ \beta & \alpha & \beta & 0 \\ 0 & \beta & \alpha & \beta \\ 0 & 0 & \beta & \alpha \end{pmatrix}$$

Here, all the diagonal elements correspond to the energy of an electron in a 2p orbital of carbon atom and are equal to α, and we have placed β's where the two carbon atoms are bonded, and zero elsewhere for the off-diagonal elements. We follow the procedure adopted for ethylene, i.e., we find the eigenvalues ε_i by diagonalizing this matrix, and set $x = (\alpha - \varepsilon)/\beta$ to obtain

$$\begin{vmatrix} x & 1 & 0 & 0 \\ 1 & x & 1 & 0 \\ 0 & 1 & x & 1 \\ 0 & 0 & 1 & x \end{vmatrix} = 0$$

i.e., in the topological matrix, we substitute x for α and 1 for β. Note that the secular determinant is symmetrical. We now have a prescription for writing the secular determinant for any conjugated system. The size of the determinant is determined by the number of carbon atoms in conjugation. Thus, propene (CH_3-$CH{=}CH_2$) has the same secular determinant as ethylene because the methyl carbon is sp^3 hybridized and is not part of the conjugated π system. Write x for all the diagonal elements. The off-diagonal elements are equal to zero if the concerned carbons are not bonded, and one if they are bonded to one other. Expansion of the determinant for butadiene by the method of minors gives

$$\begin{vmatrix} x & 1 & 0 & 0 \\ 1 & x & 1 & 0 \\ 0 & 1 & x & 1 \\ 0 & 0 & 1 & x \end{vmatrix} = 0 \Rightarrow \underbrace{x\begin{vmatrix} x & 1 & 0 \\ 1 & x & 1 \\ 0 & 1 & x \end{vmatrix} - \begin{vmatrix} 0 & 1 & 0 \\ 0 & x & 1 \\ 0 & 1 & x \end{vmatrix}}_{\text{expansion of the 4×4 determinant by minors, using the first row}} = x(x^2 - 2x) - (x^2 - 1)$$

$$= \underbrace{x^4 - 3x^2 + 1 = 0}\ \underset{x^2 = u}{\Rightarrow}\ \underbrace{u^2 - 3u + 1 = 0}_{\text{quadratic equation}}$$

$$u_{1,2} = \frac{3 \pm \sqrt{5}}{2} \Rightarrow u_1 = 2.618, u_2 = 0.382$$

$$x_{1,2} = \pm\sqrt{u_1} = \pm 1.618, x_{3,4} = \pm 0.618$$

The relationship $\varepsilon_i = \alpha - x_i\beta$ turns the x-values into orbital energies and we end up with the HMO orbital energy scheme for butadiene shown in Figure 9.9.

Starting with the lowest, the first two energy levels are filled up with the four electrons. The lowest energy (longest wavelength) transition is from ε_2 [called the highest occupied molecular orbital (HOMO)] to ε_3 (LUMO) corresponding to a transition energy of -1.236β or 2.966 eV and a transition wavelength of 418 nm. The observed wavelength is 217 nm. Though the agreement is not good, the HMO method correctly predicts that increasing conjugation shifts the longest wavelength transition to longer wavelengths; i.e., a red or bathochromic shift with respect to ethylene is observed.

Solving for the wave functions, we obtain the following linear combinations for the four MOs for butadiene

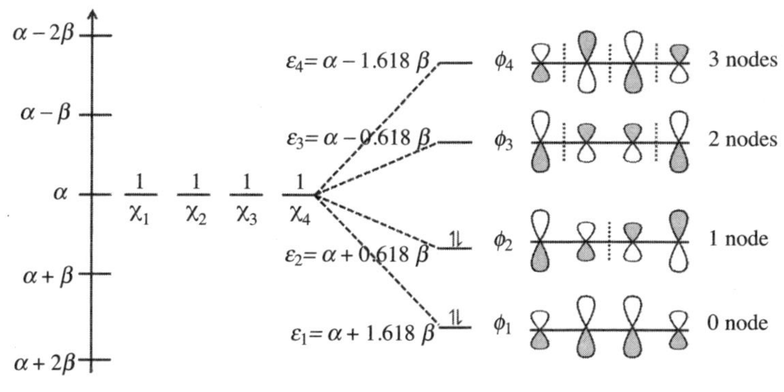

Figure 9.9. HMO energy scheme for butadiene

$$\begin{aligned}\phi_1 &= 0.37\chi_1 + 0.60\chi_2 + 0.60\chi_3 + 0.37\chi_4 \\ \phi_2 &= 0.60\chi_1 + 0.37\chi_2 - 0.37\chi_3 - 0.60\chi_4 \\ \phi_3 &= 0.60\chi_1 - 0.37\chi_2 - 0.37\chi_3 + 0.60\chi_4 \\ \phi_4 &= 0.37\chi_1 - 0.60\chi_2 + 0.60\chi_3 - 0.37\chi_4\end{aligned} \tag{9.15}$$

All coefficients in the lowest energy π orbital, ϕ_1, are of the same sign. As a consequence, ϕ_1 has no nodes between neighbouring AOs, i.e., places at which the orbital changes sign (Figure 9.9). As a rule, in quantum mechanics, the larger the number of nodes of an orbital, the higher is the energy (e.g., the 2s and 2p atomic orbitals have one node each and higher energy than a 1s atomic orbital, which has no nodes). The coefficients c_{21} and c_{22} in ϕ_2 are of the same sign and so are c_{23} and c_{24}, but there is a sign change (node) on passing from c_{22} to c_{23} (Figure 9.9). The higher-energy orbitals ϕ_3 and ϕ_4 have two and three nodes, respectively. The number of nodes in the MO ϕ_j is therefore $j - 1$ and varies from 0 to $N - 1$, i.e., 0 to 3 for butadiene. Also, orbitals with an even number of nodes are symmetrical to reflection in a plane passing through the centre of the molecule.

We can therefore make the following generalizations:

1. The number of nodes increases from 0 to $N - 1$, where N is the number of atoms in conjugation. The energy increases as the number of nodes increases because of greater localization of charge.
2. The bonding and antibonding orbitals are symmetrically placed above and below $E = \alpha$ (or nonbonding level).
3. The coefficients are symmetrical with respect to the atomic orbitals' coefficients.
4. Alternate levels are symmetrical or unsymmetrical with respect to reflection.

Moving next to hexatriene, we find the complexity increasing and it becomes difficult to solve a 6×6 determinant without the aid of computers. In the case of butadiene, we were fortunate that the problem reduced to a quadratic equation in u $(= x^2)$. For the higher systems, we may use another model, the Free Electron Model (FEM), which we shall discuss later. We first discuss the selection rules for the transitions between various energy levels.

Though the HMO method predicts the energies of molecular orbitals and the associated wave functions, one cannot make predictions regarding the allowedness of transitions between different levels. In order to make such predictions, we make use of symmetry.

9.4.1.2 Selection rules

We start with the ethylene molecule. It belongs to the D_{2h} molecular point group. The coordinate system is as follows:

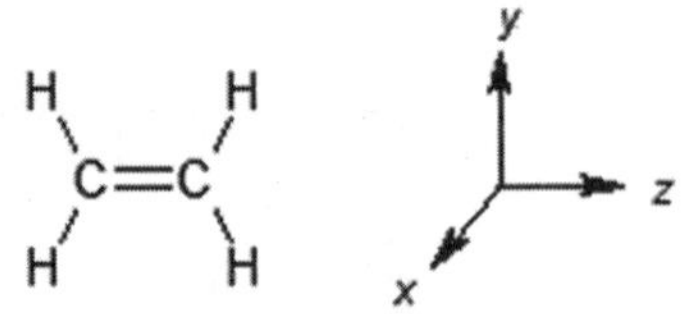

Since χ_1 and χ_2 refer to the $2p_x$ orbitals on the two carbon atoms, we examine their behaviour under the various operations of the D_{2h} point group (Table 9.2).

Table 9.2. Determination of symmetries of the ethylene wave functions

D_{2h}	E	$C_2(z)$	$C_2(y)$	$C_2(x)$	i	σ_{xy}	σ_{xz}	σ_{yz}
χ_1	χ_1	$-\chi_1$	$-\chi_2$	χ_2	$-\chi_2$	χ_2	χ_1	$-\chi_1$
χ_2	χ_2	$-\chi_2$	$-\chi_1$	χ_1	$-\chi_1$	χ_1	χ_2	$-\chi_2$
$\Gamma(R)$	2	-2	0	0	0	0	2	-2
b_{2g}	1	-1	1	-1	1	-1	1	-1
b_{3u}	1	-1	-1	1	-1	1	1	-1

Thus,

$$E\begin{pmatrix}\chi_1\\ \chi_2\end{pmatrix}=\begin{pmatrix}1 & 0\\ 0 & 1\end{pmatrix}\begin{pmatrix}\chi_1\\ \chi_2\end{pmatrix}$$

and $\Gamma(E) = 2$, which is the trace of the matrix. Similarly,

$$C_2(z)\begin{pmatrix}\chi_1\\ \chi_2\end{pmatrix}=\begin{pmatrix}-1 & 0\\ 0 & -1\end{pmatrix}\begin{pmatrix}\chi_1\\ \chi_2\end{pmatrix}$$

and $\Gamma(C_2(z)) = -2$. Under the $C_2(z)$ operation, the two carbons do not change position, but the sign of their orbital reverses since rotation is about z, and the orbital $2p_x$ lies along the x-axis. We observe that the atoms that interchange position do not contribute to the character. Similarly, we obtain the other $\Gamma(R)$'s, as shown in the table. On resolving $\Gamma(R)$, we obtain the following irreducible representations

$$\Gamma(R) = b_{3u} \oplus b_{2g}$$

Using the projection operator method (multiply each entry in the first row of Table 9.2 by the corresponding entry in the last row), we find for b_{3u}

$$\psi(b_{3u}) = 1\chi_1 + (-1)(-\chi_1) + (-1)(-\chi_2) + 1\chi_2 + (-1)(-\chi_2) + 1\chi_2 + 1\chi_1 + (-1)(-\chi_1) = (1/\sqrt{2})(\chi_1 + \chi_2)$$

after normalization, i.e., this is the bonding orbital (ϕ_1) of equation (9.14a). Similarly, it can be shown that ϕ_2 [equation (9.14b)] has b_{2g} symmetry. We may thus write the ground state electron configuration as $(b_{3u})^2$, i.e., ^1A_g. The first excited state results from the $\phi_1 \rightarrow \phi_2$ transition leading to an excited state configuration $(\phi_1)^1(\phi_2)^1$ and two states, a singlet and a triplet. The states are $b_{3u} \otimes b_{2g} = {}^1B_{1u}$ and $^3B_{1u}$. The $^1A_g \rightarrow {}^3B_{1u}$ transition is forbidden by the spin selection rule, but the $^1A_g \rightarrow {}^1B_{1u}$ transition is allowed by the electric dipole selection rules since z transforms as b_{1u}. Thus, the transition dipole moment $\int \psi_m^* \hat{\mu} \psi_l \, d\tau \equiv b_{3g} \otimes b_{1u} \otimes b_{2u} = b_{2u} \otimes b_{2u} = a_g$.

We next consider the doubly excited configuration $(\phi_2)^2$, which again corresponds to ^1A_g. The transition $^1A_g \rightarrow {}^1A_g$ is forbidden by the Laporte selection rule, since it is a $g \leftrightarrow g$ transition. Hence, the only allowed $\pi \rightarrow \pi^*$ transition for ethylene is $^1A_g \rightarrow {}^1B_{1u}$.

(i) *Trans* 1,3-Butadiene

For butadiene, symmetry distinguishes between the *cis* and *trans* isomers, since they belong to different point groups. Taking the *trans* isomer first, we find the irreducible representations for the MOs according to its point group (C_{2h}). The C_2-axis is perpendicular to the plane of the molecule and becomes the z-axis.

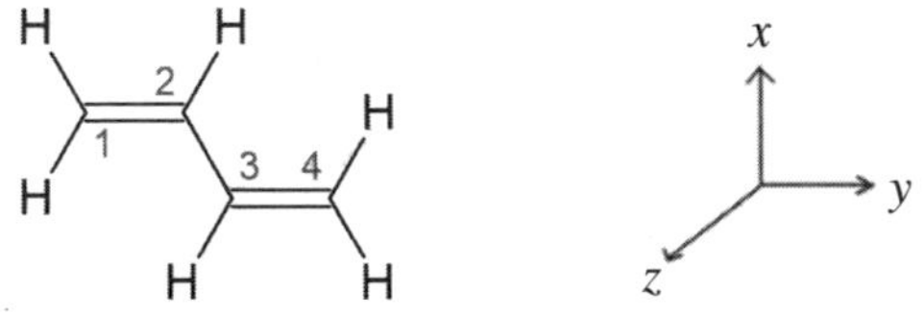

As for ethylene, we look at how the 2p$_z$ atomic orbitals transform under the various operations of the C_{2h} point group.

C_{2h}	E	$C_2(z)$	i	σ_h
χ_1	χ_1	χ_4	$-\chi_4$	$-\chi_1$
χ_2	χ_2	χ_3	$-\chi_3$	$-\chi_2$
χ_3	χ_3	χ_2	$-\chi_2$	$-\chi_3$
χ_4	χ_4	χ_1	$-\chi_1$	$-\chi_4$
$\Gamma(R)$	4	0	0	–4

Using the Great Orthogonality Theorem, we can show

$$n(a_g) = (1/4)(4 - 4) = 0$$

$$n(a_u) = (1/4)(4 + 4) = 2$$

$$n(b_g) = (1/4)\ (4 + 4) = 2$$

$$n(b_u) = (1/4)\ (4 - 4) = 0$$

Therefore, $\Gamma(R) = 2a_u \oplus 2b_g$

We again use the projection operator method to find that $\phi(a_u) = (\chi_1 - \chi_4)$ using the first row and $(\chi_2 - \chi_3)$ using the second. Similarly, $\phi(b_g) = (\chi_1 + \chi_4)$ and $(\chi_2 + \chi_3)$. From equation (9.15), we see that ϕ_1 can be written as $0.37(\chi_1 + \chi_4) + 0.60(\chi_2 + \chi_3)$. Similarly ϕ_3 can be written as $0.60(\chi_1 + \chi_4) - 0.37(\chi_2 + \chi_3)$. Hence, these two molecular orbitals transform as b_g. Likewise, ϕ_2 and ϕ_4 transform as a_u. The ground state configuration of *trans*-butadiene is therefore $(\phi_1)^2(\phi_2)^2$, i.e., ^1A_g. The first excited state gives rise to the configuration $(\phi_1)^2(\phi_2)^1(\phi_3)^1$, i.e., $(b_g)^2(a_u)^1(b_g)^1$ and the product $a_u \otimes b_g = b_u$. The resulting states are ^1B_u and ^3B_u. If we perform the various operations on the x, y and z coordinates, we find that z behaves as a_u and x and y as b_u. Thus, the $^1A_g \rightarrow {}^1B_u$ transition is allowed and xy polarized, i.e., polarized in the plane of the molecule. The $^1A_g \rightarrow {}^3B_u$ transition is forbidden by the spin selection rule.

Other singly excited transitions that are possible are $\phi_2 \rightarrow \phi_4$, $\phi_1 \rightarrow \phi_3$ and $\phi_1 \rightarrow \phi_4$. Neglecting the triplet states, it can be shown that these correspond to $^1A_g \rightarrow {}^1A_g$, $^1A_g \rightarrow {}^1A_g$ and $^1A_g \rightarrow {}^1B_u$ transitions. Only the last is allowed by the electric dipole selection rules and is xy polarized.

(ii) *Cis* 1,3-Butadiene

The *cis* isomer has C_{2V} symmetry.

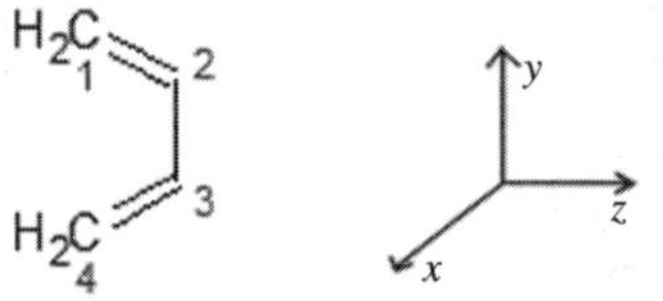

The C_2-axis is in the plane of the molecule and is labelled z. The plane of the molecule is yz. Hence, the π electrons are in the $2p_x$ orbitals.

C_{2v}	E	$C_2(z)$	$\sigma_v(xz)$	$\sigma_v(yz)$
χ_1	χ_1	$-\chi_4$	χ_4	$-\chi_1$
χ_2	χ_2	$-\chi_3$	χ_3	$-\chi_2$
χ_3	χ_3	$-\chi_2$	χ_2	$-\chi_3$
χ_4	χ_4	$-\chi_1$	χ_1	$-\chi_4$
$\Gamma(R)$	4	0	0	−4

$\Gamma(R)$ can be resolved as $2a_2 \oplus 2b_1$. Using the projection operator method, we find that $\phi(a_2) = (\chi_1 - \chi_4)$ and $(\chi_2 - \chi_3)$. Similarly, $\phi(b_1) = (\chi_1 + \chi_4)$ and $(\chi_2 + \chi_3)$. From equation (9.15), we see that ϕ_1 and ϕ_3 transform as b_1. Likewise, ϕ_2 and ϕ_4 transform as a_2. The ground state configuration of *cis*-butadiene is therefore $(b_1)^2(a_2)^2$, i.e., $^1A_1$. The first excited state gives rise to the configuration $(\phi_1)^2(\phi_2)^1(\phi_3)^1$, i.e., $(b_1)^2(a_2)^1(b_1)^1$ and the product $a_2 \otimes b_1 = b_2$. The resulting states are $^1B_2$ and $^3B_2$. If we perform the various operations on the x, y and z coordinates, we find that they are bases for b_1, b_2 and a_1, respectively. Thus, the $^1A_1 \rightarrow {}^1B_2$ transition is allowed and y polarized. The $^1A_1 \rightarrow {}^3B_2$ transition is forbidden by the spin selection rule.

Other single electron excitations that are possible are $\phi_2 \rightarrow \phi_4$, $\phi_1 \rightarrow \phi_3$ and $\phi_1 \rightarrow \phi_4$. Neglecting the triplet states, it can be shown that these transitions correspond to $^1A_1 \rightarrow {}^1A_1$, $^1A_1 \rightarrow {}^1A_1$ and $^1A_1 \rightarrow {}^1B_2$ transitions, respectively. All the transitions are allowed and polarized in z, z and y directions, respectively. We therefore observe that, although all singlet–singlet single electron excitations are permitted by the electric dipole selection rules for *cis*-butadiene, this is not the case for *trans*-butadiene, for which some of these transitions are forbidden by the Laporte selection rules, which are not applicable to the *cis* isomer which lacks a centre of symmetry.

Thus, we see that Group Theory allows us to predict which of the transitions are permitted by the electric dipole selection rules. However, it is unable to predict how allowed a transition is, i.e., the magnitude of the transition dipole moment. We introduce another theory, the free electron model (FEM) and attempt to provide an answer to this. We first describe the FEM model.

9.5 FREE ELECTRON MODEL OF LINEAR POLYENES

The Free Electron Model is suitable for calculating the wave functions of conjugated π electron systems. In this model, the π electrons are not considered to belong to specific nuclei, but are delocalized over the entire system.

Consider a conjugated polyene (molecules with alternating single and double bonds, like the butadiene example, are called polyenes). Each carbon atom is sp^2 hybridized. The three sp^2 hybrid orbitals are in the plane of the molecule and form σ bonds with either two carbon atoms and one hydrogen, or two hydrogens and one carbon (in case of the terminal carbons). This constitutes the σ bond framework. The remaining 2p orbital, perpendicular to the plane of the molecule, on each carbon atom in conjugation, contains one electron each. Thus, the number of electrons is determined by the number of carbon atoms in conjugation, because each carbon contributes one electron to the π electron system in a neutral molecule. In FEM, the molecule is first stripped off of these electrons, leaving positive charges in a zig-zag fashion on the carbon atoms. In this model, it is assumed that the positive charge is smeared out rather than localized at different nuclei, so that there is a constant uniform potential in the region of the molecule, and outside this region, the potential is infinite. This is similar to the particle-in-a-one-dimensional box problem, since the motion that interests us is in one dimension along the length of the conjugated chain.

Recall that, in the particle-in-a-one-dimensional box, the wave function goes to zero at the boundaries. This requires the boundaries to be placed beyond the end carbon atoms, else the wave function would be identically zero at these carbon atoms in each energy level, i.e., the electron density at the terminal carbon atoms would be zero, which cannot be true. We therefore extend the walls of the box by one bond in each direction.

The rationale for adding one bond length is that this gives the same wave functions as those obtained from HMO theory. All linear conjugated polyenes having N carbon atoms in conjugation have an HMO determinant in which x's occupy the diagonal, on each side of the diagonal there is 1, and all other elements of the determinant are zero. Such determinants lead to similar characteristic equations (Chebyshev polynomials), and the solutions can be expressed in a closed form. For a straight chain of N unsaturated carbons numbered sequentially,

$$\begin{aligned} x_j &= -2\cos\left(\frac{j\pi}{N+1}\right),\ j = 1, 2, ..., N \\ c_{jr} &= \sqrt{\frac{2}{N+1}}\sin\left(\frac{jr\pi}{N+1}\right) \end{aligned} \tag{9.16}$$

where r is the atom index and j is the MO index. Notice that the coefficients are similar to the particle-in-a-box eigenfunctions

$$\left(\psi_n = \sqrt{\frac{2}{L}}\sin\left(\frac{n\pi x}{L}\right)\right)$$

If the bond lengths are taken as dimensionless, an N carbon atom polyene has $N-1$ bonds. Add one bond on either side, the length of the chain is $N-1+2=N+1$, which is taken as the length of the box. Since the atoms are numbered sequentially, x refers to the position of an atom, i.e., r, and we observe a 1:1 correspondence between the coefficients in equation (9.16) and the particle-in-a-box wave functions.

9.5.1 Calculation of the wave functions

The amplitude of the FEM function at the position of each carbon atom should give the wave function in the FEM approximation. For example, for butadiene, $L = 5$. For the lowest energy orbital, $n = 1$, and x for C_1 is equal to 1, so that

$$c_{11} = \sqrt{\frac{2}{5}}\sin\frac{(1)(\pi)(1)}{5} = 0.632\sin\frac{\pi}{5} = 0.632\times\sin 36° = 0.632\times 0.588 = 0.372$$

For C_2, $x = 2$, whence

$$c_{12} = \sqrt{\frac{2}{5}}\sin\frac{(1)(\pi)(2)}{5} = 0.632\sin\frac{2\pi}{5} = 0.632\times\sin 72° = 0.632\times 0.951 = 0.602$$

Similarly, it can be shown that $c_{13} = 0.602$ and $c_{14} = 0.372$ (Figure 9.10). In a similar fashion, the wave functions for the other levels can be determined.

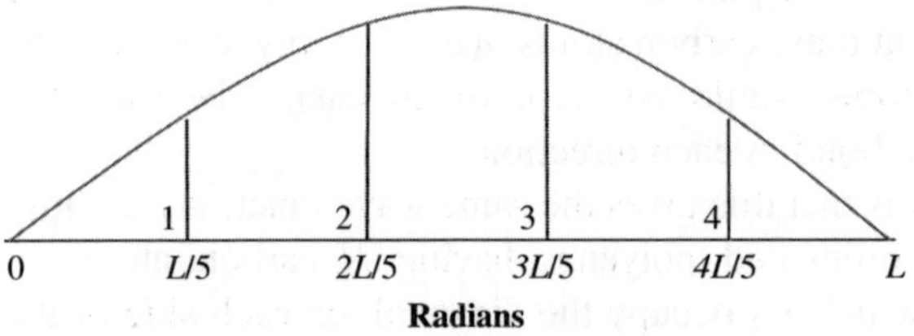

Figure 9.10. Wave function for the lowest π orbital of butadiene; it has no nodes

9.5.2 Energy of the longest wavelength transition in polyenes

The energy of the jth energy level for an electron assumed as a particle in a one-dimensional box is

$$E_j = \frac{j^2h^2}{8m_eL^2} \tag{9.17}$$

If the number of carbon atoms in conjugation is N, they contribute N delocalized electrons, which will fill the lowest energy $N/2$ FEM 'orbitals'. Therefore, the highest occupied molecular orbital number is $k = N/2$, which is also the number of double bonds in conjugation. The longest wavelength (lowest energy) transition therefore corresponds to the $k \rightarrow k+1$ transition, as shown for butadiene.

The transition energy is

$$\Delta E = E_{k+1} - E_k = \frac{(k+1)^2 h^2}{8m_e L^2} - \frac{k^2 h^2}{8m_e L^2} = \frac{(2k+1)h^2}{8m_e L^2} \tag{9.18}$$

Here, m_e represents the mass of the electron (not the molecule!) because it is the electron that undergoes the transition. According to the Bohr condition,

$$\begin{aligned} \Delta E &= h\nu = \frac{hc}{\lambda} \\ \Rightarrow \lambda &= \frac{hc}{\Delta E} = \frac{8m_e L^2 c}{h(2k+1)} \end{aligned} \tag{9.19}$$

The factor $8m_e c/h$ is a constant, and, substituting the values for m_e, c and h, we obtain $3.297 \times 10^{12}\ \text{m}^{-1}$ for this constant. Thus,

$$\lambda[\text{m}] = 3.297 \times 10^{12} L^2/(2k+1)$$

where L is also expressed in metre. Since L is usually expressed in pm and λ in nm, we can write

$$\lambda[\text{nm}] \approx 3.3 \times 10^{-3} L^2/(2k+1) \tag{9.20}$$

with the length of the box expressed in pm. In order to apply the formula to a molecule such as butadiene, we need to compute the length of the box. Since a polyene consists of alternating double and single bonds, we take the average C–C bond length (l_C) as the mean of a single (154 pm) and double (133 pm) bond length, ~140 pm. Substituting L = 700 pm (5 bond lengths) and k = 2 in equation (9.20), we get λ = 323 nm. The experimental value is 217 nm. For hexatriene, the calculated value is 452 nm, but experiment gives $\lambda_{max} \approx 250$ nm. Similarly, for octatetraene, the calculated and experimental values are 582 and 286 nm, respectively.

Clearly, one or more of our assumptions is in error. First, we have used an average bond length based on the bond lengths of ethylene and ethane, which is obviously incorrect. Second, we multiplied this by the number of bonds, assuming a linear chain. However, in the stable all-*trans* configuration, assuming a ...C–C=C... bond angle of 120° (because of sp^2 hybridization), we obtain $L = l_C\,[(N-1)\cos 30° + 2]$ because of the zig-zag nature of the carbon chain. This makes the box length shorter and hence λ smaller, in better agreement with experiment. Third, there is no justification for adding a full bond length on each side apart from the fact that this gives agreement with HMO wave functions. In fact, adding half a wavelength on either side gives better agreement between the calculated λ values and the experimental values.

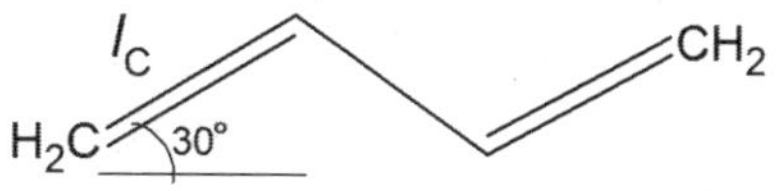

Example 9.2 Calculate the $\pi \rightarrow \pi^*$ transition wavelength for ethylene, given that the C=C bond length is 134 pm.

Solution

We assume that the π electrons are free to move approximately half bond length beyond each outermost carbon. i.e., $L = 268$ pm. The number of double bonds $k = 1$.

Using equation (9.20), $\lambda_{max} = 3.3 \times 10^{-3} \times (268)^2 / 3 = 79$ nm. The experimental value is 180 nm.

Example 9.3 Calculate the $\pi \rightarrow \pi^*$ transition wavelength for (a) 1,3-butadiene, (b) 1,3,5-hexatriene. Assume an average bond length of 140 pm.

Solution

(a) We assume that the π electrons are free to move approximately half bond length beyond each outermost carbon. i.e., $L = 4 \times 140 = 560$ pm. The number of double bonds $k = 2$. Using equation (9.20), $\lambda_{max} = 3.3 \times 10^{-3} \times (560)^2 / 5 = 207$ nm. The experimental value is 217 nm.

(b) For hexatriene, $L = 6 \times 140 = 840$ pm. The number of double bonds $k = 3$. Hence, $\lambda_{max} = 3.3 \times 10^{-3} \times (840)^2 / 7 = 333$ nm. The experimental value is 258 nm.

We see that only for 1,3-butadiene there is a fructuous agreement between the experimental and calculated values. In other cases, there is a large deviation between the computed and the experimental values. Besides the arbitrary nature of the bond length used in the calculations, there is a more fundamental argument for the failure to secure agreement between experimental and theoretical values from FEM. We had assumed a constant potential along the chain, which is certainly not the case, since the bond length fluctuates along the chain and, second, the potential is more negative near the positively charged carbon nuclei (Figure 9.11). We cannot expect better agreement between theory and experiment for these reasons.

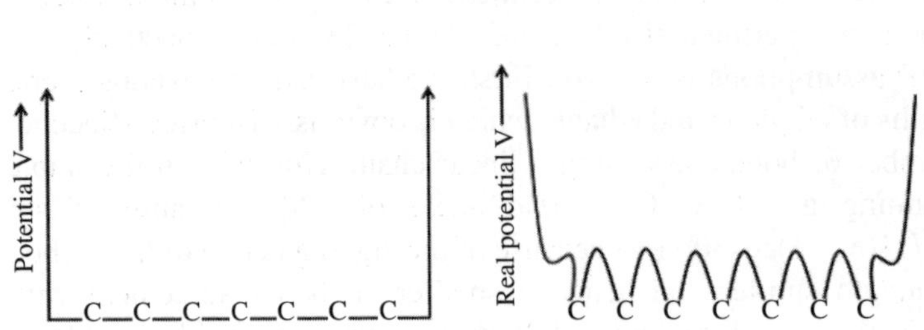

Figure 9.11. Assumed and actual potential

As the number of carbon atoms in conjugation increases, according to equation (9.20), so do L and $(2k+1)$, but because of the square dependence on L, there should be a simple proportionality between λ and the number of carbon atoms in conjugation. Such a proportionality holds good for the first few polyenes. However, as the number of double bonds increases, the wavelength of absorption approaches 550 nm as a limiting value.

This is because, as the chain length increases, the polyene twists out of plane, reducing the overlap between the participating p orbitals, and this decreases the conjugation and, consequently, $\lambda_{max.}$ Clearly, the model does not have quantitative validity, although the trend of increasing band wavelength with increasing k is correctly predicted. In longer polyenes, the λ_{max} increases to the visible range (400 – 700 nm), and the polyene becomes coloured. Thus, retinol (vitamin A), which contains a polyene chain with $k = 5$, has a pale yellow colour, and lycopene, responsible for the red colour of tomatoes, and β-carotene, responsible for the colour of carrots, have $k = 11$ (Figure 9.12).

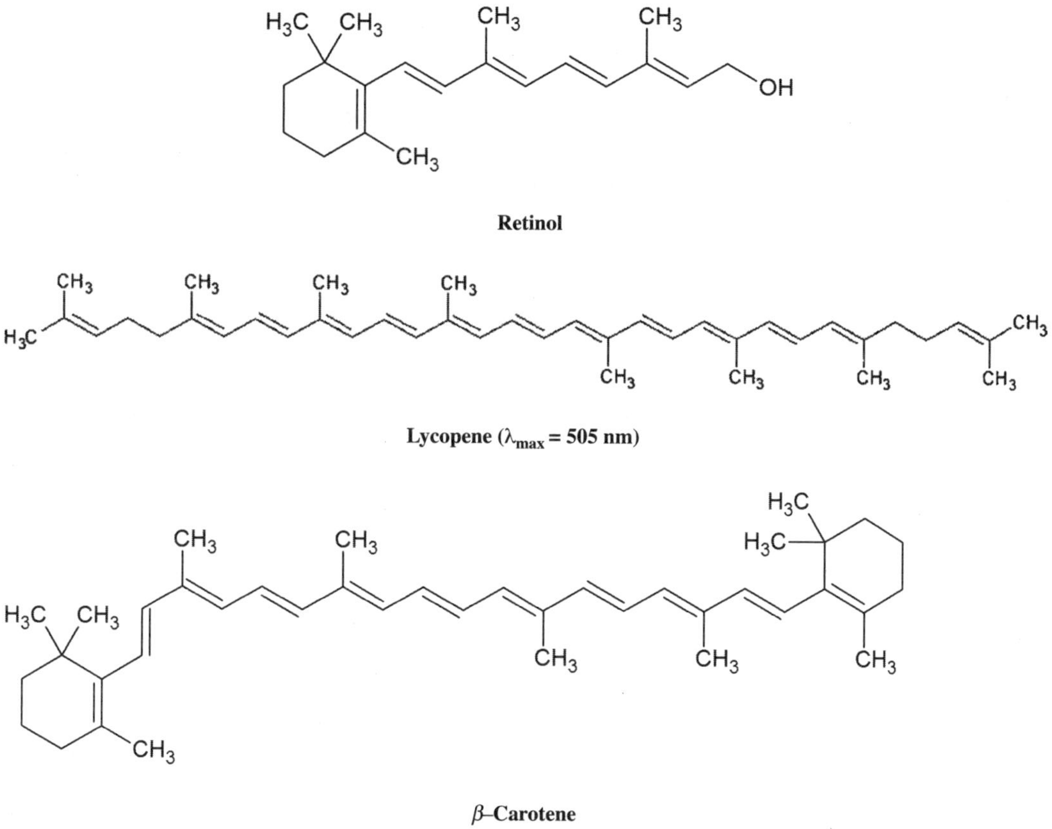

Figure 9.12. Some conjugated polyenes

Example 9.4 Calculate the $\pi \rightarrow \pi^*$ transition wavelength for lycopene.

Solution

Lycopene (Figure 9.12) has 11 double bonds in conjugation. Hence, $k = 11$. The number of bonds in conjugation is therefore 22 and $L = 22 \times 140$ pm = 3080 pm. Substituting in equation (9.20), we obtain $3.3 \times 10^{-3} \times (3080)^2 / 23 = 1361$ nm, much longer than the observed value.

9.5.3 Selection rules for particle-in-a-box wave functions

Consider an electron in a box of length L. Let the neutralizing positive charge be located at the centre of the box, i.e., at $L/2$. The dipole moment is then $\mu = e(x - L/2)$, where x is the coordinate of the electron. The transition dipole moment for a transition from state l to state m is then given by

$$\left|\vec{M}_{lm}\right| = \int_0^L \psi_m^* e\left(x - \frac{L}{2}\right)\psi_l dx = e\int_0^L \psi_m x \psi_l\, dx - \frac{eL}{2}\int_0^L \psi_m \psi_l\, dx \tag{9.21}$$

where we have dropped the asterisk because the particle-in-a-box wave functions are always real. Because of the orthogonality of the wave functions, the second term vanishes. The first term is a triple product in which x is obviously antisymmetric with respect to reflection. We also note that the wave function is odd for odd n and even otherwise. Therefore, in order to make the overall product even and the integral non-vanishing, one of the functions ψ_l and ψ_m must be odd and the other even. The particle-in-a-box is one of the cases where inspection alone allows us to decide when a transition is forbidden. Thus, odd $\leftrightarrow$ odd n and even $\leftrightarrow$ even n transitions are forbidden.

9.5.3.1 Quantitative deduction of selection rules

Insertion of the expressions for the wave functions into equation (9.21) gives

$$\left|\vec{M}_{lm}\right| = e\int_0^L \sqrt{\frac{2}{L}}\sin\frac{m\pi x}{L}\, x \sqrt{\frac{2}{L}}\sin\frac{l\pi x}{L}\, dx$$

Making use of the trigonometric relation

$$\sin A \sin B = \frac{1}{2}[\cos(A-B) - \cos(A+B)]$$

we get

$$\left|\vec{M}_{lm}\right| = \frac{e}{L}\left(\int_0^L x\cos\frac{(m-l)\pi x}{L}dx - \int_0^L x\cos\frac{(m+l)\pi x}{L}dx\right) \tag{9.22}$$

Both the integrals are of the type $\int y\cos y\,dy = y\sin y + \cos y$. Taking first the first integral and making the substitution $y = \frac{(m-l)\pi x}{L}$, $dx = \frac{L}{(m-l)\pi}dy$, we get

$$\int_0^L x\cos\frac{(m-l)\pi x}{L}dx = \frac{L^2}{(m-l)^2\pi^2}\int_0^{(m-l)\pi} y\cos y\,dy = y\sin y + \cos y\Big|_0^{(m-l)\pi}$$

$$= \frac{L^2}{(m-l)^2\pi^2}(\cos(m-l)\pi - 1)$$

where we have used $\sin(m - l)\pi = 0$, since sine of an integral multiple of π is zero. Similarly, for the other term, it can be shown that

$$\int_0^L x\cos\frac{(m+l)\pi x}{L}dx = \frac{L^2}{(m+l)^2\pi^2}(\cos(m+l)\pi - 1)$$

Substitution of both expressions in equation (9.22) gives

$$\left|\vec{M}_{lm}\right| = \frac{eL}{\pi^2}\left(\frac{\cos(m-l)\pi - 1}{(m-l)^2} - \frac{\cos(m+l)\pi - 1}{(m+l)^2}\right) \tag{9.23}$$

The cosine of an integral multiple of π, say $n\pi$, is given by $(-1)^n$. For the first integral in equation (9.23) to be non-zero, $m - l$ must be odd, else the numerator will vanish. Similarly, the second integral requires $m + l$ to be odd in order to survive. Both conditions are met if one of the integers m and l is odd and the other even. This is the same condition that we had derived from our elementary analysis using symmetry considerations alone.

We are now in a position to calculate the intensity of the π electron transitions of conjugated polyenes. Since the numerators of both terms are equal to -2 when this condition is satisfied, we may write equation (9.23) as

$$\left|\vec{M}_{lm}\right| = \frac{2eL}{\pi^2}\left(\frac{1}{(m-l)^2} - \frac{1}{(m+l)^2}\right) \tag{9.24}$$

Taking our example of hexatriene, the longest wavelength transition corresponds to the HOMO–LUMO transition for which $l = 3$ and $m = 4$. We note that, for the longest wavelength transition, $m - l$ is always 1 and $m + l$ is always large; in this case 7. The denominator of the second expression is 49, which is much larger than that of the first term, so we may neglect the second term. We therefore have $\left|\vec{M}_{lm}\right| \approx 2eL/\pi^2$ for all large polyenes. Substitution in the expression for the integrated absorption coefficient (2.36) gives

$$\bar{A} = 9.784\times10^{60}\tilde{\nu}_{lm}\frac{4e^2L^2}{\pi^4} = 1.031\times10^{22}\tilde{\nu}_{lm}L^2 \tag{9.25}$$

with the constant term expressed in $dm^3\ mol^{-1}\ cm^{-1}\ m^{-2}$. Here, $\tilde{\nu}$ is expressed in cm^{-1} and L in metre. We may alternatively write this equation as

$$\bar{A} = 1.031\times10^5 L^2/\lambda$$

with L expressed in pm and λ in nm.

For butadiene, using $\lambda = 260$ nm and $L = 73$ pm, one gets

$$\bar{A} = 2.1\times10^6\ dm^3\ mol^{-1}\ cm^{-2}$$

The observed value is

$$\bar{A} = \int_{\text{band}} \varepsilon(\tilde{\nu})d\tilde{\nu} = 3.5\times10^6\ dm^3\ mol^{-1}\ cm^{-2}$$

Considering the simplicity of the method, the agreement is quite surprising.

9.6 CHANGE OF MOLECULAR SHAPE ON ABSORPTION

We now focus on the changes in geometry brought about by electronic excitation because these determine the shape of the absorption band and the photoelectron spectra as a result of the Franck–Condon principle. It is well known that molecules change their equilibrium geometries upon excitation. In the case of polyenes, it is found that electronic excitation brings about twisting about a double bond resulting in nonplanarity and decreasing the conjugation. Similarly, formaldehyde is planar in its ground state, but becomes pyramidal in both its excited singlet and triplet $n\pi^*$ states. Such changes in geometry can be rationalized by a simple diagram called the Walsh diagram. An example of a Walsh diagram is that for a dihydride, H_2A, where A is a polyvalent atom.

9.6.1 AH_2 molecules

Consider the BeH_2 and H_2O molecules. The first is linear and the other bent though both Be and O are second row atoms. What determines whether a molecule is bent or linear? Clearly, it is the orbital energies and their relative order. As usual, we consider only the valence orbitals 2s, $2p_x$, $2p_y$ and $2p_z$ for A and 1s for the hydrogens. The 1s orbital of A is the lowest energy nonbonding orbital and transforms as the totally symmetric representation ($1a_1$) for bent, and $1\sigma_g^+$ for the linear molecule.

First consider the bent molecule with $\angle HAH = 90°$. The molecular point group is C_{2v}. The 2s and $2p_z$ orbitals transform as a_1, $2p_y$ as b_2 and $2p_x$ transforms as b_1. However, the hydrogens exchange positions under some of the operations of the C_{2v} point group, and hence suitable linear combinations have to be formed. The linear combination $H_A + H_B$ transforms as a_1 and $H_A - H_B$ as b_2. The latter is higher in energy because of the decrease in electron density between the two hydrogens. We have also seen that orbitals of similar energy and symmetry combine. Thus, the $2p_x$ orbital is nonbonding because there is no corresponding orbital of the same symmetry in the hydrogen linear combinations. The a_1 2s and $2p_z$ orbitals of A interact with the $H_A + H_B$ combination of hydrogens. The 2s orbital is lower in energy and its interaction with the hydrogens produces the $2a_1$ molecular orbital as the lowest energy bonding orbital and an antibonding orbital $4a_1$. The $2p_z$ orbital combines with the hydrogens to produce the bonding orbital $3a_1$ (see Figure 9.13). The three a_1 orbitals can mix with each other, causing s-p mixing. The $4a_1$ orbital mixes best with $3a_1$ because the energy separation between these two MOs is much smaller than that between $3a_1$ and $2a_1$. The MO contours shown in Figure 9.13 confirm these expectations.

In a similar fashion, the $2p_y$ orbital (b_2) combines with the H_A-H_B linear combination of hydrogens to form a bonding $1b_2$ and antibonding $2b_2$ orbital. The molecular orbital $1b_2$ is lower in energy than $3a_1$ because of the larger effective overlap in the former. As stated previously, the $2p_x$ orbital remains nonbonding.

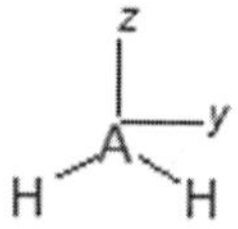

C_{2v}	E	$C_2(z)$	$\sigma_v(xz)$	$\sigma_{v'}(yz)$
H_A	H_A	H_B	H_B	H_A
H_B	H_B	H_A	H_A	H_B
$\Gamma(R)$	2	0	0	2
A_{2s}	A_{2s}	A_{2s}	A_{2s}	A_{2s}
A_{2px}	A_{2px}	$-A_{2px}$	A_{2px}	$-A_{2px}$
A_{2py}	A_{2py}	$-A_{2py}$	$-A_{2py}$	A_{2py}
A_{2pz}	A_{2pz}	A_{2pz}	A_{2pz}	A_{2pz}
A_1	1	1	1	1
A_2	1	1	–1	–1
B_1	1	–1	1	–1
B_2	1	–1	–1	1

Consider next the linear configuration (∠HAH = 180°). The 2s and $2p_z$ orbitals of A transform as σ_g and σ_u, respectively of the $D_{\infty h}$ point group. $2p_x$ and $2p_y$ together transform as π_u. Just as in the hydrogen molecule, the linear combination H_A+H_B transforms as σ_g and H_A-H_B as σ_u. Thus, the doubly degenerate π_u orbitals are nonbonding because of no matching contribution from the hydrogens. The σ_g 2s orbital of A interacts with the H_A+H_B combination of hydrogens to produce a bonding orbital $2\sigma_g$ and antibonding orbital $3\sigma_g$, respectively (see Figure 9.13). Likewise, the $2p_z$ orbital (σ_u) combines with the H_A-H_B combination of hydrogens to form a bonding $1\sigma_u$ and antibonding $2\sigma_u$ orbital. The molecular orbital $1\sigma_u$ is higher in energy than $2\sigma_g$ because the 2p orbital is higher in energy than 2s. Of the two antibonding orbitals, the $3\sigma_g$ orbital is lower in energy because it has bonding interactions between the two hydrogens.

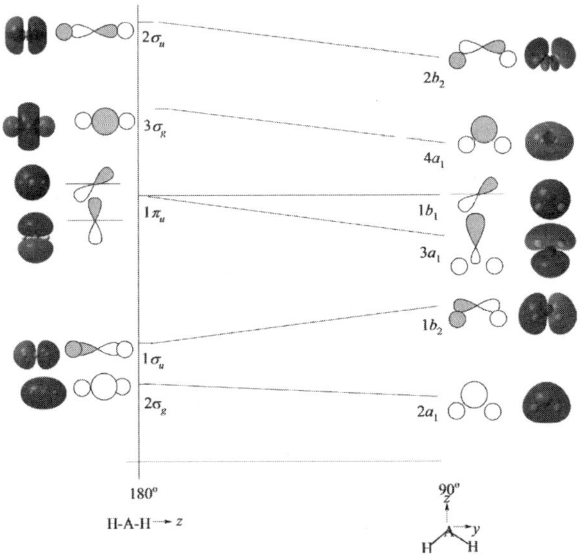

Figure 9.13. Walsh diagram for dihydrides of second row atoms

The orbital correlation diagram is shown in Figure 9.13. Note the change in the Cartesian coordinate system for the two geometries. Thus, the $2p_z$ and $2p_y$ orbitals interchange positions on bending. In what follows, we consider from the point of view of the linear molecule. Starting from the linear configuration (left), as the molecule is bent, the two hydrogens get closer, and the additional positive (bonding) overlap reduces the energy of the $2\sigma_g$ orbital, which transforms into the $2a_1$ orbital of the C_{2v} point group. The effect is opposite for the next higher orbital $1\sigma_u$, where bending brings the two hydrogens closer, but the overlap is negative. Moreover, the lobes of the $2p_z$ orbitals of the linear molecule no longer point directly at the hydrogen atoms and their overlap decreases on bending. Reduction of the symmetry of the molecule from linear to bent lifts the degeneracy of the nonbonding π_u orbital. On bending, the $2p_y$ orbital can now overlap with the hydrogen atoms, and this orbital can also mix with the 2s orbital, further lowering its energy. The π_u orbital now splits into the nonbonding $1b_1$ orbital and a low-energy $3a_1$ bonding orbital on bending. The two antibonding orbitals $3\sigma_g$ and $2\sigma_u$ stabilize on bending, the former because of the approach of the two hydrogens which form positive overlap and the latter because of the reduced antibonding overlap between the lobes of the $2p_z$ orbitals and the hydrogens on bending.

We may now use Figure 9.13 to predict whether a given AH_2 molecule will be linear or bent. The principal feature that determines this is the number of valence electrons in the molecule and the highest energy orbital they occupy. Thus, molecules with 1 or 2 valence electrons prefer a bent geometry, since the $2\sigma_g$ orbital, which they occupy, gets stabilized on bending. Examples are H_3^+ and LiH_2^+, both of which have two valence electrons (we exclude the Li $1s^2$ electrons from consideration) and are bent. If there are 3 or 4 valence electrons, they occupy $1\sigma_u$. Figure 9.13 shows that this orbital is destabilized on bending, and hence, such molecules will prefer a linear geometry. Examples are LiH_2 and BeH_2^+ (3 electrons), and BeH_2 and BH_2^+ (4 electrons). For 5 or 6 valence electrons, the preferred geometry is bent, with the electron(s) occupying the $3a_1$ orbital. For 5 electrons (BH_2), the electron configuration is $(2a_1)^2(1b_2)^2(3a_1)^1$ and the H-B-H bond angle is 131°. Increasing the occupancy of the $3a_1$ orbital to two makes the molecule more bent and the bond angle in CH_2 decreases to 102.4°.We expect similar behaviour when there are 7 or 8 electrons, since occupancy of $1b_1$ should not have any effect on the bond angle. NH_2 with 7 valence electrons and H_2O with 8 valence electrons have bond angles of 103° and 105°, respectively. Moreover, excitation of an electron from $3a_1$ to $1b_1$ in NH_2, resulting in an electronic configuration $(2a_1)^2(1b_2)^2(3a_1)^1(1b_1)^2$ increases the bond angle to 144°, as expected from the reduced occupancy of the $3a_1$ orbital. We now consider some AH_2 molecules in more detail.

BeH_2

BeH_2 is our choice of a linear molecule. Be has the ground state electron configuration $(1s)^2(2s)^2$, with two outer bonding electrons. Each hydrogen atom contributes a further electron, so that BeH_2 has four electrons to place into two molecular orbitals. The most stable state will be for two electrons to go into $2\sigma_g$ and two into $1\sigma_u$, thus producing a *linear* molecule ($^1\Sigma_g^+$). The bent geometry is not preferred, though bending the molecule lowers the energy of the $2\sigma_g$ orbital. The $1\sigma_u$ orbital, however, gets destabilized. According to qualitative molecular orbital theory, the effect is larger for higher energy orbitals, and hence, the destabilizing effect outweighs the stabilization, and the molecule prefers a linear geometry. However, when the molecule is electronically excited, the next available orbital for the excited electron is $1\pi_u$. If the molecule is bent, this electron will go to the $3a_1$ orbital, which is lower in energy. Thus, the excited state of BeH_2 is *bent*.

Methylene

Now consider singlet methylene (CH_2). The number of electrons to be filled in is six. The highest occupied level is $1\pi_u$ (linear) or $3a_1$ (bent). Clearly, the molecule prefers the bent geometry because the latter is lower in energy. Promotion of an electron from $3a_1$ to $1b_1$ results in both singlet and triplet states, where the molecule is still bent but with a larger angle. The triplet state of CH_2, $\tilde{X}^3B_1$, lies lower in energy (by 37.75 kJ mol^{-1}) than the singlet state, $\tilde{a}^1A_1$. So the former (triplet) is the ground state, and the latter a low-lying excited state.

Here, being a computational chemist, I cannot but quote from the 1981 Nobel Prize winner Roald Hoffmann on methylene:

> A sad reflection on the state of theoretical chemistry is the rarity of the situation where a computational prediction is in disagreement with an experimental result and the calculator has the courage to question the observation. I think such a rare situation presents itself in the case of the triplet ground state of methylene, CH_2.
>
> The search for the methylene spectrum is an interesting story in itself. The lowest triplet and likely ground state has been assigned $^3\Sigma_g^-$, and it has been concluded that the molecule is linear or nearly linear. On the other hand nearly every theoretical calculation, semi-empirical or ab initio, has given a bent triplet, with an HCH angle of 135° to 150°. The best of these calculations approach the true wave function so closely that, in my opinion, it is safe to say that they are correct and the experimental conclusion must be questioned.

Water

Going next to H_2O with eight valence electrons, the highest occupied level would have been $1\pi_u$ in case the molecule were linear, but the $3a_1$ level of the bent structure is greatly stabilized by bending and hence the water molecule is bent ($^1A_1$), and the HOH bond angle is 105°. If an electron is removed from the $1b_1$ orbital, it does not make any difference to its angle of bending, as shown in Figure 9.13, but if it is removed from the $3a_1$ orbital, the molecule will become less bent and its bending vibrational frequency would also decrease. Thus, we have the following electron configurations for the cations of water and the corresponding bending vibrational frequencies:

$$H_2O\ldots\ldots\ {}^1A_1\ldots\ldots 1a_1^2 1b_2^2 2a_1^2 1b_1^2 \ldots 1595\ \text{cm}^{-1}$$

$$H_2O^+\ldots\ldots\ {}^2B_1\ldots\ldots 1a_1^2 1b_2^2 2a_1^2 1b_1^1 \ldots.1431\ \text{cm}^{-1}$$

$$H_2O^+\ldots\ldots\ {}^2A_1\ldots\ldots 1a_1^2 1b_2^2 2a_1^1 1b_1^2 \ldots.873\ \text{cm}^{-1}$$

In our earlier study of photoelectron spectroscopy, we had observed that if the cation resulting from a photoelectron experiment has a geometry very different from the neutral molecule, a lot of vibrational structure is observed and the band is broad. This is demonstrated by the photoelectron spectrum of water, which shows a narrow band due to removal of an electron from the nonbonding $1b_1$ orbital (~13 eV), while the middle band (~15 eV) corresponding to the removal of an electron from the bonding $3a_1$ orbital, which mostly affects the bending, is broad (Figure 9.14), and the vibrational fine structure corresponds to the bending mode. The left-hand band (~19 eV) corresponds to the removal of an electron from the $1b_2$ bonding orbital and shows a progression in the stretching vibration which is not very well resolved.

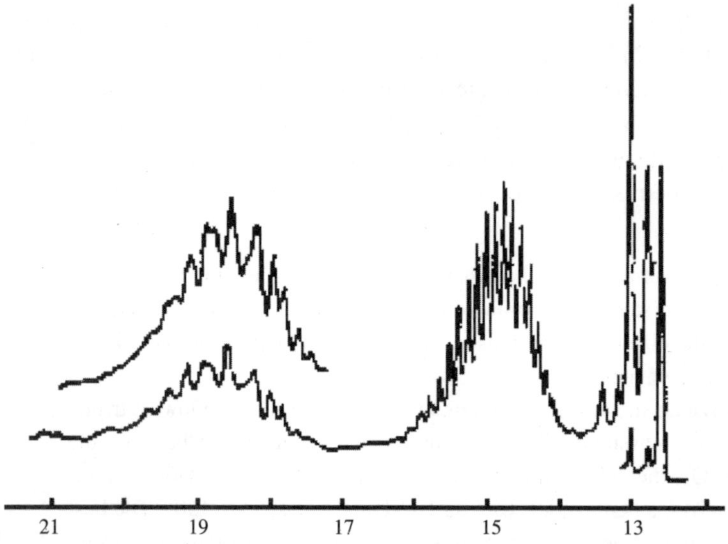

Figure 9.14. The photoelectron spectrum of water

Third row hydrides

We can extend our Walsh diagram to include third row dihydrides and consider the valence 3s and 3p electrons of the A atom only. The third row prototype of methylene is SiH_2. The H-Si-H bond angle in the ground state is 97°. As for CH_2, promotion of an electron from $3a_1$ to $1b_1$ results in singlet and triplet excited states of SiH_2 with larger bond angles 124° and 126°, respectively. Similarly, PH_2 shows behaviour similar to NH_2, its bond angle increasing from 92° to 123° on exciting the molecule from the $\ldots(3a_1)^2(1b_1)^1$ to the $\ldots(3a_1)^1(1b_1)^2$ configuration.

We note that the third row dihydrides are more bent than their second row counterparts. We have observed that the major factor that influences the shape of an AH_2 molecule having more than four valence electrons is the occupancy of the $3a_1$ orbital. This orbital has much lower energy in the angular molecule than a linear one. The energy lowering also depends on the extent of s-p mixing, which in turn depends on the proximity of the energy levels $3a_1$ and $4a_1$ in the bent molecule. These energy levels originate from the $1\pi_u$ and $3\sigma_g$ levels of the linear molecule. Hence, the extent of mixing and consequent stabilization of the bent structure depends on the energy gap between these two energy levels.

The energy gap decreases with increasing electronegativity of the atom A. Consider the Group 16 dihydrides for which the HOMO and LUMO, respectively, are the $1\pi_u$ and $3\sigma_g$ MOs for the linear geometry, since there are a total of 8 valence electrons. As we go down the group, the electronegativity of the A atom decreases and the HOMO gets destabilized and the LUMO stabilized, reducing the energy gap. This is because of two factors: The energy of the HOMO

($1\pi_u$) increases because this is a nonbonding MO localized on the A atom and its energy depends on the electronegativity of A. Lower the electronegativity of A, higher is the HOMO energy, and hence more destabilized it is. At the same time, the antibonding $3\sigma_g$ orbital becomes less antibonding as we move down the group, since the *ns* orbitals of A become more diffuse with increasing principal quantum number. Therefore, as the energy gap decreases, the mixing of the $1\pi_u$ and $3\sigma_g$ MOs increases on bending, leading to greater stabilization of the $3a_1$ orbital, providing a larger driving force for bending. At the same time, the $4a_1$ orbital is destabilized to a larger extent. For BeH_2, in which the $1\pi_u$–$3\sigma_g$ energy gap is small, the $4a_1$ level goes so high on bending that it becomes higher in energy than the $2b_2$ molecular orbital and has the appearance shown below

which clearly demonstrates p mixing.

Some other linear molecules such as HCN, CO_2 and HC≡CH also become bent on excitation. In the next section, we consider another molecule, formaldehyde, which is the simplest molecule that can show all kinds of electronic transitions.

9.6.2 The electronic structure of formaldehyde

Using the MO diagram for methylene, we can now proceed to build the molecular orbital diagram for formaldehyde by studying its interaction with an oxygen atom. The triplet ground state of methylene has two unpaired electrons, one in the $2a_1$ orbital and the other in $1b_1$. The MO diagram for formaldehyde can be constructed by assigning symmetry species to all of the AOs of oxygen used to construct the MOs:

$:CH_2$	Symmetry	Occupancy	O	Symmetry	Occupancy
C(2s)	a_1	2	2s	a_1	2
$\sigma(C_{2py}H)$	b_2	2	$2p_y$	b_2	2
$\sigma(C_{2pz}H)$	a_1	1	$2p_z$	a_1	1
n_{C2px}	b_1	1	$2p_x$	b_1	1
$\sigma^*(C_{2py}H)$	b_2	0			
$\sigma^*(C_{2pz}H)$	a_1	0			

The oxygen orbitals are lower in energy than those of methylene because of the higher electronegativity of oxygen. Thus, the O(2s) orbital cannot effectively bond with the methylene orbitals and is hence nonbonding in character. The oxygen $2p_z$ and $2p_x$ orbitals are, however, of the right energy and symmetry to bond, respectively, with the $2a_1$ and $1b_1$ orbitals of the methylene group containing one electron each. These comprise the σ(CO) and π(CO) molecular orbitals, respectively. The O($2p_y$) orbital houses the lone pair of electrons ($2b_2$).

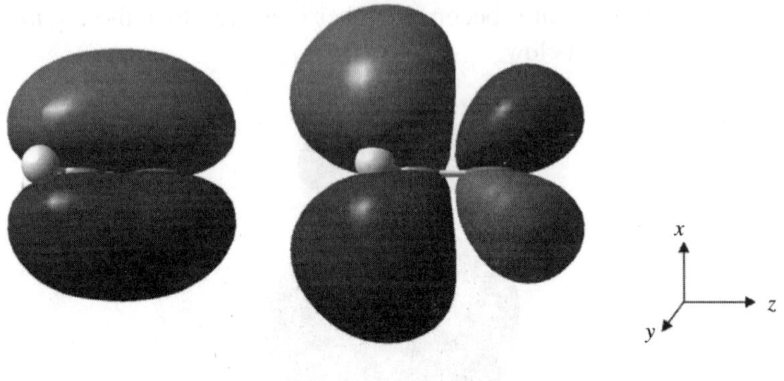

$1b_1$ (π)
Bonding

$2b_1$ (π*)
Antibonding

The molecular orbitals in order of increasing energy are O(2s) < $\sigma(C_{2s}H)$ < $\sigma(C_{2py}H)$ < σ(CO) < π(CO) < n_O < π*(CO) < σ*(CO) or $1a_1 < 2a_1 < 1b_2 < 3a_1 < 1b_1 < 2b_2 < 2b_1 < 4a_1$. The O(2$s$) AO is mixed with the C–O and C_{2pz}-H σ bonding orbitals, which all have a_1 symmetry, so these three MOs are a mixture of σ-bonding and lone pair character. The O($2p_y$) lone pair of electrons ($2b_2$) falls between the $1b_1$ and $2b_1$ (π and π*) MOs, and the electrons in these orbitals are the main concern of electronic spectroscopy.

nπ* excited state

The 12 valence electrons can occupy the first 6 MOs, resulting in a ground state electron configuration $(1a_1)^2(2a_1)^2(1b_2)^2(3a_1)^2(1b_1)^2(2b_2)^2$. The $2b_2$ n MO is the HOMO, and the $2b_1$ π* MO is the LUMO. The ground state is $\tilde{X}^1A_1$. Promotion of an electron from the nonbonding (n) $2b_2$ MO to the antibonding (π*) $2b_1$ MO gives an excited state configuration $\ldots(1b_1)^2(2b_2)^1(2b_1)^1$, which has the states $\tilde{a}^3A_2$ and $\tilde{A}^1A_2$ since $b_2 \otimes b_1 = a_2$ and two unpaired electrons can give rise to singlet and triplet states. These $\tilde{a}$ and $\tilde{A}$ states of formaldehyde and other similar molecules are known more commonly as nπ* states.

$\pi\pi$* excited state

The next excited state is formed by the promotion of an electron from the bonding π state to the antibonding π* state, resulting in the electronic configuration $\ldots(1b_1)^1(2b_2)^2(2b_1)^1$. The resulting states are $\tilde{b}^3A_1$ and $\tilde{B}^1A_1$ and are known as $\pi\pi$* excited states.

Two other transitions that are possible are n→σ* and σ→π*, resulting in excited state electron configurations $\ldots(1b_1)^2(2b_2)^1(2b_1)^0(4a_1)^1$ and $\ldots(3a_1)^1(1b_1)^2(2b_2)^2(2b_1)^1$ and states $^{1,3}B_2$

and $^{1,3}B_1$, respectively. The $\sigma\sigma^*$ excited state results from a $3a_1 \rightarrow 4a_1$ transition, resulting in $^{1,3}A_1$ excited states. However, this transition involves a large amount of energy (Figure 9.15) and is usually not observed.

Hence, four transitions are important for this chromophore (n→π^*, $\pi\rightarrow\pi^*$, n→σ^* and $\sigma\rightarrow\pi^*$) (Figure 9.15). The HOMO-LUMO transition involves transfer of the in-plane nonbonding O(2p$_y$) electron to the out-of-plane antibonding C-O π^* orbital. This n→π^* transition is characteristic of –C=O, -C=S, -N=O, -NO_2 and O-N=O chromophores. The transition $\tilde{X}^1A_1 \rightarrow \tilde{A}^1A_2$, known more commonly as n–π^* or n-to-π^* transition, is electric dipole forbidden because none of the Cartesian components corresponds to A_2. However, it is one of the most important electronic transitions (353 – 230 nm). The other three transitions are allowed by the electric dipole selection rules and have polarizations z ($\pi\rightarrow\pi^*$), y (n→σ^*) and x ($\sigma\rightarrow\pi^*$).

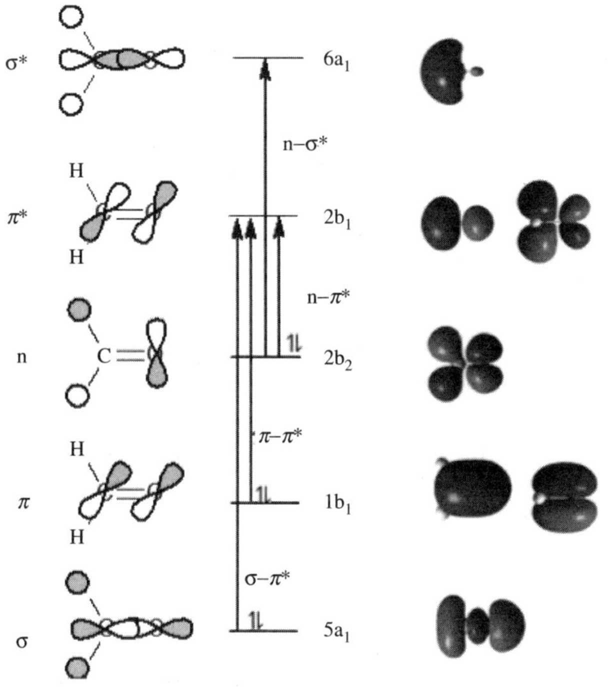

Figure 9.15. Possible transitions for formaldehyde. Two views of the π orbitals are shown for clarity. Note that they are singly degenerate

The n→π^* transition is observed at ~290 nm for formaldehyde. As stated previously, the n→π^* transitions are easily distinguished from the more common $\pi\rightarrow\pi^*$ transitions, since these are *blue* shifted in hydrogen bonding solvents, such as ethanol or water. This is due to interaction of the hydrogen 1s orbital of –OH with the *n* orbital, thereby stabilizing it and increasing the energy of the n→π^* transition. The two transitions can also be distinguished by the low intensities of the forbidden n→π^* transitions (Figure 9.16) and the fact that these usually occur at longer wavelengths.

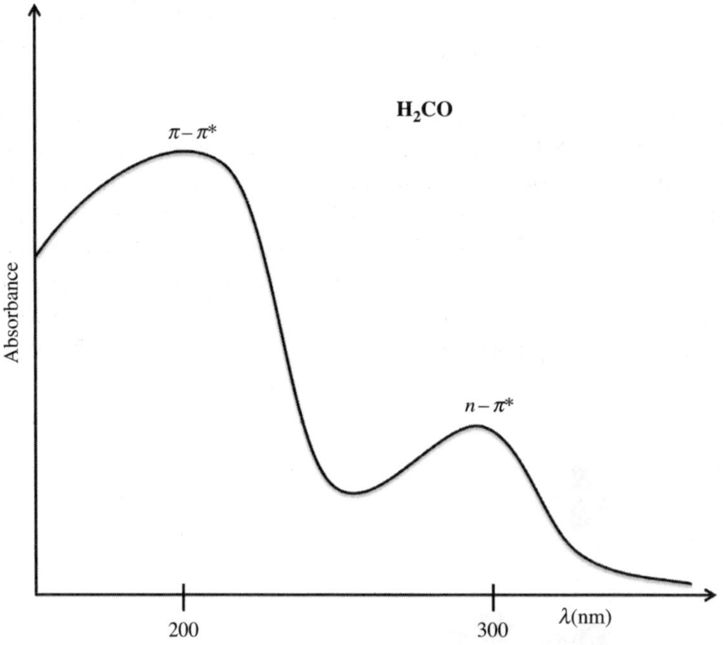

Figure 9.16. $\pi \rightarrow \pi^*$ and $n \rightarrow \pi^*$ transitions of formaldehyde

The question that arises now is how can some electronic-forbidden transitions such as $n \rightarrow \pi^*$ be observed as weak bands in the spectrum? This can be explained by the interaction between the electronic and vibrational wave functions. This is another instance of the breakdown of the Born–Oppenheimer rule, according to which we had considered the two as independent, but, apparently, there is some interaction between the two.

9.7 VIBRONIC COUPLING

The word 'vibronic' is a combination of the words 'vibrational' and 'electronic'. Because the energy required for electronic transitions, usually in the UV-Vis region, is larger than that for vibrational transitions, usually in the infrared region, sometimes a photon can excite a molecule to an excited electronic and vibrational state, called a vibronic state. If a vibration brings the molecule out of plane, the *n* electrons can interact with the π orbitals and the $n \rightarrow \pi^*$ transition can then occur. In the case of formaldehyde, vibronic coupling occurs with the ν_4 out-of-plane bend (b_1). For the $\tilde{X}^1A_1(\upsilon''=0) \rightarrow \tilde{A}^1A_2(\upsilon'=1)$ transition, the ground vibronic state remains $^1A_1$ because the $\upsilon''=0$ vibrational state is always totally symmetric. However, the excited vibronic state has $^1B_2$ symmetry since $a_2 \otimes b_1 = b_2$. The transition now becomes electric dipole allowed because the *y*-component of the electric field of the radiation transforms as b_2. It borrows some intensity from the $\tilde{X}^1A_1 \rightarrow \tilde{B}^1B_2$ $(n \rightarrow \sigma^*)$ transition near 175 nm. The $n \rightarrow \pi^*$ transition occurs at ~290 nm in formaldehyde. The other transitions shown occur at shorter wavelengths (higher energies), e.g., the $\pi \rightarrow \pi^*$ at 187 nm (Figure 9.16).

9.8 RE-EMISSION OF ENERGY BY AN EXCITED MOLECULE

A coloured solution continuously absorbs visible radiation, which means that the excited state molecules are somehow returning to the ground state and getting re-excited. We have also seen earlier that the excited state has a finite lifetime. In this section, we review the various processes by which the excited state molecules return to the ground state. These processes are usually depicted in a diagram, known as a Jablonski diagram (Figure 9.17).

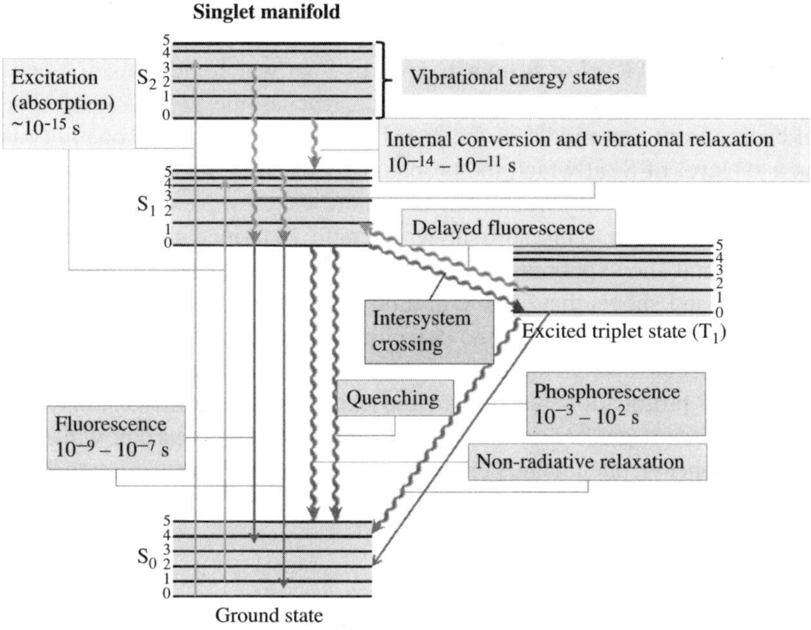

Figure 9.17. Jablonski diagram, showing electronic and vibrational energy levels

In the Jablonski diagram, a simplified notation is used to label states. The majority of molecules have singlet ground states, which are labelled S_0. When such a molecule is excited, depending on the energy absorbed from the radiation, the molecule may go to one of the singlet excited states, since the selection rules do not permit a change of multiplicity. Collectively, the singlet states are known as the *singlet manifold* and the excited singlet states are labelled S_1, S_2, ... in order of increasing energy. However, apart from the first excited state S_1, the higher energy singlet states have extremely small lifetimes, and immediately fall to the S_1 state. Hence, we need concern ourselves only with the S_1 excited state.

Excitation takes the molecule to a high vibrational level of the excited state S_1 as a result of the Franck–Condon principle. It quickly relaxes to the $\upsilon' = 0$ level of S_1 by a process known as *vibrational relaxation* (VR), also called *vibrational cascade.* This process is very fast, $<10^{-12}$ s, by which time all the excited molecules are in the $\upsilon' = 0$ level of S_1. Once in this state, the molecule is relatively stable because the lifetime of the S_1 state ($\sim 10^{-8}$ s) is comparatively larger than $\sim 10^{-13}$ s for the higher S_2, S_3, etc. excited singlet states. From this state, the molecule may return to the ground state by a process such as *external conversion*, which involves collision with another molecule or with solvent molecules or the wall of the container, and dissipation of energy as heat. Another deactivating process is *internal conversion*, by which the molecule in an excited singlet state may cross over to a higher vibrational level of S_0 if there is a crossing of potential energy curves of the two states. All these processes are radiationless and are represented as wavy arrows in Figure 9.17.

However, if the excited molecule survives these processes, then it may emit energy in the form of radiation, and this process is known as *fluorescence*, which can be considered as the reverse of absorption. However, the two transition wavelengths do not coincide because of the restrictions placed by the Franck–Condon principle.

Excitation occurs from the $\upsilon'' = 0$ level of S_0 to a υ' level of S_1 at which the transition ends vertically. From here, by vibrational relaxation the molecule reaches the $\upsilon' = 0$ level. The fluorescence transition begins from the middle of the $\upsilon' = 0$ level (at which the probability is maximum) and ends at a υ'' level of S_0 at which the probability of finding the system is large. As Figure 9.18 shows, except for the (0,0) transition, the fluorescence wavelength is always longer than the absorption wavelength. This shift in wavelength is termed the *Stokes shift.* Furthermore, the absorption band shows vibrational fine structure characteristic of the excited state, while the fluorescence band shows that for the ground state. However, there is a mirror image relationship between the two spectra, with the (0,0) lines coinciding. This mirror image relationship is observed for molecules which are fairly rigid, such as polyaromatic systems, and which have no dissociation or protonation reaction in the excited state. Some examples of molecules that show good mirror images are anthracene, rhodamine and fluorescein, while the spectra of molecules such as biphenyl, phenol and heptane show poor mirror image relationships.

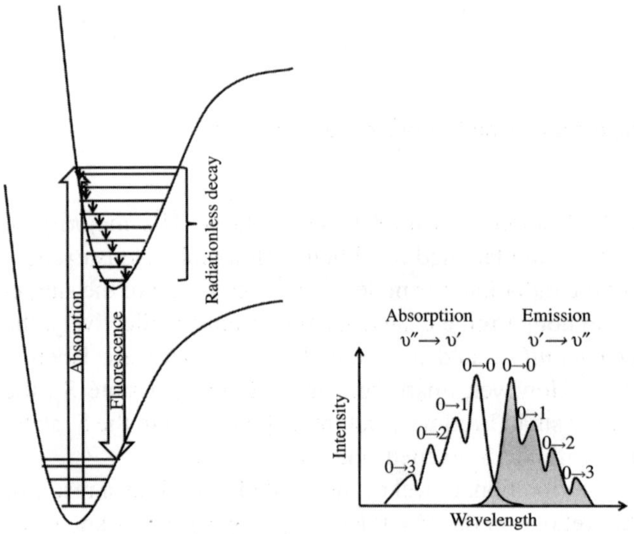

Figure 9.18. Fluorescence

Sometimes the (0,0) band is not observed if the geometries of the two states are very different, or due to solvent relaxation, which leads to a shift of this band.

Fluorescence is only observed if the excited state survives other dissipative processes like internal conversion. Fluorescein is an example of a molecule that exhibits fluorescence, but a related molecule phenolphthalein is non-fluorescent. By comparing the two structures, we find that phenolphthalein is flexible and can undergo internal rotation and lose its excited state energy, whereas fluorescein is rigid with planar rings. Hence, it cannot undergo internal conversion and the only option left for it is to undergo fluorescence to lose its excitation energy.

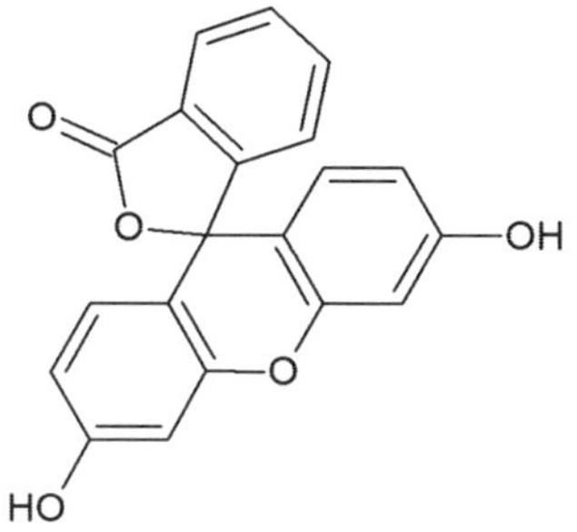

Fluorescein

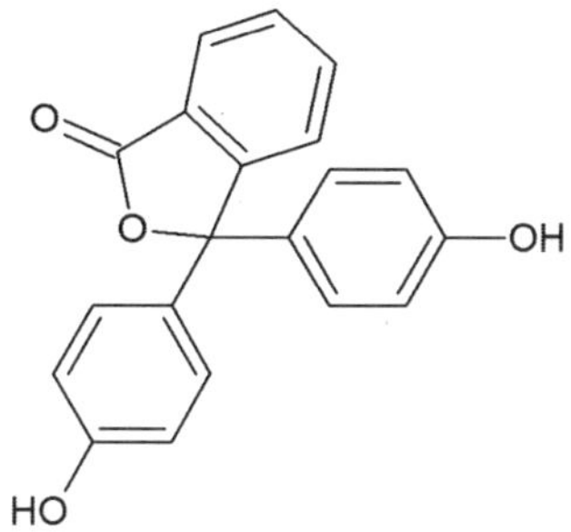

Phenolphthalein

When the excited singlet state is formed as a result of a forbidden transition, e.g., n→π*, the reverse transition (fluorescence) is also forbidden since the transition dipole moment for a transition and its reverse is the same. In such a case, the excited state has a longer lifetime since fluorescence is forbidden. The molecule stays long enough in the excited state to undergo another radiationless process, called *intersystem crossing (ISC)* to a triplet state. Since it is radiationless, ISC is not governed by the selection rules which forbid singlet-triplet transitions. This can happen if there is a crossing of the singlet S_1 and triplet T_1 potential energy curves. Since T_1 is lower in energy, the molecule ends up in a higher vibrational level of T_1 from where it goes down to the zero vibrational level of T_1 by vibrational relaxation (Figure 9.19).

From the lowest vibrational level of T_1, the molecule can only return to S_0. This is a spin-forbidden process known as *phosphorescence* (Figure 9.19). Phosphorescence refers to the emission of radiation from some excited state to one of different multiplicity, in contrast to fluorescence in which the transition is between states of the same multiplicity. Since phosphorescence is spin forbidden, it is very slow, and triplet states are characterized by longer lifetimes of the order $>10^{-3}$ s. Phosphorescence persists even after the source radiation is switched off and is hence different from fluorescence which stops immediately (Figure 9.20). As stated earlier, generally nπ* singlet excited states are long lived and show phosphorescence. On the other hand, $\pi\pi$* singlet excited states exhibit fluorescence, which is an allowed process. An example is anthracene, a highly fluorescent compound.

Figure 9.19 shows that, since the triplet state is lower in energy than the excited singlet, the phosphorescence frequency is lower than the fluorescence frequency, which in turn is lower than the absorption frequency ($\nu_p < \nu_f < \nu_a$).

Phosphorescence is enhanced by the presence of paramagnetic species like O_2 or by the presence of a paramagnetic centre in the molecule, which promotes singlet–triplet ISC. Triplet sensitizers containing heavy elements such as UO_2^+ also help in increasing phosphorescence. As learnt in the chapter on atomic spectroscopy, S is not a good quantum number for the heavier elements since Russell–Saunders coupling does not hold in such cases. Instead, the *jj*

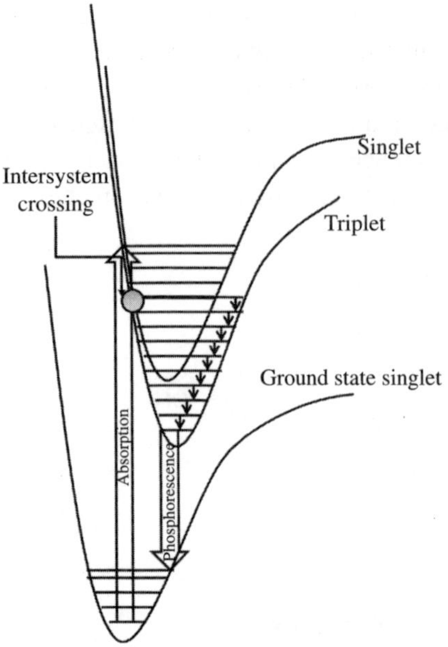

Figure 9.19. Intersystem crossing from S_1 to T_1

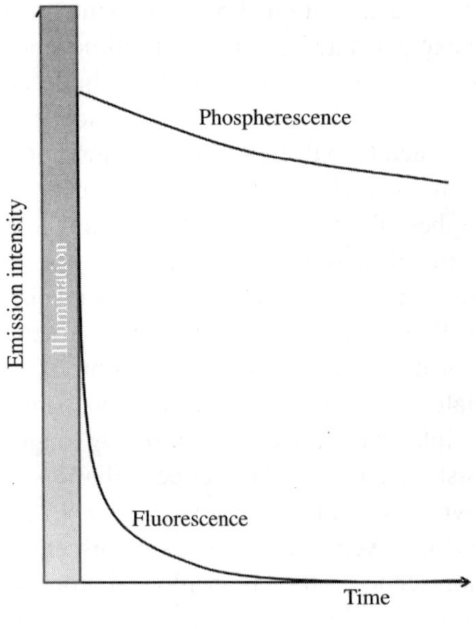

Figure 9.20. Fluorescence and phosphorescence emission intensity as a function of time

coupling scheme operates for these elements and the $\Delta S = 0$ selection rule has no meaning. An increase in phosphorescence obviously decreases fluorescence and we say that fluorescence is *quenched*.

Photochemistry is the study of processes initiated by light. Besides the processes discussed above, many other processes take place photochemically–such as dissociation and predissociation, ring opening, isomerization, excimer (excited dimer) formation as in pyrene, which decrease the quantum efficiency of fluorescence and phosphorescence. One important example of a photochemical reaction is the mechanism of vision. The retina of the eye contains rhodopsin, also called 'visual purple', a combination of the protein opsin with 11-*cis*-retinal, which acts as a chromophore and is a receptor of photons entering the eye. Opsin itself does not absorb visible light and 11-*cis*-retinal absorbs at 380 nm, but their combination, rhodopsin, has a very broad absorption spectrum in the visible which tails into the blue. The absorption maximum shifts to ~500 nm, which matches the output of the sun closely. The conjugated double bonds are responsible for the ability of the molecule to absorb over the entire visible region. They play another important role: in its electronically excited state, the conjugated chain can isomerize by twisting about an excited C=C bond, forming 11-*trans*-retinal. The primary step in vision is photon absorption followed by isomerization. The shape of the molecule changes as a result of this isomerization. The molecule changes from a bent structure to a more or less linear structure. The molecule no longer fits into the protein, which changes shape, and triggers a nerve impulse to the brain. The photochemistry of vision was only understood in the last century, when the identity of the retinals was established in 1933 by Wald, for which he shared the Nobel Prize in 1967.

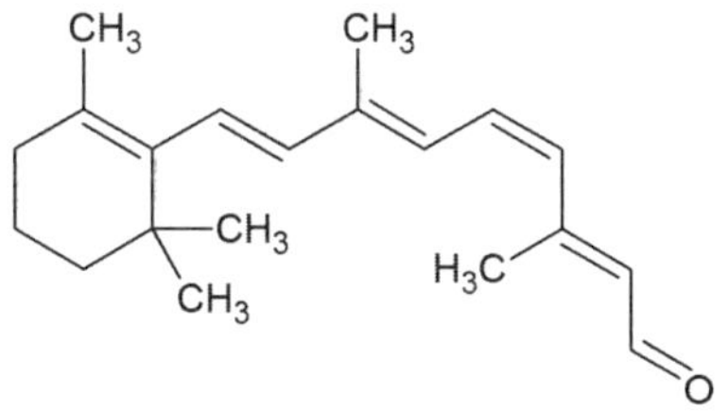

11-*cis*-retinal

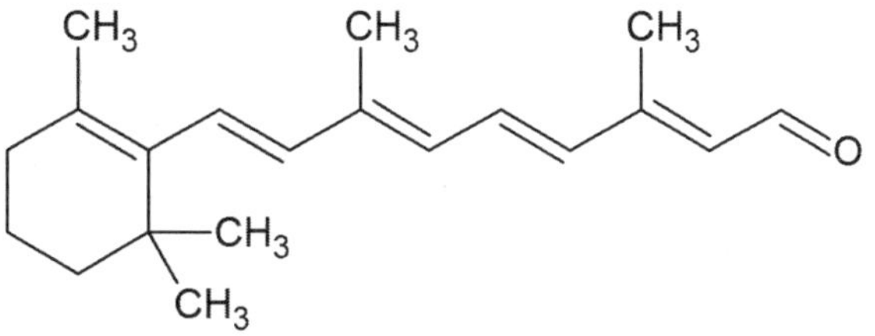

11-*trans*-retinal

Photosynthesis is another example of a photochemical process, which involves the absorption of light by the chlorophyll in plants to produce carbohydrates from carbon dioxide and water. Photography uses the action of light on grains of silver chloride or silver bromide to produce an image. Ozone formation in the upper atmosphere results from action of light on oxygen molecules. Solar cells, which are used to power satellites and space vehicles, convert light energy from the sun to chemical energy and then release that energy in the form of electrical energy.

9.9 KINETICS OF PHOTOCHEMISTRY

The first step in photochemistry is absorption of radiation by a molecule in the ground S_0 state. The rate of this process is given by the intensity of the absorbed radiation, $I_a \equiv k_a[S_0]$. We thus have the following processes

$$S_0 + h\nu_a \xrightarrow{I_a} S_1$$

$$S_1 \xrightarrow{I_f} S_0 + h\nu_f$$

The rate of disappearance of the excited state S_1 molecules is given by

$$\frac{d[S_1]}{dt} = I_a - k_f[S_1] \tag{9.26}$$

where k_f is the fluorescence rate constant and is a first-order rate constant ($I_f = k_f[S_1]$).

We had earlier observed in Chapter 1 that $N_m = N_m^0 e^{-A_{ml}t}$, where $N_m^{\,0}$ refers to the population of the upper level when the radiation is just switched off (i.e., I_a is zero). We thus infer that the fluorescence rate constant is equivalent to the Einstein coefficient of spontaneous emission A_{ml}, which is equal to [equation (2.40)] $A_{ml} = \frac{\ln 10}{N_A} 8\pi\tilde{\nu}^2 c \frac{g_l}{g_m} \int_{\text{band}} \varepsilon(\tilde{\nu})d\tilde{\nu}$. Inserting the values of all constants, we obtain

$$A_{ml} = 2.88 \times 10^{-9} \tilde{\nu}^2 \frac{g_l}{g_m} \int_{\text{band}} \varepsilon(\tilde{\nu})d\tilde{\nu} \tag{9.27}$$

Thus, A_{ml}, which governs the decay of the excited state, can be determined from the shape of the absorption curve. In the absence of all other processes, A_{ml} is equivalent to the fluorescence rate constant k_f. The lifetime of the upper state is defined as the time taken for N_m to fall to $1/e$ of its initial value $N_m^{\,0}$, i.e., $\tau_0 = 1/A_{ml} = 1/k_f$ *if fluorescence is the only process taking place*. Normally $k_f \approx 10^8$ s^{-1}. If there are no competing processes, the lifetime is equal to the true radiative lifetime τ_0.

The excited state may also get deactivated by intersystem crossing (rate constant k_{ISC}) or internal conversion (rate constant k_{IC}). In the presence of these competing processes, the set of equations gets modified to

$$S_0 + h\nu_a \xrightarrow{I_a} S_1$$

$$S_1 \xrightarrow{I_f} S_0 + h\nu_f$$

$$S_1 \xrightarrow{k_{\text{ISC}}} T_1 \xrightarrow{I_p} S_0 + h\nu_p$$

$$S_1 \xrightarrow{k_{\text{IC}}} S_0$$

Using the steady state approximation for the excited state S_1, we write

$$\frac{d[S_1]}{dt} = I_a - k_f[S_1] - k_{\text{ISC}}[S_1] - k_{\text{IC}}[S_1] = 0$$

$$\Rightarrow I_a = (k_f + k_{\text{ISC}} + k_{\text{IC}})[S_1] \tag{9.28}$$

Efficiencies of photochemical processes are described by quantum yields (branching ratios into that decay channel). The quantum yield is defined as

$$\phi = \frac{\text{Rate of a specified process}}{\text{Rate of photon absorption}}$$

The quantum yield for fluorescence is defined as the fraction of molecules undergoing fluorescence and is given by

$$\phi_f = \frac{\text{Rate of fluorescence}}{\text{Rate of all processes}} = \frac{k_f[S_1]}{(k_f + k_{\text{ISC}} + k_{\text{IC}})[S_1]} = \frac{k_f}{k_f + k_{\text{ISC}} + k_{\text{IC}}} \tag{9.29}$$

In the absence of any other process, or if fluorescence is the only process taking place, the lifetime of the excited state is given by $\tau_0 = 1/k_f = 1/A_{ml}$. This is also called the *intrinsic* or *natural* lifetime of the excited state.

In the presence of other processes, the observed lifetime is reduced to

$$\tau = \frac{1}{k_f + k_{\text{ISC}} + k_{\text{IC}} + \ldots} \tag{9.30}$$

since fluorescence is not the only process occurring. The denominator is a sum of first-order rate constants, and hence, the quenching process remains first order.

Thus, we may write equation (9.29) as

$$\phi_f = \frac{k_f}{\sum_i k_i} = \frac{\tau}{\tau_0} \tag{9.31}$$

Hence, since other energy wastage processes reduce the lifetime of the excited species, $\phi_f < 1$ and $\tau < \tau_0$. The lifetime τ is related to the intrinsic or natural lifetime by $\tau = \tau_0\phi_f$. The values of τ obtained from this equation agree with those determined by direct observation or calculation.

Besides these processes, quenching of fluorescence also occurs due to external molecules. For example, dissolved oxygen promotes intersystem crossing and increases phosphorescence. Solvents such as DMSO contain large atoms (sulphur) and can effectively quench fluorescence. Similarly, other large anions like the halides, pseudohalides, etc. can also quench fluorescence. In the presence of a quencher, an additional second-order process has to be taken into account:

$$S_1 + Q \xrightarrow{k_q} S_0 + Q$$

This modifies equation (9.29) to

$$\phi_f = \frac{k_f[S_1]}{(k_f + k_{\text{ISC}} + k_{\text{IC}} + k_q[Q])[S_1]} = \frac{k_f}{k_f + k_{\text{ISC}} + k_{\text{IC}} + k_q[Q]} \tag{9.32}$$

If we designate the quantum yield in the absence of quencher as ϕ_f^0, we may write

$$\begin{aligned}\frac{\phi_f^0}{\phi_f} &= \frac{k_f}{k_f + k_{\text{ISC}} + k_{\text{IC}}} \, \frac{k_f + k_{\text{ISC}} + k_{\text{IC}} + k_q[Q]}{k_f} = 1 + \frac{k_q[Q]}{k_f + k_{\text{ISC}} + k_{\text{IC}}} \\ &= 1 + K_{\text{SV}}[Q]\end{aligned} \tag{9.33}$$

This is called the *Stern–Volmer equation* and K_{SV} the Stern–Volmer constant, which is equal to $\dfrac{k_q}{k_f + k_{ISC} + k_{IC}} = k_q\tau$.

If ϕ_f^0 / ϕ_f, the relative quantum yield, is plotted against the concentration of the quencher $[Q]$, a straight line is obtained with intercept equal to '1' and slope equal to K_{SV}. The relative quantum yield can be replaced by the relative fluorescence intensity (I_f^0 / I_f), which is easily measured. If τ, the fluorescence lifetime in the absence of quencher is known, k_q can be determined from the slope of the Stern–Volmer plot. The lifetime can be easily measured by observing the first-order rate of decay of the excited state (Figure 9.20). Since the fluorescence intensity is proportional to the excited state concentration, a plot of $\ln(I_f)$ versus time gives a straight line with slope $-k$, where k is the combined rate constant of all the processes and is the reciprocal of the lifetime, τ. Larger the value of k_q, more efficient is the quencher. Thus, among the halide ions, the quenching efficiency increases as $Cl^- < Br^- < I^-$. Figure 9.21 shows the Stern–Volmer plot for the measured data on the quenching of fluorescence of riboflavin (in 0.02 M acetic acid) by halide ions. For this, the value of K_{SV} obtained is 6.3 dm^3 mol^{-1} for quenching by chloride ions. The literature value of τ is ~5 ns. This gives a value of (6.3 dm^3 mol^{-1}) / (5×10^{-9} s) $\approx 1.3 \times 10^9$ dm^3 mol^{-1} s^{-1} for k_q. Notice that since τ is very small (of the order of ns), very large values of k_q can be determined by this procedure. It should also be mentioned that the presented data are students' data obtained using a crude photofluorimeter.

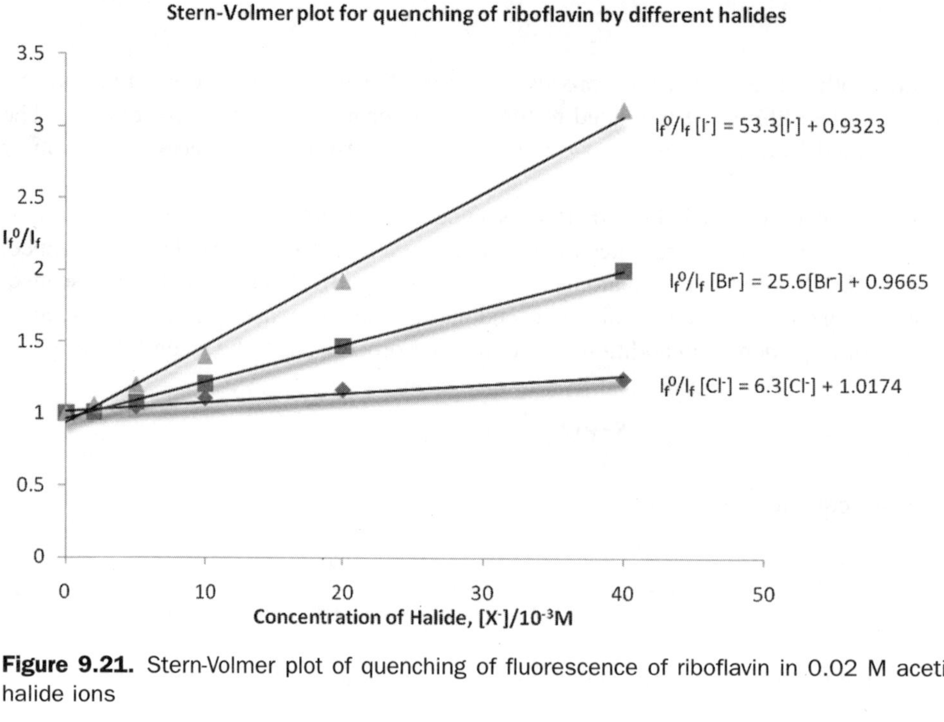

Figure 9.21. Stern-Volmer plot of quenching of fluorescence of riboflavin in 0.02 M acetic acid by halide ions

There are two main types of quenching mechanisms – *static* (due to complex formation before excitation) and *dynamic* or *collisional*. We shall focus on the latter. A diffusion controlled reaction is a bimolecular reaction in which the time taken by the reactant molecules to diffuse towards and collide with each other controls the rate of reaction. If we assume that every collision is effective, the rate constant k_D is given by the Stokes–Einstein relation

$$k_D = \frac{8000RT}{3\eta} \text{ dm}^3 \text{ mol}^{-1} \text{ s}^{-1} \qquad (9.34)$$

where R is the gas constant (8.314 J K^{-1} mol^{-1}), T is the temperature in Kelvin, and η is the viscosity in kg m^{-1} s^{-1} (usually called poise). The viscosity of water at 20 °C is 1 cP = 10^{-3} kg m^{-1} s^{-1}. Inserting this value in equation (9.34), we obtain a value of 6.5×10^9 dm^3 mol^{-1} s^{-1} for k_D. Though the order of the quenching rate constant is of the correct magnitude, it is much smaller than the diffusion rate constant, suggesting that only a fraction of the collisions is effective in quenching. For collisional quenching, we also have the relation $I_f^0/I_f = \tau_0/\tau$. Deviations from the Stern–Volmer equation and values larger than k_D indicate some other mechanism of quenching also operating, such as complex formation (static quenching).

9.10 SUMMARY

- Polyatomic molecules can have σ, π and n orbitals. Saturated hydrocarbons can show $\sigma\rightarrow\sigma^*$ transitions only, but these occur in the vacuum UV region. Molecules with heteroatoms can also have nonbonding orbitals and $n\rightarrow\pi^*$ transitions.
- $n\rightarrow\pi^*$ transitions are forbidden but are observed with low intensity due to vibronic coupling.
- $n\rightarrow\pi^*$ transitions are shifted to shorter wavelengths in hydrogen bonding solvents.
- The opposite effect is observed for $\pi\rightarrow\pi^*$ transitions.
- Increasing conjugation shifts $\pi\rightarrow\pi^*$ transitions to longer wavelengths.
- Excited state molecules return to the ground state by a variety of processes, of which fluorescence and phosphorescence involve emission of photons.

9.11 EXERCISES

1. The *oscillator strength*, f, is defined as the ratio of the observed intensity of a transition to that of a fully allowed transition,

$$f = \frac{|\vec{M}_{lm}|^2}{|\vec{M}_{01}|^2} = \frac{8\pi^2 \nu_{lm} m_e}{3he^2} |\vec{M}_{lm}|^2$$

   where ν_{lm} is the frequency of the transition (in Hz).
   Consider an electron oscillating harmonically along the x-axis. Show that the oscillator strength for the transition from the ground state is exactly 1/3.
2. (a) Is the transition $A_{1g} \rightarrow E_{2u}$ forbidden for electric dipole transitions in benzene?
   (b) The ClO_2 molecule (C_{2v}) was trapped in a solid matrix. Its ground state is known to be of B_1 symmetry. Light polarized parallel to the y-axis (parallel to the O-O separation) excited the molecule to an upper state. What is the symmetry of that state?

(c) What states of (a) anthracene (**1**) and (b) coronene (**2**) may be reached by electric dipole transitions from their (totally symmetric) ground states?

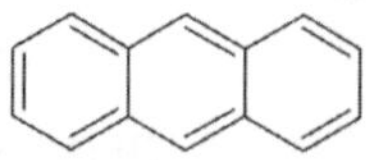

1

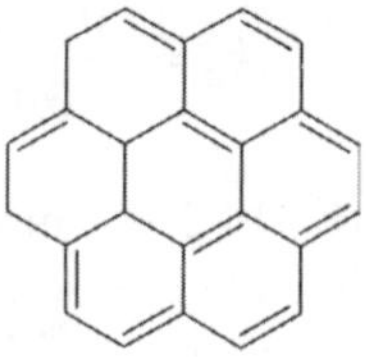

2

(d) Show that the lowest energy electronic transition of *p*-benzoquinone ($^1A_g \rightarrow {}^1B_{1g}$) is formally forbidden. The transition becomes 'vibronically allowed' in combination with certain vibrational excitations. What are the symmetries of the vibrational modes that can be excited in this way?

3. (a) Use the Walsh diagram for an AH_2 molecule (Figure 9.13) to explain why the electronic spectrum of NH_2 ($^2B_1 \rightarrow {}^2A_1$) shows a long progression of lines in the spectrum corresponding to excitation of several quanta in the bending vibrational mode.
   (b) In the first excited state, S_1, ethyne (acetylene) has a *trans* bent configuration with $\angle HCC = 120°$, compared to 180° in the linear ground state, S_0. In addition, the C-C bond length increases to 1.34 Å in S_1 from 1.21 Å in S_0. What can you say about the vibrational coarse structure that you would expect for the S_1-S_0 absorption spectrum?
4. Naphthalene contained in an alcohol–ether low-temperature glass absorbs light at $\lambda = 315$ nm, and two different emission spectra are observed. The quantum yields and observed lifetimes for the two emissions are (a) $\phi = 0.3$, $\tau = 3 \times 10^{-7}$ s; (b) $\phi = 0.03$, $\tau = 2.3$ s. How can these observations be interpreted?
5. Calculate the energy separations in Joule, kilojoule per mole, electron volt and reciprocal centimetre (wavenumbers) between the levels (a) $n = 3$ and $n = 1$, (b) $n = 7$ and $n = 6$ of an electron in a box of length 1.50 nm.
6. The photoelectron spectra of methyl isocyanide and deuteromethyl isocyanide in the gas phase were observed by Lake and Thompson (1971). Peaks were observed at the ionization potentials listed below.

C^1H_3NC		C^2H_3NC	
IP/eV	**Relative peak height**	**IP/eV**	**Relative peak height**
11.24	210	11.25	225
11.41	25	11.38	30
11.52	45	11.53	55
12.23	170	12.27	180
12.36	185	12.38	165
12.46	235	12.5	205
12.6	170	12.61	180
12.69	210	12.73	185

C^1H_3NC		C^2H_3NC	
IP/eV	Relative peak height	IP/eV	Relative peak height
12.83	130	12.84	130
12.92	120	12.95	110
15.57	40	15.72	90
15.67	80	15.81	115
15.76	110	15.9	155
15.86	140	15.99	185
15.95	165	16.08	200
16.05	180	16.17	220
16.14	170	16.26	210

Deduce the adiabatic and vertical ionization potentials, and state whether these refer to ionization from nonbonding or other molecular orbitals. Use the vibrational fine structure to deduce wavenumbers of fundamental vibration modes of the various molecule–ions formed. A selection rule for a vibration of a molecule–ion to be observed in a photoelectron spectrum is that the vibration is totally symmetric. An observed line may be due to simultaneous excitation of two vibration modes. The infrared spectrum of gaseous C^1H_3NC shows bands centred on the wavenumbers listed below.

CH_3, sym. stretch: 2951 cm^{-1}.

N=C, sym. stretch: 2166 cm^{-1}

CH_3, sym. bend: 1455 cm^{-1}

N–C, stretch: 945 cm^{-1}

With the aid of these data, and by comparison of the wavenumbers for C^1H_3NC and C^2H_3NC, assign the observed vibrations to the fundamental modes of the molecule–ions. Hence establish whether the photoelectrons have been ionized from bonding or anti-bonding orbitals.

7. (a) When 30.4 nm radiation is used to produce the photoelectron spectrum of benzene, the highest energy photoelectrons have kinetic energy of 31.5 eV. Find the ionization energy of the highest energy MO in benzene (which is a π MO).
   (b) What would be the kinetic energy of the highest-energy photoelectrons emitted from benzene if 58.4 nm radiation were used?
8. (a) Derive the selection rules for the transitions of an electron in a one-dimensional box.
   (b) Calculate the integrated absorption coefficient for the longest wavelength transition of the electrons of hexatriene. Take λ_{max} as 268 nm and the box length as 0.73 nm.
   (c) Calculate the frequency and hence the integrated absorption coefficient for the lowest π electron transition of octatetraene. (Use the square well potential with a total box length of 0.95 nm.)

9. Lutein, also known as food colouring E161b, is found in egg yolks, fat cells and green leaves. It is often used in poultry feed to enhance the colour of egg yolks.

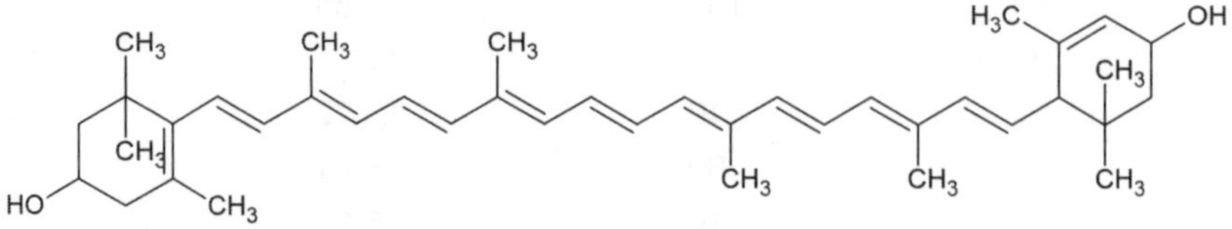

   (a) How many π electrons are there in this molecule?
   (b) If the average bond length in this molecule is 140 pm, approximately over what length are the electrons delocalized?
   (c) Employing the 'particle-in-a-box' approximation, calculate the energy of the HOMO–LUMO transition in eV. Can the transition be responsible for the observed yellow colouration? Explain.
   (d) Identify two other transitions, with $\Delta n > 1$, which give rise to absorption in the visible region and calculate the wavelengths of light absorbed by molecules undergoing these transitions. What colours are absorbed by this molecule?

10. (a) Identify the ground and excited states of the same spin multiplicity for the *cis*-butadiene molecule.
    (b) Which spin-allowed transitions are electric-dipole allowed and with what polarization?

11. (a) Derive the selection rules for the $n\pi^*$, $\pi\pi^*$, $n\sigma^*$ and $\pi\sigma^*$ transitions of formaldehyde.
    (b) How would you distinguish between $n\pi^*$ and $\pi\pi^*$ transitions?

12. Water has an $^1A_1$ ground state and an excited state with a configuration $\ldots(3a_1)^2(1b_1)^1(2b_2)^1$. What states result from this configuration? Is it an allowed transition from the ground state to any of these states?

13. Use simple Group Theoretical arguments to decide which of the following transitions are allowed:
    (a) the ($\pi\rightarrow\pi^*$) transition in ethylene
    (b) the ($n\rightarrow\pi^*$) transition in a C_{2v} carbonyl group.

14. Indicate, with the help of the Jablonski diagram, the various processes by which the excited electronic states get deactivated. What is the difference between fluorescence and phosphorescence?

15. (a) Derive the Stern–Volmer equation

$$\frac{\phi_0}{\phi} = 1 + \frac{k_q}{k_f + k_i}[Q] = 1 + \tau k_q[Q]$$

    where τ is the lifetime of the excited state = $(k_f + k_i)^{-1}$ and k_f and k_i are, respectively, the first-order rate constants for fluorescence and internal conversion, and k_q is the second order rate constant for quenching. Explain the significance of the three rate constants.
    (b) Hann (1974) found that the fluorescence intensity from a solution of 1-methyl anthracene in methanol (2.5×10^{-5} mol dm^{-3}) was decreased by the addition of

1-methylpyridinium toluene-*p*-sulphonate. The intensity of fluorescence, at 25 °C, in arbitrary units, is tabulated below as a function of 1-methylpyridinium toluene-*p*-sulphonate concentration.

10^2 (1-methylpyridinium toluene-*p*-sulphonate)/mol dm^{-3}	Intensity (arbitrary units)
0	7.41
2	3.28
2.9	2.67
3.8	2.13
4.7	1.92
5.6	1.72

The lifetime of the upper singlet state of 1-methyl anthracene in methanol at 25 ºC is 4.9 ns. Calculate the rate of quenching of fluorescence by 1-methylpyridinium toluene-*p*-sulphonate, by use of the Stern–Volmer equation.

16. (a) The long-wavelength absorption band of benzophenone is relatively weak (ε_{max} = 10 m^2 mol^{-1}) and the band is shifted to shorter wavelength when the solvent is changed from cyclohexane to ethanol. For 4,4′-bis(dimethylamino) benzophenone, the long-wavelength band is much stronger (ε_{max} = 1000 m^2 mol^{-1}) and there is a solvent effect in the opposite direction. Give an explanation for these observations.
    (b) The phosphorescence emission spectrum of benzophenone shows a vibrational progression with a separation of 1600–1700 cm^{-1} (C=O stretching), but the phosphorescence spectrum of 2-acetylnaphthalene resembles that of 1-chloronaphthalene and shows a series of vibrational bands separated by ~ 1400 cm^{-1} (aromatic ring vibration). Can you accord for this?
    (c) The fluorescence spectrum of anthracene vapour shows a series of peaks of increasing intensity with individual maxima at 440, 410, 390, and 370 nm, and followed by a sharp cut-off at shorter wavelengths. The absorption spectrum rises sharply from zero to a maximum at 360 nm with a trail of peaks of lessening intensity at 345, 330, and 305 nm. Account for these observations.
17. Answer the following:
    (a) The intrinsic lifetime and the observed lifetime of a state usually differ. Explain.
    (b) For an allowed transition of wavelength 600 nm, the lifetime of the excited state is 10 s. Estimate the natural line-width of this band.
18. Consider
    (a) a π electron in a long conjugated molecule of length 1 nm,
    (b) an oxygen molecule in a container of length 5 cm

    Treating both systems as particles in a one-dimensional box, calculate the spacing between the two lowest energy levels. Quote your answers in J, kJ mol^{-1}, eV and cm^{-1}.
19. The MOs of naphthalene (D_{2h}) give the following ground state configuration:

$$(1b_{1u})^2(1b_{2g})^2(1b_{3g})^2(2b_{1u})^2(1a_u)^2$$

    The next (empty) orbital is $2b_{2g}$.

(a) What are the symmetries of the ground and first excited states?
(b) Is this an allowed transition for absorption of a photon by electric dipole rules?
(c) If it is an allowed transition, what is its polarization?

20. (a) List some important differences between fluorescence and phosphorescence.
(b) The luminescent first-order decay of a certain organic molecule yields the following data:

t(s)	0	1	2	3	4	5	10
I	100	43.5	18.9	8.2	3.6	1.6	0.02

where I is the radiative intensity. Calculate the mean lifetime τ for the process. Is it fluorescence or phosphorescence?

21. Many aromatic hydrocarbons are colourless, but their anion and cation radicals are often strongly coloured. Give a qualitative explanation for the phenomenon. (*Hint:* Consider only the π molecular orbitals.)
22. The lowest triplet (T_1) in naphthalene molecule is about 11,000 cm^{-1} below the excited singlet (S_1) electronic level at 77 K. Calculate the ratio of populations in these two states. (*Hint:* Do not forget the degeneracies.)
23. How would you employ the particle-in-a-box model to qualitatively explain the difference in the appearance of stilbene (colourless) and C_6H_5-$(CH{=}CH)_4$-C_6H_5 (brown)?
24. Assuming that the average carbon–carbon bond length in hexatriene is 150 pm, use the particle-in-a-one-dimensional box model to calculate the longest wavelength peak in its absorption spectrum.
25. The fluorescence of a protein is due to tryptophan, tyrosine and phenylalanine (assuming that the protein does not contain a prosthetic group that is fluorescent). Iodide ions are known to quench the fluorescence of tryptophan. If a protein is known to contain only one tryptophan group and iodide fails to quench the fluorescence, what can you conclude about the location of the tryptophan residue?
26. Anthracene is colourless, but tetracene is bright orange. Explain.
27. A phosphorescent crystal was illuminated with a strong light. At $t = 0$, the light beam was turned off and the intensity of the phosphorescent radiation was measured in intervals of milliseconds. The results were as shown in the accompanying table. What is the intensity of the phosphorescence radiation at $t = 0$? What is the lifetime of the phosphorescent state?

t(ms)	Intensity (arbitrary units)
1	65.0
2	43.6
3	29.2
4	19.6
5	13.1

29. A molecule that has been excited to a higher electronic state can lose its excess energy by emitting a light quantum (fluorescence) or by transferring its energy by collision with another molecule (quenching). The competition between the two modes of energy loss

is often studied to test the significance of encounter-controlled events (quenching by collision and transfer of energy into kinetic energy requires an encounter with another body). Different molecules have different quenching efficiencies. More complex molecules are usually more efficient quenchers since they have larger effective size and a greater number of degrees of freedom. The competing process can be represented by the following mechanism, where F denotes the fluorescing molecule and Q the quenching molecule (Q may be the same as F):

$$\mathrm{F} + h\nu \rightarrow \mathrm{F}^* \qquad k_a$$

$$\mathrm{F}^* \rightarrow \mathrm{F} + h\nu' \qquad k_f$$

$$\mathrm{Q} + \mathrm{F}^* \rightarrow \mathrm{F} + \mathrm{Q}^* \qquad k_q$$

The *fluorescent yield* is defined as the ratio of the intensity of the light absorbed. Show that the fluorescent yield is

$$\frac{I_f}{I_a} = \frac{k_f[\mathrm{F}^*]}{k_f[\mathrm{F}^*] + k_q[\mathrm{F}^*][Q]}$$

which can be rearranged to get

$$\frac{I_a}{I_f} = 1 + \frac{k_q[Q]}{k_f}$$

This last equation is known as the *Stern–Volmer equation*. Outline an experiment by which you can determine k_q using the Stern–Volmer equation and any other information you may need.

is often studied to test the [illegible] collision and transfer of energy [illegible] body. Different molecules have different [illegible] efficiencies. More complex molecules are usually more efficient quenchers since they have larger [illegible] and a greater number of degrees of freedom. The quenching process can be represented by the following mechanism, where [illegible] denotes the fluorescing molecule and Q the quenching molecule. Q may be [illegible]

[illegible]

[illegible]

[illegible]

[illegible] that the [illegible]

[illegible]

which can be [illegible]

[illegible]

[illegible] equation [illegible]

Appendix 1

THE SPECTRUM OF THE NON-RIGID ROTATOR

The stretching effect of the centrifugal forces on the bond lengths, and hence the moment of inertia of a rotating molecule, can be calculated classically as follows. Consider a single particle of mass m rotating about a fixed point with an angular velocity of ω. Assume that, in the absence of rotation, the distance of the particle from the fixed point is r_e, and, as a result of the rotation, the distance increases to r when the particle rotates. The bond stretching is accompanied by a restoring force, given by $k(r - r_e)$, which balances the centrifugal force given by $mr\omega^2$. Therefore,

$$k(r-r_e) = mr\omega^2$$

$$\Rightarrow kr - mr\omega^2 = kr_e$$

$$\Rightarrow r = \frac{k}{k - m\omega^2} r_e \tag{1}$$

Therefore,

$$r - r_e = \frac{mr\omega^2}{k} \tag{2}$$

The total energy of the rotating system is given by the sum of its kinetic and potential energies, i.e.,

$$E = T + V = \frac{1}{2} I\omega^2 + \frac{1}{2} k(r - r_e)^2$$

Substitution of equation (2) gives

$$\begin{aligned}
E &= \frac{1}{2} I\omega^2 + \frac{1}{2} k\left(\frac{mr\omega^2}{k}\right)^2 = \frac{1}{2} I\omega^2 + \frac{1}{2} k\left(\frac{mr^2\omega^2}{kr}\right)^2 \\
&= \frac{1}{2} I\omega^2 + \frac{1}{2} k\left(\frac{I\omega^2}{kr}\right)^2 = \frac{1}{2}\frac{(I\omega)^2}{I} + \frac{1}{2} k\left(\frac{I^2\omega^2}{krI}\right)^2 \\
&= \frac{L^2}{2I} + \frac{L^4}{2kr^2I^2} \\
&= \frac{L^2}{2mr^2} + \frac{L^4}{2km^2r^6}
\end{aligned} \tag{3}$$

The first term in equation (3) is the major term. Substituting equation (1) for r and approximating $r \approx r_e$ in the second, minor term, we obtain

$$E = \frac{L^2}{2mr_e^2}\left(1 - \frac{m\omega^2}{k}\right) + \frac{L^4}{2km^2r_e^6}$$

Since $\frac{m\omega^2}{k} << 1, 1 - \frac{m\omega^2}{k}$ can be approximated by $1 - \frac{2m\omega^2}{k}$. Therefore,

$$\begin{aligned}
E &= \frac{L^2}{2mr_e^2} - \frac{L^2}{2mr_e^2} \times \frac{2m\omega^2}{k} + \frac{L^4}{2km^2r_e^6} \\
&= \frac{L^2}{2mr_e^2} - \frac{\omega^2 L^2}{kr_e^2} + \frac{L^4}{2km^2r_e^6} \\
&= \frac{L^2}{2mr_e^2} - \frac{(I\omega)^2 L^2}{kr_e^2 I^2} + \frac{L^4}{2km^2r_e^6} \\
&= \frac{L^2}{2mr_e^2} - \frac{L^4}{kr_e^2 m^2 r_e^4} + \frac{L^4}{2km^2r_e^6} \\
&= \frac{L^2}{2mr_e^2} - \frac{L^4}{2km^2r_e^6} = \frac{L^2}{2mr_e^2} - \frac{L^4 m}{2km^3r_e^6} \\
&= \frac{L^2}{2I} - \frac{L^4 m}{2kI^3}
\end{aligned}$$

To convert this classical result to a quantum mechanical result, we apply the quantum restrictions on the angular momentum squared operator. The allowed energy values are, therefore,

$$E_J = \frac{\hbar^2}{2I} J(J+1) - \frac{m\hbar^4 J^2 (J+1)^2}{2kI^3}$$

In terms of the rotational terms, the result becomes $F(J) = E_J/hc$

$$\begin{aligned}
F(J) &= \frac{h^2}{4\pi^2 \times 2Ihc} J(J+1) - \frac{mh^4 J^2 (J+1)^2}{16\pi^4 \times 2kI^3 \times hc} \\
&= \frac{h}{8\pi^2 Ic} J(J+1) - \frac{mh^3}{32\pi^4 kI^3 c} J^2 (J+1)^2 \text{ cm}^{-1}
\end{aligned}$$

Since $\tilde{B} = h/8\pi^2 Ic, I = h/8\pi^2 \tilde{B}c$,

$$F(J) = \tilde{B}J(J+1) - \frac{16m\pi^2 \tilde{B}^3 c^2}{k} J^2 (J+1)^2 \text{ cm}^{-1}$$

Since $\omega = \frac{1}{2\pi c}\sqrt{\frac{k}{m}}$, it follows that $\frac{k}{m} = 4\pi^2 c^2 \omega^2$,

$$F(J) = \tilde{B}J(J+1) - \frac{4\tilde{B}^3}{\omega^2} J^2 (J+1)^2 \text{ cm}^{-1}$$

Appendix 2

CHARACTER TABLES OF SOME IMPORTANT SYMMETRY GROUPS

$\mathbf{C}_1$	E
A	1

$\mathbf{C}_s$	E	σ_h		
A	1	1	x, y, R_z	x^2, y^2, z^2, xy
B	1	–1	z, R_x, R_y	yz, xz

$\mathbf{C}_i$	E	i		
A_g	1	1	R_x, R_y, R_z	$x^2, y^2, z^2, xy, xz, yz$
A_u	1	–1	x, y, z	

$\mathbf{C}_2$	E	C_2		
A	1	1	z, R_z	x^2, y^2, z^2, xy
B	1	–1	x, y, R_x, R_y	yz, xz

$\mathbf{C}_3$	E	C_3	C_5^2		$\varepsilon = \exp(2\pi i/3)$
A	1	1	1	z, R_z	x^2, y^2, z^2, xy
E	$\{^1_1$	ε^*	$\varepsilon\}$	$(x, y)(R_x, R_y)$	$(x^2 - y^2, xy)(yz, xz)$

$\mathbf{C}_4$	E	C_3	C_4^3			
A	1	1	1	1	R_z	$x^2 + y^2, z^2$
B	1	–1	1	–1		$x^2 - y^2, xy$
E	$\{^1_1$	$-i$	1	$i\}$	$(x, y)(R_x, R_y)$	(yz, xz)

$\mathbf{C}_5$	E	C_5	C_5^2	C_5^3	C_5^4		$\varepsilon = \exp(2\pi i/5)$
A	1	1	1	1	1	z, R_z	$x^2 + y^2, z^2$
E_1	$\{^1_1$	ε^*	ε^{2*}	ε^2	$\varepsilon\}$	$(x, y)(R_x, R_y)$	(yz, xz)
E_2	$\{^1_1$	ε^{2*}	ε	ε^*	$\varepsilon^2\}$		$(x^2 - y^2, xy)$

$\mathbf{C_6}$	E	C_6	C_3	C_2	C_3^2	C_6^5		$\varepsilon = \exp(2\pi i/6)$
A	1	1	1	1	1	1	z, R_z	$x^2 + y^2, z^2$
B	1	−1	1	−1	1	−1		
E_1	$\{^1_1$	ε^*	$-\varepsilon$	−1	$-\varepsilon^*$	$\varepsilon\}$	(R_x, R_y)	(yz, xz)
E_2	$\{^1_1$	$-\varepsilon^*$	$-\varepsilon^*$	1	$-\varepsilon$	$\varepsilon^*\}$		$(x^2 - y^2, xy)$

$\mathbf{D_2}$	E	$C_2(z)$	$C_2(y)$	$C_2(x)$		
A_1	1	1	1	1		$x^2 + y^2, z^2$
B_1	1	1	−1	−1	z, R_z	xy
B_2	1	−1	1	−1	y, R_y	xz
B_3	1	−1	−1	1	z, R_x	yz

$\mathbf{D_3}$	E	$2C_3$	$3C_2$		
A_1	1	1	1		$x^2 + y^2, z^2$
A_2	1	1	−1	z, R_z	xy
E	2	−1	0	$(x, y)(R_x, R_y)$	$(x^2 - y^2, xy)(xz, yz)$

$\mathbf{D_4}$	E	$2C_4$	C_2	$2C_2'$	$2C_2''$		
A_1	1	1	1	1	1		$x^2 + y^2, z^2$
A_2	1	1	1	−1	−1	z, R_z	
B_1	1	−1	1	1	−1		$x^2 - y^2$
B_2	1	−1	1	−1	1		xy
E	2	0	−2	0	0	$(x, y)(R_x, R_y)$	(xz, yz)

$\mathbf{D_5}$	E	$2C_5$	$2C_5^2$	$5C_2$		
A_1	1	1	1	1		$x^2 + y^2, z^2$
B_1	1	1	1	−1	z, R_z	
B_2	2	2cos 72°	2cos 144°	0	$(x, y)(R_x, R_y)$	(xz, yz)
B_3	2	2cos 144°	2cos 72°	0		$(x^2 - y^2, xy)$

$\mathbf{D_6}$	E	$2C_6$	$2C_3$	C_2	$3C_2'$	$3C_2''$		
A_1	1	1	1	1	1	1		$x^2 + y^2, z^2$
A_2	1	1	1	1	−1	−1	z, R_z	
B_1	1	−1	1	−1	1	−1		
B_2	1	−1	1	−1	−1	1	$(x, y)(R_x, R_y)$	
E_1	2	1	−1	−2	0	0		(xz, yz)
E_2	2	−1	−1	2	0	0		$(x^2 - y^2, xy)$

$\mathbf{C_{2v}}$	E	C_2	$\sigma_v(xz)$	$\sigma_v'(yz)$		
A_1	1	1	1	1	z	x^2, y^2, z^2
A_2	1	1	–1	–1	R_z	xy
B_1	1	–1	1	–1	x, R_y	xz
B_2	1	–1	–1	1	y, R_x	yz

$\mathbf{C_{3v}}$	E	$2C_3$	$3\sigma_v$		
A_1	1	1	1	z	$x^2 + y^2, z^2$
A_2	1	1	–1	R_z	
E	2	–1	0	$(x, y)\ (R_x, R_y)$	$(x^2 - y^2, xy)(xz, yz)$

$\mathbf{C_{4v}}$	E	$2C_4$	C_2	$2\sigma_v$	$2\sigma_d$		
A_1	1	1	1	1	1	z	$x^2 + y^2, z^2$
A_2	1	1	1	–1	–1	R_z	
B_1	1	–1	1	1	–1		$x^2 - y^2$
B_2	1	–1	1	–1	1		xy
E	2	0	–2	0	0	$(x, y)(R_x, R_y)$	(xz, yz)

$\mathbf{C_{5v}}$	E	$2C_5$	$2C_5^2$	$5\sigma_v$		
A_1	1	1	1	1	z	$x^2 + y^2, z^2$
B_1	1	1	1	–1	R_z	
B_2	2	2cos 72°	2cos 144°	0	$(x, y)(R_x, R_y)$	(xz, yz)
B_3	2	2cos 144°	2cos 72°	0		$(x^2 - y^2, xy)$

$\mathbf{C_{6v}}$	E	$2C_6$	$2C_3$	C_2	$3\sigma_v$	$3\sigma_d$		
A_1	1	1	1	1	1	1	z	$x^2 + y^2, z^2$
A_2	1	1	1	1	–1	–1	R_z	
B_1	1	–1	1	–1	1	–1		
B_2	1	–1	1	–1	–1	1		
E_1	2	1	–1	–2	0	0	$(x, y)(R_x, R_y)$	(xz, yz)
E_2	2	–1	–1	2	0	0		$(x^2 - y^2, xy)$

$C_{\infty v}$	E	C_{∞}^{Φ}	...	$\infty\sigma_v$		
$A_1 \equiv \Sigma^+$	1	1	...	1	z	$x^2 + y^2, z^2$
$A_2 \equiv \Sigma^-$	1	1	...	−1	R_z	
$E_1 \equiv \Pi$	2	$2\cos\Phi$	...	0	$(x, y)(R_x, R_y)$	(xz, yz)
$E_2 \equiv \Delta$	2	$2\cos 2\Phi$	...	0		$(x^2 - y^2, xy)$
$E_3 \equiv \Phi$	2	$2\cos 3\Phi$	...	0		
⋮	⋮	⋮	⋱	⋮		

Bibliography

TEXTBOOKS

Atkins, P. W. & de Paula, J. (2010) *Physical Chemistry.* Oxford University Press; 10th edition.

Banwell, C. & McCash, E. (1994) *Fundamentals of Molecular Spectroscopy*, 4th Ed., McGraw-Hill Higher Education: England.

Barrow, G. M. (1962) *Introduction to Molecular Spectroscopy.* McGraw-Hill Book Co., Inc.: N.Y.

Brand, J.C.D. & Speakman, J. C. (1975) *Molecular Structure: Physical Approach.* Hodder & Stoughton Educational; 2nd edition.

Chang, R. (1970) *Basic Principles of Spectroscopy.* McGraw-Hill Book Co., Inc.: N.Y.

Cotton, F. A. (1990) *Chemical Applications of Group Theory.* 3rd Ed. Wiley Interscience: New York.

Hollas, J. M. (2003) *Modern Spectroscopy,* 4th Ed., Wiley-Blackwell: England.

Ladd, M. (1998) *Symmetry and Group Theory in Chemistry.* Horwood chemical science series, Horwood Publishing: Chichester, England.

Schonland, D. S. (1965) *Molecular Symmetry: An Introduction to Group Theory and Its Uses in Chemistry.* van Nostrand: London.

OTHER REFERENCES

Bethune, D. S., Meijer, G., Tang, W. C., Rosen, H. J., Golden, W. G., Seki, H., Brown, C. A. & de Vries, M. S. (1991) Vibrational Raman and infrared spectra of chromatographically separated C_{60} and C_{70} fullerene clusters. *Chem. Phys. Lett.* **179:** 181–186.

DeLucia, F. & Gordy, W. (1969) Molecular-beam maser for the shorter-millimeter-wave region: Spectral constants of HCN and DCN. *Phys. Rev.* **187:** 58–65.

Field, G. B., Somerville, W. B. & Dressler, K. (1966) Hydrogen molecules in astronomy. *Annu. Rev. Astron. Astrophys.* **4:** 207–244.

Fleming, J.W. & Chamberlain, J. (1974) High resolution far infrared Fourier transform spectrometry using Michelson interferometers with and without collimation, *Infrared Physics.* **14(4)**: 277–292, ISSN 0020-0891, http://dx.doi.org/10.1016/0020-0891(74)90034-7.

Gaussian 09, Revision **C.01**, Frisch, M. J.; Trucks, G. W.; Schlegel, H. B.; Scuseria, G. E.; Robb, M. A.; Cheeseman, J. R.; Scalmani, G.; Barone, V.; Mennucci, B.; Petersson, G. A.; Nakatsuji, H.; Caricato, M.; Li, X.; Hratchian, H. P.; Izmaylov, A. F.; Bloino, J.; Zheng, G.; Sonnenberg, J. L.; Hada, M.; Ehara, M.; Toyota, K.; Fukuda, R.; Hasegawa, J.; Ishida, M.; Nakajima, T.; Honda, Y.; Kitao, O.; Nakai, H.; Vreven, T.; Montgomery, J. A., Jr.;

Peralta, J. E.; Ogliaro, F.; Bearpark, M.; Heyd, J. J.; Brothers, E.; Kudin, K. N.; Staroverov, V. N.; Kobayashi, R.; Normand, J.; Raghavachari, K.; Rendell, A.; Burant, J. C.; Iyengar, S. S.; Tomasi, J.; Cossi, M.; Rega, N.; Millam, N. J.; Klene, M.; Knox, J. E.; Cross, J. B.; Bakken, V.; Adamo, C.; Jaramillo, J.; Gomperts, R.; Stratmann, R. E.; Yazyev, O.; Austin, A. J.; Cammi, R.; Pomelli, C.; Ochterski, J. W.; Martin, R. L.; Morokuma, K.; Zakrzewski, V. G.; Voth, G. A.; Salvador, P.; Dannenberg, J. J.; Dapprich, S.; Daniels, A. D.; Farkas, Ö.; Foresman, J. B.; Ortiz, J. V.; Cioslowski, J.; Fox, D. J. Gaussian, Inc., Wallingford CT, 2009.

Gilliam, O. R., Johnson, C. M. & Gordy, W. (1950) Microwave spectroscopy in the region from two to three millimeters. *Phys. Rev.* **78:** 140–144.

Hann, R. A. (1974) Remarkable temperature shift of the fluorescence maximum of a crystalline anthracene derivative. *Chem. Phys. Lett.* **25:** 271–273.

Krupenie, P. H. (1972) The spectrum of molecular oxygen, *J. Phys. Chem. Ref. Data* **1:** 423–543.

Kuzmany, H., Pfeiffer, R., Hulman, M. & Kramberger, C. (2004) Raman spectroscopy of fullerenes and fullerene–nanotube composites. *Phil. Trans. R. Soc. Lond. A.* **362**: 2375–2406.

Lake, R. F. & Thompson, H. W. (1971) The photoelectron spectra of methyl isocyanide and trideutero-methyl isocyanide. *Spectrochim. Acta A* **27:** 783–786.

Linstrom, P. J. W. G. Mallard, W.G. (2005) Eds., *NIST Chemistry WebBook, NIST Standard Reference Database Number 69*, National Institute of Standards and Technology: Gaithersburg MD, 20899, http://webbook.nist.gov

Nayler, P. & Whiting, M. C. (1955) Researches on polyenes. Part III. The synthesis and light absorption by dimethylpolyenes. *J. Chem. Soc.* 3037–3047.

Rosenberg, A. & Ozier, I. (1974) The forbidden pure rotational spectrum of SiH_4: the *R* branch. *Can. J Phys.*, **52**: 575.

Rosenblum, B., Nethercot, Jr., A. H. & Townes, C. H. (1958) Isotopic mass ratios, magnetic moments and the sign of the electric dipole moment in carbon monoxide. *Phys. Rev.* **109:** 400–412.

Siegbahn, K., Nordling, C., Johannson, G., Hedman, J. Hedén, P.F., Hamrin, K., Gelius, U., Bergmark, T., Werme, L.O., Manne, R. & Baer, Y. (1969) ESCA applied to free molecules, North Holland Publ. Co.: Amsterdam, London.

Stern, O. & Volmer, M. (1919) Über die abklingzeit der fluoreszenz. *Physik Z.* **20:** 183–188.

Index